Lecture Notes in Electrical Engineering   1550

The book series *Lecture Notes in Electrical Engineering* (LNEE) publishes the latest developments in Electrical Engineering—quickly, informally and in high quality. While original research reported in proceedings and monographs has traditionally formed the core of LNEE, we also encourage authors to submit books devoted to supporting student education and professional training in the various fields and applications areas of electrical engineering. The series cover classical and emerging topics concerning:

- Communication Engineering, Information Theory and Networks
- Electronics Engineering and Microelectronics
- Signal, Image and Speech Processing
- Wireless and Mobile Communication
- Circuits and Systems
- Energy Systems, Power Electronics and Electrical Machines
- Electro-optical Engineering
- Instrumentation Engineering
- Avionics Engineering
- Control Systems
- Internet-of-Things and Cybersecurity
- Biomedical Devices, MEMS and NEMS

For general information about this book series, comments or suggestions, please contact leontina.dicecco@springer.com.

To submit a proposal or request further information, please contact the Publishing Editor in your country:

China

Jasmine Dou, Editor (jasmine.dou@springer.com)

India, Japan, Rest of Asia

Swati Meherishi, Editorial Director (Swati.Meherishi@springer.com)

Southeast Asia, Australia, New Zealand

Ramesh Nath Premnath, Editor (ramesh.premnath@springernature.com)

USA, Canada

Michael Luby, Senior Editor (michael.luby@springer.com)

All other Countries

Leontina Di Cecco, Senior Editor (leontina.dicecco@springer.com)

**** This series is indexed by EI Compendex and Scopus databases. ****

Patrick Siarry · Simon King Sing Cheung ·
M. A. Jabbar · Xiaolong Li
Editors

Proceedings of the 2024 International Conference on Wireless Communications, Networking and Applications

 Springer

Editors
Patrick Siarry
Laboratory of Images, Signals and Intelligent
Systems
Paris-Est Créteil University
Vitry-sur-Seine, France

M. A. Jabbar
Department of CSE
Vardhaman College of Engineering
Hyderabad, Telangana, India

Simon King Sing Cheung
Info Technology Office
Hong Kong Metropolitan University
Kowloon, Hong Kong

Xiaolong Li
Electronics and Computer Engineering
Technology
Indiana State University
Terre Haute, IN, USA

ISSN 1876-1100　　　　　ISSN 1876-1119 (electronic)
Lecture Notes in Electrical Engineering
ISBN 978-981-95-6945-8　　　ISBN 978-981-95-6946-5 (eBook)
https://doi.org/10.1007/978-981-95-6946-5

This work was supported by Shenzhen Qianhai Scieway Innovation Technology Co., Ltd.

This Springer imprint is published by the registered company Springer Nature Singapore Pte Ltd.
The registered company address is: 152 Beach Road, #21-01/04 Gateway East, Singapore 189721, Singapore

If disposing of this product, please recycle the paper.

Preface

The 8th Annual 2024 International Conference on Wireless Communications, Networking and Applications (WCNA2024) was held during December 20–22, 2024 at Shenzhen, Guangdong, China.

It aimed to engage researchers, engineers, and students in the fields of wireless communications, networking, and applications. The conference provided a forum for sharing experiences and original research contributions on those topics. Researchers and practitioners were invited to submit their contributions to WCNA2024.

WCNA2024 proceeding tended to compile the most up-to-date, comprehensive, and globally representative state-of-the-art research in wireless communications, networking, and applications. All the accepted papers have been submitted to strict peer-review by 3-4 expert referees, and selected based on originality, significance and clarity for the purpose of the conference. The conference program was rich, profound and featuring high-impact presentations of selected papers and additional late-breaking contributions. We sincerely hoped that the conference not only offered participants a broad overview of the latest research results in related fields, but also provided them with a valuable platform for academic networking and exchange.

The Technical Program Committee members worked very hard to meet the review deadline. There were four editors, all internationally recognized leading experts in their respective research fields. They demonstrated outstanding proficiency and achieved distinction in their professions. The proceedings were scheduled to be published on Springer Book Series Lecture Notes in Electrical Engineering as a volume-quickly, informally and in high quality.

We would like to express our sincere gratitude to all the members of International Program Committee and organizers for their hard work, precious time and endeavor preparing for the conference. Our deepest thanks also went to the volunteers and staffs for their long-hours work and generosity they've given to the conference. The last but not least, we would like to thank each and every of the authors, speakers and participants for their great contributions to the success of WCNA2024.

Patrick Siarry
M. A. Jabbar
Simon King Sing Cheung
Xiaolong Li

Committees

General Chair

Siarry Patrick — Universite Paris-Est Creteil, France

Editors

Siarry Patrick — Universite Paris-Est Creteil, France
M. A. Jabbar — Vardhaman College of Engineering, India
Simon King Sing Cheung — Hong Kong Metropolitan University, China
Xiaolong Li — Indiana State University, USA

Technical Program Committee

Sanjiv Rao Godla — Aditya University, India
Jun Tao — Jianghan University, China
Qiang (Shawn) Cheng — University of Kentucky, USA
Samir Ladaci — Ecole Nationale Polytechnique Algiers, Algeria
Yinglei Song — Jiangsu University of Science and Technology, China
Sumit Kushwaha — Chandigarh University, India
Trong-Minh Hoang — Posts and Telecommunication Institute of Technology, Vietnam
Daniela Litan — Hyperion University, Bucharest, Romania
Sachin Kumar — Kyungpook National University, South Korea
Valentina Emilia Balas — Aurel Vlaicu University of Arad, Romania
Jose Anand — KCG College of Technology, India
Alexei G. Shishkin — Moscow State University, Russia
Ranjith Kumar Gatla — Institute of Aeronautical Engineering, India
Anand Nayyar — Duy Tan University, Vietnam
Abdel Ghani Aissaoui — University of Tahri Mohamed of Bechar, Algeria
Tao Yu — China Electronics Technology Group Corporation, China
Zaid Ameen Abduljabbar — University of Basrah, Iraq
Isabel S. Jesus — Institute Superior of Engineering of Porto, Portugal

Vilson Luiz Dalle Mole	Federal Technological University of Paraná Brazil, Brazil
Umesh C. Pati	National Institute of Technology, India
Weiguo Ma	Nantong University, China
Sampan Rittidech	Mahasarakham University, Thailand
Ningzhi Wang	Anhui University, China
Jagadeesha R. Bhat	Indian Institute of Information Technology, India
Raouyane Brahim	Hassan II University, Morocco
Atanaska Bosakova-Ardenska	University of Food Technologies, Bulgaria
Klimis Ntalianis	University of West Attica, Greece
Suman Kr. Dey	National Institute of Technology Rourkela, India
Prof.Abdel-Badeeh M. Salem	Ain Shams University, Egypt
Daniela Litan	Hyperion University, Romania
Amina Omrane	University of Sfax, Tunisia
Gheorghe Grigoras	"Gheorghe Asachi" Technical University of Iasi, Romania
Tan-Jan Ho	Chung Yuan Christian University, Taiwan
Noor Zaman	Taylor's University, Malaysia
Ajay B Gadicha	P R Pote Patil College of Engineering & Management Amravati, India
Mikulas Alexik	Slovak Society for Applied Cybernetics and Informatics, Slovak
Abhishek Shukla	Dr. A.P.J. Abdul Kalam Technical University (AKTU), India

Contents

About the Editors

Patrick Siarry was born in France in 1952. He received the PhD degree from the University Paris 6, in 1986 and the Doctorate of Sciences (Habilitation) from the University Paris 11, in 1994. He was first involved in the development of analog and digital models of nuclear power plants at Electricité de France (E.D.F.). Since 1995 he is a professor in automatics and informatics. His main research interests are computer-aided design of electronic circuits, and the applications of new stochastic global optimization heuristics to various engineering fields. He is also interested in the fitting of process models to experimental data, the learning of fuzzy rule bases, and of neural networks.

Dr. M. A. Jabbar is a distinguished professional in the field of Computer Science and Engineering, with a specialization in Artificial Intelligence (AI) and Machine Learning (ML). He serves as the Professor and Head of the Department of Computer Science and Engineering (AI&ML) at Vardhaman College of Engineering, Hyderabad, Telangana, India. His work primarily focuses on advancing AI and ML, and his dedication and expertise have earned him numerous accolades. These include the Best Faculty Researcher Award from the CSI Mumbai Chapter, the Best HOD of the Year Award from CSI, the Outstanding Leadership Award from IEEE Hyderabad Section, and the Distinguished Contributor Award from the IEEE CS Chapter in 2021.

His remarkable research contributions span artificial intelligence, machine learning, and their interdisciplinary applications, solidifying his reputation as a leader in these fields. In addition to his research, Dr. M.A. Jabbar holds senior memberships in globally recognized professional organizations such as the Association for Computing Machinery (ACM) and the Institute of Electrical and Electronics Engineers (IEEE). He is also a lifetime member of the Computer Society of India (CSI) and the Indian Science Congress Association (ISCA). His diverse roles in editorial and research capacities reflect his unwavering commitment to advancing computer science and its transformative applications.

Dr. Cheung is currently the Director of IT at the Hong Kong Metropolitan University. He received his BSc and PhD in Computer Science from the City University of Hong Kong, and was admitted as IET fellow, IMA fellow, BCS fellow, HKIE fellow, HKCS fellow. He is active in research, with near 200 publications in the form of books, book chapters, journal articles and conference papers in two distinct areas, namely, innovation and technology in education, and software and system engineering. He won the Outstanding Research Publication Award from the Open University of Hong Kong in 2016, the 1st class Achievement in Computer and IT from Shenzhen Science and Technology Association in 2016, the 1st class Outstanding CIO Award from the Hong Kong IT Joint Council in 2015, and an Honoree for IT Excellence from the CIO Asia in 2015. He has been listed in the Who's Who in Science and Engineering since 2007.

Xiaolong Li obtained a Bachelor's degree in Electronic and Information Engineering from Huazhong University of Science and Technology in 1999, followed by a Master's degree in Electronic and Information Engineering from the same university in 2002, and a PhD in Electronic and Computer Engineering from the University of Cincinnati (2006). Since 2008, he has been an assistant professor in the Department of Electronic and Computer Engineering Technology at Indiana State University.

Main Research Interests: Wireless and mobile networks, wireless ad hoc networks, wireless Internet, modeling and performance analysis, QoS in communication networks, wireless mesh networks, microcontroller-based applications, security in wireless networks.

Generative Artificial Intelligence for Wireless Communications: Algorithms, Applications, and Beyond

Kwok Tai Chui[1], Simon K. S. Cheung[1(✉)], Jiaqi Liu[1], and Kwan Keung Ng[2]

[1] Hong Kong Metropolitan University, Hong Kong SAR, China
{jktchui,kscheung}@hkmu.edu.hk, s1304012@live.hkmu.edu.hk
[2] Ming-Ai (London) Institute, London, UK
s.ng@ming-ai.org.uk

Abstract. Generative models have been immersed in various artificial intelligence applications. This field of research is also known as generative artificial intelligence (GenAI). Any form of data (e.g., videos, images, audio, text, and time-series data) can be generated, providing additional and useful data to enhance the performance of machine learning models. GenAI features the ability to address various issues, such as rare event collection, high cost of data labelling, bias mitigation, privacy, and data scarcity. Wireless communications have become a vital part of today's digital era, offering flexibility, accessibility, ease of maintenance, and cost-effectiveness. Smartphones, tablets, notebooks, and smartwatches are common tools that rely on wireless communications. In this research, we conducted a literature review to analyze the 1180 latest relevant works (published in 2015–2024) employing GenAI for wireless communications. The algorithms, applications, and results were discussed and compared. The analysis focused on fundamental characteristics such as the yearly publication count, the various subject areas of these publications, the leading ten journals and conferences, as well as a visual representation of keywords in a word cloud. This was succeeded by a thorough examination of the ten most frequently cited publications. Additionally, two significant open challenges were highlighted, along with recommendations for possible future research avenues.

Keywords: Data Augmentation Algorithms · Data Generation Algorithms · Generative Adversarial Networks · Generative Artificial Intelligence · Wireless Communications

1 Introduction

Generative artificial intelligence (GenAI) creates new paths to breakthroughs in artificial intelligence. In early history, artificial intelligence planning [1] and data augmentation [2] led to the development of GenAI. In 2014, the advancement of deep learning algorithms emerged in the family of variational autoencoder [3] and generative adversarial networks [4], which demonstrated effectiveness in synthesizing additional training data to enhance

© The Author(s) 2026
P. Siarry et al. (Eds.): WCNA 2024, LNEE 1550, pp. 1–13, 2026.
https://doi.org/10.1007/978-981-95-6946-5_1

the model's performance. In 2017, the proposal of a generative pre-trained transformer [5] (GPT-1) formed the foundation of the generative era. GenAI booms at full speed since the release of ChatGPT in 2022 [6].

GenAI benefits to various issues (i) rare event collection [7]: The data distribution of various practical applications comprises majority class(es) and minority class(es). Minority classes are sometimes rare events that do not commonly exist in the populations. Thus, collecting sufficient samples of rare events usually requires extensive efforts and high cost, along with a large sample of majority classes. This may also be not feasible in some applications, such as equipment failure events, disaster events, and financial crashes; (ii) privacy [8]: Some types of data contain personal information, which some owners choose to anonymise the data before sharing subject to proper usage or even refuse to use the data. This challenge usually appears in biomedical applications; and (iii) data scarcity [9]: Many artificial intelligence algorithms are deep learning algorithms that require sufficient (often a massive amount of) training data to achieve robust and accurate models. The data distribution in the feature space of the machine learning problems leads to sparsity, creating another challenge.

Attention is drawn to GenAI in wireless communications, which has become crucial to offer flexibility, accessibility, ease of maintenance, and cost-effectiveness [10, 11]. If the question is whether wired communication or wireless communication is to be used, given that the same performance is ensured, the answer to wireless communication will be dominated because users witness the advantages of wireless communications.

In the literature, some recent publications conducted surveys on GenAI for specific topics of wireless communications, such as 6th-generation mobile networks [12], physical layer communications [13], semantic communication networks [14], space-air-ground integrated networks [15], and networking [16]. In contrast, our research work conducted a thorough review of the algorithms and applications of GenAI for wireless communications from a macroscopic perspective (not specific topics of interest as in the literature). Insights and future research directions will also be presented.

The rest of this paper covers (i) a search process of relevant research works in Sect. 2; (ii) a quantitative analysis and basic characteristics in Sect. 3; (iii) a qualitative analysis and key insights in Sect. 4; an exploration of future research directions in Sect. 5; and a conclusion in Sect. 6.

2 Search Process

In the search process to select relevant articles for GenAI for wireless communications, Scopus (one of the popular scientific abstract and citation databases) was used with the TITLE-ABS-KEY function on 1 November 2024. The query comprised: TITLE-ABS-KEY("GenAI" OR "generative artificial intelligence" OR "generative adversarial networks" OR "GANs" OR "variational autoencoder" OR "data generation" OR "data augmentation" OR "synthetic data") AND TITLE-ABS-KEY("wireless communication" OR "telecommunications" OR "cellular networks" OR "mobile networks" OR "sixth generation" OR "fifth generation" OR "fourth generation" OR "radio communication" OR "satellite communication" OR "Wi-Fi" OR "Bluetooth" OR "ZigBee" OR "Z-wave" OR "wireless sensor networks" OR "wireless systems"). 1657 documents were returned.

1180 documents were shortlisted (28.8% of documents were excluded) based on the inclusion and exclusion criteria below:

- As a common language, English articles were included. 30 documents (1.81%) were excluded for other languages, including 27 Chinese articles, 2 Korean articles, and 1 Turkish article.
- Including document types: articles, conference papers, and book chapters. 137 documents (8.27%) were excluded, including 101 conference reviews, 24 reviews, 6 books, 3 retracted, 1 erratum, 1 editorial, and 1 data paper.
- Select the year range from 2015 to 2024. 310 documents (18.7%) were excluded.

Figure 1 summarizes the workflow of the inclusion and exclusion criteria.

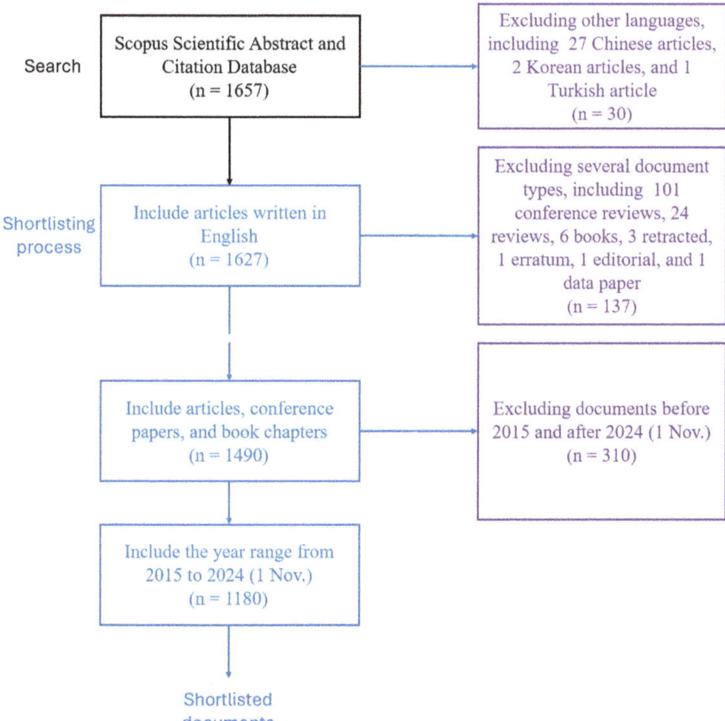

Fig. 1. Shortlisting process of relevant articles on GenAI for wireless communications with inclusion and exclusion criteria.

3 Quantitative Analysis and Basic Characteristics

The quantitative analysis and basic characteristics of the 1180 shortlisted articles are conducted. To begin, Fig. 2 shows the trends in the number of publications from 2015 to 2024. At the beginning of the last decade, the number of related publications in the field of GenAI for wireless communications received limited attention, with a small fluctuation in the annual number of publications of 25 to 37 in 2015–2017. The numbers reached a new level of about 70 publications in 2018–2019 and another level of hundred in 2020–2021. These numbers increased from 179 to 296 in the last two years (2022–2024). The trends aligned with the emergence of new technologies, including deep learning algorithms and generative pre-trained transformers.

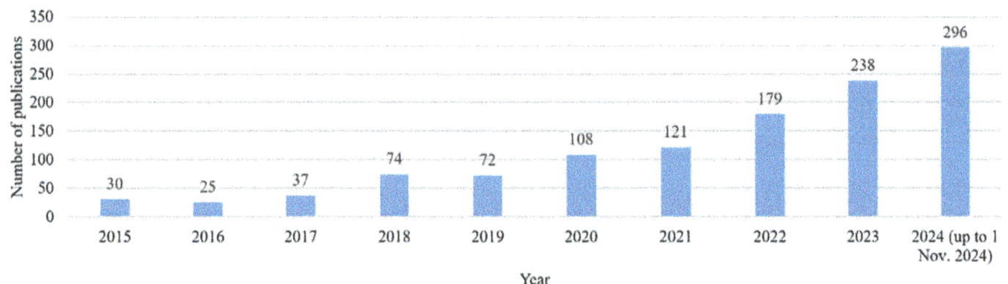

Fig. 2. Number of publications by years from 2015 to 2024.

To analyze the disciplines of wireless communication applications using GenAI, Fig. 3 shows the total number of publications in 2015–2024 in the top 24 subject areas in ascending order.

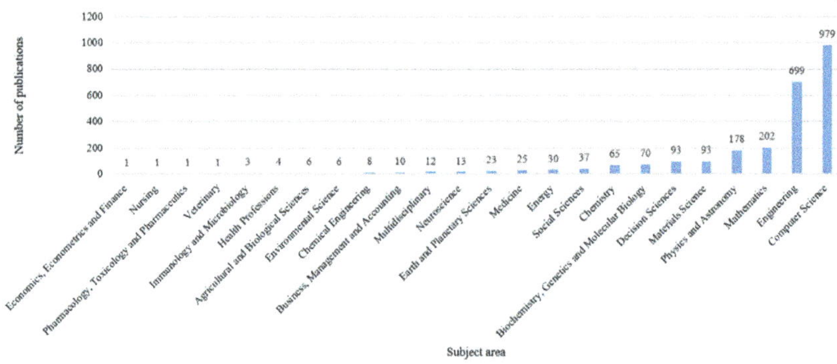

Fig. 3. Number of publications in 24 subject areas.

Computer science (n = 699) and engineering (n = 979) dominated the contributions, with a sum of 65.5% of publications. There are eight other areas exceeding 1% of the overall publications, namely mathematics (n = 202, 7.89%), physics and astronomy (178, 6.95%), materials science (n = 93, 3.63%), decision sciences (93, 3.63%), biochemistry, genetics, and molecular biology (70, 2.73%), chemistry (65, 2.54%), social sciences (37, 1.45%), and energy (30, 1.17%), summing to 30%. It is worth noting that science-related subject areas also played important roles in GenAI for wireless communications, which are not fully dominated by computer science, engineering, and mathematics.

Among the 1180 shortlisted articles, there are 672 journal papers (56.9%), 488 conference papers (41.4%), and 20 book chapters (1.69%). In total, these works were published in 249 journals and 311 conferences. To investigate the sources of publications better, Table 1 highlights the top 10 highest number of publications in journals and conferences, with the percentage (out of the total number of publications). These contributed to 38.7% (n = 260) for the journal side and 25% (n = 122) for the conference side, reflecting a broad coverage of scopes in research works in GenAI for wireless communications.

Table 1. Top ten sources of publications of 114 journals and 44 conferences.

Name of journal	Count	Percentage (%)	Name of conference	Count	Percentage (%)
IEEE Access	62	9.23	IEEE International Conference on Communications	20	4.10
IEEE Internet of Things Journal	47	6.99	Lecture Notes in Computer Science Including Subseries Lecture Notes in Artificial Intelligence and Lecture Notes in Bioinformatics	20	4.10
Sensors	46	6.85	ACM International Conference Proceeding Series	18	3.69
IEEE Transactions on Wireless Communications	23	3.42	Proceedings IEEE Global Communications Conference Globecom	15	3.073770492
IEEE Sensors Journal	17	2.53	IEEE Vehicular Technology Conference	11	2.254098361
IEEE Transactions on Mobile Computing	17	2.53	ICASSP IEEE International Conference on Acoustics Speech and Signal Processing Proceedings	10	2.049180328
Sensors Switzerland	15	2.23	Proceedings of SPIE The International Society for Optical Engineering	9	1.844262295

(continued)

Table 1. (continued)

Name of journal	Count	Percentage (%)	Name of conference	Count	Percentage (%)
IEEE Transactions on Cognitive Communications and Networking	12	1.79	IEEE Wireless Communications and Networking Conference WCNC	7	1.43442623
IEEE Transactions on Vehicular Technology	12	1.79	Lecture Notes in Networks and Systems	7	1.43442623
Electronics	9	1.34	Advances in Intelligent Systems and Computing	5	1.024590164

To examine the scopes of the existing works, a word cloud of the top 200 keywords is presented in Fig. 4. The consideration is directed to two perspectives (i) The total count of keywords is 6935. Fifteen keywords exceed 100 counts, listing in descending order, generative adversarial networks ($n = 457, 6.59\%$), deep learning ($n = 315, 4.54\%$), machine learning ($n = 196, 2.83\%$), wireless communications ($n = 188, 2.71\%$), learning systems ($n = 157, 2.26\%$), wireless sensor networks ($n = 146, 2.11\%$), data augmentation ($n = 138, 1.99\%$), internet of things ($n = 135, 1.95\%$), adversarial networks ($n = 134, 1.93\%$), convolutional neural networks ($n = 132, 1.90\%$), mobile telecommunication systems ($n = 129, 1.86\%$), neural networks ($n = 122, 1.76\%$), channel state information ($n = 115, 1.66\%$), deep neural networks ($n = 107, 1.54\%$), and location-based services ($n = 107, 1.54\%$). (ii) Recall there are 1180 shortlisted publications. Twelve keywords are selected in more than 10% of publications; they are generative adversarial networks ($n = 457, 38.7\%$), deep learning ($n = 315, 26.7\%$), machine learning ($n = 196, 16.6\%$), wireless communications ($n = 188, 15.9\%$), learning systems ($n = 157, 13.3\%$), wireless sensor networks ($n = 146, 12.4\%$), data augmentation ($n = 138, 11.7\%$), internet of things ($n = 135, 11.4\%$), adversarial networks ($n = 134, 11.4\%$), convolutional neural networks ($n = 132, 11.2\%$), mobile telecommunication systems ($n = 129, 10.9\%$), and neural networks ($n = 122, 10.3\%$). There is a special remark that some researchers prefer not to repeat the theme in the paper title or keywords. This may help widen the coverage of topics [17, 18]. As a result, the word counts returned in Scopus are generally less than the ground truth.

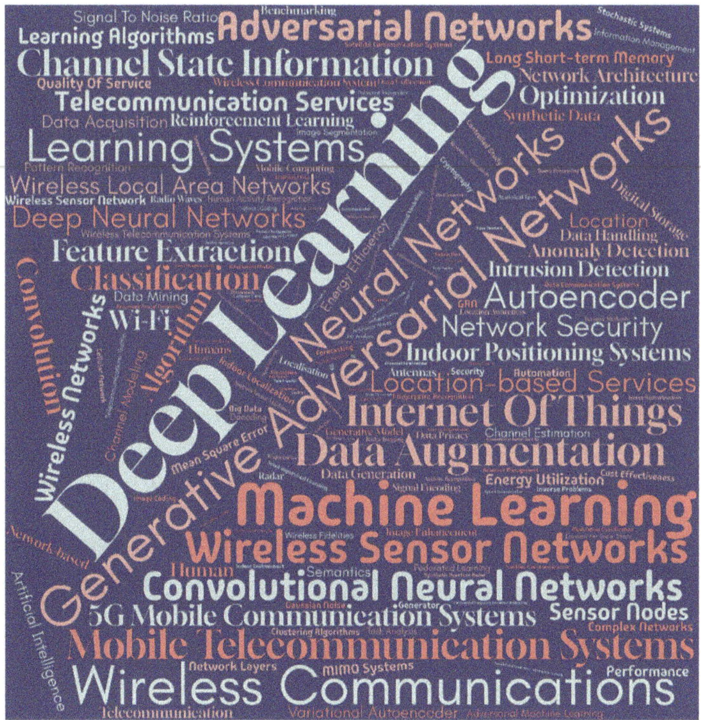

Fig. 4. A word cloud of the top 200 keywords.

4 Qualitative Analysis of Most Cited Publications

Yet, there are 1180 shortlisted articles which cannot be fully discussed. Therefore, the top 10 most cited publications in the recent 5-year (2020–2024) range [19–28] and previous 5-year (2015–2019) range [29–38] are studied to investigate the popular and impactful research works.

Table 2 summarizes the citation counts, algorithms, and applications of each work. Seventeen of twenty works (85%) are journal publications, providing in-depth proposals, performance evaluations, and analyses of various GenAI for wireless communications. Many diverse applications are seen from the list in Table 2, except some receiving attention to a larger extent, including network traffic, activity recognition, and attach detection. Regarding data generation algorithms, most works adopted traditional data augmentation approaches and baseline/variants of generative adversarial networks.

5 Key Open Research Challenges

5.1 Proportion of Ground Truth Data and Synthetic Data

Similar to GenAI for other applications, the knowledge about the appropriate proportion of ground truth and synthetic data is limited.

Some general ideas are (i) improve the imbalanced ratio between majority classes and minority classes so that models become less biased toward the majority classes, preserving the model's ability and fairness in all classes [39, 40], (ii) not limit the amount of generated data so that the data diversity can be enhanced [41, 42], and (iii) applying both data augmentation and data generation to maximize the potential high-quality data [43, 44].

5.2 Limited Studies in the Investigation of Appropriate Variants of GenAI Algorithms

In the keyword analysis and most cited publications, it can be seen that there were limited studies on an in-depth and extensive performance evaluation and comparison of various GenAI algorithms with benchmarking in wireless communication applications. Particularly, one may be interested in some generic algorithms that yield satisfactory performance so that domain knowledge of the variants of GenAI and time-consuming customization and model training can be avoided.

There are some recommended readings on performance evaluation and analysis using different variants of GenAI algorithms, such as generative adversarial networks [45, 46]. Detailed architectures and hyperparameter tuning are desired for fair judgement on the effectiveness of individual algorithms. In addition, ablation experiments should also be performed to examine the benefits of individual techniques [47, 48].

Table 2. The top 10 most cited publications in the recent 5-year (2020–2024) range and previous 5-year (2015–2019) range.

Work	Year of pub	Citation counts	Applications	Data generation algorithms
Recent 5-year (2020–2024) range				
[19]	2020	269	Modelling the effect of wireless communication channels	Conditional generative adversarial networks
[20]	2020	129	Radio modulation classification	Data augmentation via rotation, flip, and Gaussian noise insertion
[21]	2021	108	A WiFi-based human activity recognition	Eight types of transformation using deformation, data-independent, and task-specific methods
[22]	2021	95	Blockage prediction and proactive handoff in vision-based 6G communications	ViWi data generation

(*continued*)

Table 2. (*continued*)

Work	Year of pub	Citation counts	Applications	Data generation algorithms
[23]	2020	94	Denial of service attack detection	Average dependence estimator
[24]	2020	83	Network load prediction	Anomaly insertion to network traffic data
[25]	2021	82	Indoor localization	Amplitude feature deep convolutional generative adversarial networks
[26]	2021	77	Blockchain security framework	Estimation algorithm
[27]	2020	75	Constructing and adapting WiFi indoor radio map	Gaussian process regression conditioned least-squares generative adversarial networks
[28]	2021	69	A model-free resource allocation for ultra-reliable and low-latency 6G communications	Generative adversarial networks
Previous 5-year (2015–2019) range				
[29]	2019	226	3D object detection with LiDAR	Rotational perturbations on pose generation
[30]	2019	167	Preventing wireless jamming attacks	Generative adversarial networks
[31]	2018	125	Allocating tasks in heterogeneous spatial crowdsourcing environments	Randomize experimental scenarios
[32]	2018	105	Iris segmentation of near-infrared images	Data augmentation by lowering the quality of images
[33]	2018	97	Optimizing caching policy in device-to-device communications	Synthesizing user preference using content popularity
[34]	2016	90	Spatial co-location pattern mining	Data synthesis with varying data size and the neighbor distance variable

(continued)

Table 2. (*continued*)

Work	Year of pub	Citation counts	Applications	Data generation algorithms
[35]	2018	85	Telecom fraud detection	Generative adversarial networks
[36]	2019	80	Activity recognition with channel state information	Semi-supervised generative adversarial network
[37]	2017	79	Network traffic classification	Auxiliary classifier generative adversarial networks
[38]	2019	70	Generating network traffic data at the internet protocol packet layer	Generative adversarial networks

6 Conclusion

In conclusion, this research investigates the application of GenAI in wireless communications, highlighting its transformative potential across various applications. Through a comprehensive literature review of 1,180 recent studies published from 2015 to 2024, we identified a significant upward trend in GenAI research, showcasing an increasing interest in algorithms and applications that enhance wireless communication systems. By examining key characteristics such as publication frequency, subject areas, and influential sources (journals, conferences, and book chapters), we provide an overview of the field. To the best of our knowledge, this review is the first of its kind to focus on the general consideration of GenAI in wireless communications. The findings indicate that GenAI effectively addresses several critical challenges within wireless communications, including the collection of rare event data, privacy concerns, and data scarcity, which are particularly pertinent in today's data-driven landscape. Furthermore, our analysis of the top-cited publications underscores the pivotal role of GenAI in evolving wireless technologies, revealing opportunities for innovation across diverse domains.

Despite significant advancements, notable challenges persist, which pave the way for future research directions. These include improving algorithmic efficiency, addressing privacy concerns, and developing methods for enhanced data augmentation. Our research serves as a foundation for upcoming studies, emphasizing the necessity for interdisciplinary collaboration and innovative methodologies in leveraging GenAI to push the boundaries of wireless communication capabilities. Ultimately, as the digital landscape continues to evolve, the synergy between GenAI and wireless technologies will play an integral role in shaping the future of connectivity, efficiency, and user experience.

References

1. Bonet, B., Geffner, H.: Planning and control in artificial intelligence: a unifying perspective. Appl. Intell. **14**(3), 237–252 (2001)

2. Shorten, C., Khoshgoftaar, T.M.: A survey on image data augmentation for deep learning. J. Big Data **6**(1), 1–48 (2019)
3. Zou, C., Yang, F., Song, J., Han, Z.: Channel autoencoder for wireless communication: state of the art, challenges, and trends. IEEE Commun. Mag. **59**(5), 136–142 (2021)
4. Chakraborty, T., KS, U.R., Naik, S.M., Panja, M., Manvitha, B.: Ten years of generative adversarial nets (GANs): a survey of the state-of-the-art. Mach. Learn.: Sci. Technol. **5**(1), 011001 (2024)
5. Yenduri, G., et al.: Gpt (generative pre-trained transformer)–a comprehensive review on enabling technologies, potential applications, emerging challenges, and future directions. IEEE Access **12**, 54608–54649 (2024)
6. Wu, T., et al.: A brief overview of ChatGPT: the history, status quo and potential future development. IEEE/CAA J. Automatica Sinica **10**(5), 1122–1136 (2023)
7. Chui, K.T., Gupta, B.B., Arya, V., Torres-Ruiz, M.: Selective and adaptive incremental transfer learning with multiple datasets for machine fault diagnosis. Comput. Mater. Continua **78**(1), 1363–1379 (2024)
8. Janssen, M., Brous, P., Estevez, E., Barbosa, L.S., Janowski, T.: Data governance: organizing data for trustworthy Artificial Intelligence. Gov. Inf. Q. **37**(3), 101493 (2020)
9. Alzubaidi, L., et al.: A survey on deep learning tools dealing with data scarcity: definitions, challenges, solutions, tips, and applications. J. Big Data **10**(1), 46 (2023)
10. Oberascher, M., Rauch, W., Sitzenfrei, R.: Towards a smart water city: a comprehensive review of applications, data requirements, and communication technologies for integrated management. Sustain. Cities Soc. **76**, 103442 (2022)
11. Khanh, Q.V., Hoai, N.V., Manh, L.D., Le, A.N., Jeon, G.: Wireless communication technologies for IoT in 5G: Vision, applications, and challenges. Wirel. Commun. Mob. Comput. **2022**(1), 3229294 (2022)
12. Celik, A., Eltawil, A.M.: At the dawn of generative AI Era: a tutorial-cum-survey on new frontiers in 6G wireless intelligence. IEEE Open J. Commun. Soc. **5**, 2433–2489 (2024)
13. Van Huynh, N., et al.: Generative AI for physical layer communications: A survey. IEEE Trans. Cogn. Commun. Netw. **10**(3), 706–728 (2024)
14. Liang, C., et al.: Generative AI-driven semantic communication networks: Architecture, technologies and applications. IEEE Trans. Cogn. Commun. Netw. (Early Access)
15. Zhang, R., et al.: Generative AI for space-air-ground integrated networks. IEEE Wireless Commun. (Early Access)
16. Navidan, H., et al.: Generative Adversarial Networks (GANs) in networking: a comprehensive survey & evaluation. Comput. Netw. **194**, 108149 (2021)
17. Uddin, S., Khan, A.: The impact of author-selected keywords on citation counts. J. Informet. **10**(4), 1166–1177 (2016)
18. Waltman, L.: A review of the literature on citation impact indicators. J. Informet. **10**(2), 365–391 (2016)
19. Ye, H., Liang, L., Li, G.Y., Juang, B.H.: Deep learning-based end-to-end wireless communication systems with conditional GANs as unknown channels. IEEE Trans. Wireless Commun. **19**(5), 3133–3143 (2020)
20. Huang, L., Pan, W., Zhang, Y., Qian, L., Gao, N., Wu, Y.: Data augmentation for deep learning-based radio modulation classification. IEEE Access **8**, 1498–1506 (2019)
21. Zhang, J., et al.: Data augmentation and dense-LSTM for human activity recognition using WiFi signal. IEEE Internet Things J. **8**(6), 4628–4641 (2021)
22. Charan, G., Alrabeiah, M., Alkhateeb, A.: Vision-aided 6G wireless communications: blockage prediction and proactive handoff. IEEE Trans. Veh. Technol. **70**(10), 10193–10208 (2021)

23. Baig, Z.A., Sanguanpong, S., Firdous, S.N., Nguyen, T.G., So-In, C.: Averaged dependence estimators for DoS attack detection in IoT networks. Futur. Gener. Comput. Syst. **102**, 198–209 (2020)
24. Sevgican, S., Turan, M., Gökarslan, K., Yilmaz, H.B., Tugcu, T.: Intelligent network data analytics function in 5G cellular networks using machine learning. J. Commun. Netw. **22**(3), 269–280 (2020)
25. Li, Q., et al.: AF-DCGAN: amplitude feature deep convolutional GAN for fingerprint construction in indoor localization systems. IEEE Trans. Emerging Top. Comput. Intell. **5**(3), 468–480 (2021)
26. Rathore, S., Park, J.H., Chang, H.: Deep learning and blockchain-empowered security framework for intelligent 5G-enabled IoT. IEEE Access **9**, 90075–90083 (2021)
27. Zou, H., et al.: Adversarial learning-enabled automatic WiFi indoor radio map construction and adaptation with mobile robot. IEEE Internet Things J. **7**(8), 6946–6954 (2020)
28. Kasgari, A.T.Z., Saad, W., Mozaffari, M., Poor, H.V.: Experienced deep reinforcement learning with generative adversarial networks (GANs) for model-free ultra reliable low latency communication. IEEE Trans. Commun. **69**(2), 884–899 (2021)
29. Manhardt, F., Kehl, W., Gaidon, A.: Roi-10d: Monocular lifting of 2d detection to 6d pose and metric shape. In: Proceedings of the IEEE/CVF Conference on Computer Vision and Pattern Recognition, pp. 2069–2078. IEEE, California (2019)
30. Erpek, T., Sagduyu, Y.E., Shi, Y.: Deep learning for launching and mitigating wireless jamming attacks. IEEE Trans. Cogn. Commun. Netw. **5**(1), 2–14 (2019)
31. Wang, L., Yu, Z., Han, Q., Guo, B., Xiong, H.: Multi-objective optimization based allocation of heterogeneous spatial crowdsourcing tasks. IEEE Trans. Mob. Comput. **17**(7), 1637–1650 (2018)
32. Bazrafkan, S., Thavalengal, S., Corcoran, P.: An end to end deep neural network for iris segmentation in unconstrained scenarios. Neural Netw. **106**, 79–95 (2018)
33. Chen, B., Yang, C.: Caching policy for cache-enabled D2D communications by learning user preference. IEEE Trans. Commun. **66**(12), 6586–6601 (2018)
34. Yu, W.: Spatial co-location pattern mining for location-based services in road networks. Expert Syst. Appl. **46**, 324–335 (2016)
35. Zheng, Y.J., Zhou, X.H., Sheng, W.G., Xue, Y., Chen, S.Y.: Generative adversarial network based telecom fraud detection at the receiving bank. Neural Netw. **102**, 78–86 (2018)
36. Xiao, C., Han, D., Ma, Y., Qin, Z.: CsiGAN: Robust channel state information-based activity recognition with GANs. IEEE Internet Things J. **6**(6), 10191–10204 (2019)
37. Vu, L., Bui, C.T., Nguyen, Q.U.: A deep learning based method for handling imbalanced problem in network traffic classification. In: Proceedings of the 8th International Symposium on Information and Communication Technology, pp. 333–339. ACM, Viet Nam (2017)
38. Cheng, A.: PAC-GAN: Packet generation of network traffic using generative adversarial networks. In: 2019 IEEE 10th Annual Information Technology. Electronics and Mobile Communication Conference (IEMCON), pp. 0728–0734. IEEE, Canada (2019)
39. Chui, K.T., Gupta, B.B., Arya, V., Bansal, R., Colace, F.: A lightweight generative adversarial network for imbalanced malware image classification. In: ACM International Conference Proceeding Series, pp. 1–4. Association for Computing Machinery, India (2023)
40. Ding, H., Sun, Y., Huang, N., Shen, Z., Cui, X.: TMG-GAN: generative adversarial networks-based imbalanced learning for network intrusion detection. IEEE Trans. Inf. Forensics Secur. **19**, 1156–1167 (2023)
41. Liu, Y., et al.: Deep generative model and its applications in efficient wireless network management: A tutorial and case study. IEEE Wirel. Commun. **31**(4), 199–207 (2024)
42. Iglesias, G., Talavera, E., González-Prieto, Á., Mozo, A., Gómez-Canaval, S.: Data augmentation techniques in time series domain: a survey and taxonomy. Neural Comput. Appl. **35**(14), 10123–10145 (2023)

43. Chui, K.T., Lee, L.K., Wang, F.L., Cheung, S.K., Wong, L.P.: A review of data augmentation and data generation using artificial intelligence in education. In: International Conference on Technology in Education, pp. 242–253. Springer Nature Singapore (2023)
44. Jiang, X., Ge, Z.: Data augmentation classifier for imbalanced fault classification. IEEE Trans. Autom. Sci. Eng. **18**(3), 1206–1217 (2021)
45. Müller-Franzes, G., et al.: A multimodal comparison of latent denoising diffusion probabilistic models and generative adversarial networks for medical image synthesis. Sci. Rep. 13(1), 12098 (2023)
46. Boroujeni, S.P.H., Razi, A.: Ic-gan: An improved conditional generative adversarial network for rgb-to-ir image translation with applications to forest fire monitoring. Expert Syst. Appl. **238**, 121962 (2024)
47. Wang, Y., Xu, X., Hu, L., Fan, J., Han, M.: A time series continuous missing values imputation method based on generative adversarial networks. Knowl.-Based Syst. **283**, 111215 (2024)
48. Lateef, F., Kas, M., Chahi, A., Ruichek, Y.: A two-stream conditional generative adversarial network for improving semantic predictions in urban driving scenes. Eng. Appl. Artif. Intell. **133**, 108290 (2024)

Design of Intelligent Calibration System for Medical Device Testing Based on Wireless QA Control

Jing Sun[✉], Dong Shen, Wenqing Zhu, Yannan Ren, Yang Feng, and Ziyan Zhao

Shandong Province Institute of Measurement, Jinan 250014, Shandong, China
`sunjing136@email.cn`

Abstract. In response to the problems of errors, low efficiency, and insufficient utilization in manual calibration of medical devices, this study adopts wireless communication technology to achieve real-time data exchange and remote operation between the device and the central system, effectively reducing manual participation. By combining intelligent calibration algorithms, dynamic analysis and automatic adjustment of device operation are implemented to ensure autonomous calibration when it deviates from the standard. The system also integrates a real-time Quality Assurance (QA) monitoring module to monitor key indicators of devices, trigger alarms in case of abnormalities, and initiate fault warning and preventive maintenance processes to reduce downtime. The experimental data shows that the average calibration time of the device has been reduced by 66%, and the utilization rate has been increased to 79.8%, fully verifying the significant effectiveness of the system in shortening the calibration cycle and enhancing device utilization, which is conducive to the comprehensive improvement of medical device efficiency and operation level.

Keywords: Medical Devices · Intelligent Calibration · Wireless Communication Technology · Real-time Monitoring · Device Utilization

1 Introduction

The accuracy and stability of medical devices in modern healthcare organizations are directly related to the effectiveness of medical diagnosis and treatment. It is important to ensure the calibration accuracy and reliability of medical devices [1, 2]. Traditional calibration models rely heavily on manual operations, which face problems of inefficiency, high errors, and poor device utilization. In today's hospital environment, the number of medical devices is huge and it is difficult for technicians to perform effective calibration and maintenance of medical devices during peak patient times. Manual calibration-induced device downtime increases the workload and interferes with healthcare services, thereby delaying patient care [3, 4]. In this regard, the development of an intelligent and efficient calibration system is important for the medical community.

To compensate for the weaknesses of traditional calibration systems in terms of real-time response, degree of automation and remote manipulation, this study designs

P. Siarry et al. (Eds.): WCNA 2024, LNEE 1550, pp. 14–24, 2026.
https://doi.org/10.1007/978-981-95-6946-5_2

an intelligent calibration and inspection system for medical devices based on wireless QA control technology. The system utilizes wireless QA control technology to collect and transmit device status information in real-time, ensuring that the operational data is rapidly fed back to the central control platform [5, 6]. Combined with intelligent calibration algorithms, the system can autonomously monitor the state of the device, and once deviations from standard parameters are detected, the calibration process is initiated to quickly restore the accuracy of the device. Fault early warning and predictive maintenance modules are integrated into the system design. These modules detect potential faults in advance through in-depth analysis of the device's historical operating data, effectively reducing the device failure rate. This greatly improves the efficiency of device usage, optimizes the hospital's resource allocation, and reduces economic losses caused by device failures. The application of intelligent calibration systems has greatly improved the utilization and operational safety of medical devices [7, 8]. The significant reduction in manual operation time and labor costs has improved the overall efficiency of the hospital, ensuring that patients receive timely diagnosis and treatment services. At the same time, the remote monitoring function of the system enables technicians to grasp the parameter status of hospital devices at any time.

2 Related Works

In previous studies, many scholars have proposed different solutions for the calibration and quality control of medical devices. In order to improve the accuracy and automation level of device calibration, Dowrick T [9] designed a customized calibration device for rapid calibration in surgical environments. He compared the device with manual calibration and evaluated the calibration quality using metrics such as stereo reprojection, stereo reconstruction, tracking stereo reprojection, and tracking stereo reconstruction error. Sabir A [10] studied the performance of several sensors in low-cost hardware and high acceleration environments. He used accelerometers, gyroscopes, and magnetometers to estimate Euler angles and proposed a robust and easy to implement calibration method that can calibrate sensors without any external device. Meanwhile, Lu Y H [11] introduced a new method for device calibration using custom driven laser beam sensors, which solved the problem of tool center point calibration in robot manipulators used in the automation industry. However, these systems still have problems such as poor real-time performance and difficulty in remote operation, and have not effectively solved pain points such as scattered devices and complex manual operations. Therefore, traditional systems are still inadequate in handling large-scale device management, real-time monitoring, and calibration. However, existing methods still have limitations in automatic calibration and fault warning of devices, with insufficient intelligence level, and cannot automatically calibrate when the device deviates. By combining wireless communication technology and intelligent algorithms, it is expected to solve these shortcomings of the current system. Therefore, this article proposes an intelligent calibration system based on wireless QA control to improve the efficiency and accuracy of device management.

3 System Implementation Based on Wireless QA Control

3.1 Design of Wireless Communication Module

The core of wireless communication module design is to achieve real-time data transmission and remote control between medical devices and central control systems. This module adopts a dual-mode communication method of Wi-Fi and Bluetooth Low Energy (BLE) to meet the requirements of large capacity data transmission and low power consumption of devices. Wi-Fi modules (such as ESP8266) are used for transmitting device status and calibration data, such as operating time, temperature, pressure, etc., to the central control system through wireless networks for remote monitoring and real-time data analysis. The BLE module is used for short distance, low-power communication, suitable for data interaction between devices and mobile terminals (such as technicians' phones or tablets), improving portability and flexibility. During the data transmission process, MQTT protocol is used for lightweight data encapsulation and transmission, ensuring efficiency in low bandwidth environments. The MQTT (Message Queuing Telemetry Transport) protocol transmits device identification, status, and calibration results to the control system via Wi-Fi or BLE, enabling parallel data processing across multiple devices. To ensure data security, the system uses AES-256 encryption algorithm to encrypt the data, and combines a two factor authentication mechanism to ensure the security and reliability of remote operations and data transmission. The central control system uses an MQTT server to receive and process device data. In combination with edge computing, some data and calibration instructions are processed locally on the device side to reduce delay and improve system response speed and real-time performance.

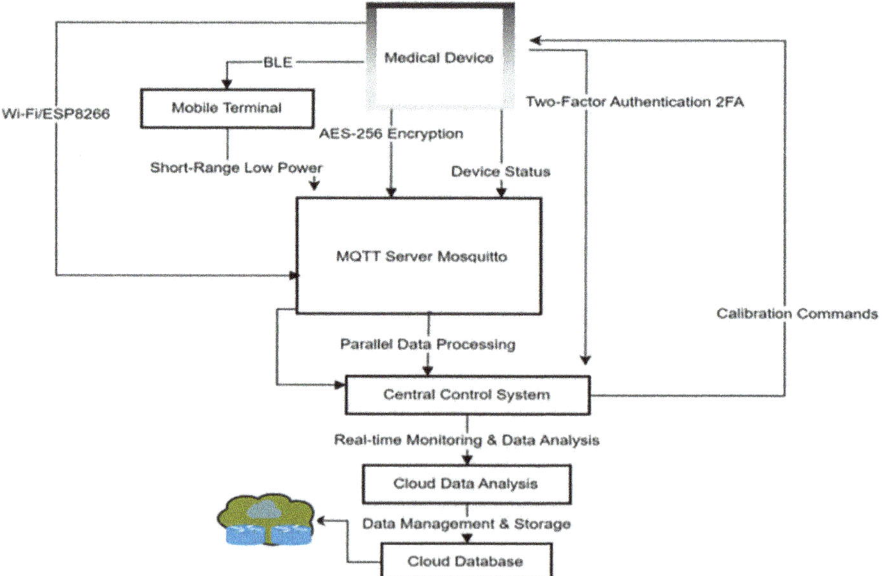

Fig. 1. Workflow of intelligent calibration system for medical device testing

In Fig. 1, the medical device transmits its status and calibration data in real-time to the MQTT server Mosquitto through the Wi-Fi (ESP8266) module, which acts as a lightweight protocol intermediary to ensure parallel data processing across multiple devices. The data transmission of the device includes important parameters such as operating time, temperature, pressure, etc., while using AES-256 encryption algorithm to ensure data security. The central control system receives device data through MQTT servers and monitors and analyzes it in real-time. The system integrates edge computing technology, and some data is directly processed locally on the device side to reduce delay and improve system response speed. When the device encounters an abnormality, the central control system sends calibration instructions to the device through the MQTT server, and the device performs automatic calibration upon receiving the instructions. On the other hand, Bluetooth low-power modules are used for short-range data transmission between devices and mobile terminals (such as technicians' phones or tablets), improving operational convenience. Figure 1 also shows that the system enhances the security of remote operations through two factor authentication, and all data is finally stored in a cloud database to support subsequent data management and analysis. This process achieves real-time monitoring and remote calibration of medical devices through the lightweight MQTT protocol and wireless communication methods such as Wi-Fi and BLE, significantly improving the efficiency and security of the system.

3.2 Integration of Intelligent Calibration Algorithms

In order to achieve self-testing and automatic calibration of medical devices, an intelligent calibration algorithm based on machine learning is adopted. This algorithm automatically determines whether the device deviates from the standard by analyzing the real-time operating data of the device, and triggers the calibration program when necessary. The status evaluation function is used to comprehensively evaluate the operational status of the device, as shown in Formula 1:

$$S_{eval} = \sum_{i=1}^{n} w_i \cdot f_i(x_i) \tag{1}$$

The performance of the device under different working conditions is calculated using the weighting function S_{eval}. w_i represents the weight of each parameter x_i, and $f_i(x_i)$ is the feature function related to the parameter. By adjusting the weight w_i, the parameters that have the greatest impact on the device status can be effectively highlighted, achieving a comprehensive evaluation of the device status. The information gain in feature selection is shown in Formula 2:

$$IG(X,Y) = H(Y) - H(Y|X) = -\sum_{v \in Y} P(y)\log P(y) + \sum_{r \in X} P(x) \sum_{v \in Y} P(y|x)\log P(y|x) \tag{2}$$

The information gain $IG(X,Y)$ measures the degree to which the uncertainty of the target variable is reduced after applying feature X. Among them, $H(Y)$ is the entropy of the target variable, representing pure randomness without any information, while

$H(Y|X)$ is the conditional entropy of the target variable when the feature X is known. By selecting features with higher information gain, the predictive performance of the model can be effectively improved. The cost function for optimizing the calibration model is shown in Formula 3:

$$J(\theta) = \frac{1}{m} \sum_{j=1}^{m} (y_j - h_\theta(x_j))^2 \tag{3}$$

Formula 3 defines the cost function for calibrating the model to evaluate its predictive performance. The cost function $J(\theta)$ represents the mean square error between the predicted and actual values of the model, where m is the number of samples and y_j is the actual observed value. $h_\theta(x_j)$ is the predicted value calculated by the model based on the parameter θ. By minimizing the cost function and optimizing the calibration model, it can more accurately adjust the device parameters to ensure that the device operates in a standard status.

Firstly, the system collects key data from the sensors of the device, such as temperature, pressure, vibration, etc. These data are denoised and normalized for model training. To improve model performance, time-domain and frequency-domain analysis methods are used to extract data features, and recursive feature elimination algorithms are used for feature selection to ensure that the model can effectively capture device status changes.

Next, supervised learning models such as random forests or support vector machines are used to train based on historical device data. The model learns the operating status of the device under different working conditions, establishes standard working intervals, and defines the calibration accuracy requirements of the device. When operating in real-time, the system inputs the current data of the device into a trained model to predict whether the device deviates from the standard. When the data deviates from the normal working range, the system determines whether calibration is needed based on the evaluation results of the model. The calibration program automatically starts, and the system adjusts the operating parameters of the device to ensure that the accuracy of the device is restored to the preset range. The core of this method lies in the adaptive learning ability of machine learning algorithms to achieve accurate prediction of device operating deviations, combined with device self-calibration mechanism, reducing the need for manual intervention, and improving the efficiency and accuracy of calibration.

Figure 2 shows the curve of the training error and validation error in the machine learning-based intelligent calibration algorithm after 100 iterations. The blue solid line represents the training error. As the number of iterations increases, the training error gradually decreases from 33 to nearly 0, indicating that the model has optimized the fitting of the device's operating data and captured the operating characteristics of the medical device during the learning process. The iterative trend of validation error is roughly the same as that of training error, and its decrease is usually lower than that of training error, indicating that the intelligent calibration algorithm performs well on training data.

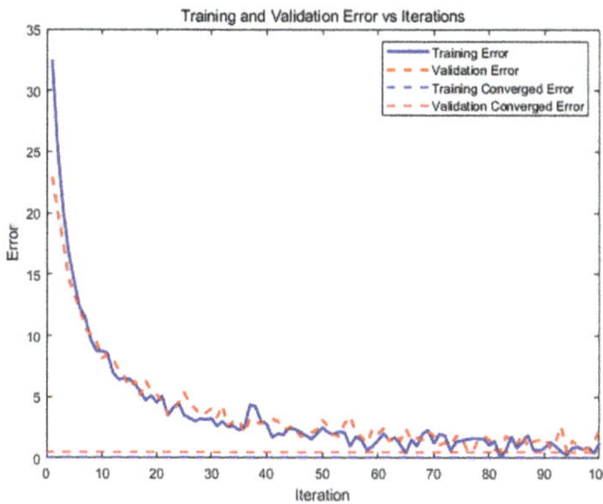

Fig. 2. Iterative process of training error and validation error in intelligent calibration algorithm

3.3   Real-Time QA Monitoring System

Real-time Quality Assurance (QA) system continuously monitors the temperature, pressure, vibration, and operating status of medical devices to ensure its condition. The system collects operational data from devices through an integrated sensor network, and then transmits this data in real-time to the central control system. After data collection, the raw data is denoised and smoothed using the sliding average algorithm.

Table 1. Monitoring results of key parameters

Parameter Name	Raw Data	Processed Data	Monitoring Result
Temperature	25.5 °C	25.4 °C	Normal
Pressure	101.3 kPa	101.2 kPa	Normal
Vibration	0.02 g	0.015 g	Slightly Deviating
Operating Status	Running	Running	Normal
Humidity	40%	39.50%	Normal
Power Consumption	150 W	145 W	Normal

Table 1 shows the monitoring results of the real-time quality assurance system for medical devices on key parameters, including temperature, pressure, vibration, operating status, humidity, and power consumption. The raw data in Table 1 reflects the current working status of the device, with a temperature of 25.5 °C and a pressure of 101.3 kPa, indicating that it is within the normal range. However, after processing with the sliding average algorithm, the values of temperature and pressure slightly change, dropping to 25.4 °C and 101.2 kPa respectively, further verifying that the device remains stable during the monitoring process. In addition, the vibration parameters show a slight deviation, with the original value of 0.02 g decreasing to 0.015 g after processing, indicating that there may be a slight mechanical issue with the device that requires further attention. The operating status remains "running", indicating that the device has not malfunctioned. The humidity and power consumption data are 39.5% and 145 W, respectively, both displayed within the normal range. Overall, these data fully demonstrate the importance of intelligent calibration systems in real-time monitoring of device status, ensuring its operational stability.

The real-time QA system adopts anomaly detection algorithm and analyzes the operating status of devices based on Z-score detection method. The system learns the historical operating data of the device and sets standard thresholds based on the normal operating range of the device. When the real-time parameters of the device deviate from the normal range, the anomaly detection model recognizes the abnormal status and automatically triggers an alarm. Meanwhile, the system determines the severity of deviations through its built-in intelligent decision-making module. When the deviation is within the tolerance range, the system continues to monitor; once the deviation exceeds the preset threshold, the automatic calibration process is triggered.

4   Results and Discussion

4.1   Calibration Accuracy

To evaluate the accuracy changes before and after device calibration, standard deviation and mean square error (MSE) are used as evaluation indicators for the calibration effect. Firstly, the system collects key parameters through the device's sensors and records the data before calibration. Subsequently, the system launches an intelligent calibration algorithm based on machine learning to calibrate the device, adjust its parameters, and restore it to its standard working status.

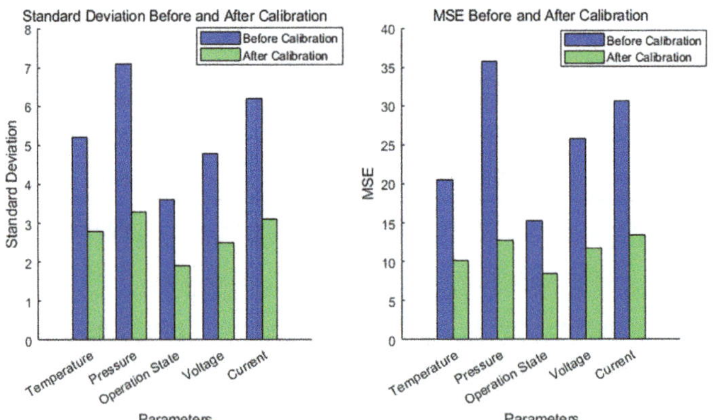

Fig. 3. Comparison of standard deviation and mean square error (MSE)

Figure 3 compares the standard deviation and mean square error (MSE) of the device before and after calibration on five key indicators: temperature, pressure, operating conditions, voltage, and current. The standard deviation of temperature before calibration in the left figure of Fig. 3 is 5.2, which decreases to 2.8 after calibration, indicating that temperature fluctuations are reduced and the stable operation of the device is improved; the standard deviation of pressure is reduced from 7.1 to 3.3, indicating an improvement in the precision of pressure control of the calibrated device. The figure on the right shows the change in mean square error. The MSE of the voltage before calibration is 25.8, but decreases to 11.7 after calibration, indicating a significant reduction in the degree of deviation of the voltage parameters from the standard values, and the operation of the device is more in line with the expected standards. In addition, the MSE of the operating status decreases from 15.3 to 8.4, indicating a significant improvement in the overall performance of the device through intelligent calibration algorithms. These data fully demonstrate that machine learning-based intelligent calibration algorithms effectively improve the accuracy and stability of devices.

After calibration, the same parameters of the device are collected again, and the data before and after calibration is compared and analyzed. The fluctuation range of device operating parameters is evaluated by calculating the standard deviation of data before and after calibration. Meanwhile, the mean square error is used to measure the degree of deviation of device parameters from standard values before and after calibration. Finally, the system uploads the error comparison results before and after calibration to the central control system, so that technicians can further analyze the calibration effect and optimize the device's operational performance.

4.2 Calibration Time Evaluation

The calibration time of the device is recorded. The average calibration time is compared with traditional manual calibration time, and the percentage of time reduction is calculated.

Table 2 compares the calibration efficiency of four devices under traditional manual calibration and intelligent calibration methods. In traditional manual calibration, the average calibration time for device A is 35 min, and the total calibration time is 350 min; device B takes 28 min, with a total time of 280 min; device C takes 40 min, with a total time of 400 min; device D takes 25 min, with a total time of 250 min, indicating that manual calibration takes a long time. After using intelligent calibration, the average calibration time of device A is shortened to 12 min, and the total time is reduced to 120 min, a reduction of 65.71%. Device B is reduced from 28 min to 9 min, with a total time of 90 min and a reduction rate of 67.86%. Device C is reduced to 15 min, with a total time of 150 min and a reduction rate of 62.50%. Device D decreases from 25 min to 8 min, for a total time of 80 min, with a reduction rate of 68.00%. These results indicate that the intelligent calibration system significantly improves calibration efficiency, reducing the average calibration time by 66%, demonstrating its importance in medical device management.

4.3 System Response Time Evaluation

The response time of the measurement system from device anomaly detection to calibration program initiation is measured, and the average response time is calculated. In order to analyze and compare the average response time of 20 devices and compare it with the response time of traditional methods, this article collects response data from

Table 2. Calibration time of devices

Device Name	Calibration Method	Average Calibration Time (minutes)	Calibration Count	Total Calibration Time (minutes)	Time Reduction Percentage
Device A	Manual Calibration	35	10	350	-
Device A	Wireless CalibrationControl	12	10	120	
Device B	Manual Calibration	28	10	280	
Device B	Wireless CalibrationControl	9	10	90	
Device C	Manual Calibration	40	10	400	
Device C	Wireless CalibrationControl	15	10	150	
Device D	Manual Calibration	25	10	250	
Device D	Wireless CalibrationControl	8	10	80	

the devices, as shown in Fig. 4. One group is the average response time of an intelligent calibration system based on wireless QA control, and the other group is the average response time of traditional manual calibration methods.

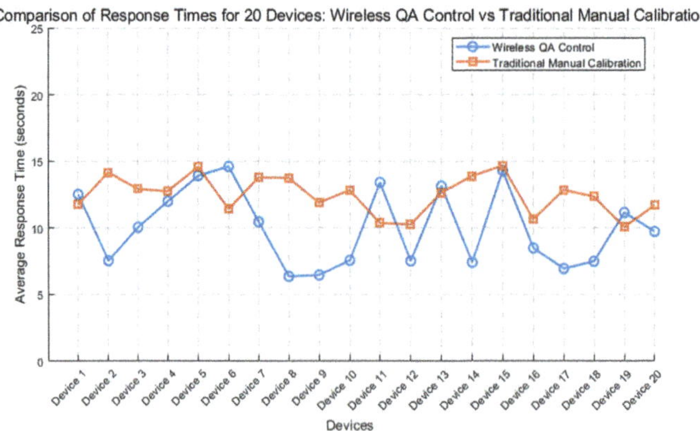

Fig. 4. Comparison of response time on 20 medical devices

Figure 4 compares the response time of an intelligent calibration system based on wireless QA control with traditional manual calibration methods on 20 medical devices. The blue line in the figure represents the wireless QA control method, which has a response time between 5 and 15 s. Devices 8 and 9 are particularly fast, overall showing excellent flexibility and stability. This indicates that wireless QA control can quickly initiate the calibration process after detecting device abnormalities, effectively reducing device downtime and improving device utilization. In contrast, the traditional manual calibration method represented by the red line has a response time that fluctuates between 10 and 15 s, revealing the lag in monitoring and response of this method, which is not conducive to the smooth operation of the device. In summary, the data clearly indicates the significant advantages of wireless QA control in improving device management efficiency, reducing failure rates, and enhancing security.

5   Conclusions

In this article, an intelligent calibration and detection system for medical devices based on wireless QA control is constructed, which integrates wireless communication technology, intelligent calibration algorithm and real-time QA monitoring mechanism to realize remote calibration and dynamic monitoring of medical devices. Through wireless networks, devices can communicate with the central control platform for data exchange, and calibration can be completed without the need for technical personnel to be present on site, greatly improving operational efficiency. Based on machine learning technology and intelligent calibration algorithms, the system can autonomously identify device deviations and accurately regulate their status, ensuring optimal operation of the device. Calibration data is stored uniformly in the cloud, providing convenience for subsequent data analysis and management work. Although there is still room for improvement in the performance of the system in complex networks and high concurrency scenarios, future research can focus on enhancing the robustness of the system, optimizing data processing flows, and expanding the adaptive range of intelligent algorithms to adapt to more diverse devices and application scenarios.

References

1. Petti, P.L., Rivard, M.J., Alvarez, P.E., et al.: Recommendations on the practice of calibration, dosimetry, and quality assurance for gamma stereotactic radiosurgery: report of AAPM Task Group 178. Med. Phys. **48**(7), e733–e770 (2021)
2. Liang, Y., Li, C., Liu, S., et al.: A 14-b 20-MS/s 78.8 dB-SNDR energy-efficient SAR ADC with background mismatch calibration and noise-reduction techniques for portable medical ultrasound systems. IEEE Trans. Biomed. Circ. Syst. **16**(2), 200–210 (2022)
3. binti Abdullah, N., Noor, N.M., Dolah, M.T., et al.: Precision and reliability: Calibration coefficients and long-term stability analysis of radiotherapy dosimeters calibrated by SSDL, Nuklear Malaysia. Asian J. Med. Technol. **3**(2): 15–32 (2023)
4. Jindal, D., Kaur, H., Patil, R.K., et al.: Validation–In pharmaceutical industry: equipment validation: a brief review. Adesh Univ. J. Med. Sci. Res. **2**(2), 94–98 (2020)
5. Özgüner, O., Shkurti, T., Huang, S., et al.: Camera-robot calibration for the da vinci robotic surgery system. IEEE Trans. Autom. Sci. Eng. **17**(4), 2154–2161 (2020)
6. Jiang, J., Luo, X., Luo, Q., et al.: An overview of hand-eye calibration. Int. J. Adv. Manuf. Technol. **119**(1), 77–97 (2022)
7. Jannah, N., Syaifudin, S., Soetjiatie, L., et al.: Simple and low cost design of infusion device analyzer based on Arduino. Indonesian J. Electron. Electromed. Eng. Med. Inform. **2**(2), 80–86 (2020)

8. Yu, H., Li, Q., Wang, R., et al.: A deep calibration method for low-cost air monitoring sensors with multilevel sequence modeling. IEEE Trans. Instrum. Meas. **69**(9), 7167–7179 (2020)
9. Dowrick, T., Xiao, G., Nikitichev, D., et al.: Evaluation of a calibration rig for stereo laparoscopes. Med. Phys. **50**(5), 2695–2704 (2023)
10. Sabir, A., Zakriti, A.: Simplification of calibration of low-cost MARG sensors without high-precision laboratory equipment. Int. Rev. Appl. Sci. Eng. **14**(2), 170–182 (2023)
11. Lu, Y.H., Lin, Y.W., Hsu, S.C., et al.: Equipment calibration with a laser interruption sensor. Int. J. iRobotics **7**(1), 8–13 (2024)

Optimization of RFID Tag Recognition and Data Security System in Smart Libraries Integrating Embedded Technology

Yuanyuan Sun[✉]

Shandong Vocational College of Light Industry, Zibo, Shandong Province, China
Sunyysd@email.cn

Abstract. In order to solve the problems of low efficiency in borrowing and returning books and poor storage environment in libraries, the writer puts forward an RFID Smart Library Management System. Firstly, introduce key technologies such as RFID technology, ultra-high frequency electronics, and radio frequency technology. Secondly, the library management was combined with RFID technology, embedded technology, and network communication technology to complete the design and implementation of three sub modules: book management, computer anti-theft, and environmental monitoring. The above three sub modules are organically integrated into a new generation of intelligent library management system, fundamentally releasing the pressure of personnel when borrowing and returning books, enhancing library security, and improving the environmental quality of book preservation. Finally, test the system performance. The results show that under the application background of RFID technology, the system has the characteristics of simple book return process, high book positioning accuracy, convenient and fast search, and safe and reliable, fully meeting the expected design standards and requirements.

Keywords: Library · RFID · Book management · Computer anti-theft · environmental monitoring

1 Introduction

With the rapid development of information technology, various new information technologies emerge one after another and have been widely applied in society. At the same time, with the rise of technological revolutions in various fields of society, the current information management of libraries is no longer able to continue to adapt to the needs brought about by the continuous expansion of library scale, the increasing number of literature, and the growing demand for readers. How to adopt more advanced information technology to improve the management and service quality of libraries has become a new challenge. At present, the management of libraries has gone through multiple stages such as manual management, data informatization, and barcode [1]. The traditional manual management method was that before the popularization of computer technology, the service process of the library had to be manually managed; In the stage

P. Siarry et al. (Eds.): WCNA 2024, LNEE 1550, pp. 25–35, 2026.
https://doi.org/10.1007/978-981-95-6946-5_3

of data informatization, libraries use information technology to manage the literature and reader information in the library, greatly improving the efficiency of library management; In the stage of barcode management, the library combines barcode technology and computer technology to convert manual input into manual barcode scanning in book data operations, completely freeing library management personnel from manual operations and further improving the efficiency and service quality of library management [2]. Compared with barcode technology, RFID technology has the following advantages: RFID electronic tags store more abundant information; The information stored in RFID electronic tags can be modified; RF tags are difficult to replicate and have higher security; Through wireless transmission, the range of information transmission is wider and not limited by field of view [3]. RFID technology initially emerged in the field of logistics warehouse management. With the improvement of the performance of RFID electronic tag reading and writing devices and the reduction of tag costs, RFID technology has been widely applied in various fields of society [4]. The author proposes to combine the library intelligent management system with advanced RFID technology, embedded technology, and network communication technology to complete the design of a new generation of library intelligent management system, which mainly includes subsystems such as book management, computer anti-theft, and environmental monitoring [5].

2   Literature Review

The construction and application of smart libraries can help people intelligently and accurately collect, organize, and collect books, providing people with a highly interactive, free, and flexible reading experience. Radio Frequency Identification (RFID) as an automatic recognition technique in general, mainly includes components such as antennas, electronic tags, and card readers. It has the characteristics of fast scanning, strong readability, and safety and reliability. By applying this technology to the design of library intelligent management systems, it can not only ensure the 24-h self-service book return function, but also intelligently and efficiently organize disorderly books, simplify the workload of librarians, shorten the waiting time of readers, and effectively improve the level of library intelligent management. Recently, many researches have been made on how to apply RFID to smart library systems: Ude, E. N., and other researchers put forward an RFID system reliability model that considers the dependence of failure count on two procedures. Finally, when assessing the reliability of the software, we incorporate the failure rate and the test coverage as a parameter, which allows us to perform better estimation and prediction of software reliability than current NHPP Software Reliability Growth Model (SRGMs) [6]. Vedanth, V. S. and others has adopted barcodes to identify, register, steal, store, and classify. The proposed system is able to store the data in electronic form so that the reader can read it. Using RFID labels, it is possible to issue and accept the books. It can also compute related elements like expiry, renewal, and penalties from the library database [7]. Andhare et al. proposed a LMS for Automatic Library Management System (RMI), which is based on RFID (RFID). Their design allows for the identification and tracing of lots of tagged books by means of radio waves. The LMS is designed to offer an interactive gateway to show the available information about the library, to distribute the books, to follow up on the published books, and to bring back the books. The Internet of Things makes it easier to collect data on a cloud platform so that it can be remotely accessible in a library system. This will make it easier to borrow, update, and return RFID labels [8]. Based on this research, the author proposes a design of an RFID based library intelligent management system. Through the deep integration of multiple technologies, this intelligent library management system provides libraries with more comprehensive management functions and safer management modes.

3 Research Methods

3.1 Introduction to RFID Technology

RFID technology was first applied during World War II, when it was used to determine whether an aircraft flying into a country's airspace was an enemy aircraft or a British aircraft. By the 1880s, RFID technology was mostly used in national military affairs and had not been widely popularized. By 1977, the National Laboratories of the United States began to apply RFID technology to daily life and it had been widely used. At that time, RFID applications were mostly in the low-frequency stage, with high costs, limited information transmission, and short communication distances. Then, as high frequency RFID technique developed, RFID could be used extensively in the community. RFID is a kind of noncontact auto-recognition technique, which makes use of the space coupling property of RF signal to realize information communication. RFID is one of the most common sensing techniques [9]. A complete and sound RFID system requires not only readers and electronic tags, but also storage and host systems. An RFID system architecture is illustrated in Fig. 1.

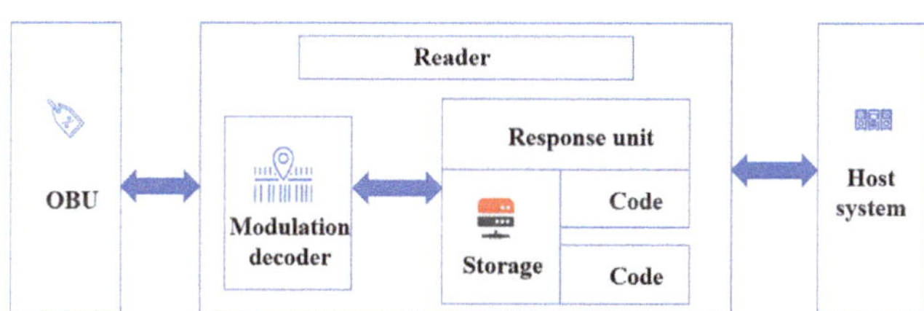

Fig. 1. RFID system structure

The working principle of the system is as follows: using a reader and antenna transmission, it completes the unified transmission of a certain frequency RF signal. When an electronic tag appears in the identification area of the RF signal, the electronic tag can be successfully activated and generate corresponding induced current. Subsequently, the electronic tag sends the RF signal uniformly to the reader through the antenna. The reader then uses modulation and demodulation to decode and process the electronic tag information, and sends and transmits it to the host system for unified processing. Ultra high frequency electronics mainly refer to electronic technology located in the frequency range of 300 MHz~3 GHz. Radio frequency technology mainly refers to electronic technology located in the frequency range of 30 kHz~300 GHz. In specific applications, ultra-high frequency electronics and radio frequency technology mainly involve wireless design, radio frequency circuit design, signal transmission, etc. By applying this technology to the field of wireless communication, it can help people to communicate wirelessly anytime and anywhere without leaving their homes [10].

3.2 Composition of RFID System

According to the usage of RFID, there are two categories: Passive Label and Active Label. The active label is mostly used to follow precious objects and to monitor the surroundings, whereas the passive label is usually applied to stock an asset with a moving reader or a stationary reader. There are many factors that affect the read and write range of passive ultra-high frequency RIFD: antenna transmission power, reader read and write frequency, and environmental interference. The card reading rate is given in Table 1. A comparative study of passive RFID techniques is presented. LF or HF are used as the basis of electromagnetic induction or near field coupling. The low-frequency operation is performed in a low-band (e.g., 134 kHz) and therefore has a smaller reading range. But low frequency labels are less susceptible to environmental disturbances, so they can be used more often, for example, in an environment where there is water or metallic material. Thus, they are able to be embedded within a biological organism to identify and locate it, as well as to efficiently collect information about the stock in a plant [11].

Table 1. Operating frequency and reading range of passive ultra-high frequency RFID

Frequency range	Typical frequency	Read range	Tag Price
Low Frequency, LF	125–134 kHz	<0.5 m	1
High Frequency, HF	13–56 MHz	<1.0 m	0.5
Ultra High Frequency, UHF	865–928MHz	1–10 m	0.15

In general, ultra-high-frequency RFID is a method of transmission of power by means of wireless communication. Different from the induction near field RFID, the remote UHV RFID (also called far field RFID) is based on the backscatter modulation. More specifically, an RFID antenna transmits an electromagnetic field produced by a reader, and because the signal is gradually attenuated in the course of freedom transmission, only a little power is transmitted to the label antenna. The label's absorption of the incoming EM waves has two main functions: a part of this power is used to supply power to the circuitry within the label, and the rest is reflected. The reader's design is to capture and decipher this signal by modulation of the label's message (that is, identification) into a backscattered signal of the label [12]. Apart from the identification of labels, RFID readers can also offer a signal intensity, i.e., the power of the label. Some of the readers, for instance, offer a signal strength in dBm units (e.g., Impinj's Speedway reader), while others offer signal strength without units (e.g. ThingMagic's Mercury 5e reader). Passive labels should be driven by the electromagnetic field transmitted by the reader. Based on the Friis Equation in Formula (1), the incoming power of the tag is extremely small. Two conditions have to be satisfied for successful read-out of the passive UHF RFID label.

Firstly, the label has to get enough power from the reader to activate the inner circuitry; and secondly, the reader has to be sufficiently sensitive to pick up the label's reaction.

$$P_r = G_r G_t \left(\frac{\lambda}{4\pi \pi R} \right)^2 \cdot P_t \qquad (1)$$

3.3 Overall Design of Library Intelligent Management System

The Smart Management System of Library is an integrated scheme of RFID and Embedded Technology, and network communication technology, including sub modules such as book management, computer anti-theft, and environmental monitoring. The book management module utilizes RFID technology to embed RFID tags in all books, achieving automated management of book borrowing and returning in a unified borrowing and returning mode; The computer anti-theft module is a billing based computer anti-theft system developed to address the phenomenon of theft of computers and other items caused by leaving one's seat in public places. It uses mobile booth anti-theft technology to supplement the campus one card billing function; The environmental monitoring module uses sensors such as temperature, humidity, and gas monitoring to detect the overall environmental conditions of the borrowing room in real time, and is combined with an air purification processor to provide an excellent environment for readers to borrow and store books.

3.4 Book Management Submodule Design

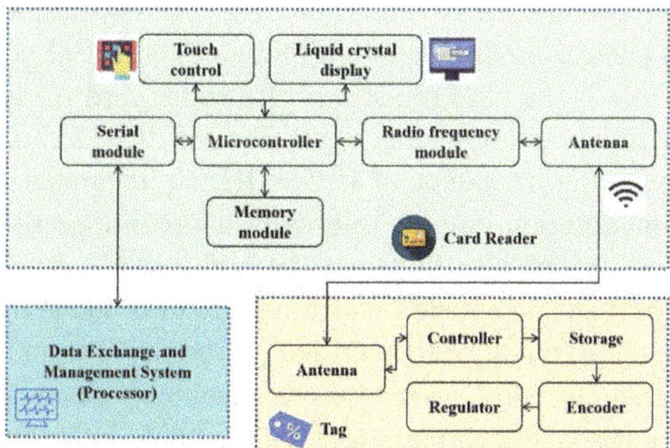

Fig. 2. Design Block Diagram of Book Management Module

Library administration sub-module is realized by RFID system, which can be used for self-use. RFID, also called RF recognition, is a kind of noncontact auto recognition technique. The RFID hardware is generally composed of 4 components: electronic tags, system design antennas, readers, and data exchange and management systems (processors). Electronic tags, also known as electronic responders, are the carriers of RFID systems, consisting of electronic chips that store information and built-in wireless antennas. Each tag has its unique code, and the reader/writer is responsible for reading or writing tag information and displaying and processing data. Data Interchange and Management System is used to store and manage the data, as well as the read and write control of cards. All books will be identified using RFID radio frequency tags, which record information such as title, author, publisher, call number, and ISBN/ISSN. Set up borrowing and returning areas in the library, using high-performance radio frequency card readers as automatic borrowing and returning machines. After verifying the identity of the campus card reader, the card reader sends electromagnetic waves to detect valid radio frequency tags around the books. After reading and processing the radio frequency tag data, the data is sent to the data exchange and management system through RS232 serial port, and eventually it will write the data to a database, which will finish self-lending and reposting. The design diagram of the book management module is shown in Fig. 2.

3.5 Hardware Design

The radio frequency tag adopts the library's 13 56 MHz high-frequency passive tag, which is powered by a card reader during operation. The radio frequency tag working at 13.56 MHz is mainly suitable for multi tag recognition and short-range recognition, with a recognition distance of 0–1.2 m. When operating at a baud rate of 2 kHz–4 kHz, it can simultaneously read data from 6–8 radio frequency tags. The RF tag is not damaged by human factors and has a read and write frequency of no less than 100000 times. Each book in the library is identified by a 64 bit UID radio frequency tag, which can meet the needs of the vast majority of campus libraries' book collections.

In practical applications, the control function of the microcontroller is slightly different from the communication of the data exchange and management system. The card reader is mainly composed of a main controller, RF communication module, LCD display module, touch control module, storage module, and serial port module. The RF communication module is the core module of the entire card reader, consisting of the MFRC632 chip, matching circuit, and antenna. The MFRC632 chip is suitable for non-contact tags with a working frequency of 13.56 MHz. It integrates encoding, decoding, modulation, and demodulation circuits internally and can directly drive antennas for close range operation. In order to filter out the high-order harmonics generated during the modulation process, an antenna matching circuit is added to the module. When MFRC632 operates, the signal to be transmitted is modulated from the data in the transmission buffer and sent out in the form of electromagnetic waves through the antenna. The RF tag responds with load modulation of the RF field. After the antenna detects the response signal of the RF tag, the signal is processed by the antenna matching circuit and sent to the receiving pin. The internal receiving buffer detects and demodulates the signal, and the processed data is sent to the parallel data interface for reading by the main controller.

3.6 Software Implementation

The software implementation of the book management module mainly consists of two parts: the control program of the card reader module and the database management program on the PC. The self borrowing software process in the book management module is shown in Fig. 3, and the self returning software process is similar to it.

The control program in the card reader module includes subroutines such as communication with RF tags, communication with databases, touch screen control, LCD display, and receipt printing. All programs are designed in C language.

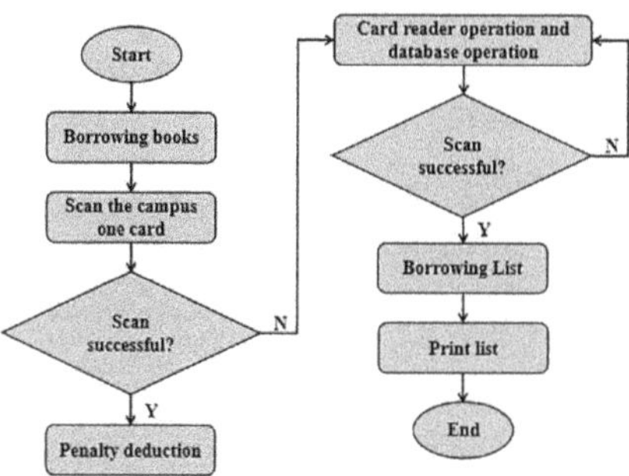

Fig. 3. Self borrowing software process

Anti collision design is an important part of the communication subroutine between the card reader module and the RF tag. When the card reader is working, when borrowing and returning multiple books at the same time, that is, when multiple RF tags need to communicate with the card reader at the same time, there will be mutual interference between the tags, which may cause the card reader to be unable to identify any RF tag, which is called tag collision in RF communication. In order to enable the card reader to read multiple tags simultaneously, the ISO 18000 TypeA standard specifies the use of time slot Aloha algorithm. The author adopts an adaptive time slot Aloha anti-collision algorithm, and its principle will not be further described. The data sent by the card reader to the data exchange and management system is stored in a program buffer on the PC. The database management program allocates a background thread to constantly check whether the program buffer is empty. Once the program buffer is checked for data, the data is read and processed before being written to the database.

3.7   Computer Anti-Theft Submodule Design

Computer anti-theft is a submodule that integrates the existing campus one card management system (including document management, account management, and other public information management, as well as the classification, summary, statistics, and query of accounts) with the anti-theft system. Mainly including balance display, deduction and billing management systems, as well as general service fee management systems. This module is fully integrated with the existing campus one card and connected to the campus network information center database. The anti-theft platform can be rented on a per use or on time basis through the one card system to prevent theft of items after personnel leave. The system adopts one card control and automatically locks when entering the anti-theft state. No device can be disarmed until an alarm is triggered. The anti-theft device adopts a universal "mouse hat" design, which can adapt to different types of laptops. The anti-theft module is equipped with an independent alarm, which can achieve the effect of circuit breaker alarm. Dual color indication of work and alarm status, disconnection alarm, which means cutting off any connection of the platform to trigger an alarm, and three contact anti-theft is safer (base, machine back, and mouse cap), which means that any contact will trigger the alarm when it bounces up, and the alarm is an independent type. At the same time, the embedded devices inside the anti-theft device push alarm information to the user's mobile phone through communication methods such as GPRS/3G, achieving timely detection and prevention of theft incidents.

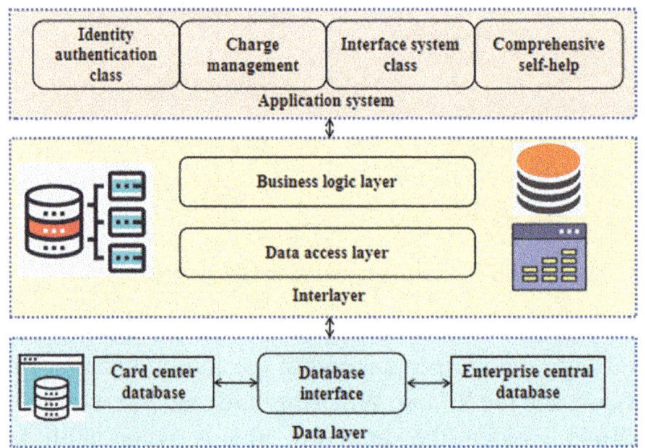

Fig. 4. Software Block Diagram of Computer Anti theft Module

The main technical points of the computer anti-theft module lie in the software framework of the anti-theft system, including identity authentication, fee management, and interface system. With the help of the existing one card database, seamless integration with other one card devices can be achieved. Figure 4 shows the software framework of this module. The identity authentication category adopts the existing one card authentication method, while the fee management category allows the school to configure and set charges on a per use or per time basis. The comprehensive self-service category refers to providing value-added services such as SMS alarms, GPS tracking, and network communication.

4   Results Analysis

4.1   Test Plan

The purpose of system testing is to verify the effectiveness and reliability of the system. Starting from load performance testing and security testing, the focus is on testing the implementation effects of the functions of the circulation workstation, self-service borrowing and returning, and book anti-theft detection modules. Firstly, set up a testing environment. Set the server hard disk memory to 3000 GB and the workstation CPU memory to 128 MB, using the Windows 2010/XP operating system, and build a SQLServer2000 database. Secondly, determine the scope of system testing. Focus on testing the response time, load performance, and reliability of the three major modules mentioned above. Finally, clarify the system detection content. Using manual testing, focus on testing the above three modules, and comprehensively observe and record the performance of each module, such as response time and server response speed.

4.2 Test Cases and Analysis of Test Results

Circulation Workstation Module Testing: During the testing of the circulation workstation module, 60 sample books are randomly selected from a large number of books, and the circulation workstation is comprehensively tested to see if it meets the relevant design standards and requirements. The test results are shown in Table 2:

Table 2. Test Results of Circulation Workstation

Test Items	Test Result
Response speed (tags/second)	Greater than 20
Reading range radius (satellite communication)	110
Conversion speed (volume/hour)	900
Tag readout rate (%)	Greater than 98
Power supply voltage (satellite communication)	5
Power consumption of power supply (satellite communication)	Less than 16
Operating frequency (MHz)	925~930

From the data in Table 2, it can be seen that the circulation workstation module provides readers with normal borrowing functions, and the book return operation is simple and smooth. Multiple books can be operated simultaneously. In addition, the module has a high response speed.

Self borrowing and returning module testing: Randomly select 1000 sample books from a massive collection of books and conduct self borrowing and returning module testing on them. The test results are shown in Table 3:

Table 3. Self service borrowing and returning test results

Testing Item	Test Result
Response speed (tags/second)	Greater than 20
Reading range radius (mm)	120
Conversion speed (volume/hour)	900
Signature reading rate (%)	Greater than 98

From the data in Table 3, it can be seen that readers are allowed to borrow and return multiple books at once, and the touch screen soft keyboard operation is smooth, with a signing rate of over 98%.

5 Conclusion

The author conducted a careful analysis and research on the problems that exist in the practical application of RFID technology, embedded technology, and other technologies in intelligent library management. They completed the design of three sub modules: book management, computer anti-theft, and environmental monitoring in the library, realizing the automation and intelligence of library management, improving and perfecting the existing operating mode of the library, and providing readers with better and more convenient services. In summary, under the application background of RFID technology, the system has the characteristics of simple operation, high book recognition rate, and high level of intelligent book management, fully meeting the practical application needs and providing important platform support for promoting the continuous development of libraries towards intelligence, digitization, and informatization. Although the feasibility and necessity of promoting library intelligent management systems have been proven in practice, their application is not very widespread in the current situation of libraries in China. This is mainly due to the existence of constraints such as inconsistent standards for RFID technology application in libraries and high initial investment costs. But with the rapid development of RFID technology today, the intelligent library management system will inevitably become more perfect, and the prospects for application and promotion are very broad.

References

1. Oluranti, F.E., Olorunleke, A.R., Abiodun, A.: Digital preservation skills of electronic information resources' management among library personnel in selected universities in nigeria. Eur. J. Comput. Sci. Inf. Technol. **11**(3), 75–87 (2023)
2. Sakamoto, R.: E-resource management situation and future challenges: the case study of seinan gakuin university library. Pharm. Libr. Bull. **67**(3), 96–101 (2022)
3. Liu, P.: Design and implementation of library seating management system. J. Comput. Commun. **12**(8), 15 (2024)
4. Kamath, A.K., Kumar, B., Aggarwal, S., et al.: An automated validation framework for power management and data retention logic kits of standard cell library **29**(06), 121–122
5. Mckinley, M.: Cultural adaptation of a text message library designed to support diet, activity and weight management behaviour in the postpartum period in the UK: the supporting mums (sms) study. Proceedings **91**, 111–114 (2024)
6. Ude, E.N., Nwamaka, E.A., Guha, K., et al.: Rfid library management software dependability through reliable fault-detection and fault correction procedures. Microsyst. Technol. **30**(5), 647–659 (2024)
7. Vedanth, V.S., Rohan, M.S., Sai, S.V., et al.: Rfid based automated library management system. Int. J. Res. Appl. Sci. Eng. Technol. **40**(09), 48–53 (2023)
8. Andhare, M., Bhangale, K., Kumbhar, V.S., et al.: Iot-enabled rfid-based library management and automatic book recommendation system using collaborative learning **23**(05), 135–137 (2023)

9. Wang, W.: Retracted article: optimization of book information search in intelligent library system management based on cellular network. Opt. Quant. Electron. **56**(10), 1 (2024)
10. Takahashi, C.: Aiming for library management on the university idea "higher education providing brilliant prospects." Nurs. Inf. J, Japan Nurs. Libr. Assoc. **29**, 56–57 (2022)
11. Glory, K.B., Venkatesan, D., Devi, G., Kiran, C.S., et al.: An effective storage management for university library using weighted k-nearest neighbor algorithm. IEEE North Karnataka Subsection Flagship Int. Conf. (NKCon) **2022**, 1–5 (2022)
12. Ji, Y.: Design and application of intelligent library intelligence authentication system based on information entropy early warning models. Appl. Math. Nonlinear Sci. **9**(1), 66–69 (2024)

Multi Layer Anti Intrusion Algorithm for Information Emergency Network Under Active Passive Combination

Huaikai Chen[✉]

Yunnan Medical Health College, Yunnan 650033, Kunming, China
18687161752@163.com

Abstract. In response to the problem of multiple types of intrusion data in information emergency networks, a multi-layer anti intrusion algorithm for information emergency networks is proposed, which combines active and passive methods. This strategy combines the technological advantages of active defense and passive detection, and constructs a multi-level and all-round protection system. Passive detection adopts a network coding method based on information theory security, which encodes and decodes data packets through linear operations to ensure the confidentiality of information during transmission; Active defense uses support vector machine algorithm to achieve intrusion detection. By introducing kernel functions to map data to high-dimensional space, the optimal separation hyperplane is found in the high-dimensional space and mapped back to the original space, thus achieving non-linear classification of intrusion data. The experimental results show that this method can accurately detect various types of network intrusion data, and it also has the ability of self-learning and optimization, which can continuously adapt to changes in the network environment and improve the protection effect.

Keywords: Information emergency network · Multi-layer anti intrusion algorithm · Network coding · Support Vector Machine · Active passive combination

1 Introduction

The information emergency network refers to an emergency communication network constructed using information technology to respond quickly, coordinate efficiently, and command accurately in response to natural disasters, accidents, public health emergencies, and other emergencies [1]. The construction of emergency communication networks includes the comprehensive use of various communication methods such as satellite communication, private network communication, and public network communication, establishing a unified emergency information platform to achieve rapid collection, processing, analysis, and sharing of information, and providing strong support for command and decision-making [2]. However, while the openness of the internet provides convenience for information sharing, it also makes emergency networks a target of malicious

P. Siarry et al. (Eds.): WCNA 2024, LNEE 1550, pp. 36–46, 2026.
https://doi.org/10.1007/978-981-95-6946-5_4

attacks and faces serious security threats. Especially in the context of the information age, the computer network operating environment is becoming increasingly complex, and various network attack methods are emerging, such as illegal access, virus infection, network vulnerability exploitation, etc., all of which pose serious challenges to the secure and stable transmission of information emergency networks [3]. To address these security threats, it is particularly important to build an effective network defense system.

Reference [4] proposes an optimization mode for an Industry 4.0 network intrusion detection system based on a bidirectional long short-term memory interpretable artificial intelligence framework. The interpretable artificial intelligence framework is used to perform preprocessing tasks such as data cleaning and normalization, in order to improve the data quality of network intrusion detection. Use Krill swarm optimization algorithm to select important features from the database. This method can accurately detect intrusions and provide better security and privacy within industry network systems, but its defense capability against unknown attacks has certain limitations. Reference [5] proposes an unsupervised domain adaptive triplet network based on hierarchical attention for network intrusion detection, introducing joint loss to force the hierarchical attention triplet network to learn compact and discriminative embeddings of benign network traffic, while staying away from representations of known attacks. Then, a class of support vector machine models is trained on the basis of benign embedding for unknown attack detection tasks. This method improves the detection rate of unknown attacks, but the detection efficiency is poor.

Traditional network security defense measures, such as firewalls and intrusion detection systems, although able to resist external attacks to a certain extent, are gradually becoming inadequate in the face of increasingly complex and diverse network attack methods. Passive defense measures typically only detect and respond to attacks after they occur, and cannot effectively prevent them before they occur. Active defense measures can quickly detect and respond to potential intrusion behaviors through real-time monitoring and analysis of network traffic or system logs, but their defense capabilities against unknown attacks still have certain limitations. Therefore, this article proposes a multi-layer anti intrusion algorithm for information emergency networks that combines active and passive approaches. By integrating the advantages of active and passive defense technologies, a multi-level and all-round defense system is constructed to improve the security and stability of network information transmission.

2 Multi Layer Network Intrusion Prevention Algorithm Combining Active and Passive Approaches

In response to the security vulnerabilities in emergency networks, this article proposes a security protection mechanism that combines active and passive measures. The purpose is to organically combine different levels of security protection measures, and implement network security protection from multiple aspects such as information protection, network attack monitoring, and emergency response through deep protection security configuration, ensuring the confidentiality, reliability, and availability of the network and transmitted information.

2.1 Passive Safety Strategy

Due to the openness of the information emergency network channel, the network is highly susceptible to attacks such as eavesdropping and interception, and the confidentiality of information faces enormous challenges. Traditional security strategies based on mathematical puzzles are used to complete security operations such as data encryption, identity authentication, and message authentication. Each end node needs to perform complex encryption and decryption operations, while ad hoc network nodes have limited energy and computing power, and are not suitable for complex encryption and decryption operations. At the same time, ad hoc networks are a decentralized network, making it difficult to find a reliable central node to complete key distribution, authentication, and other tasks.

Adopting network coding based on information theory security to ensure the confidentiality of information. Compared to traditional security strategies, network coding based on information theory security only needs to introduce a small amount of random information and mix it with the source message (such as not requiring random data according to weak security standards). The mixed information is meaningless to eavesdroppers and cannot obtain any meaningful information from it. Although network coding was initially proposed primarily to improve network data throughput, it allows for encoding operations on data packets, and the process of network coding is a data mixing process. This design principle increases the degree of coupling between data packets, so that the correct and complete decoding of transmitted information no longer depends on the successful reception of one or several data packets. In other words, the difficulty of deciphering transmitted information after being eavesdropped by attackers will increase, thus improving the security of network data communication. Different secure network encoding methods can be used in the design based on the number of encoded data packets [6]. When there are a large number of encoded data packets, network coding packet header compression method is used; When the number of encoded data packets is small, network encoding source construction method is used.

The encoding node maintains a unified finite field body. When performing encoding operations, the encoding node randomly selects multiple elements from the finite field to perform network encoding operations on the data part of the input data packet, and uses the selected coefficients as encoding coefficients, placing them in the global encoding vector field at the header of the data packet for decoding by the sink.

The commonly used finite field in random linear network coding contains 0 elements. If 0 elements are selected during the coding process, it is equivalent to reducing the number of data packets involved in coding, which is not conducive to secure network coding [7]. Due to the random selection of encoding coefficients by encoding nodes, there may be a linear correlation between the encoding coefficients before and after, which reduces the encoding efficiency and affects the anti eavesdropping effect. For example, assuming x, y, z is the original information sent by the source, three pieces of raw data are encoded by a random linear network and sent out for analysis using equation analysis to analyze the data packet obtained by the eavesdropper, as shown in formula

(1).

$$\begin{cases} x + y + z = A \\ x + y + 2z = B \end{cases} \tag{1}$$

For eavesdroppers, although they cannot obtain all the source information through decoding, they can obtain the value of z through simple linear operations. This encoding method can cause local information leakage and does not meet the security network encoding standards. To solve the above problems, optimization and improvement are proposed for the process of random linear network coding.

Firstly, optimize the finite field F in random linear network coding by excluding the zero elements in the finite field. To address the issue of local information leakage caused by randomly selecting global encoding vectors, secure source coding is used at the source for encoding operations. Assuming the data to be sent by the source is $S = (a, b, c)$.

(1) The source node needs to attach a random data packet r at the end of sending data s, forming a data packet as shown in formula (2).

$$S' = \begin{pmatrix} a \\ b \\ c \\ r \end{pmatrix} \tag{2}$$

(2) Select random number p as the source and generate the matrix shown in formula (3) with p as the seed:

$$L = \begin{bmatrix} p & p+1 & p+2 & p+3 \\ p^2 & (p+1)^2 & (p+2)^2 & (p+3)^2 \\ p^3 & (p+1)^3 & (p+2)^3 & (p+3)^3 \\ p^4 & (p+1)^4 & (p+2)^4 & (p+3)^4 \end{bmatrix} \tag{3}$$

(3) By multiplying L and s, the result is sent out as the source information. Through this source construction, the output of the source can ensure information security and prevent information leakage.

2.2 Active Security Strategy

The active safety strategy adopted in this article is the Support Vector Machine (SVM) method, which aims to find the optimal separation hyperplane in the feature space to maximize the positive and negative sample intervals on the training set. This separating hyperplane can separate different data, thereby achieving the purpose of classification. This method is a new type of learning machine based on the VC dimension theory of statistical learning theory and the principle of structural risk minimization. Its theory is complete, and it not only has good nonlinear processing ability, generalization ability, and learning performance, but also shows unique advantages and good application prospects in solving small sample, nonlinear, and high-dimensional recognition problems in pattern recognition. It is suitable for intrusion detection and other applications. In network intrusion detection, SVM algorithm can identify potential intrusion behaviors by analyzing network traffic or system logs and other data. This recognition is based on SVM's ability to classify data, which can effectively distinguish between normal behavior and abnormal behavior (i.e. intrusion behavior). Adopting SVM active defense strategy can monitor the real-time operation status of the network or system, detect and respond to security incidents in a timely manner.

2.2.1 Network Information Feature Extraction Module

The network information collected by the network information collection module belongs to high-dimensional data and contains a large amount of redundant data. In order to simplify the process of network security intrusion detection, network information features are extracted [8].

This study uses the 2-v-gram language model of n-gram to extract effective information from data packets, thereby obtaining the frequency matrix of the occurrence of effective network information. The frequency of the n-gram language model is mainly measured by sliding a window (length n). The window sliding step is 1 byte, and sequential sliding is performed on the data packet, while calculating the frequency of 256^n valid network information occurrences. From the above description, it can be seen that the larger the value of n, the greater the amount of structural information in the network information features [9]. In order to solve the problem of large information content in the network information feature structure mentioned above, the frequency of byte pairs separated by three positions in the effective information is measured to extract network information features more effectively. Set the sliding window size to 6, the byte pair spacing to 4, and the first and last bytes together form a 2v-gram byte pair, denoted as $\{G, g\}$, and convert it to decimal form, denoted as $\{71, 103\}$. Based on this, construct a network information appearance matrix and calculate the network information appearance frequency matrix [10].

Assuming the effective information of the network is $B = [b_1, b_2, \cdots, b_l]$ and the n-gram language model is $\beta = [\beta_1, \beta_2, \cdots, \beta_n]$, it should be noted that $n < l$. The formula for calculating the frequency of occurrence of effective information in the network is:

$$f(\beta/B) = \frac{\beta}{l - n + 1} \tag{4}$$

In the equation, $l - n + 1$ represents the total number of times the window slides on the set of valid information in the network.

Calculate the frequency of occurrence of 2v-gram byte pairs based on formula (1), expressed as:

$$f(\{\beta_1, \beta_{v+2}\}/B) = \sum_{\beta_2, \cdots, \beta_{v+1}} p(\{\beta_1, \beta_2, \cdots, \beta_{v+1}, \beta_{v+2}\}/B) \tag{5}$$

If the byte pair spacing is $v > 0$, the calculation result of formula (5) can be seen as the marginal probability of the $(v + 2) - gram$ distribution starting from β_1 and ending at β_{v+2}.

2.2.2 SVM Based Network Intrusion Detection

Construct a single class SVM intrusion detection classifier based on the network information features obtained in the previous section, fuse the single class SVM intrusion detection classifiers according to the combination rules, and execute the combination intrusion detection classifier to obtain the network security intrusion detection results.

The single class SVM intrusion detection classifier separates the required information from the original feature space and constructs a hyperplane using Gaussian function mapping, satisfying the following formula:

$$\begin{cases} Min_{w,\xi,\rho}\left(\frac{1}{2}\|W\|^2 - \rho + \frac{1}{hC}\sum_i \xi_i\right) \\ W \cdot \phi(x_i) \geq \rho - \xi_i, \xi_i \geq 0, \forall i = 1, 2, \cdots, h \end{cases} \tag{6}$$

In the equation, W represents the orthogonal vector separating the hyperplanes; ρ represents the boundary; h represents the total number of training modes; C represents the unseparated part of the original network information feature space; ξ_i represents the slack variable for punishing the rejected mode; $\phi(x_i)$ represents the i training mode.

By solving formula (7), the separation hyperplane can be obtained, and the decision function of the single class SVM intrusion detection classifier is:

$$\begin{cases} f_{svc}(z) = I\left(\sum_i a_i K(x_i, z) \geq \rho\right) \\ \sum_{i=1}^h a_i = 1 \end{cases} \tag{7}$$

In the equation, I represents the indicator function; a_i represents the calculation coefficient, provided by formula (6).

If $f_{svc}(z) = 1$, it indicates that pattern z is the target class; If $f_{svc}(z) = 0$, it indicates that pattern z is an intrusion class.

The single class SVM intrusion detection classifier cannot reliably calculate the probability distribution of intrusion classes. Therefore, based on constraint $\int_{R^d} p(x/C_t)dx = 1$, L different single class SVM intrusion detection classifiers are fused to obtain:

$$y_{avg}(x) = \frac{1}{L}\sum_{i=1}^L p_i(x/C_t) \tag{8}$$

According to formula (8), the criteria for determining network security intrusion detection are as follows: if $y_{avg}(x) < \theta$, then network information x is classified as an intrusion; If $y_{avg}(x) \geq \theta$, then network information x is the target class. Among them, θ refers to the predefined threshold.

Construct a hyperplane using the obtained network information features, design a decision function for a single class SVM intrusion detection classifier, and fuse different single class SVM intrusion detection classifiers to obtain detection classifier $y_{avg}(x)$. Set a network security intrusion threshold of θ. If $y_{avg}(x) < \theta$, it indicates that the network information is intrusion information, otherwise it is target information.

In summary, based on the combination of active and passive strategies, the implementation of multi-layer intrusion prevention in the information emergency network has been completed.

3 Experimental Analysis

3.1 Experimental Environment

In order to verify whether the proposed method meets the excellent standards in practical application, simulation comparison is used to verify its feasibility, and the performance of the proposed method is analyzed based on the comparison results.

Approximately 1 million web log data were collected from the CSIC2020 dataset and Github open source data, of which 80000 logs contained SQL injection attacks and 20000 logs contained XSS injection attacks. During the injection process, directly injecting code may be blocked by firewalls and filters. Therefore, in the data preprocessing stage, URL decoding and Unicode decoding operations need to be performed on the data to expose injection behavior.

To ensure the accuracy of the experimental process, the average value of multiple experiments was selected. The formula for the missed recognition rate, false recognition rate, and recognition success rate of intrusion data is:

$$DR = \frac{TP}{TN} \times 100\% \tag{9}$$

$$ACC = \frac{FN}{PN} \times 100\% \tag{10}$$

$$FAR = \frac{FN}{FP} \times 100\% \tag{11}$$

In the above equation, DR represents the recognition success rate of intrusion data, TP represents the accurate recognition of intrusion sample data, TN represents the total number of samples, ACC represents the misidentification rate of intrusion data, FN represents the correct number of misidentified intrusion samples, TN represents the total number of normal samples, FAR represents the missed recognition rate of intrusion data, and TN represents the total number of intrusion samples.

3.2 Experimental Results and Analysis

Extract the characteristics of information emergency network intrusion data using the proposed method, as shown in Table 1.

According to Table 1, it can be seen that the proposed method can effectively extract information emergency network intrusion and achieve feature pattern extraction of intrusion data.

The proposed method was compared experimentally with the methods in reference [4] and [5] to identify intrusion data from 1 million web log data. The specific recognition success rate is shown in Fig. 1.

Observing Fig. 1, it can be seen that the proposed method experienced fluctuations in recognition success rate at approximately 600000 data points. After observing experimental records and investigating, it was found that network users frequently output information during this period, resulting in a decrease in recognition rate. However, when the network returned to normal, the recognition rate also returned to normal until

Table 1. Extraction Results of Abnormal Intrusion Feature Patterns in Train Communication Networks

Number	Intrusion data description
1	Network vulnerabilities
2	Script remote execution of arbitrary command vulnerability
3	Script execution rejection vulnerability
4	Parameter Format String Vulnerability
5	The Hacker Crackdown
6	Local Area Network Denial of Service Attack
7	Malicious software attack
8	Denial of Service Attack
9	Security server error
10	Distributed Denial of Service Attack
11	SQL injection
12	XSS injection attack

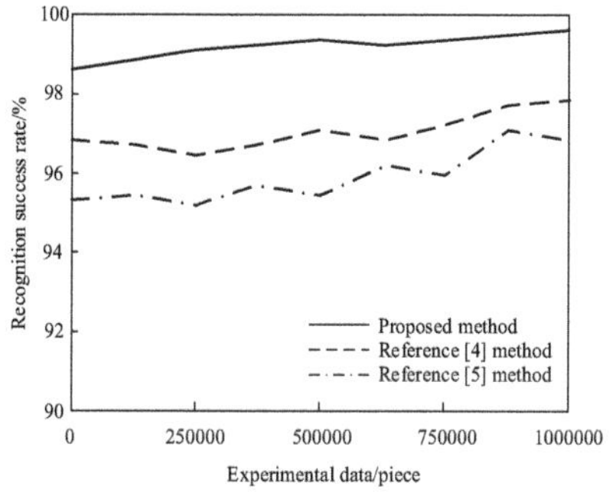

Fig. 1. Success rate of data identification for information emergency network intrusion

the end of the experiment. Although the proposed method showed a decrease in recognition rate, the average success rate of identifying intrusion data in information emergency networks reached 98.8%, which is higher than the literature comparison methods. This indicates that the proposed method has a high success rate of identifying intrusion data and meets the standards of practical applications.

In the process of identifying intrusion data, the accuracy of identification may decrease due to the possibility of misidentification. So it is necessary to calculate the misidentification rate to ensure the security of normal data. The misidentification rate of intrusion recognition on 1 million web log data was tested, and the comparison results are shown in Fig. 2.

Observing Fig. 2, it can be seen that the proposed method has a low misidentification rate for information emergency network intrusion data, and the misidentification level is relatively stable, with the highest not exceeding 0.5%. When there is limited experimental data, the literature comparison method has a lower false recognition rate. However, as the data increases, the false recognition rate gradually decreases, but still remains higher than the proposed method. This is because the proposed method adopts an active passive multi-layer defense mechanism, and each layer can identify and defend against different network threats. The SVM method used has the ability of real-time updating and self-learning, which can continuously adapt to changes in the network environment and new security threats, helping to reduce the false recognition rate caused by changes in the network environment.

The missed recognition rate can affect the recognition accuracy. If a missed recognition occurs, the intrusion is successful, resulting in data loss or damage to the information emergency network, causing serious harm. It is necessary to strictly control the missed recognition rate and set an extremely low threshold (1%) to avoid missed recognition as much as possible. The missed recognition rate of intrusion detection on 1 million web log data was tested, and the comparison results are shown in Fig. 3.

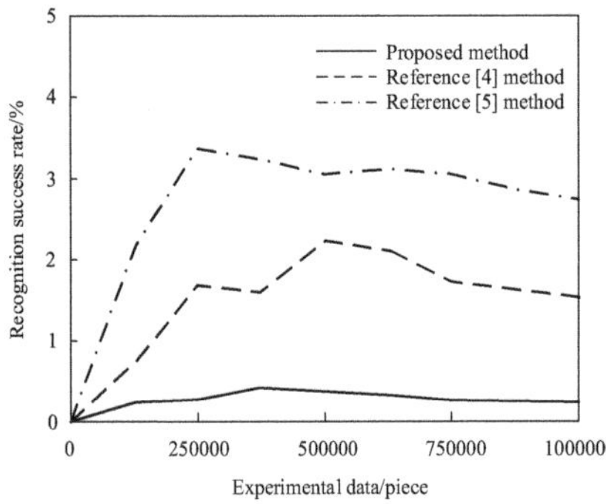

Fig. 2. Misidentification rate of information emergency network intrusion data

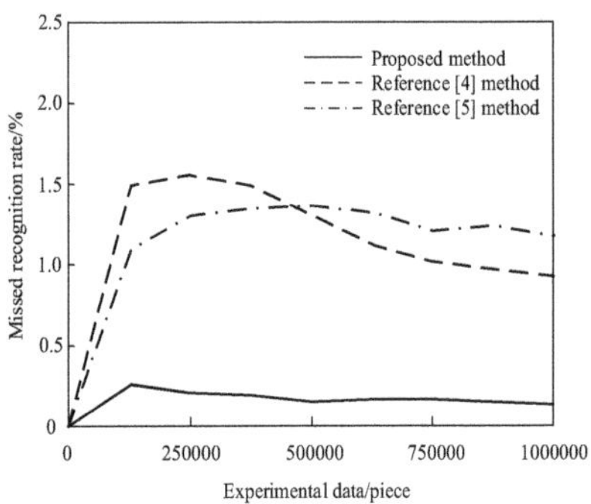

Fig. 3. Identification rate of intrusion data leakage

Observing Fig. 3, it can be seen that the missed recognition rate of the proposed method does not exceed 0.25%, and it can comprehensively identify information emergency network intrusion data. This is mainly because the proposed method can timely detect and respond to potential intrusion behaviors based on the classification results of SVM algorithm after extracting the features of network intrusion data. Meanwhile, with the continuous changes in the network environment, SVM algorithm can also achieve intrusion data recognition through continuous learning and updating, reducing the missed recognition rate to an extremely low level and avoiding the impact on normal sample data.

4　Conclusion

The multi-layer anti intrusion algorithm of information emergency network under the combination of active and passive provides new ideas and solutions for network security protection. By integrating the technological advantages of active defense and passive detection, this algorithm significantly improves the security and stability of the network. Specifically, the multi-layer anti intrusion algorithm of the information emergency network combines multiple security technologies to form a comprehensive network security protection system. These technologies can monitor network traffic and behavior in real-time, issue real-time alerts and intercept abnormal behavior, effectively preventing potential network intrusion threats. The experimental results show that this method can not only improve the real-time and accuracy of network defense, but also effectively respond to unknown attacks and complex attack scenarios, thereby comprehensively enhancing the security protection capability of information emergency networks.

References

1. Huang, Z.: Hybrid device-to-device and device-to-vehicle networks for energy-efficient emergency communications. IEEE Syst. J. **17**(4), 5429–5440 (2023)
2. Bu, B.: Network coding data distribution technology between streams of emergency system based-wireless multi-hop network. Comput. Commun. **186**(3), 22–32 (2022)
3. Kumar, N., Kumar, U.: Artificial intelligence for classification and regression tree based feature selection method for network intrusion detection system in various telecommunication technologies. Comput. Intell. **40**(1), 1–23 (2024)
4. Sridhar, S.S.S.S.: An optimized model for network intrusion detection systems in industry 4 0 using xai based bi 1stm framework. Neural Comput. Appl. **35**(15), 11459–11475 (2023)
5. Jinghong, L., Xudong, L., Jun, B.Z.: A novel hierarchical attention-based triplet network with unsupervised domain adaptation for network intrusion detection. Appl. Intell.: Int. J. Artif. Intell. Neural Netw. Complex Probl.-Solving Technol. **53**(10), 11705–11726 (2023)
6. Zhang, X., Tian, L., Guo, S., et al.: STF-Net: sparsification transformer coding guided network for subcortical brain structure segmentation. Biomed. Eng./Biomed. Technik **69**(5), 465–480 (2024)
7. Fu, T., Lin, Y., Lin, L., et al.: Network architecture of non-coding RNAs provides insights into the pathogenesis of upper tract urothelial carcinoma. Urol. Oncol. **40**(8), 11–21 (2022)
8. Yang, H., Zhang, Z., Zhang, L.: Network security situation assessments with parallel feature extraction and an improved BiGRU. J. Tsinghua Univ. (Sci. Technol.) **62**(5), 842–848 (2022)
9. Mokkapati, R., Dasari, V.L.: Embedded signal artificial neural network based intelligent non-dependent feature selection for cyber attack classification in signal-based networks. Traitement du Signal: Signal Image Parole **40**(3), 905–914 (2023)
10. Ma, Y., Li, R.: Simulation of network information security encrypted transmission based on DFT-S-OFDM. Comput. Simul. **39**(1), 358–361, 393 2022)

Application of an Augmented Reality-Based Assisted Inspection System in Airport

Weijia Ye[1]([✉]), Guomin Sun[2], Ning He[1], and Sunle Lan[3]

[1] R&D Center, The Second Research Institute of CAAC, Chengdu, China
yeweijia@Caacsri.Com, hening@caacsri.com
[2] Chongqing Airport Group Co., Ltd., Chongqing, China
[3] Shenzhen Airport (Group) Co., Ltd., Shenzhen, China
lansl98@foxmail.com

Abstract. With the rapid development of augmented reality (AR) technology, its application in airport operations has shown significant potential. This paper explores an AR-based auxiliary inspection system designed to enhance efficiency and accuracy in airport facility maintenance and security inspections. By leveraging AR devices such as HoloLens, the system provides real-time guidance, overlays critical information, and enables remote collaboration. The implementation of this system reduces human error, optimizes resource allocation, and improves decision-making processes, demonstrating its value in advancing the digitization and intelligence of airport operations.

Keywords: Augmented Reality · Airport · Civil Aviation · Equipment inspect

1 Introduction

The complexity of modern airport operations requires regular inspections to maintain safety and efficiency. Ensuring efficient operations and maintaining high safety standards are critical priorities. These inspections cover a wide range of areas, including infrastructure, mechanical systems, and security equipment. However, traditional inspection methods often face significant challenges, such as reliance on manual processes, limited real-time data access, and susceptibility to human error. These limitations can lead to inefficiencies and missed maintenance issues, ultimately affecting airport performance and safety.

Therefore, we use the AR technology apply for routine inspect work in airport. The advent of augmented reality (AR) technology offers a transformative solution to these challenges. AR enables the overlay of digital information onto the physical environment, providing real-time, context-aware guidance. This capability is particularly valuable in airport inspections, where accuracy and timely decision-making are paramount. By equipping inspection teams with AR glasses, personnel can visualize critical data, access step-by-step instructions, and interact with virtual information directly within their field of view [1].

P. Siarry et al. (Eds.): WCNA 2024, LNEE 1550, pp. 47–54, 2026.
https://doi.org/10.1007/978-981-95-6946-5_5

Moreover, AR facilitates remote collaboration, allowing experts located off-site to guide on-ground inspection teams in real-time. This reduces the need for physical presence, enhances knowledge sharing, and accelerates problem resolution. Additionally, the systems can integrate with IoT sensors and digital twins, enabling dynamic updates and predictive maintenance by analyzing live data streams [2].

This paper presents the development and application of an AR-based auxiliary inspection system tailored for the airport environment. The system aims to enhance inspection accuracy, optimize operational workflows, and support proactive decision-making. Through case studies and performance evaluations, we demonstrate how AR technology can significantly improve inspection outcomes, contributing to safer and more efficient airport operations.

2 Related Work

The inspection and troubleshooting auxiliary system consist of four parts, the Management platform, AR wearable device, Remote Assistant Service and WeChat applet.

Currently, research on the application of augmented reality (AR) in airport security focuses on several key areas: 1) Enhanced Screening: AR improves efficiency by overlaying real-time data, helping identify threats during baggage and passenger checks. 2) Advanced Surveillance: Integrating AR with security cameras allows for real-time tracking and visualization of critical data, enhancing situational awareness. 3) Training Simulations: AR provides immersive scenarios for security staff, simulating potential threats and improving emergency responses. Some studies also explore combining AR with digital twins and IoT for predictive maintenance and operational optimization.

We have previously conducted research on applying AR technology in the field of airport safety. For instance, we developed a pre-flight inspection assistance system based on AR technology, which uses AR glasses to help airport staff improve the efficiency of pre-flight inspections.[3][4] Additionally, we explored integrating augmented reality with visual SLAM (VSLAM) technology. By utilizing visual recognition to extract feature points from the spatial environment and combining this with augmented reality spatial registration, we achieved the overlay of virtual elements onto the real environment. This fusion of virtual and real elements enhances inspection efficiency during the inspection process [5, 6].

The implemented features include inspection process assistance and remote spatial registration. By optimizing our previous technical research and aligning with the practical needs of airport operational departments, we developed the inspection assistance system. This system enhances the efficiency of airport personnel inspections and diversifies fault troubleshooting methods.

2.1 Introduction of Inspection and Troublesh Shooting Auxiliary System

The system mainly helps with airport inspection personnel during daily inspection tasks. For example, when staff are uncertain about the displayed parameters of a device in the inspection process, the AR glasses can display the correct diagram information to assist with the inspection. Additionally, when staff need to repair faulty equipment, the system offers two methods: first, personnel can search documents through the AR glasses and follow the instructions to repair the equipment; second, they can connect remotely with an expert, who can make spatial annotations that appear on the AR glasses to help the staff quickly troubleshoot the equipment.

The system is mainly composed of four parts: Inspection Assistance System Service Platform，the glasses interface, a WeChat mini-program, and the expert service interface. Their respective functions are as follows: the inspection assistance system backend primarily manages basic system information, communicates with the airport database, transmits data to the glasses, and stores files generated during the inspection process. The glasses interface is responsible for displaying the UI of the inspection workflow and showing inspection data transmitted from the backend. Additionally, staff can use the glasses to call remote experts and establish a connection with the expert interface. After completing the inspection work using the glasses, personnel can upload their signatures through the WeChat mini program. The entire workflow is shown in Fig. 1.

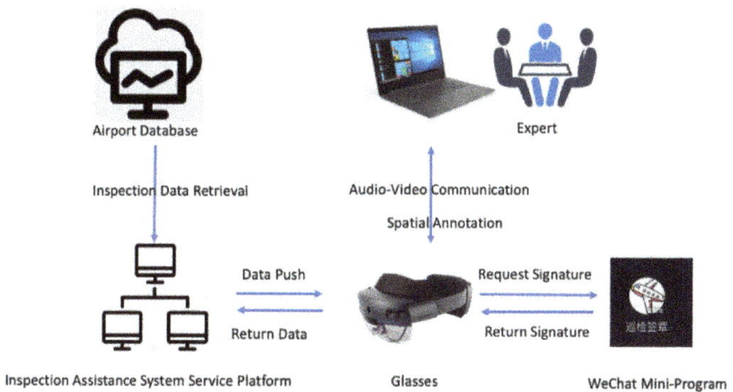

Fig. 1. Inspection auxilary system flow chart

2.2 Inspection Assistance System Service Platform

First, the Inspection Assistance System Service Platform includes a personnel login management feature. It retrieves inspection data from the airport and allows the addition of correct diagram information for each step of the inspection process. After the inspection is completed, the platform provides access to view on-site photos taken during the inspection, personnel signatures, and other related inspection details. Managers can use this platform to review and monitor the inspection work.[7].

Figure 2 below shows the completed inspection data displayed on the system service platform. You can see the pre-uploaded equipment assistance images as well as the on-site photos taken by staff wearing the glasses during the equipment inspection.

2.3 AR Glass UI Design and Function Realized

The AR glasses we use are Microsoft's HoloLens 2. Airport staff mainly complete the inspection process on-site using the program we developed for the AR glasses. The entire system's UI design and main functions can be seen in the figure below. To clearly display

50   W. Ye et al.

the UI, we used screen-sharing software to synchronize the display with the AR glasses and then recorded a video. Figure 3 is a screenshot taken from the recorded video. And the following images displayed in the glasses are screenshots that we obtained through screen-sharing software.

As shown in the image, the left column is the function menu, which includes Follow, Inspection, Search, Scan, and Call Remote Expert. The right side is the display panel, where the current panel shows inspection information. The entire process supports voice and gesture interactions.

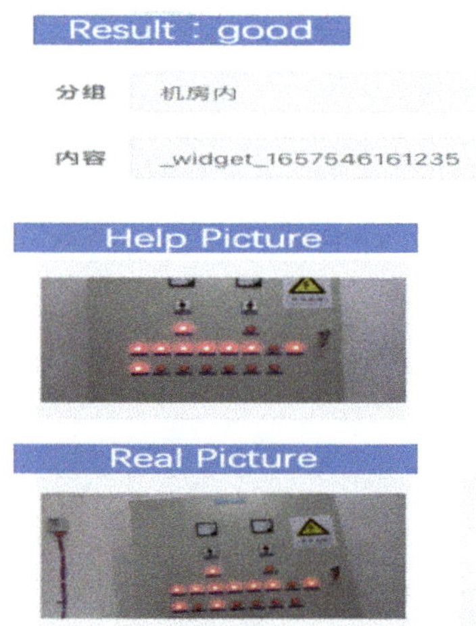

Fig. 2. The inspection workflow displayed on the backend shows the correct diagram information for the equipment and the real equipment information captured through on-site photos

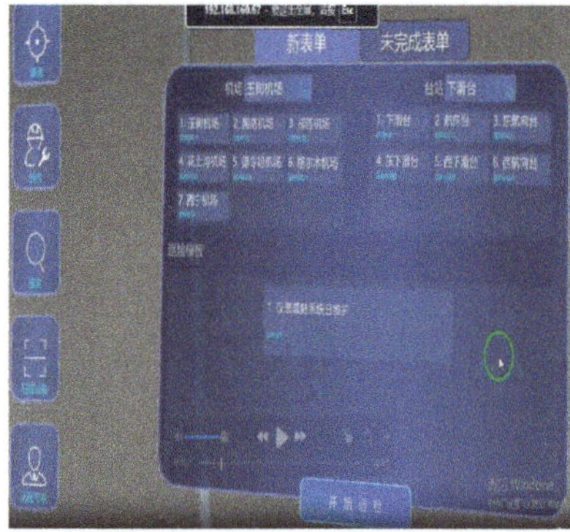

Fig. 3. The system's main interface consists of two sections: the left column displays the system's function options, while the right side contains the information display panel for the selected function. Currently, the inspection function's information panel is being displayed.

1) The "Follow" function is primarily designed to help airport staff quickly operate the panel even when they move to a distant location. Its purpose is that no matter where the staff move, by clicking or using a voice command to activate the follow function, the entire panel information will quickly move in front of the wearer. This reduces the need for the user to adjust the panel's position and improves the overall efficiency of the workflow.

2) The "Inspection" function is the main feature of our system. Before starting the inspection, the inspector needs to select the airport, inspection station, and inspection template, as shown in the image. Each button has text below it, which serves as a voice prompt [8].

Figure 4 shows the information panel of the inspection function, where the corresponding information and inspection template need to be selected before starting the inspection.

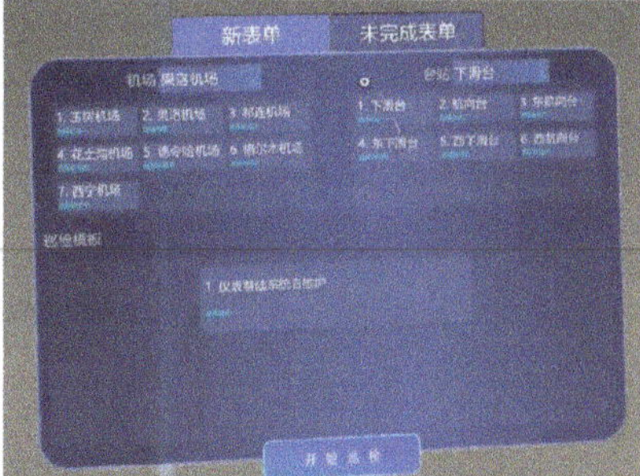

Fig. 4. Inspection Function Information Panel

The first image in Fig. 5 illustrates the inspection workflow after beginning the inspection. On the left side of the screen is the inspection content, where users can choose to take on-site photos, while the right side displays a correct example image for the current inspection step.

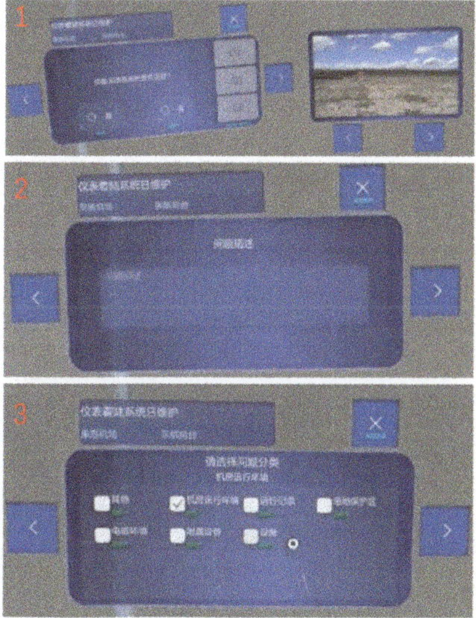

Fig. 5. The Daily Equipment Inspection Workflow Panel, Issue Description Panel, and Equipment Issue Classification Panel

3) The "Search" function allows inspectors to look up manuals when encountering issues that require additional reference during the inspection process. By entering keywords related to the problem on the glasses, the system performs a fuzzy search in the knowledge base to display the relevant manual files. This enables inspectors to troubleshoot and repair issues directly on-site.

4) The "Scan" function primarily enables the visualization of equipment parameters. Each inspection device on-site is assigned a unique asset QR code [9]. By scanning the QR code with the AR glasses' camera, the system provides a visual display within the AR glasses, assisting inspection personnel in checking the equipment on-site.

In the image below, the left side displays the basic information of the scanned equipment, including its type, location, and model. The right side shows an image of the equipment's exterior [10] (Figs. 6 and 8).

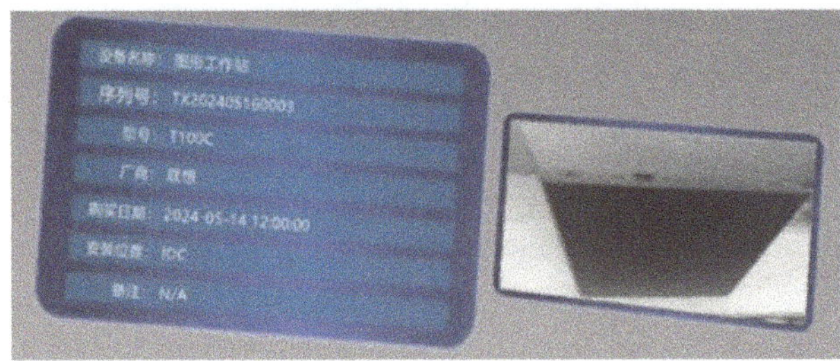

Fig. 6. Panel Information Displayed in AR Glasses After Equipment Scanning

2.4   The Wechat Mini Program

The WeChat mini program primarily enables the handwritten signature functionality for staff. After completing the inspection tasks on the AR glasses, due to the limitations of the glasses, handwritten input cannot be performed directly. Therefore, by entering the 5-digit code displayed on the glasses (as shown in the first image of Fig. 7) into the mini-program, a communication link between the glasses and the mini-program is established. This allows the inspector to provide their handwritten signature, completing the entire inspection process.

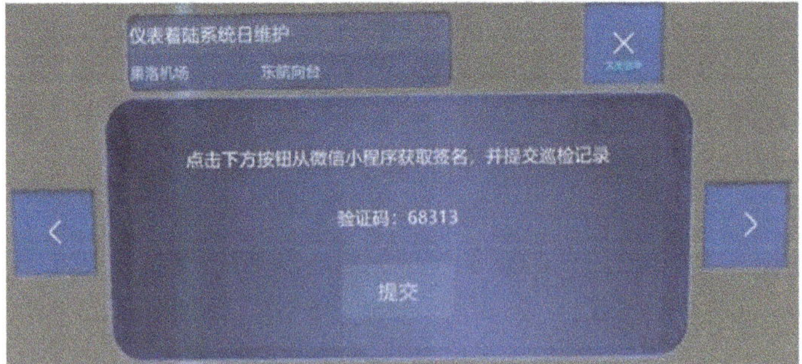

Fig. 7. After completing the inspection process on the glasses, a verification code is displayed on the glasses to establish communication with the mini-program.

The inspection process is considered complete only after the personnel signature is submitted through the mini program. Airport management can then log into the Inspection Assistance System Service Platform to review completed inspection records, which include the details of each inspection item, on-site photos, and personnel signatures.

2.5   Remote Expert Sytem

As mentioned earlier, when inspection personnel encounter a malfunction in the inspection equipment, they have several options: 1) refer to the correct diagram of the equipment, or 2) search for the relevant troubleshooting manual. If neither method resolves the issue, a third option becomes necessary. The inspector can use the AR glasses to call a remote expert on the computer [10]. Once the connection is established, the expert can guide the on-site staff to troubleshoot the equipment by providing spatial annotations.

First, data exchange between HoloLens and the computer is achieved using the WebRTC network protocol [11]. On the HoloLens side, the Unity AR development platform is used to overlay annotation information received from the computer onto the real-world scene. On the computer side, the expert can create or edit annotations through the user interface, and this data is synchronized in real time over the network for display on the HoloLens.

The image below shows a test scene in Unity. On the left is the camera feed captured after the program starts. By selecting the line annotation tool in the UI and drawing a circle on the camera feed, the red circle in the image represents the annotated information. This illustrates the workflow on the expert's computer interface. Once the annotation is completed, the inspection personnel can see the annotated information synchronized in their AR glasses, thereby improving the efficiency of equipment fault troubleshooting [2].

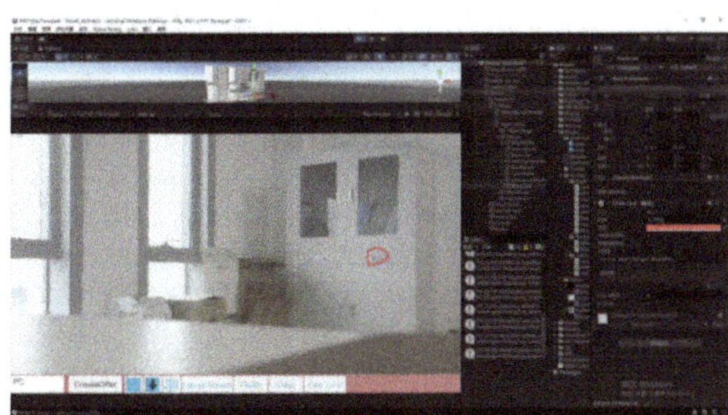

Fig. 8. A demo running in Unity demonstrating the line-drawing annotation feature.

3   Conclusion

In this article, we introduced an airport equipment inspection assistance system based on AR technology. This system enhances the efficiency of airport staff during daily inspection tasks. Additionally, when faced with troubleshooting equipment issues, it provides effective support through methods such as offering correct example comparisons, searching for tool manuals, and connecting with remote experts. By leveraging augmented reality to seamlessly integrate virtual and real-world information, the system brings inspections to life. The fully electronic workflow significantly improves the overall efficiency of the inspection process.

References

1. Makris, S., Pintzos, G., Rentzos, L., Chryssolouris, G.: Assembly support using AR technology based on automatic sequence generation. CIRP Ann. Manuf. Technol. **62**, 9–12 (2013)
2. Fussell, S., Setlock, L., Setlock, L.D., Kraut, R.: Effects of head-mounted and scene-oriented video systems on remote collaboration on physical tasks. CHI '03: Proceedings of the SIGCHI Conference on Human Factors in Computing Systems, pp. 513–520 (2003)
3. Weijia, Y., Qingyang, Y., Ning, H., Lifei, P.: Augmented reallity for assist apron operation safety supervisor in aircraft inspection process
4. Weijia, Y., Jin, W., Ning, H.: Application of civil aviation apron operation safety inspector auxiliary system based on augmented reality technology in airport
5. Salas-Moreno, R.F., Glocken, B., Kelly, P.H., et al.: Dense planner slam, in Mixed and Augement Reality (ISMAR). 2014 IEEE International Symposium on, pp. 157–164, IEEE (2014)
6. Qin, T., Li, P., Shen, S.: VINS-Mono: a robust and versatile monocular visual-inertial state estimator. IEEE Trans. Rob. **34**(4), 1004–1020 (2018)
7. Fiorentino, M., Uva, A.E., Gattullo, M., Debernardis, S., Monno, G.: Augmented reality on large screen for interactive maintenance instructions. Comput. Ind. **65**, 270–278 (2014)
8. Laha, B., Bowman, D.A., Socha, J.J.: Effects of VR system fidelity on analyzing isosurface visualization of volume datasets. IEEE Trans. Vis. Comput. Graph. **20**, 513–522 (2014)
9. Wagner, D., Langlotz, T., Schmalstieg, D.: Robust and unobtrusive marker tracking on mobile phones. IEEE International Symposium on Mixed and Augmented Reality (ISMAR), pp 121–124 (2008)
10. Tecchia, F., Alem, L., Huang, W.: 3D helping hands: a gesture based MR system for remote collaboration. In: Proc. ACM VRCAI. pp. 323–328 (2012)
11. Kurillo, G., Bajcsy, R., Nahrsted, K., Kreylos, O.: Immersive 3D environment for remote collaboration and training of physical activities. In: Proc. IEEE VR. pp. 269–270 (2008)

CMSFF-UNet: A Network for Underwater Image Segmentation to Boost Segmentation Accuracy

Huiming Li and Chaobing Huang[✉]

School of Information Engineering, Wuhan University of Technology, Wuhan, China
`huangcb@whut.edu.cn`

Abstract. Underwater image segmentation is currently a challenging issue in underwater image technology and has gradually attracted more attention from researchers. Despite significant advancements in this field, the complex underwater environment continues to pose difficulties, leaving room for improvement in segmentation accuracy. To enhance underwater image segmentation performance, the Multi-Scale Feature Fusion Module (MSFFM) proposed in this paper is designed for upsampling feature fusion. It strengthens the incorporation of shallow semantic information. Additionally, the feature fusion process incorporates Channel prior Convolutional attention(CPCA). Finally, the optimization process utilizes a composite loss function incorporating Dice Loss and Focal Loss. Results of experiments carried out on the SUIM dataset demonstrate that the model we proposed achieves excellent segmentation accuracy for underwater imagery.

Keywords: Underwater Image Segmentation · Attention Mechanism · Multi-Scale Feature Fusion · Combined Loss Function

1 Introduction

Underwater image segmentation represents a crucial domain within computer vision research. It denotes the procedure of separating the target from the background within underwater images. The segmented targets can be further used for tasks such as recognition, classification, localization, and tracking, which are of significant importance for the exploration and utilization of marine resources. Underwater images feature light attenuation and low contrast. These characteristics make processing and analyzing underwater images more arduous compared to traditional image processing.

Currently, underwater image segmentation methodologies are principally classified into two techniques: conventional algorithm-based solutions and neural network-oriented frameworks. Traditional image segmentation methods include thresholding methods [1], edge detection [2], clustering [3], and others. While traditional methods can complete image segmentation tasks more quickly, they lack generality and require the selection of appropriate algorithms based on the characteristics of the image. Therefore, they are not suitable for handling complex scene segmentation tasks.

© The Author(s) 2026

P. Siarry et al. (Eds.): WCNA 2024, LNEE 1550, pp. 55–63, 2026.

https://doi.org/10.1007/978-981-95-6946-5_6

In recent years, deep learning-based image segmentation algorithms have gained significant attention, leading to the development of numerous new approaches. Patil et al. [4] designed a unique Generative Adversarial Network (GAN) to segment underwater motion targets. Their approach combines some initial video frames and the present frame for motion saliency prediction. This estimated saliency is then input into the network for foreground estimation. Jahidul et al. [5] established the SUIM benchmark for underwater semantic segmentation. Concurrently, they proposed the SUIM-Net model, which balances performance and computational efficiency. Hambarde et al. [6] introduced the underwater Generative Adversarial Network. This end-to-end network can estimate depth from a single underwater image. Chicchon et al. [7] proposed a method that uses image contours along with a joint loss function to improve segmentation quality and spatial resolution. However, the above methods do not address the issue of boundary blur in underwater image segmentation, leading to poor segmentation accuracy. Therefore, when tackling underwater image semantic segmentation tasks, a network with clear, precise, and high-accuracy boundaries is required.

ResNet50 is adopted as the feature extraction network of U-Net in this paper to address the issues of blurry boundaries and segmentation accuracy in underwater image semantic segmentation. First, the proposed Multi-Scale Feature Fusion Module (MSFFM) is used for upsampling feature fusion to strengthen the input of shallow semantic information. Next, the feature fusion process incorporates Channel Prior Convolutional Attention(CPCA). Finally, the optimization process utilizes a composite loss function incorporating Dice Loss and Focal Loss. Results of experiments carried out on the SUIM dataset demonstrate that the model we proposed achieves excellent segmentation accuracy for underwater imagery.

2 Propose Methods

2.1 Channel Prior Convolutional Attention Module

Figure 1 depicts the Channel Prior Convolutional Attention(CPCA) module [8]. Given an input feature map $F \in R^{C \times H \times W}$, it initially passes through a channel attention module [9]. This generates a 1D channel attention feature map $M_C \in R^{C \times 1 \times 1}$. The resulting M_C is then element-wise multiplied with F, yielding the channel-refined feature map $F_C \in R^{C \times H \times W}$. Subsequently, F_C undergoes processing by a spatial attention module [10] to generate a 3D spatial attention map $M_S \in R^{C \times H \times W}$. Finally, M_S is element-wise multiplied with F_C, producing the final output feature map $\hat{F} \in R^{C \times H \times W}$.

$$F_C = CA(F) \otimes F \tag{1}$$

$$\hat{F} = SA(F_C) \otimes F_C \tag{2}$$

where $\otimes$ in Eqs. 1 and 2 denotes element-wise multiplication.

The calculation of channel attention is shown in Eq. 3:

$$CA(F) = \sigma(MLP(AvgPool(F)) + MLP(MaxPool(F))) \tag{3}$$

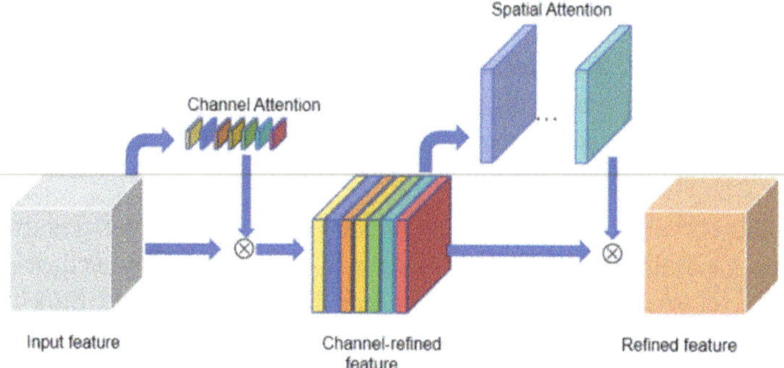

Fig. 1. Channel prior convolutional attention module

The calculation of spatial attention is shown in Eq. 4:

$$SA(F) = Conv_{1\times1}\left(\sum_{i=0}^{3} Branch_i(DwConv(F))\right) \tag{4}$$

2.2 Multi-Scale Feature Fusion Module

The CMSFF-UNet uses ResNet50 as the backbone and features a U-shape architecture. During the upsampling process, bilinear interpolation is applied five times. High-level feature maps effectively capture contextual information, whereas low-level ones preserve fine spatial details. Simple feature fusion techniques usually overlook this information diversity in feature maps, leading to reduced segmentation accuracy. Therefore, this research employs the multi-scale feature fusion method(MSFFM). The designed of MSFFM draws on the PANet and FPN feature fusion architectures. Part(c) of Fig. 2 shows the MSFFM. The MSFFM fuses the upsampled features with the encoder features, enabling the model to grasp understand the contextual information of the input feature map.

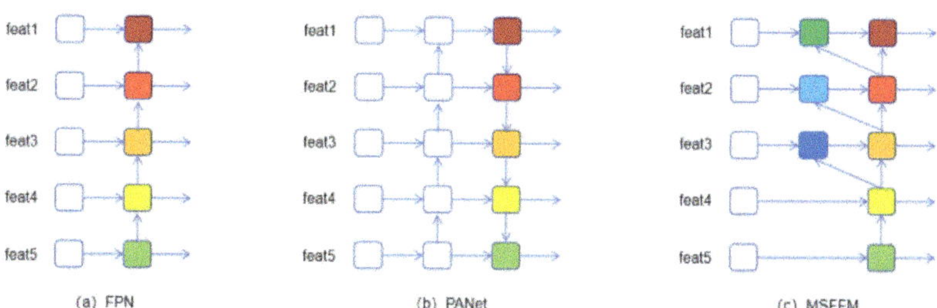

Fig. 2. Convergence of network architecture features

2.3   Network Model

The model uses ResNet50 as the backbone network and extracts multi-scale features through MSFFM, incorporating the CPCA attention mechanism to obtain better semantic information. Figure 3 shows the architecture of CMSFF-UNet [11]. The complete network architecture includes the following key steps:

- Feature Extraction: ResNet50 is used as the encoder for U-Net. The deep feature extraction capability of ResNet50, along with its rich shallow and deep features, enables U-Net to accurately capture both fine details and semantic information from images.
- Multi-Scale Feature Fusion: MSFFM uses linear interpolation to blend upsampled feature maps with the output features from upper encoder layers. By expanding the receptive field, this approach empowers the model to more effectively capture the contextual information within input feature maps.
- CPCA: By adaptively weighting features across both channel and spatial locations, the neural network can emphasize important features, thereby improving the precision of the segmentation task.

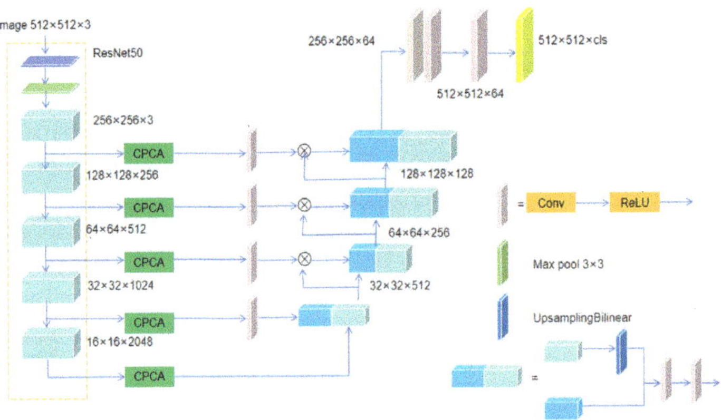

Fig. 3. CPCA Multi-Scale Feature Fusion Network

2.4   Loss Function

The loss function of CMSFF-UNet combines Focal Loss [12] and Dice Loss [13]. Equation 5 represents Focal Loss, and Eq. 7 represents Dice Loss. Dice Loss, due to its robustness against class imbalance and sensitivity to small target segmentation, ensures that the model can accurately identify and segment different categories, even maintaining high precision in complex underwater environments. Focal Loss dynamically adjusts the weight of easily segmented samples, encourages the model to focus on hard-to-segment ones, thereby enhancing the model's generalization ability and training efficiency. Therefore, Focal Loss and Dice Loss are combined in this paper to improve the convergence speed of the model when facing imbalances in dataset classes and segmentation difficulty.

$$L_{Focal} = -\alpha_t (1 - p_t)^\gamma \, log(p_t) \tag{5}$$

$$\text{Dice coefficent} = \frac{2TP}{2TP + FP + FN} \tag{6}$$

$$L_{Dice} = 1 - Dicecoefficent \tag{7}$$

The composite loss function of CMSFF-UNet is defined as follows:

$$L = L_{Focal} + L_{Dice} \tag{8}$$

Table 1. Comparison of experimental results

Model	IoU(%)								mIoU(%)
	BW	HD	PF	WR	RO	RI	FV	SR	
U-Net	79.46	32.25	21.85	33.94	23.65	50.28	38.16	42.16	39.85
U-Net(ResNet)	87.75	77.20	13.67	66.98	38.13	70.12	73.21	63.52	61.32
U-Net(Vgg)	86.61	76.69	1.94	61.90	69.76	61.06	75.30	55.14	61.05
SegNet [14]	80.63	45.67	17.45	32.24	55.72	47.62	43.92	51.51	46.85
SUIMNet [5]	80.64	63.45	23.27	41.25	60.89	53.12	46.02	57.12	53.22
PSPNet [15]	82.51	65.04	28.54	46.56	62.88	55.80	46.78	55.98	55.51
CMSFF-UNet(our)	83.54	77.46	26.02	62.62	68.70	68.57	70.57	62.62	65.01

3 Experimental

3.1 Datasets

The SUIM dataset for underwater image semantic segmentation is utilized in this paper
to evaluate the model's segmentation accuracy. It comprises 1,525 natural underwater
images with ground truth semantic labels, along with 110 test images also annotated with
ground truth labels. This dataset provides pixel-level annotations for 8 object categories:
Fish and Vertebrates (FV), Human divers (HD), Wrecks and Ruins (WR), Reefs and
Invertebrates (RI), Robots (RO), Background (BW), Sea-floor and Rocks (SR), Aquatic
plants and Sea-grass (PF).

3.2 Evaluation Metrics

As presented in Table 2, the evaluation metrics for semantic segmentation are derived
from the confusion matrix.

mIoU, mPA, and Accuracy are used as evaluation metrics to evaluate the segmentation performance of the network architecture in this paper. The following are the formulas
for Accuracy and mIoU.

$$\text{Accuracy} = \frac{TP + TN}{FP + FN + TP + TN} \tag{9}$$

$$\text{mIoU} = \frac{1}{N} \times \left(\frac{\text{TP}}{\text{FN} + \text{FP} + \text{TP}} \right) \tag{10}$$

Table 2. Segmentation confusion matrix results

state	Predicted results	
	Negative	Positive
Negative	TN	FP
Positive	FN	TP

Here, N represents the number of classes. TP indicates correctly predicted positives, TN represents correctly predicted negatives, FP refers to incorrectly predicted positives, and FN denotes incorrectly predicted negatives.

To calculate mPA, we first determine the classification ratio for each class and then calculate their average. Denote the accuracy of each class pixel as P. Below is the formula for computing mPA.

$$\text{mPA} = \frac{\text{Sum(P)}}{N} \tag{11}$$

3.3 Implementation Details

The experimental environment is configured with an NVIDIA GeForce RTX 3090 GPU, Intel(R) Xeon(R) Platinum 8255C CPU, and Torch 1.7.0, while the experimental hyperparameters are set as follows: 100 epochs, a learning rate of 0.0004, a batch size of 8, an input size of 512×512, a weight decay of 0.0001, and the Adam optimizer. Transfer learning is employed using ImageNet pre-trained weights to shorten training time on the SUIM dataset.

3.4 Comparison Experiment

The SUIM dataset was used to assess the performance of CMSFF-UNet. Its performance was then benchmarked against several other approaches. Table 1 represents the comparison results. CMSFF-UNet achieved an mIoU of 65.01%, demonstrating a clear advantage over classic networks such as U-Net, SegNet, and PSPNet, when tested on the SUIM test dataset.

3.5 Ablation Study

To validate the proposed method, the baseline is a UNet with ResNet50 as the backbone network. In the ablation study, the channel prior convolutional attention (CPCA) module, multi-scale feature fusion module (MSFFM), and combined loss function (CLS) proposed in this paper were sequentially added to the baseline model. Table 3 outlines the setup of these ablation experiments. It demonstrates that as the three modules were incrementally added to the model, its performance improved step-by-step. Finally, the Accuracy, mPA, and mIoU reached 85.78%, 79.03%, and 65.01%, respectively, which are improvements of 2.28%, 9.46%, and 3.69% compared to the baseline model.

Table 3. Ablation experiment results

NO	CPCA	MSFFM	CLF	mIoU(%)	mPA(%)	Accuracy(%)
1				61.32	69.57	83.50
2	√			62.19	71.87	84.42
3	√	√		62.61	73.03	85.40
4	√	√	√	65.01	79.03	85.78

3.6 Results Visualization

Predictions are performed on the SUIM test set using both the UNet and CMSFF-UNet models. Figure 4 illustrates the segmentation results, and CMSFF-UNet is able to segment underwater scenes with higher recognition and detection accuracy. Additionally, the boundary precision and accuracy achieved in segmenting underwater objects demonstrate superior performance compared to traditional algorithms.

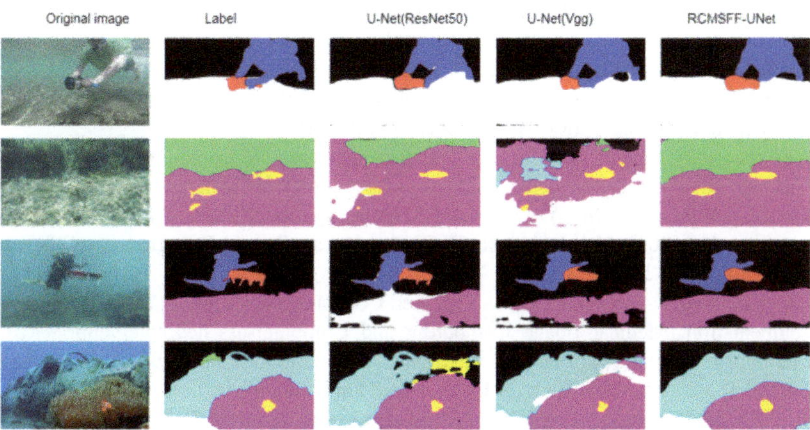

Fig. 4. Prediction results

4   Conclusion

We propose CMSFF-UNet, a network for underwater image segmentation to boost segmentation accuracy. Firstly, the MSFFM module is proposed to address the issue of semantic information disparity between deep and shallow features in this network. Next, the CPCA module is incorporated into the feature fusion process to avoid the loss of important feature information. Finally, a composite loss function is applied to address the problem of sample imbalance, thereby improving the model's accuracy.

On the SUIM dataset, the proposed CMSFF-UNet model achieves an mIoU of 65.01% and an mPA of 79.03%. CMSFF-UNet tackles the issues of low performance and coarse segmentation boundaries in existing underwater scene segmentation models, making it of significant importance for the study of underwater image semantic segmentation.

References

1. Senthilkumaran, N., Vaithegi, S.: Image segmentation by using thresholding techniques for medical images. Comp. Sci. Eng. Int. J. **6**(1), 1–13 (2016)
2. Al-Amri, S.S., Kalyankar, N.V., Khamitkar, S.D.: Image segmentation by using edge detection. Int. J. Comp. Sci. Eng. **2**(3), 804–807 (2010)
3. Dhanachandra, N., Manglem, K., Chanu, Y.J.: Image segmentation using K-means clustering algorithm and subtractive clustering algorithm. Procedia Comp. Sci. **54**, 764–771 (2015)
4. Patil, P.W., Thawakar, O., Dudhane, A., et al.: Motion saliency based generative adversarial network for underwater moving object segmentation. 2019 IEEE International Conference on Image Processing (ICIP). IEEE, pp. 1565–1569 (2019)
5. Islam, M.J., Edge, C., Xiao, Y., et al.: Semantic segmentation of underwater imagery: Dataset and benchmark. 2020 IEEE/RSJ International Conference on Intelligent Robots and Systems (IROS). IEEE, pp. 1769–1776 (2020)
6. Hambarde, P., Murala, S., Dhall, A.: UW-GAN: Single-image depth estimation and image enhancement for underwater images. IEEE Trans. Instrum. Meas. **70**, 1–12 (2021)
7. Chicchon, M., Bedon, H., Del-Blanco, C.R., et al.: Semantic segmentation of fish and underwater environments using deep convolutional neural networks and learned active contours. IEEE Access **11**, 33652–33665 (2023)
8. Huang, H., Chen, Z., Zou, Y., et al.: Channel prior convolutional attention for medical image segmentation. Comput. Biol. Med. **178**, 108784 (2024)
9. Bastidas, A.A., Tang, H.: Channel attention networks. /Proceedings of the IEEE/CVF Conference on Computer Vision and Pattern Recognition Workshops, pp. 0–0 (2019)
10. Guo, C., Szemenyei, M., Yi, Y., et al.: Sa-unet: Spatial attention u-net for retinal vessel segmentation. 2020 25th International Conference on Pattern Recognition (ICPR). IEEE, pp. 1236–1242 (2021)
11. Ronneberger, O., Fischer, P., Brox, T.: U-net: Convolutional networks for biomedical image segmentation. Medical Image Computing and Computer-assisted Intervention–MICCAI 2015: 18th international conference, Munich, Germany, October 5–9, 2015, proceedings, part III 18, pp. 234–241. Springer International Publishing (2015)
12. Ross, T.Y., Dollár, G.: Focal loss for dense object detection. Proceedings of the IEEE Conference on Computer Vision and Pattern Recognition, pp. 2980–2988 (2017)
13. Soomro, T.A., Afifi, A.J., Gao, J., et al.: Strided U-Net model: Retinal vessels segmentation using dice loss. 2018 Digital Image Computing: Techniques and Applications (DICTA). IEEE, pp. 1–8 (2018)
14. Badrinarayanan, V., Kendall, A., Cipolla, R.: Segnet: a deep convolutional encoder-decoder architecture for image segmentation. IEEE Trans. Pattern Anal. Mach. Intell. **39**(12), 2481–2495 (2017)
15. Zhao, H., Shi, J., Qi, X., et al.: Pyramid scene parsing network. Proceedings of the IEEE Conference on Computer Vision and Pattern Recognition, pp. 2881–2890 (2017)

Research on Text Recognition in the Automotive Field Based on XGBoost and Feature Engineering

Yishu Zhao[1,2], Meng Zhang[1,2], Fan Zhang[1,2], and Jing Yang[1,2(✉)]

[1] China Automotive Technology and Research Center Co., Ltd., Tianjin, China
yangjing2012@catarc.ac.cn
[2] China Auto Information Technology (Tianjin) Co., Ltd., Tianjin, China

Abstract. This study focuses on the application and research of XGBoost algorithm and feature engineering in text recognition in the automotive field. First, data sources from multiple parties are obtained to form multi-source heterogeneous data sources. Secondly, feature engineering is constructed, which includes removing noise using regular expressions, text normalization, word segmentation and part-of-speech tagging of sentences, and the construction of text feature space. Then, the principle of the XGBoost algorithm and its advantages in text feature processing are analyzed, and an efficient text recognition model is built. Finally, multiple models are selected for parameters. The results show that the rational application of feature engineering and the XGBoost algorithm can significantly improve the accuracy and stability of text recognition, providing new ideas and methods for the development of text recognition technology. This method provides a new theoretical framework and practical path for the innovation of text recognition technology in the automotive field and contributes to the development research of China's automotive industry.

Keywords: Multi-source heterogeneity · Feature engineering · XGBoost algorithm · Text recognition · Parameter selection

1 Background and Introduction

In the wave of digital transformation, text data is growing exponentially. How to achieve efficient parsing and precise cognition of massive text information has become an important subject of interdisciplinary research. Text recognition, as a fundamental technology for natural language understanding, plays a key role in frontier fields such as knowledge graph construction, public opinion monitoring and analysis, cross-language interaction, and intelligent dialogue systems. Its technological advancements not only drive the development of core algorithms in artificial intelligence but also provide key support [1–3] for the intelligent transformation of industries such as finance, healthcare, and education.

XGBoost, an optimized distributed Gradient Boosting library, has achieved great success in the field of machine learning thanks to its outstanding performance and efficiency. Not only has it performed well in various data science competitions, but it has also

P. Siarry et al. (Eds.): WCNA 2024, LNEE 1550, pp. 64–73, 2026.
https://doi.org/10.1007/978-981-95-6946-5_7

demonstrated strong advantages in practical applications, becoming one of the preferred algorithms for many enterprises and research institutions. For example, literature [4] has proposed an automotive internationalization system platform based on a multi-database system and XGBoost. The permutation and combination join algorithm among 10 underlying text databases was studied, and the optimization function for database selection and join was proposed. The machine learning model XGBoost was used to train and simulate the parameters, and the results showed that the optimization method could basically meet the user's requirements. Literature [5]: To accurately predict car sales, it is proposed to predict car sales based on multi-source heterogeneous big data and the XGBoost algorithm. To eliminate the problem of multicollinearity, principal component analysis is used to analyze sample features with a cumulative weight ratio of over 90%. The trained model was used to test the validation set, and the results showed that the average accuracy between predicted sales and actual sales over multiple months was 88.52 percent, with an average difference of only 11.48 percent. This provides a reference for businesses to predict car sales. However, to fully leverage the performance advantage of the XGBoost algorithm in regression prediction, high-quality feature engineering is key. As a core component of the machine learning process, feature engineering significantly enhances the model's ability to capture data patterns by systematically preprocessing, constructing and screening the original text data into numerical features with strong representational power. Scientific and reasonable feature engineering not only enhances the recognition accuracy of feature importance by XGBoost, but also effectively improves the generalization performance and prediction accuracy of the model.

To sum up, this paper presents a text recognition method that combines the XGBoost algorithm with feature engineering technology. This method, through systematic theoretical derivation and empirical research, delves into the core construction elements of efficient text recognition models, aiming to provide a new theoretical framework and practical path for technological breakthroughs in the field of text recognition.

2   The Method of This Paper

The research method in this paper is divided into five steps. The first step is data source selection. Selecting the appropriate data source can further improve the reliability of the results. The second step is feature engineering of the data, which mainly involves preprocessing the original data to form a standardized, dimensionally reasonable, and information-complete feature database. The third is an introduction to the XGBoost model, which explains the operation steps of each step of the model through mathematical formulas. The fourth is model parameter selection, which mainly involves choosing the best set of model parameters to achieve the highest accuracy of the model. The fifth is results and analysis, analyzing the strengths and weaknesses of the model and the next research direction. The roadmap of the research methods in this paper is shown in Fig. 1 below.

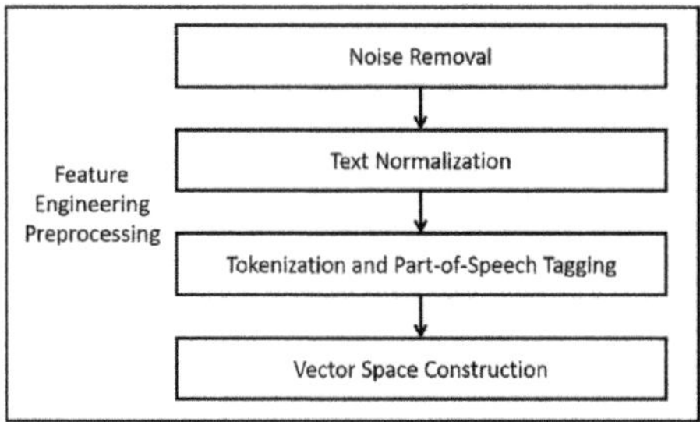

Fig. 1. Roadmap of research methods in this paper

2.1 Data Source Selection

Data is mainly collected through automated web crawler technology, using a distributed crawler architecture, equipped with IP rotation mechanisms and request frequency control systems to comply with the anti-crawling strategies of each platform. All data collection strictly adheres to the Cybersecurity Law and the platform's Robots protocol, and only processes non-private information that is fully public and has no access restrictions. The main dimensions of data collection are as shown in Table 1.

Table 1. Main Dimensions of Data collection

Serial numbers	Data source	Fields
1	Social media matrix	Short content platforms: Weibo (including trending topics, influencer updates)
		Image-text communities: Xiaohongshu (grass-planting notes, product reviews), Zhihu (question-and-answer knowledge base)
2	Video and live streaming platforms	Professional video sites: Bilibili (including bullet-screen data, UP host ecosystem), Tencent Video
		Short video platforms: Xigua Video, Weishi
3	News and information cluster	Aggregation: Toutiao, NetEase News, Sina News
		Portal sites: Tencent.com, ifeng.com, The Paper
		Vertical fields: Autohome, Dongchedi, Pacific Auto Network, 36Kr, Huxiu.com, Xueqiu

2.2 Feature Engineering Processing

Text data has rich semantic information and complex grammatical structures. How to extract valuable features from the original text is the key to text recognition. The feature engineering preprocessing in this paper is carried out in four steps. The first step is to remove noise using regular expressions, the second step is to normalize the text, the third step is to perform word segmentation and part-of-speech tagging on the statements, and the fourth step is to construct the text feature space. The preprocessing steps of feature engineering in this paper are shown in Fig. 2 below.

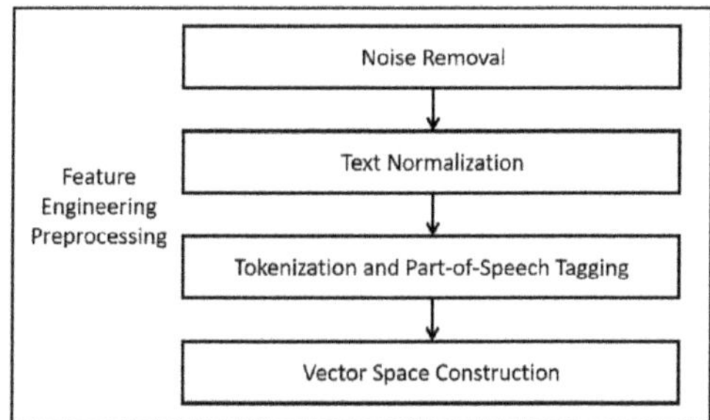

Fig. 2. Figure of the feature engineering preprocessing steps in this paper

(1) Noise removal

In modern data acquisition and text analysis processes, the original text is often mixed with a variety of noisy data such as HTML tags, encoded error characters, invisible control characters, etc. These impurities not only affect the readability of the text, but also seriously interfere with the results of subsequent natural language processing tasks. Noise removal mainly involves using regular expressions to remove special characters and redundant information. As follows:

1. Special character processing: Removing HTML tags (BeautifulSoup), eliminating garbled text and other processing operations. The key code is as follows.

```
# Remove the <script> and <style> tags and their contents
text    =    re.sub(r'<(script|style).*?    >.    *?    </\1>',    ",    text,
flags=re.DOTALL|re.IGNORECASE)
    # Remove other HTML tags
    text = re.sub(r'<[^>]+>', ", text)
    # Remove HTML entity characters
    text = re.sub(r'&[a-z0-9]+; ', ' ', text, flags=re.IGNORECASE)
```

2. Redundant information filtering: Remove non-text content such as AD text, copyright notices, etc. The key code is as follows.

```
# Handling AD content (deleting the entire line)
    for keyword in ad_keys:
        text        =        re.sub(rf'^.*{keyword}.*$\n?        ',        ",        text,
flags=re.MULTILINE|re.IGNORECASE)
    # Handle copyright notices
    for pattern in copyright_regex:
        text = re.sub(pattern, ", text, flags=re.IGNORECASE)
```

(2) Text normalization.

Text normalization refers to the process of converting a text to a standard consistent form, which mainly includes: eliminating randomness and variability in the text, retaining core semantic content, providing consistent input for downstream tasks such as standardizing full-width/half-width characters,standardizing currency symbols. Key codes are as follows:

```
# Traditional to Simplified
text = convert_to_simplified(text)
# Full Angle to half Angle
text = full_to_half(text)
# Standardizing numbers
text = re.sub(r'[Zero One two three four Five Six Seven Nine]', lambda x: str('
zero one two three four Five Six Seven Nine '.index(x.group()))), text)
```

Word segmentation and part-of-speech tagging
1. Word segmentation is the process of splitting consecutive text into meaningful language units to provide input for subsequent NLP tasks such as part-of-speech tagging and syntactic analysis. The key code in the jieba toolkit is as follows [6, 7].

```
seg_list = jieba.cut(text, cut_all=False)print(" Exact pattern: "+ "/".join(seg_list))
# Add a custom dictionary
jieba.load_userdict("user_dict.txt") # Format per line: "word frequency part-of-speech"
# Keyword extraction import jieba.analyse
keywords = jieba.analyse.extract_tags(text, topK=5)
```

2. The purpose of part-of-speech tagging is to identify the grammatical role of words in context, help distinguish synonyms (such as differentiating between "Apple" as a company and as a fruit), identify noun entities, verb actions, etc., and improve syntactic analysis [8, 9]. The key code is as follows:

```
words = pseg.cut(" The intelligent driving of this car is quite good, but I dare not use it when driving at high speed ")for word, flag in words:
    print(f"{word}({flag})", end=" ")
```
Vector space construction

The construction of the feature vector space is the process [10] of converting text data into numerical vectors, and the bag-of-words model is one of the common methods for constructing the feature vector space. The bag-of-words model regards each word in the text as a feature and takes the occurrence frequency of each word in the text as the value of this feature. It is simple and intuitive but ignores the order and semantic relationship between words. TF-IDF, based on the bag-of-words model [11], takes into account the rarity of words in the document collection and adjusts the weights of words to better reflect the importance of words; Word embeddings map words into a high-dimensional vector space, providing a richer representation of text features by capturing semantic relationships between words. Key code for constructing spatial vectors with TF-IDF:

```
from sklearn.feature_extraction.text import TfidfVectorizer
tfidf_vectorizer = TfidfVectorizer()
X_tfidf = tfidf_vectorizer.fit_transform(corpus)
```

2.3   Introduction to the XGBoost Model

2.3.1   Introduction to the XGBoost Algorithm

Since its proposal, the XGBoost algorithm has received extensive attention from both academia and industry. Numerous studies have shown that XGBoost excels in handling structured data, with advantages such as efficiency, accuracy, and scalability. It has been widely used in fields such as finance, healthcare and marketing, but there is relatively little research on XGBoost in text recognition. Although text data has characteristics such as unstructuredness and high dimensionality, which are quite different from traditional structured data, the XGBoost algorithm itself has strong model capabilities and flexibility, and through reasonable feature engineering processing, it has achieved good results in text recognition tasks.

2.3.2   Theoretical Basis of the XGBoost Model

This study uses the XGBoost algorithm for modeling, which is based on the gradient boosting framework and optimizes model performance by iteratively constructing a regression classification tree (CART). In the process of generating the tree structure, XGBoost uses the Greedy Algorithm for optimal feature selection [5], whose core splitting criterion is as shown in formula (1).

$$Gain = \frac{1}{2}\left[\frac{G_L^2}{H_L + \lambda} + \frac{G_R^2}{H_R + \lambda} - \frac{(G_L + G_R)^2}{G_L + G_R + \lambda}\right] - \gamma \tag{1}$$

The four parts of Formula (1) are: the score of the left subtree (evaluating the prediction results of the left subset (sample set IL) after feature segmentation), the score of the right subtree (evaluating the prediction results of the right subset (sample set IR) after feature segmentation), the unpartitioned score, and the model complexity resulting from adding new leaves. Assuming IL and IR are the sample sets to the left and right of the segmentation points respectively, calculate the Gini coefficient $I = IL \cup IR$ [5]

for each segmentation scheme. The partition with the smallest Gini coefficient is the optimal partition of the node, and its model is shown in Eq. (2).

$$y_i = \phi(x_i) = \sum_{k=1}^{k} f_k(x_i) \, , fk \in F \tag{2}$$

where K is the number of CART trees and F is the CART regression classification tree model, on this basis, assuming that the errors of the learners of the base model are independent of each other [5], when a new function F(t) is added to the model, the learning objective of the XGBoost algorithm is to find $f_t(x)$ and minimize the objective function, as shown in formula (3) below:

$$\hat{y}_i^{(t)} = y_i^{(t-1)} + f_t(x_i) \tag{3}$$

Learning the i-th sample in the t-th round model requires combining the results of the previous t-1 rounds to learn the current round model, so the objective function of the i-th sample during the t-th round is formula (4).

$$L^{(t)} = \sum_{i=1}^{n} l\left(y_i, \hat{y}_i^{(t)}\right) + \Omega(f_t)$$

$$L^{(t)} = \sum_{i=1}^{n} l\left(y_i, y^{(t-1)} + f_t(x_i)\right) + \Omega(f_t) \tag{4}$$

2.3.3 XGBoost Model Prediction

The prediction function of the XGBoost Model is constructed through the Additive Model, whose core mathematical expression is Formula (5).

$$\hat{y}_i = \varphi(x_i) = \sum_{k=1}^{K} f_k(x_i), f_k \in \Gamma \tag{5}$$

In Formula (5), K is the total number of ensemble trees (iterations), f_k is the kth CART regression tree belonging to the function space Γ. The output of each f_k output is the weight of the leaf node,which is $w_q(x_i)$.

2.4 Model Parameter Selection

When choosing parameters for the XGBoost model, it is necessary to combine data characteristics, task objectives, and computing resources, and find the optimal parameter combination through experiments and tuning. The parameter explanations and values are shown in Table 2 below.

Table 2. Parameter explanation and values

Serial numbers	Parameters	Instructions	Default value
1	eta	Control how much each tree contributes to the final prediction	0.1
2	n_estimators	Control the number of base learners (trees)	100
3	max_depth	Control the complexity of the tree to prevent overfitting	5
4	subsample	Control the proportion of samples used for each tree to prevent overfitting	1.0
5	colsample_bytree	Control the proportion of features used by each tree to prevent overfitting	1.0
6	lambda	Control model complexity and prevent overfitting	1
7	min_child_weight	Control the minimum sample weight sum of leaf nodes to prevent overfitting	1
8	gamma	Control the growth of the tree and split only when the loss function drops more than gamma	0

2.5 Model Building and Evaluation

Input the text data processed by feature engineering into the XGBoost model for training, set the hyperparameters such as the learning rate, the maximum depth of the tree, and the subsampling ratio, select the optimal combination of hyperparameters through cross-validation, and use the XGBoost API to implement model training and prediction.

Model evaluation: Evaluate the performance of the model using metrics such as accuracy, precision, recall, and F1 value. Accuracy indicates the proportion of samples that the model predicts correctly to the total samples; precision indicates the proportion of samples predicted to be positive that are actually positive; recall indicates the proportion of samples that are actually positive that are correctly predicted to be positive; the F1 value is the harmonic mean of precision and recall, taking precision and recall into account. Select the optimal combination of feature engineering methods and model parameters by comparing the model performance under different feature engineering methods and different model parameters.

3 Experimental Results and Analysis

3.1 Dataset

This study uses a self-collected dataset, which includes multiple categories of text samples, each with a corresponding label. During the experiment, the dataset was divided into a training set and a test set in a specific proportion for the model's training and performance evaluation, respectively.

3.2 Test Results

The impact of feature engineering methods: Bag-of-words models are simple and intuitive, but ignore the order and semantic relationship between words, resulting in higher sparsity of feature representations and affecting the performance of the model. TF-IDF takes into account the rarity of words based on the bag-of-words model and adjusts the weights of words to better reflect the importance of words, but still has the problem of insufficient feature representation. The TF-IDF method combined with information gain feature selection reduces feature dimensions and enhances feature representativeness by choosing the most valuable features, thereby improving model performance.

The influence of model parameters: The learning rate controls the step size of model updates in each iteration. A learning rate that is too small leads to slow convergence of the model, while a learning rate that is too large leads to model oscillation. The maximum depth of the tree determines the complexity of the model. A tree that is too deep can lead to overfitting, while a tree that is too shallow can lead to underfitting. Subsampling ratios increase the generalization ability of the model by randomly selecting some samples for training. By adjusting these hyperparameters reasonably, the optimal balance point of the model's performance can be found.

4 Conclusions and Analysis

This study combines the XGBoost algorithm with feature engineering to study text recognition. The experiments show that reasonable feature engineering can effectively extract and select text features, provide high-quality input for the model, improve recognition accuracy and stability, and further optimize the model performance by tuning the XGBoost hyperparameters.

There are still some deficiencies in this study. The scale and diversity of the experimental dataset are limited, so it may be difficult to fully reflect the practical application; there is room for improvement in feature engineering methods to explore more advanced word embedding and deep learning feature extraction techniques; model evaluation metrics need to be more comprehensive, with a focus on robustness and interpretability in addition to existing metrics. Therefore, future research can be carried out from the following four aspects. One is to expand the scale and diversity of the experimental dataset, covering more domains and types of text data, in order to improve the generalization ability of the model. Second, delve into more advanced feature engineering methods and combine them with deep learning techniques to further enhance the performance of text recognition. Third, intensify research on the interpretability of the model to explore how the XGBoost model can provide more intuitive and understandable prediction results in text recognition tasks, providing better support for practical applications. Fourth, study how to combine the XGBoost algorithm with other machine learning algorithms and natural language processing techniques to build more complex and powerful text recognition systems and promote the development and application of text recognition technology.

References

1. Chen, Z., Li, D., Tan, Y.: Beijing Geodesy **39**(02), 127–134 (2025)
2. Hu, R., He, C., Zhang, W., et al.: Handwritten Chinese text recognition based on convolutional recurrent neural networks. Sci. Technol. Eng. **25**(04), 1547–1554 (2025)
3. Feng, X., Yao, W.: Bill text recognition method based on bidirectional long Short-Term memory network and sparse self-attention. J. Sensing Technol. **37**(11), 1946–1951 (2024)
4. Shengqiang, H., Hui, W., Fan, Z., Kai, K.: Construction and research of automotive internationalization system based on multi-database system and XGBoost. 2024 International Conference on Electronics and Devices, Computational Science (ICEDCS), pp. 527–532. Marseille, France (2024)
5. Yi, W., Jia, Z., Fan, Z.: Sales forecast based on multi-source heterogeneous and XGBOOST models. China Automotive **01**, 9–15 (2021)
6. Baoga, G., Caireng, A.J.: A study on Tibetan word segmentation based on an improved Hidden Markov model. Info. Technol. Information. (03), 64–67 (2025)
7. Xiyu, W., Dongbo, W.: Across language texts based on language model automatic segmentation study. Library Journal 1–19 (2025)
8. Tang, M., Zhao, H., Li, W.: Modern Information Technology **8**(22), 85–91 (2024)
9. Yang, L., Qiankun, X., Chang, L., et al.: Research on a large language model with automatic part-of-speech tagging across languages based on classics. Info. Data Work **46**(02), 82–90 (2025)
10. Liu, S., Liu, J., Guo, B., et al.: Text classification based on vector space model. Comp. Prog. Skills and Maint. **6**, 44–47 (2024). https://doi.org/10.16184/j.cnki.com.prg.2024.06.003
11. Li, Z.: Research on optimization of neural machine translation model based on continuous bag-of-words model. Auto. Instrument. **11**, 48–52 (2024)

Visual Optimization Design of Web Page Interface Text Based on Computer Interaction Technology

Hongmei Wang[1] and Xiaoou He[2]([✉])

[1] College of Fine Arts and Design, University of Jinan, Jinan, Shandong, China
[2] School of Art of Soochow University, Suzhou, China
`Xiaoou_He@163.com`

Abstract. In the current visual optimization design of web page interface text, there are many problems such as the long loading time of non-text content such as pictures and videos, which leads to delayed display of text content, and the lack of effective hierarchical design such as titles and subtitles in text presentation. This article aimed to use computer interaction technology to solve the problems of slow interface response time and unclear web page content hierarchy in the visual optimization design of web page interface text. The study used media queries in Cascading Style Sheets (CSS) to optimize web pages for specific screen widths and device types, and dynamically adjust the layout and style of the page based on different conditions. Using fluid layout and CSS's flexible box mode, the text and other elements in the web page can automatically adjust their width and arrangement according to the size of the browser window. Combining flexible layout and fluid layout technology, the arrangement of text blocks on different devices can be automatically adjusted. Computer interaction technology and real-time response mechanism are used to interact with dynamic text and analyze user interaction behavior, thereby optimizing the visual effect of web page text. Experiments show that the response time of the text visual design of the web page interface in this article on desktops of different sizes is within 1s; the page response time for users to perform interactive operations on the web page is within 0.5s, and the scrolling response time is within 0.2s; when the network speed is not less than 10Mbps, the page loading time is within 2s, the image loading time is within 1.5s, and the video loading time is within 2.5s. The study verified the effectiveness of computer interaction technology in the visual optimization design of web page interface text and improved users' overall satisfaction with the web page.

Keywords: Computer Interaction Technology · Web Interface · Text Visual Optimization · Cascading Style Sheets · Real-Time Response Mechanism

1 Introduction

With the rapid development of Internet technology, web design has gradually become a key link in improving user experience. Modern web pages not only need to have rich content and beautiful pictures, but also need to ensure the simplicity and functionality

P. Siarry et al. (Eds.): WCNA 2024, LNEE 1550, pp. 74–83, 2026.
https://doi.org/10.1007/978-981-95-6946-5_8

of the overall design to adapt to the browsing habits of different users. As users' requirements for information acquisition speed continue to increase, fast and clear content display has become an important goal of web design. As the core carrier of information transmission, the visual presentation of text is particularly critical. The current web design trend emphasizes simplicity and beauty, and advocates the effective transmission of information. The visual optimization of text is an important part of it. Text not only needs to be readable, but also needs to have a clear hierarchical structure to ensure that users can quickly capture key information. In addition, web design must not only consider aesthetics, but also compatibility and cross-platform adaptability. As an important part of web design, text visual optimization plays a positive role in areas such as search engine optimization (SEO) [1–3]. It not only affects the user's reading experience, but also affects the overall web page interaction efficiency.

In order to effectively solve the above problems, the study introduces computer interaction technology. Computer interaction technology improves the visual display effect of text in web page interfaces in many ways, including optimizing interface response time, improving web page loading speed, and dynamically adjusting text structure. Through computer interaction technology, the dynamic content of web pages can be loaded instantly, avoiding the delay in text display caused by the slow loading of non-text content. The main contribution of this article is to analyze the common problems in the text display of web page interfaces in detail and propose an optimization solution based on computer interaction technology, focusing on solving problems such as slow web page response time and unclear text hierarchy. The study verifies the effectiveness of these technical solutions through experimental design and explores text optimization strategies in different scenarios.

2   Related Work

With the popularization of the Internet and the development of technology, web design has become an important way for users to obtain information and communicate and interact. Text visual design has become the key for users to have a good experience on information-intensive websites. To study the design of variable fonts from the perspective of information dissemination design, Huang G et al. [4] studied a new interactive art expression method, which realized the interactive art expression of multimedia elements in display and control. In view of the current lack of media for publishing information related to performing arts, Mahardika I N M et al. [5] developed an interface design for a performing arts website, realizing a user-centered design. In order to enhance the importance of online learning resources in the field of education, Yuan S et al. [6] studied the visual communication design method based on deep learning and successfully improved the design speed and performance of visual communication in online learning materials. However, despite the continuous advancement of web design technology, text design still faces many challenges. Problems such as web page loading delay, unclear hierarchy of web page text, and poor text readability are still prominent.

Computer interaction technology has been widely used in web design in recent years, greatly promoting the improvement of user experience. The core of computer interaction technology is to perceive user behavior and adjust the content or layout of the web page in real time, thereby achieving a smarter and more personalized interactive experience [7–9]. Through advanced caching mechanisms, content distribution network (CDN) technology, and browser-based performance optimization technology, web pages can respond quickly during user browsing, reduce loading time, and improve the overall smoothness of the page. In order to combine aesthetics with user needs, Leiva L A et al. [10] studied a convolutional neural network model and successfully provided personalized design of web pages based on user visual preferences. In order to explore the impact of web design on the future of hotels, Chen J et al. [11] analyzed web fonts and web corner curvature, and used experiments to verify the positive impact of fonts and corner curvature on web pages. In response to the large differences in user experience of learning management systems, YuF et al. [12] studied and designed a user-centered course web page, which effectively improved the user experience. In order to study the application of augmented reality technology in the user experience of web services, scholars such as Yongkang X [13] further analyzed the advantages of combining augmented reality technology with web pages and successfully designed web pages with augmented reality features. In addition, interactive technology can also adaptively adjust the text layout and visual effects of web pages according to the user's browsing habits and device characteristics, thereby meeting the personalized needs of different user groups [14, 15]. However, current text visual optimization usually only focuses on static content and lacks sufficient consideration of dynamic content. With the continuous changes in user interaction patterns, traditional static typesetting methods have been unable to meet the needs of modern web design. Although existing text visual optimization methods have achieved certain results in some aspects, there is still much room for improvement in adaptive typesetting, dynamic optimization, and personalized text display.

3 Text Visual Optimization Design Based on Computer Interaction Technology

3.1 Adaptive Design and Text Layout Optimization

In terms of adaptive design, when the device screen width is 1200 pixels or more, the page uses a multi-column layout; when the page pixel is 600 pixels or less, it switches to a single-column layout. On wide-screen devices, the text is distributed in multiple columns, and on narrow-screen devices such as mobile devices, the text is automatically rearranged into a single column for display. After the page layout is adaptively designed, in order to ensure that users have the best reading experience on different devices, a responsive font and image adjustment method is adopted, and relative units are used to define the font size. For image content, clear images are loaded on high-resolution devices, while smaller images are loaded on low-resolution or small-screen devices. The text readability score can be used to evaluate the ease of reading text.

The formula for the text readability score R is:

$$R = \frac{\mathcal{T} \times (\delta_1 - \delta_2)}{\alpha \times d \times (\delta_1 + \delta_2)} \tag{1}$$

$\mathcal{T}$ is the total number of words in the text; δ_1 is the brightness value of the text color; δ_2 is the brightness value of the background color; α is the font size; and d is the line spacing. Formula (1) takes into account the effects of text length, brightness values of text and background colors, font size, and line spacing. The higher the readability score, the easier the text is to read.

In terms of text layout optimization, on desktop computers, web page content adopts a double-column or multi-column layout to increase the information density of the page. On mobile devices, it automatically switches to a single-column layout so that the text is arranged in a natural reading order. Relative units can be used to define font size, and line spacing can be optimized according to different screen sizes. The adjustment of line spacing depends on the size and resolution of the device, reducing the phenomenon of reduced reading experience caused by overly dense text layout on small-screen devices. On small-screen devices, traditional navigation menus take up too much screen space. To this end, responsive navigation design is used to display the navigation menu on small-screen devices in the style of a "hamburger menu" that can only expand when the user clicks on it. Responsive image loading technology is used to enable the browser to load the appropriate image size according to the screen size, and to dynamically adjust embedded videos and other non-text content according to the device screen size and resolution. In addition, in order to solve the problem that the fixed navigation bar and sidebar of the web page occupy too much screen space, through layout optimization, the fixed elements are automatically shrunk or hidden when the page is scrolled, and only expanded when the user needs it.

3.2 Dynamic Text Interaction and User Behavior Analysis

The dynamic text interaction method combines computer interaction technology and real-time response mechanism. The dynamic text interaction method in this article is mainly reflected in four aspects, namely, text dynamic loading and lazy loading, real-time text adjustment and response, text scaling and custom adjustment, and interactive text highlighting. The formula for dynamic loading efficiency is:

$$\mathbb{E} = \frac{\sum_{i=1}^{n}(\varphi_i \times \omega_i)}{t_l + \theta \times \sum_{j=1}^{m} B_j} \tag{2}$$

φ_i is the number of successfully loaded contents; ω_i is the quality score of the i-th content; t_l is the total time required for loading; B_j is the size of the j-th loaded content; θ is the weight coefficient for adjusting the impact of content loading time on efficiency; m and n are the number of loaded contents and the number of successfully loaded contents, respectively. The web page interface based on computer interaction technology is shown in Fig. 1.

From Fig. 1, it can see that with the cooperation of computer interaction technology, the visual optimization design of the web page interface text is realized. In this web page, the font can be dynamically adjusted, page format can be customized, and the text block arrangement can be customized. The web page also has multi-functional designs such as the navigation menu bar and the operation prompt bar. In order to enhance the user's browsing experience, the design uses virtualization technology, and the navigation menu bar or operation prompt bar can only be displayed when the user needs it. In addition, users can also adjust and add more functions that meet their own needs in the personalized settings.

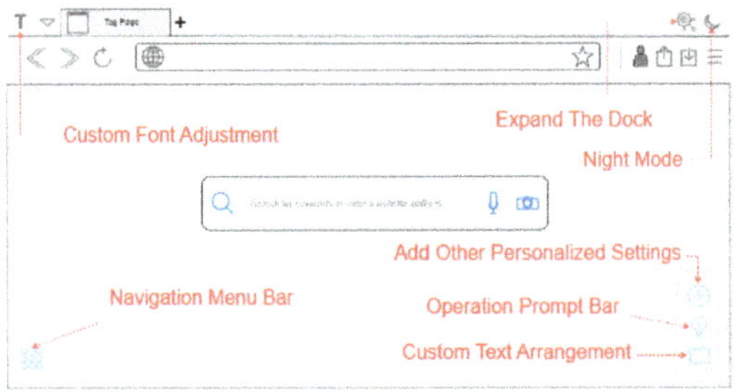

Fig. 1. Web page interface diagram

User behavior analysis provides data support for web design and content layout optimization by monitoring and analyzing user interaction on web pages. The specific user behavior analysis process is shown in Fig. 2.

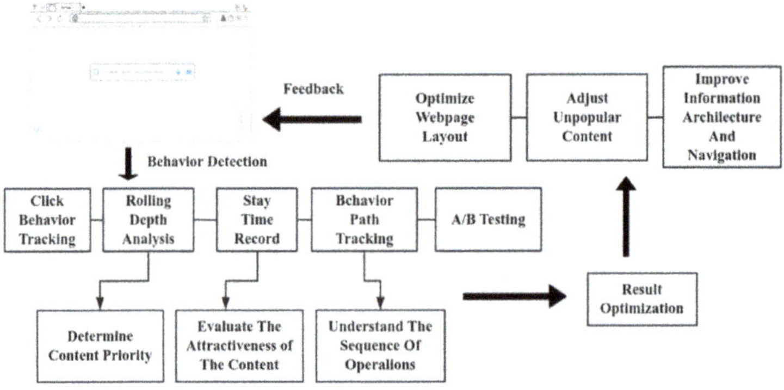

Fig. 2. User behavior analysis process

Figure 2 shows the specific process of the user behavior analysis method. As can be seen from Fig. 2, the analysis of user behavior mainly includes click-through rate tracking analysis, scroll depth analysis, dwell time analysis, interactive behavior analysis and A/B testing. By tracking the user's click behavior on the web page, a corresponding heat map is generated to show the user's attention to different areas of the web page. This article analyzes the depth of the user's scrolling page to understand the user's attention and neglect of the text content when browsing the web page. It can record the user's stay time in a specific text area, track the user's behavior path, and analyze the user's operation sequence and click flow when browsing the web page. Through the above operations, it can understand the user's reading logic and information acquisition habits. The A/B testing method is used to verify the effect of different text layouts. Users are randomly assigned to different text layouts, and the user behavior data of each version is compared to finally determine the text layout method that best suits the user. Finally, the results of the behavior analysis are fed back to the web page system for corresponding adjustments and optimizations.

The formula for user behavior response time is:

$$t_r = t_c + t_l + \beta \times t_n + \gamma \times t_d + \lambda \times N \tag{3}$$

Among them, t_r is the total response time of user behavior; t_c is the delay time after the user clicks; t_l is the content loading time; t_n is the time required for the browser to present the loaded content to the user; t_d is the network delay time; N is the number of user interactions on the web page; β, γ and λ are weight coefficients used to adjust the impact of corresponding factors on the total response time.

4 Experimental Design and Testing

4.1 Experimental Preparation and Data Collection

The experiment uses VSCode (Visual Studio Code) as the main code editor for web development, uses the Bootstrap framework for web design, and uses Vue.js as the JavaScript framework to implement dynamic text interaction and user behavior monitoring. Selenium is used as a testing platform to simulate user operations such as clicks, scrolling, input, etc., to test the dynamic interaction functions of the web page. Google Lighthouse is used to analyze and test the performance, accessibility, and best practices of the web page. The device types mainly include desktop computers, tablets, and smartphones; the operating systems include Windows, macOS, iOS, and Android. The dataset mainly comes from the actual running web page log files, user interaction records, and the loading time statistics of the test web pages. The dataset contains more than 100,000 user interaction records, which are distributed in multiple different web pages and usage scenarios. The experimental operating system is Ubuntu 20.04 LTS, the server is Intel(R) Xeon(R) CPU E5–2620 v4 @ 2.10GHz, 128GB RAM, and the network environment is 100Mbps optical broadband.

4.2 Test Scenarios and Steps

In order to evaluate the text visual optimization effect of the web page interface, the experiment designed 5 test scenarios to verify the adaptive design, dynamic interaction and text hierarchy optimization effect of the web page interface text visual. The 4 test scenarios are: responsive text layout test on the desktop, adaptive text layout test on the mobile terminal, readability test of text hierarchy, dynamic text interaction and user behavior response test.

The test steps for each test scenario are as follows.

For the responsive text layout test on the desktop (Scenario 1): open the test web page, record the initial layout effect, reduce the browser window to 1024 × 768, observe the changes in the text layout, and record whether there are elements that exceed the screen or the layout is chaotic; then adjust the window to 800 × 600, and record whether the text and pictures are still displayed according to the adaptive layout rules; finally, close the web page and record the response time and the impact on text readability each time the window is adjusted.

For the adaptive text layout test on mobile (Scenario 2): use a mobile device to open the test web page and check the initial layout effect. Read the web page content in portrait mode and record whether there are any improper layout or abnormal text scaling. Switch the device to landscape mode and observe the rearrangement of the web page content, recording whether the clarity of the layout and hierarchy is maintained. Finally, retest with different font size settings and record the impact of font size on the layout.

For the readability test of text hierarchy (Scenario 3): open the test web page, step by step browse the page content containing multiple titles, subtitles, paragraphs and lists, and record the reading order under different text hierarchies. Retest under different font sizes and contrast settings, and record the difference between the hierarchy before and after visual optimization.

For dynamic text interaction and user behavior response test (Scenario 4): Open the web page and click on various text interaction elements such as hyperlinks, buttons, dynamically loaded content, etc., on the page, and record the response speed and visual effect after each click. Perform mouse hover and other operations on the test web page to observe whether there is a responsive dynamic effect display. Simulate fast browsing and slow scrolling of web pages, record the system's response to user scrolling behavior, and pay attention to the display process of dynamically loaded content. Finally, measure the size of the click hotspot and the accuracy of the text interactive area.

4.3 Experimental Data Analysis

Through the above experimental steps, Table 1 shows the scenario test results of scenario 1, scenario 2 and scenario 3.

Table 1 clearly shows the detailed test results of scenario 1, scenario 2, and scenario 3. For scenario 1, it can see that scenario 1 mainly tests the desktop screen size, and the test indicators are layout adaptation, whether the text exceeds the screen, response time, and layout change. From Table 1, it can see that after continuous adjustment of the screen size, the web page design in this article can make the text adapt to the screen size very well. In addition, there is no situation where the text goes beyond the screen. In terms of response time, the webpage has the longest response time of 0.6s at 640x480 pixels, followed by 0.5s at 1920x1080 pixels, and the shortest response time is only 0.3s at 1024x768 pixels. In terms of layout changes, as the screen size continues to decrease, the layout of the webpage can change more and more seriously.

As for scenario 2, it can see that scenario 2 mainly tests different font sizes for horizontal and vertical screens on mobile terminals. The test indicators include whether the text layout is adapted, whether horizontal scrolling is required, and whether the image display is normal. As can be seen from Table 1, regardless of whether the screen is horizontal or vertical and whether the font size changes, the text on the web page can adapt well to the layout. In portrait mode, no horizontal scrolling is required regardless of the font size, while in landscape mode, horizontal scrolling is only required for large fonts. In terms of image display, the images in portrait mode are displayed normally, and

Table 1. Scenario 1, scenario 2 and scenario 3 test results

Scenario 1				
Window Size (pixels)	Layout Compatible?	Text Exceed The Screen?	Response Time (seconds)	Layout Changes
1920x1080	✓	×	0.5	Nothing
1024x768	✓	×	0.3	Minor Adjustment
800x600	✓	×	0.4	Significant Adjustments
640x480	✓	×	0.6	Serious Adjustment
Scenario 2				
Equipment Direction	Font Size	Text Layout Adaptability	Need Horizontal Scrolling?	Image Display Normal?
Vertical Screen	Small	✓	×	Normal
Vertical Screen	Large	✓	×	Normal
Landscape Screen	Small	✓	×	Normal
Landscape Screen	Large	✓	✓	Image Offset
Scenario 3				
Text Hierarchy	Clear?	User Understanding Time (seconds)		Found Key Information?
First Level Title	✓	3.6		✓
Secondary Title	✓	4.9		✓
Paragraph Text	×	10.1		×
Reference Paragraph	✓	5.7		✓

only in landscape mode with large fonts, the web page images are slightly offset. This situation is often caused by the size of the image itself.

For scenario 3, scenario 3 mainly tests the readability of the text hierarchy. The test indicators are clarity, user understanding time and key information search. As can be seen from Table 1, at the three text levels of the first-level title, second-level title and quoted paragraph, the web page can be displayed clearly, and users can find key information, and the user's understanding time is also within 6 s. For paragraph text, the web page display is not as clear as the other three text levels, and the user's understanding time is increased to 10.1s. This is because the general key information of the text is in the first-level title, second-level title and quoted paragraph, so the web page weakens the display of paragraph text. Figure 3 shows the test results of scenario 4.

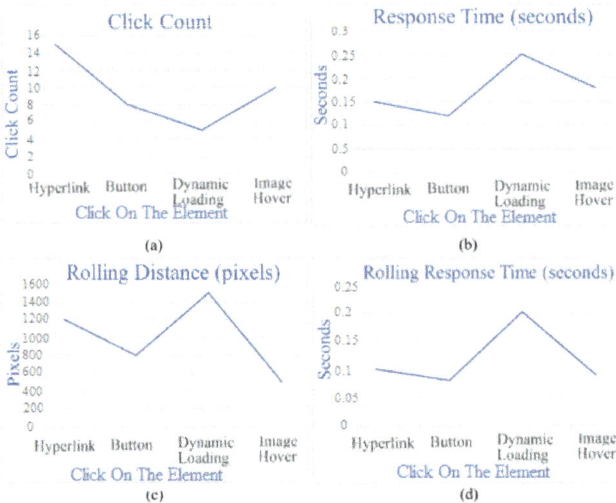

Fig. 3. Scenario 4 test results. (a) Number of clicks, Fig. 3 (b) Response time, Fig. 3 (c) Scrolling distance, Fig. 3 (d) Scrolling response time

As can be seen from Fig. 3(a), user operations on the web page include clicking hyperlinks, buttons, dynamically loading content, and hovering over images. As can be seen from Fig. 3(a), the number of clicks is 15, 8, 5, and 10, respectively. As can be seen from Fig. 3(b), the average response time is 0.15s, 0.12s, 0.25s, and 0.18s, respectively; Fig. 3(c) shows that the rounded values of the scroll distance are 1200 pixels, 800 pixels, 1500 pixels, and 500 pixels, respectively. Figure 3(d) shows the average values of the scrolling response time, which are 0.1s, 0.08s, 0.2s and 0.09s respectively. Based on Fig. 3, the webpage optimization design based on computer interaction technology in this article has a good effect in the test of scenario 4. Regardless of the operation performed on the webpage, the average response time and average scrolling time are within 0.5s.

5   Conclusions

This article proposes a series of solutions to the problems of loading delay and lack of hierarchy in text presentation in current web design. The visual design of web text is optimized through responsive design and user behavior analysis. However, the study did not test the design of web pages on specific devices. In future research, the test scenarios can be further expanded to cover more device types and performance under different network conditions. At the same time, artificial intelligence and machine learning technologies can be combined to achieve more personalized web design.

References

1. Iqbal, M., Khalid, M.N., Manzoor, A., et al.: Search Engine Optimization (SEO): A Study of important key factors in achieving a better Search Engine Result Page (SERP) Position. Sukkur IBA J. Comp. Math. Sci. **6**(1), 1–15 (2022)

2. Erdmann, A., Arilla, R., Ponzoa, J.M.: Search engine optimization: the long-term strategy of keyword choice. J. Bus. Res. **144**, 650–662 (2022)
3. Lewandowski, D., Schultheiß, S.: Public awareness and attitudes towards search engine optimization. Behaviour & information technology **42**(8), 1025–1044 (2023)
4. Huang, G.: Analysis of visual communication elements in variable font design from the perspective of digital technology. Intel. Decis. Technol. (Preprint), 1–17
5. Mahardika, I.N.M., Purwadi, J., Susanto, B.: Penerapan User Interface Design Pattern untuk Situs Web Katalog Seni Pertunjukan Indonesia. J. Terapan Teknologi Informasi **7**(2), 107–119 (2023)
6. Yuan, S.: Visual communication design of web-based learning resources in the digital era. J. Intell. Fuzzy Sys. (Preprint), 1–12 (2024)
7. Azhari, M.H.: Peran Visual dalam Meningkatkan Engagement Pengguna di Situs Web Berita. J. Komunitas Literasi **1**(1), 1–10 (2024)
8. André Costa, S.F., Moreira, J.J.: Towards an AI-driven user interface design for web applications. Procedia Comp. Sci. **237**, 179–186 (2024)
9. Aripiyanto, S., Azhari, M., Munawarohman, R., et al.: Usability improvement through user interface design with human centered design (HCD) method on junior high school websites. 2022 Seventh International Conference on Informatics and Computing (ICIC), 1–7 (2022)
10. Leiva, L.A., Shiripour, M., Oulasvirta, A.: Modeling how different user groups perceive webpage aesthetics. Univ. Access Info. Soc. **22**(4), 1417–1424 (2023)
11. Chen, J., Lehto, X.: Shaping digital luxury perception: the impact of curvature in website design. Tourism Manage. **107**, 105059–105059 (2025)
12. Yu, F., Urquhart, L.R., Sharmin, S., et al.: Evaluating and enhancing canvas course website: prioritizing user-centered design. Proc. Ass. Info. Sci. Technol. **61**(1), 1171–1173 (2024)
13. Yongkang, X., Jethro, S., Conor, F., et al.: Web XR user interface research: design 3D layout framework in static websites. Applied Sci. **12**(11), 5600–5600 (2022)
14. Jomini, N.S., L, A.C., Cynthia, P.: The effects of news site design on engagement and learning. Journalism Practice **16**(6), 1226–1246 (2022)
15. Jiyu, L., Shaopeng, S.: Design method of cross-border e-commerce website color matching for patients with anxiety disorder under cognitive psychology. Psychiatria Danubina **34**(s1), 662–663 (2022)

Intelligent Speech Recognition Based on Internet of Things Technology in Russian MOOC Systems

Lingxu Xiao[✉]

Sanya Aviation and Tourism College, Sanya, China
aa936621579@163.com

Abstract. With the development of Internet of Things technology, the application of intelligent speech recognition in educational platforms has become increasingly widespread. Traditional noise suppression techniques lack flexibility in dealing with noise with both high and low frequencies, and cannot maintain high accuracy in various noise scenarios. To solve this problem, this article collected speech and visual signals through the fourth generation Amazon Echo Dot, used a Multimodal Generative Adversarial Network (MGAN) to generate simulated speech signals under noise, and combined the U-Net model for noise suppression. Furthermore, the Wav2Vec 2.0 model was used to capture contextual information in speech signals, thereby achieving efficient and accurate speech recognition in a multi-noise environment. The experimental results indicate that the proposed method obtains a speech recognition accuracy of 94.3% in noisy environments, significantly better than existing traditional methods, effectively improving the speech interaction effect in the Russian MOOC (Massive Open Online Course) system and enhancing the user learning experience.

Keywords: Intelligent Speech Recognition · Internet of Things Devices · Multimodal Generative Adversarial Networks · Noise Reduction Techniques · Russian MOOC Systems

1 Introduction

With the rapid development of Internet of Things (IoT) technology, intelligent speech recognition plays an important role in the field of education. MOOC [1, 2] systems, as an important form of online education [3], rely on intelligent speech recognition technology [4] to achieve human-computer interaction and user command response. However, existing speech recognition methods still face many challenges in practical applications. In noisy environments, traditional speech recognition technologies often rely on convolutional neural networks (CNN) [5] and recurrent neural networks (RNN) [6], but these methods are difficult to effectively filter noise when faced with complex noise [7, 8], resulting in serious interference of speech signals, which affects the accuracy of recognition. These methods have limited ability to handle a variety of background noises and is unable to satisfy the requirements of high-precision recognition in real-world settings.

© The Author(s) 2026
P. Siarry et al. (Eds.): WCNA 2024, LNEE 1550, pp. 84–94, 2026.
https://doi.org/10.1007/978-981-95-6946-5_9

Existing models often perform poorly when processing long-term dependent speech signals due to loss of context information. Especially in speech recognition scenarios with complex dialogues or long sentences, traditional models find it difficult to maintain a good understanding of the previous and following contexts, resulting in deviations in recognition results. The resource limitation of IoT devices [9, 10] is also a major problem in the practical application of speech recognition. Due to the limited computing power of IoT devices, many highly complex speech recognition algorithms [11] cannot be effectively run on these devices, influencing the system's real-time and efficacy. In this context, how to enhance speech recognition accuracy in noisy environments and adapt to the resource limitations of IoT devices has become a major challenge facing current speech recognition technology.

For the purpose of dealing with the above problems, this article proposes an intelligent speech recognition solution based on IoT technology, which is optimized specifically for the application scenarios in the Russian MOOC system. Using MGAN technology, by combining visual information such as lip shape and expression with speech signals, multimodal input improves the system's speech recognition ability in complex background noise. Generative Adversarial Network (GAN) simulates speech signals under background noise through the generator, and continuously optimizes the generated speech through the discriminator, so that the system can maintain high speech clarity in a multi-noise environment. In addition, this article combines the noise suppression technology of the U-Net structure to further enhance the anti-noise ability of the speech signal after being transmitted through the IoT device. Compared with traditional noise suppression algorithms such as convolutional neural networks and spectral subtraction, the U-Net structure has higher flexibility and accuracy in processing complex noise. To solve the problem of context loss in long-term dependent speech recognition, this article also introduces the Wav2Vec 2.0 model, which can better capture the context information in the speech signal and improve the recognition accuracy of long-sentence speech through self-supervised learning and self-attention mechanism.

2 Related Work

The application of IoT technology in education has made significant progress in recent years. Madni et al. [12] used analytical induction to explore the elements impacting the use of IoT for e-learning in developing nations' Higher Educational Institutions (HEIs), and proposed an IoT based E-Learning adoption model. The results showed that privacy, infrastructure readiness, funding constraints, usability, teacher support, interactivity, attitude, and network and data security were key factors influencing the adoption of IoT in higher education. Sejati et al. [13] improved students' understanding and ability in electricity management practices in schools by implementing IoT technology usage education related to energy efficiency and conservation education. These studies demonstrate the flexibility of IoT technology in education, helping students master complex concepts through hands-on practice.

The research on noise suppression technology in the field of speech recognition is also remarkable. Zhu et al. [14] proposed a novel framework that combines Speech Enhancement (SE) and self-supervised pre-training to improve the performance of Automatic Speech Recognition (ASR) models in noisy environments, achieving stronger noise robustness in various noisy scenarios. Fan et al. [15] proposed a two-stage deep spectrum fusion and joint training framework that includes mask mapping fusion and gated loop fusion for noise robust end-to-end automatic speech recognition, significantly improving recognition performance. These studies provide new ideas for the application of speech recognition technology in noisy environments.

3 Speech Recognition Methods Based on MGAN

3.1 Voice and Visual Signal Collection in IoT Devices

In order to ensure that the collected data is close to real application scenarios, this article chose a typical Russian MOOC learning environment, including two conditions: quiet and noisy, and simulated the actual situations of home learning and public place learning, respectively. This setting can verify the stability and robustness of the model in a variety of loud situations.

In IoT devices, speech and visual signals are collected mainly relies on the fourth generation Amazon Echo Dot device, which has a built-in multi-microphone array and high-definition camera. Voice collection is completed through an omnidirectional microphone, which can capture voice signals in the environment from all directions. To improve the quality of speech acquisition in noisy environments, this study introduces a multi-microphone array and enhances the speech signal through beamforming technology. The core idea of beamforming technology is to use multiple microphones to simultaneously collect sound signals, and by weighting and combining the signals received by microphones at different positions, enhance the speech signal from the target direction and suppress noise signals from other directions. The mathematical principle can be described as:

$$o(t) = \sum_{i=1}^{N} w_i x_i(t - \tau_i) \tag{1}$$

$o(t)$ represents the output signal obtained through beamforming, $x_i(t)$ is the signal received by the ith microphone, w_i is the weight related to the microphone position, and τ_i is the time delay of the signal reaching different microphones. The system dynamically adjusts the "beam" direction by adjusting the weight and time delay , focusing on the speaker's voice and maximizing the suppression of interfering sound sources from other directions. This method effectively improves the signal- to -noise ratio of the voice signal and improves the accuracy of voice collection in complex background noise environments.

This article applies Fast Fourier Transform (FFT) for data processing. The collected time domain speech signal $x(t)$ is transformed into a frequency domain representation by performing FFT transformation:

$$X(f) = \int_{-\infty}^{\infty} x(t)e^{-j2\pi ft}dt \tag{2}$$

In the frequency domain, the frequency distribution of speech signals and noise signals is often different. By analyzing the spectrum, a bandpass filter can be designed to retain the main frequency band of the speech signal and filter out the noise frequency bands. The signal after filtering is:

$$Y(f) = H(f) \cdot X(f) \tag{3}$$

$H(f)$ is the transfer function of the bandpass filter, which is optimized for the main frequency range of speech. Finally, the frequency domain signal $Y(f)$ is converted back to the time domain signal through inverse FFT:

$$y(t) = \int_{-\infty}^{\infty} Y(f)e^{j2\pi ft}\,dt \tag{4}$$

This process effectively removes environmental noise while preserving key information in the speech signal, enabling subsequent speech recognition models to capture and process speech data more accurately.

3.2 Construction and Application of MGAN

MGAN is a model that uses GAN to process multimodal data such as speech and visual signals. MGAN mainly consists of two parts: generator and discriminator, which are trained together in an adversarial manner. Figure 1 is a schematic diagram of the MGAN model structure.

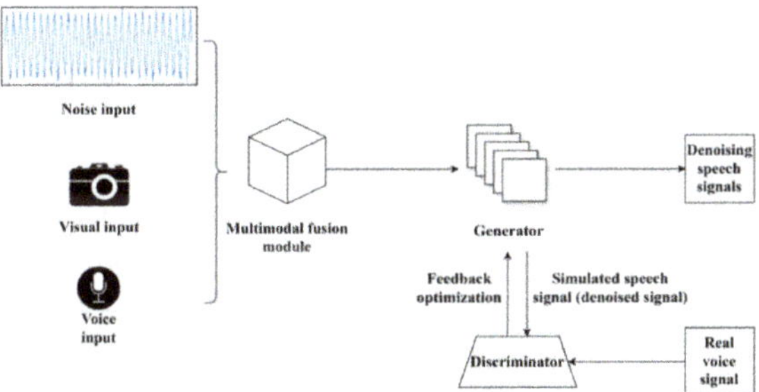

Fig. 1. MGAN Model Framework

The generator generates realistic speech by combining visual signals such as noise, mouth shape, and facial expressions, capturing speech features in different noisy environments. The discriminator uses binary classification to determine whether the speech is real or generated, guiding the generator to optimize. The MGAN model uses a cross entropy loss function, where the generator minimizes the difference between generated speech and real speech, and the discriminator maximizes discrimination accuracy to improve the quality of generated speech.

MGAN simulates speech signals in noisy environments through a generator to enhance speech clarity in noisy learning scenarios. In order to optimize the human-computer interaction effect of the Russian MOOC system, MGAN combines speech and visual signals to improve the accuracy of speech recognition in complex noisy backgrounds. During the generation and optimization process, the discriminator continuously feeds back the output of the generator to ensure that the generated speech signal is closer to the real speech and overcome the difficulty of recognizing Russian pronunciation in noisy environments. In addition, when processing long conversations or complex sentences, MGAN enhances its ability to understand the Russian context through the fusion of visual and speech signals, ensures the speech interaction effect in a variety of noise scenarios, and further improves the accuracy of speech recognition and interactive experience in the Russian MOOC system.

3.3 Noise Suppression Mechanism of U-Net Model

The noise suppression mechanism of the U-Net model in speech recognition is achieved through its unique encoder decoder structure. This model extracts the features of the input signal through a gradually shrinking encoder section, and then restores the signal through a gradually expanding decoder section while removing noise.

The encoder of the U-Net model consists of multiple convolutional layers and pooling layers, which are used to extract features from the input speech signal. Each convolutional layer captures the features of the speech signal at different scales through multiple convolution kernels. These features include various noises in the spectrum and different frequency components of the speech. Through the pooling layer, the signal is gradually downsampled, reducing the spatial dimension of the data, but retaining the main features in the signal. This layer-by-layer reduction process helps to capture global information, allowing the model to effectively extract the key features in the speech signal and gradually filter out the noise as useless information. The decoder part is symmetrical to the encoder. Its task is to restore the signal by upsampling layer by layer. At each decoding stage, the model processes the speech features retained in the previous encoding process to gradually restore the original speech signal. The decoder does not simply restore the signal, it also carefully filters the features so that the model can better distinguish between speech signals and noise. In this way, the U-Net model effectively suppresses background noise and retains clear speech signals. One of the keys to its noise suppression ability is the jump connection of U-Net. These connections directly map the features in the encoder to the corresponding layers of the decoder, thereby retaining high-resolution speech feature information. This design allows the decoder to not only rely on global information but also combine local detail information when restoring the signal, so as to better handle complex noise scenes. In this way, both high-frequency and low-frequency noise are effectively suppressed.

3.4 Wav2Vec 2.0 in Context Information Capture

Wav2Vec 2.0 captures contextual information through a self-supervised learning model, relying on pre-training a large amount of unlabeled speech data to gradually learn the features and patterns of speech signals. During the pre-training process, the speech signal is divided into several time frames, with each frame representing a segment of the speech. Convolutional networks are used to extract low-level features from these frames.

Subsequently, these features are input into a Transformer-based self-attention mechanism, where the self-attention layer is responsible for capturing the relationship between each time frame and other frames, thereby understanding the long-term dependencies of speech signals. In this way, Wav2Vec 2.0 can effectively capture and understand contextual information.

The optimization of this model is achieved through contrastive learning, identifying similar and different speech segments in different contexts. During training, some speech signals can be masked to hide several speech frames, and the model can infer the content of the hidden parts based on the context. The Transformer structure utilizes a multi-head self-attention mechanism to handle the dependency relationships between multiple speech segments, improving recognition performance in complex contexts. Compared with traditional recurrent neural networks, Wav2Vec 2.0 exhibits stronger contextual information retention ability, especially when processing long-term speech signals, significantly improving recognition accuracy. This structure relies on context inference in various noisy scenarios, successfully enhancing speech recognition's robustness and accuracy.

4 Experimental Design for Speech Recognition

4.1 Experimental Environment Configuration and Dataset Description

The experiment relies on IoT devices and high-performance computing platforms. In terms of hardware, the fourth generation Amazon Echo Dot is used to collect voice and visual signals, equipped with multiple microphone arrays and high-definition cameras, and the computing platform is NVIDIA GeForce RTX 4060 GPU (64GB video memory). The software environment includes tools such as Windows 10, Python 3.8, Pytorch 1.8, etc. The experiment collected speech and visual data in both quiet and noisy environments, simulating learning scenarios in homes and public places, covering various background noises. By using beamforming and frequency domain filtering to enhance speech signals, the dataset contains speech segments with different noise levels and multiple Russian accents, which are used to verify the robustness of the model in complex contexts.

4.2 Evaluation Indicators

The evaluation indicators in this article mainly include speech recognition accuracy (Accuracy), word error rate (WER) and signal-to-noise ratio (SNR), which are used to comprehensively evaluate the performance of the speech recognition system in a noisy environment.

Accuracy measures the proportion of correct speech recognized by the model, which is defined as the ratio of the number of correctly recognized words to the total number of words:

$$Accuracy = \frac{n}{n_t} \times 100\% \tag{5}$$

n is the number of words recognized correctly, and n_t is the total number of words.

WER is used to quantify the degree of recognition errors, and the calculation formula is:

$$WER = \frac{S + D + I}{N} \times 100\% \tag{6}$$

S represents the number of replacement errors, D represents the number of deletion errors, I represents the number of insertion errors, and N represents the total number of reference words. The lower the WER, the fewer recognition errors.

SNR is used to evaluate the system's ability to resist noise in different noise environments, and the following is the calculating formula:

$$SNR = 10 log_{10}\left(\frac{P_{signal}}{P_{noise}}\right) \tag{7}$$

P_{signal} and P_{noise} respectively represent the power of the speech signal and the noise power. The higher the SNR, the better the quality of the speech signal relative to the noise.

5   Results Analysis

5.1   Comparative Analysis of Model Performance

In order to comprehensively evaluate the performance of various models in noisy environments, this article compared the MGAN model with DeepSpeech 2, Conformer, and QuartzNet, and Table 1 displays the findings.

Table 1. Comparison of Speech Recognition Performance of Different Methods

Method	Accuracy (%)	WER (%)	SNR (dB)
MGAN	94.3	5.7	20.1
DeepSpeech 2	89.6	10.4	18.9
Conformer	91.5	8.9	19.1
QuartzNet	90.2	9.5	19.4

Table 1 shows that the accuracy, word error rate, and signal-to-noise ratio of the MGAN model reached 94.3%, 5.7%, and 20.1dB, respectively, showing significant improvements compared to DeepSpeech 2, Conform, and QuartzNet. These results demonstrate the potential of the MGAN model in the field of speech recognition.

5.2   Robustness Analysis of MGAN Model in Noisy Environment

In the research of speech recognition systems, different noise environments have a significant impact on the accuracy of the model. This study evaluated the accuracy performance of MGAN, DeepSpeech 2, Conformer, and QuartzNet models in various environments such as quiet rooms, coffee shops, wind noise, street noise, and offices. The result is shown in Figure 2.

Figure 2 shows that MGAN achieves the highest accuracy of 98.5% in quiet rooms and maintains a stable performance of 90.5% in street noise environments. In contrast, DeepSpeech 2 performs weaker in noisy environments, with an accuracy rate of only 83.7% in street noise environments. The performance of Conformer and QuartzNet is between MGAN and DeepSpeech 2, indicating that MGAN exhibits stronger robustness in different background noise environments through its multimodal input and noise suppression techniques, making it the preferred solution for speech recognition in multi-noise environments.

5.3 Performance Comparison of Speech Recognition Systems on Different Speech Lengths

In speech recognition systems, the length of the speech signal has a substantial effect on the recognition model's performance. As speech segments progress from short to long, models need to have strong capabilities in handling contextual information and long-term dependencies, especially in MOOC long dialogue scenarios where the performance of speech recognition systems directly affects the user's interactive experience. This study analyzed the recognition performance of different methods in speech of different lengths and explored the processing ability of the model for long-term speech. The result is shown in Figure 3.

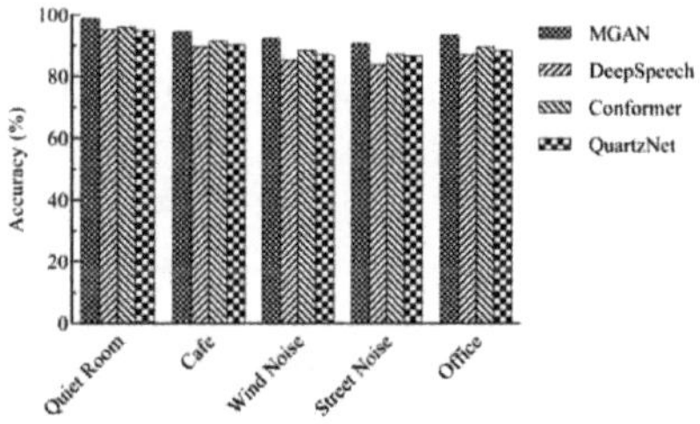

Fig. 2. Comparison of accuracy of speech recognition models in noisy environments

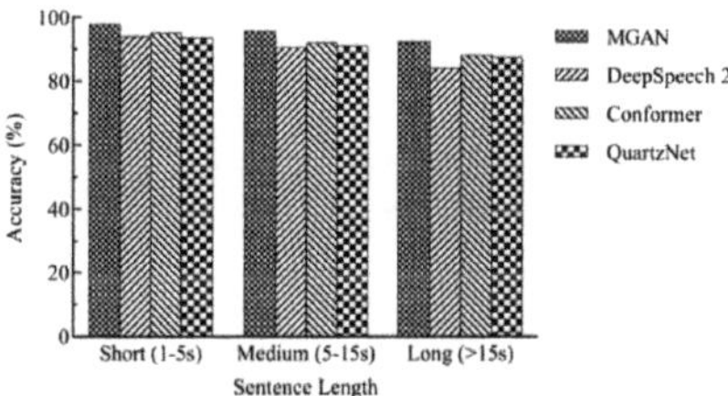

Fig. 3. Comparison of accuracy of speech recognition models under different speech lengths

In Figure 3, MGAN performs the best in processing phrase sounds with an accuracy of 97.8%, and still maintains a high accuracy of 92.3% in long speech segments. In contrast, DeepSpeech 2 performs worse than MGAN, with its recognition accuracy for long speech dropping to 84.2%. Conformer and QuartzNet have accuracies of 88.0% and 87.5% respectively when processing long speech, slightly better than DeepSpeech 2 but not as good as MGAN. This indicates that MGAN not only performs well in short-term conversations, but also has significant advantages in long-term speech recognition, making it an ideal model for handling complex speech environments.

5.4 Performance of MGAN Model on Different Resource Devices

The resource limitations of IoT devices have a potential impact on the performance of speech recognition systems, especially in complex noisy environments. To evaluate the performance of the MGAN model at different resource levels, this article set four resource conditions of 100%, 75%, 50%, and 25%, and tested the changes in speech recognition accuracy, WER, and processing delay, respectively. These data were collected through multiple experiments, demonstrating the performance differences between resource abundant and resource limited situations, as shown in Figure 4.

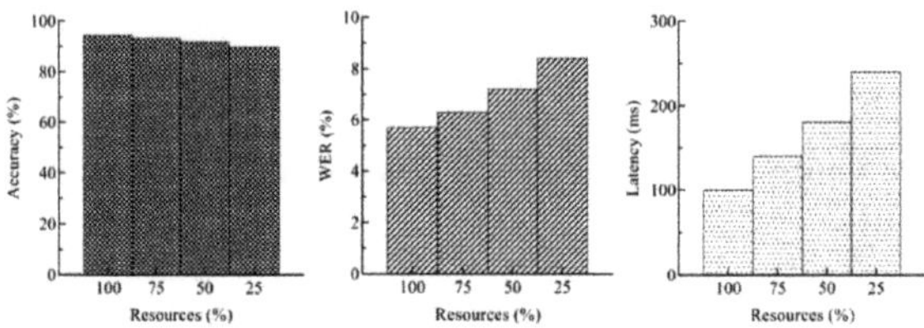

Fig. 4. Performance of MGAN model under different resource constraints

Figure 4 shows that as device resources decrease, accuracy decreases from 94.3% to 89.6%, word error rate increases from 5.7% to 8.4%, and processing latency significantly increases from 100 milliseconds to 240 milliseconds. When the resources are reduced to below 50%, the degree of performance degradation becomes more pronounced, indicating that the MGAN model has a higher dependence on resources. In the case of sufficient resources, the system maintains high accuracy and low latency. However, under resource limited conditions, although the recognition performance has declined, the MGAN model still has strong noise resistance and can maintain relatively stable performance.

6 Conclusions

This article proposes an optimization solution for Russian MOOC systems by combining IoT technology with intelligent speech recognition algorithms. MGAN and U-Net models were used to achieve noise suppression, and Wav2Vec 2.0 was used to capture contextual information, significantly improving speech recognition accuracy in noisy environments, ultimately achieving high accuracy in experiments. However, the performance of the proposed solution in handling latency and accuracy on low resource devices still needs further optimization. Future work should focus on researching resource constraints, exploring more efficient lightweight models, and further improving the generalization capacity of the model in multilingual environments to accommodate a greater variety of application situations.

Acknowledgment. Project Supported by the Education Department of Hainan Province, 2024 Hainan Provincial Higher Education Teaching Reform Research Project "Construction and Reform of 'Russian Audio-Visual Speaking' Course in Vocational Colleges" (Project Number: Hnjg2024-247).

References

1. Papadakis, S.: MOOCs 2012–2022: An overview. Adv. Mobile Learn. Educ. Res. **3**(1), 682–693 (2023)
2. Oh, E. Grace, M.-H.C., Yunjeong, C.: Learners' perspectives on MOOC design. Distance Education **44**(3), 476-494 (2023)
3. Johnson, C.C., Walton, J.B., Strickler, L., Elliott, J.B.: Online teaching in K-12 education in the united states: a systematic review. Rev. Educ. Res. **93**(3), 353–411 (2023)
4. Weng, Z., et al.: Deep learning enabled semantic communications with speech recognition and synthesis. IEEE Trans. Wirel. Comm. **22**(9), 6227–6240 (2023)
5. Shashidhar, R., Shashank, M.P., Sahana, B.: Enhancing visual speech recognition for deaf individuals: a hybrid LSTM and CNN 3D model for improved accuracy. Arab. J. Sci. Eng. **49**(9), 11925–11941 (2024)
6. Yuan, Q., Dai, Yu., Li, G.: Exploration of English speech translation recognition based on the LSTM RNN algorithm. Neural Comput. Appl. **35**(36), 24961–24970 (2023)
7. Bruntha, P.M., et al.: Application of switching median filter with L 2 norm-based auto-tuning function for removing random valued impulse noise. Aerospace Systems **6**(1), 53–59 (2023)
8. Bhaskar, S., Thasleema, T.M.: LSTM model for visual speech recognition through facial expressions. Multimedia Tools and Appl. **82**(4), 5455–5472 (2023)
9. Baucas, M.J., Spachos, P.: Improving remote patient monitoring systems using a fog-based IoT platform with speech recognition. IEEE Sensors J. **23**(15), 17611–17618 (2023)
10. Lin, Y.-B., Liao, Y.-F., Chen, S.-H., Hwang, S.-H., Wang, Y.-R.: VoiceTalk: multimedia-IoT applications for mixing Mandarin, Taiwanese, and English. ACM Trans. Internet Technol. **23**(2), 1–30 (2023)
11. Kaur, A.P., Singh, A., Sachdeva, R., Kukreja, V.: Automatic speech recognition systems: a survey of discriminative techniques. Multimedia Tools and Appl. **82**(9), 13307–13339 (2023)
12. Madni, S.H.H., et al.: Factors influencing the adoption of IoT for E-learning in higher educational institutes in developing countries. Frontiers in Psychology **13**, 915596 (2022)
13. Sejati, W., Melinda, V.: Education on the use of iot technology for energy audit and management within the context of conservation and efficiency. Int. Trans. Educ. Technol. **1**(2), 138–143 (2023)
14. Zhu, Q.-S., Zhang, J., Zhang, Z.-Q., Dai, L.-R.: A joint speech enhancement and self-supervised representation learning framework for noise-robust speech recognition. IEEE/ACM Trans. Audio, Speech, and Language Processing **31**, 1927–1939 (2023)
15. Fan, C., Ding, M., Yi, J., Li, J., Lv, Z.: Two-stage deep spectrum fusion for noise-robust end-to-end speech recognition. Appl. Acoust. **212**, 109547 (2023)

A Survey on Optimization of I/O Stack in File System Read/Write System Calls

Yang Shen, Min Xie, and Wenzhe Zhang^(⊠)

College of Computer Science and Technology, National University of Defense Technology,
Changsha, China
`{shenyang_23,xiemin,zhangwenzhe}@nudt.edu.cn`

Abstract. With the continuous advancement of computer hardware architecture, the performance of storage devices has significantly improved. The emergence of new storage devices (such as Optane SSDs) has greatly reduced the overhead introduced by hardware devices in the I/O path. However, other components within the I/O stack still present notable performance bottlenecks, limiting the full potential of storage device performance. This paper focuses on optimizing the I/O stack of file system-based software read/write system calls, addressing various bottlenecks in the existing I/O stack, such as context switch overhead, physical page allocation, cache page lookup, dirty page marking, I/O aggregation, and scheduling. The paper provides a detailed overview and analysis of the existing optimization solutions. Additionally, the paper summarizes the advantages and disadvantages of these optimization techniques and proposes future research directions, aiming to provide theoretical support and practical guidance for optimizing the I/O path in high-performance storage device environments.

Keywords: file system · I/O path optimization · High-performance storage

1 Introduction

With the continuous development of computer hardware architecture, storage media have undergone significant changes. From traditional mechanical hard drives to modern solid-state drives (SSDs), and now to low-latency, high-performance storage devices, advancements in storage technology have driven a transformation in the I/O performance of computer systems. In particular, the emergence of low-latency, high-performance storage devices (such as Intel's Optane SSD and Samsung's Z-SSD) has profoundly reshaped the I/O paths in computer systems. Traditional storage devices (e.g., mechanical hard drives) often become performance bottlenecks, whereas the widespread adoption of low-latency SSDs has significantly improved the I/O performance of storage devices and enhanced I/O path efficiency.However, other components of the I/O stack, especially at different layers of the file system (such as the cache layer and block layer), still incur noticeable performance overheads that limit the full potential of storage device performance. At the same time, with the widespread adoption of multi-core CPUs, operating system design increasingly emphasizes the parallelism, throughput, and optimization of

© The Author(s) 2026
P. Siarry et al. (Eds.): WCNA 2024, LNEE 1550, pp. 95–102, 2026.
https://doi.org/10.1007/978-981-95-6946-5_10

I/O paths under multi-core architectures. As a result, optimizing I/O paths on high-speed storage devices to reduce unnecessary overheads has become a critical issue in current I/O optimization research.

In modern computing systems, two common methods for accessing persistent data are explicit I/O (such as read/write system calls) and memory-mapped I/O (mmap). Explicit I/O relies on the operating system's system call mechanism to load and store data via the kernel. By increasing the size of the I/O buffer, the number of system calls and the overhead of context switching can be reduced, thereby improving the efficiency of I/O operations. Particularly when handling large files or continuous data ranges, explicit I/O can fully leverage the hardware's bandwidth advantages to achieve more efficient data transfers.

At the same time, file systems can generally be divided into the Virtual File System (VFS) layer, file system layer, driver layer, and device layer. During read/write system calls, the software-induced overheads mainly include: system call context switching overhead, cache page lookup overhead, physical page allocation overhead, dirty page marking overhead, LBA lookup overhead, block-level I/O aggregation overhead, block-level I/O scheduling overhead, DMA mapping and unmapping overhead, and device interrupt overhead, among others. This paper focuses on optimization strategies for various layers of the file system involved in read/write system calls.

2 Existing Research

The overhead of system calls mainly arises from the context switching between user space and kernel space. Some works have reduced this overhead by using specific interfaces, such as scatter-gather I/O, which bundles multiple I/O requests together to reduce the number of system calls. io_uring is an efficient asynchronous I/O framework in the Linux kernel that optimizes storage and network I/O performance by sharing a ring buffer and reducing system call overhead, making it particularly suitable for high-concurrency scenarios. However, the main drawback of these methods is the lack of application transparency; they are not as widely adopted or accepted as interfaces like the read/write system calls. Although non-transparent methods, while bypassing the work required to ensure application transparency, can often provide better I/O performance, the primary issue is that they necessitate a redesign of applications to use new APIs and programming models, which can impose a significant burden on developers. As early as 2008, a scheme was proposed that encouraged applications to use larger user-space buffers [1]. This approach reduces the number of system calls by transferring more data in a single operation. This idea influenced subsequent work, especially in schemes for I/O aggregation in user space [2–5]. These methods typically require modifying the application logic or common libraries to aggregate small I/O requests into larger ones to reduce the number of system calls. In such approaches, modifying common libraries ensures application transparency, but it should be noted that the execution of the aggregation algorithm may incur additional computational and memory overhead.

The Linux kernel's management of caches also introduces certain overheads. These overheads are not significant on traditional storage devices like HDDs but have become a performance bottleneck on faster, lower-latency SSDs. The Direct I/O method bypasses the Linux kernel page cache, transferring data directly to user-space buffers, thereby avoiding the extra overhead introduced by the cache layer. However, this method also loses the performance optimizations provided by the operating system, such as prefetching and write-back caching, which may lead to more cache misses and storage device accesses, thus reducing I/O performance. To address this issue, recent research [6] proposed a transparent dynamic decision mechanism integrated into the kernel. This mechanism dynamically selects either direct I/O or cached I/O based on factors such as the size of the I/O request, file lock contention, and memory pressure. By combining the advantages of both modes, this approach significantly improves I/O performance in distributed file systems.

In addition, some studies focus primarily on reducing the overhead of various steps in the cache layer, particularly the overhead of physical page allocation, cache page lookup, and dirty page marking. In Linux, physical page allocation is mainly handled by the buddy allocator. However, retrieving physical pages from the buddy allocator is often expensive because page splitting and merging introduce additional overhead, and the coarse-grained locks limit the concurrency of allocations.

The most straightforward optimization approach is to improve the memory allocator [7–11]. These studies enhance the parallelism of memory allocators by using range locks, lock-free designs, or increasing the number of instances. At the same time, by simplifying the allocator structure, the overhead of each allocation is reduced.

Other works have designed dedicated cache pools specifically for page cache [12–14]. These cache pools reduce the fragmentation overhead when acquiring physical pages from the buddy allocator by pre-allocating pages. One important improvement by Zhiyue Li and colleagues [14] involved dividing the allocation list of cache pools by file, and coupling the physical page allocation list of each file with the file's metadata structure. This reduced cache misses caused by the cache queue and improved bandwidth.

Page indexing overhead is another significant performance bottleneck in the cache layer. Currently, Linux uses xarray for memory page indexing. Xarray employs RCU locks, which can lead to serious contention issues in multi-writer scenarios. Kiet Tuan Pham and his team [15] implemented a lock-free xarray using lock-free techniques, effectively solving this issue. Additionally, early research suggested separating the management of dirty and clean pages. Some subsequent works [13, 14] adopted this approach, and its advantages are reflected in the page indexing process by eliminating insert operations by the main thread and dirty bit modification operations by offline threads on xarray, both of which usually contend for xarray's spinlock. A study of the asynchronous I/O stack [12] proposed a scheme to eliminate this overhead entirely by overlapping the time of inserting physical pages into xarray, DMA mapping time, physical page allocation time, and the time the device spends reading data from the storage media. This method effectively optimizes the overhead caused by cache misses, but provides limited performance improvement in cache hit scenarios.

Another major overhead in the cache layer comes from the need to copy data between user buffers and kernel buffers during read or write system calls. This overhead is determined by user demand and cannot be fully avoided. However, the mmap interface does not require data copying, instead mapping the virtual address directly to physical memory. Inspired by this, Jiwoong Park and colleagues [16] designed a mechanism to eliminate data copying overhead by unmapping the user-space buffer and remapping the virtual address to the kernel buffer. However, on multi-core CPUs, the unmapping operation incurs a high cost. Furthermore, if the user subsequently performs a write operation on the virtual address, a fallback (copy-on-write) is needed, further increasing the overhead. To mitigate these issues, they used batch operations, private PTE detection, and historical behavior analysis [17]. Some non-transparent solutions [18–20] achieve zero-copy by using interfaces different from read or write. However, these methods require developers to adopt entirely new interfaces, which increases programming complexity and places an additional burden on developers.

The overhead at the block layer mainly focuses on I/O aggregation and I/O scheduling. Some studies have streamlined or even completely removed I/O aggregation and I/O scheduling at the block layer [12, 21], or bypassed these two processes for certain important threads [22]. Caeden Whitaker and others tested the performance and energy efficiency of I/O scheduling algorithms on ultra-low latency (ULL) storage, concluding that I/O schedulers significantly increase the latency of individual requests while also raising energy consumption [23]. I/O scheduling shows more pronounced benefits on traditional HDDs, but on newer storage media such as SSDs, and especially on ULL devices, it tends to incur performance losses. The benefits of I/O aggregation are also significant on HDDs, as it greatly reduces disk seek overhead. However, unlike HDDs, SSDs are transparent to the operating system. Existing research indicates that, in certain specific scenarios, removing I/O aggregation can reduce access latency. However, comprehensive testing of I/O aggregation on different storage media is still lacking.

The overhead of device interrupts refers to the notifications sent by storage devices to the CPU to indicate that data transfer is complete. Polling is the primary method for eliminating device interrupt overhead. Some studies have significantly reduced latency by replacing device interrupts with polling [21, 24–26]. Bryan Harris and colleagues experimentally demonstrated that traditional polling has a higher energy efficiency advantage over interrupts on ultra-low latency storage devices [25]. Gyusun Lee and others proposed an efficient hybrid polling approach that minimizes CPU cycles used for polling without sacrificing I/O latency. By considering the I/O time characteristics of storage devices in idle and busy states, this scheme makes appropriate sleep-time decisions, optimizing I/O performance while minimizing polling CPU cycles [26]. Additionally, some research has designed I/O paths for NVMe SSDs that reduce this overhead by eliminating the bottom half of interrupt handling [27].

Some studies focus on improving the parallelism of the entire I/O stack. By introducing range locks, fine-grained locks, lock-free data structures, and increasing the number of queue instances, these approaches effectively address lock contention issues on the I/O path, such as inode locks and LRU internal locks, significantly enhancing system I/O performance and concurrency. Costa Prats designed a daemon that detects write-heavy processes and reduces their I/O priority, enabling inodes of other processes to be written and unlocked during cache flushing, thus reducing waiting time for light-load processes and alleviating inode lock issues. This mitigates prolonged system stalls and improves system response speed [28]. Chang-Gyu Lee introduced file-level range locks and node logging optimizations based on NVM (including fine-grained inode structures and precise NAT updates) to support parallel writes and metadata management optimizations in the F2FS file system, solving the inode lock issue [29]. Jiwoo Bang and colleagues proposed the Finer-LRU scheme, which splits the LRU list into multiple sublists with independent locks, optimizing the page-reclaim process to reduce lock contention, thereby improving memory management performance in multi-core HPC systems [30]. The nCache framework replaces mutexes with read-write locks to manage file metadata caches, resolving cascading tree lock issues and enhancing the concurrency and scalability of shared file I/O [31]. CFIO introduced a novel conflict-free channel design that distributes I/O requests to different PU queues to eliminate conflicts. Meanwhile, the k-RR scheduler uses batch scheduling and PU dual-registers to implement an efficient I/O pipeline, maximizing the internal parallelism of NVMe SSDs [32].

Many studies have revealed bottlenecks in the I/O path through detailed testing, providing directions for I/O stack optimization. Maria F. Borge and colleagues identified three novel performance anomalies caused by SSD hardware and their significant impact on read throughput. They also analyzed the role and limitations of existing software mechanisms in mitigating these anomalies [33]. Ruiming Lu and others analyzed logs from over a million Alibaba NVMe SSDs, revealing trends in their reliability and the challenges faced in large-scale deployments, particularly their susceptibility to slow failures [34]. Zebin Ren and colleagues characterized the performance and overhead of the Linux storage I/O stack and APIs on high-performance hardware, revealing competitiveness under light loads, bottlenecks under heavy loads, and significant limitations of the Linux block I/O scheduler [35]. Cheongjun Lee and colleagues studied the I/O characteristics of Intel Knights Landing (KNL) processors, focusing on performance bottlenecks in single-threaded buffered write operations. They identified bottlenecks in memory cgroup management and the process of storing pages in xarray structures [36].

3 Trends and Outlook

With the widespread adoption of multi-core processors and the performance improvements in storage devices, the parallelization of kernel data structures has become a major trend in enhancing I/O stack performance. By introducing parallel mechanisms, the I/O processing capacity in multi-core systems can be significantly increased, thus reducing bottlenecks and improving concurrency. Common parallelization methods include range locks, fine-grained locks, lock-free data structures, and increasing the number of queue instances. Each of these parallel schemes has its own strengths and weaknesses. However, there is currently a lack of detailed comparisons regarding the applicability of these schemes in different scenarios within the I/O stack. Moreover, in practical applications, selecting an appropriate parallelization method requires considering not only the performance bottlenecks at different layers of the I/O stack but also weighing its impact on single-threaded performance. In certain cases, excessive parallelization may result in overhead from lock management and synchronization mechanisms, which could negatively affect single-thread performance.

4 Summary

This paper explores methods for optimizing the performance of software read/write system calls in the I/O path within high-performance storage environments. With the widespread use of SSDs and low-latency storage devices, the bottleneck issues in traditional I/O stacks have gradually been exposed, limiting the full release of storage device performance. By analyzing the bottlenecks at various layers of the I/O stack (such as context switching overhead, physical page allocation, cache page lookup, dirty page marking, I/O aggregation, and scheduling), this paper reviews the strengths and weaknesses of existing optimization techniques. While many optimization solutions have achieved good results in different scenarios, there are still shortcomings in optimization strategies for various bottlenecks and different storage media. Based on current technologies, this paper proposes research directions for optimizing the I/O path in high-performance storage environments, aiming to provide theoretical support and practical guidance for I/O stack optimization in future multi-core architectures and high-performance storage devices.

References

1. Lameter, C.: Bazillions of Pages. Linux Symposium, p. 275 (2008)
2. Chowdhury, M.K.H., Tang, H., Bez, J.L., Bangalore, P.V., Byna, S.: Efficient Asynchronous I/O with Request Merging. In: 2023 IEEE International Parallel and Distributed Processing Symposium Workshops (IPDPSW), pp. 628–636. IEEE (2023)
3. Wang, X., Cherniack, M.: Improving query I/O performance by permuting and refining block request sequences. In: Proceedings of the 15th ACM International Conference on Information and Knowledge Management, pp. 652–661 (2006)
4. Joo, Y., Seo, D., Shin, D., Lim, S.S.: Enlarging i/o size for faster loading of mobile applications. IEEE Embed. Syst. Lett. **12**(2), 50–53 (2019)
5. Jensen, Q., Jagodzinski, F., Islam, T.: FILCIO: Application agnostic I/O aggregation to scale scientific workflows. In: 2021 IEEE 45th Annual Computers, Software, and Applications Conference (COMPSAC), pp. 1587–1592. IEEE (2021)
6. Qian, Y., et al.: Combining buffered I/O and direct I/O in distributed file systems. In: 22nd USENIX Conference on File and Storage Technologies (FAST 24), pp. 17–33 (2024)
7. Bhandari, K., Chakrabarti, D.R., Boehm, H.J.: Makalu: Fast recoverable allocation of non-volatile memory. ACM SIGPLAN Notices **51**(10), 677–694 (2016)
8. Coburn, J., et al.: NV-Heaps: Making persistent objects fast and safe with next-generation, non-volatile memories. ACM SIGARCH Comp. Architect. News **39**(1), 105–118 (2011)
9. Oukid, I., Booss, D., Lespinasse, A., Lehner, W., Willhalm, T., Gomes, G.: Memory management techniques for large-scale persistent-main-memory systems. Proc. VLDB Endow. **10**(11), 1166–1177 (2017)
10. Zhang, L., Swanson, S.: Pangolin: a fault-tolerant persistent memory programming library. In: 2019 USENIX Annual Technical Conference (USENIX ATC 19) (2019)
11. Wrenger, L., Rommel, F., Halbuer, A., Dietrich, C., Lohmann, D.: LLFree: scalable and optionally persistent page-frame allocation. In: 2023 USENIX Annual Technical Conference (USENIX ATC 23), pp. 897–914 (2023)
12. Lee, G., et al.: Asynchronous I/O stack: a low-latency kernel I/O stack for ultra-low latency SSDs. In: 2019 USENIX Annual Technical Conference (USENIX ATC 19), pp. 603–616 (2019)

13. Malliotakis, I., Papagiannis, A., Marazakis, M., Bilas, A.: Hugemap: Optimizing memory-mapped I/O with huge pages for fast storage. In: Euro-Par 2020: Parallel Processing Workshops: Euro-Par 2020 International Workshops, Warsaw, Poland, August 24–25, 2020, Revised Selected Papers 26, pp. 344–355 (2021)

14. Li, Z., Zhang, G.: StreamCache: revisiting page cache for file scanning on fast storage devices. In: 2024 USENIX Annual Technical Conference (USENIX ATC 24), pp. 1119–1134 (2024)

15. Pham, K.T., et al.: ScaleCache: a scalable page cache for multiple solid-state drives. In: Proceedings of the Nineteenth European Conference on Computer Systems, pp. 641–656 (2024)

16. Park, J., Min, C., Yeom, H.Y., Son, Y.: z-READ: Towards efficient and transparent zero-copy read. In: 2019 IEEE 12th International Conference on Cloud Computing (CLOUD), pp. 367–371. IEEE (2019)

17. Park, J., Min, C., Yeom, H., Son, Y.: Design and implementation of efficient and transparent zero copy read. IEEE Access 12, 174078–174093 (2024)

18. Druschel, P., Peterson, L.L.: Fbufs: A high-bandwidth cross-domain transfer facility. ACM SIGOPS Operat. Sys. Rev. 27(5), 189–202 (1993)

19. Khalidi, Y.A., Thadani, M.N.: An efficient zero-copy I/O framework for UNIX (1995)

20. Pai, V.S., Druschel, P., Zwaenepoel, W.: IO-Lite: A unified I/O buffering and caching system. ACM Trans. Comp. Sys. (TOCS) 18(1), 37–66 (2000)

21. Caulfield, A.M., et al.: Moneta: a high-performance storage array architecture for next-generation, non-volatile memories. In: 2010 43rd Annual IEEE/ACM International Symposium on Microarchitecture, pp. 385–395. IEEE (2010)

22. Zhang, J., et al.: FlashShare: punching through server storage stack from kernel to firmware for ultra-low latency SSDs. In: 13th USENIX Symposium on Operating Systems Design and Implementation (OSDI 18), pp. 477–492 (2018)

23. Whitaker, C., Sundar, S., Harris, B., Altiparmak, N.: Do we still need I/O schedulers for low-latency disks? In: Proceedings of the 15th ACM Workshop on Hot Topics in Storage and File Systems, pp. 44–50 (2023)

24. Seo, D., Joo, Y., Dutt, N.: Improving virtualized I/O performance by expanding the polled I/O path of Linux. In: Proceedings of the 16th ACM Workshop on Hot Topics in Storage and File Systems, pp. 31–37 (2024)

25. Harris, B., Altiparmak, N.: When poll is more energy efficient than interrupt. In: Proceedings of the 14th ACM Workshop on Hot Topics in Storage and File Systems, pp. 59–64 (2022)

26. Lee, G., Shin, S., Jeong, J.: Efficient hybrid polling for ultra-low latency storage devices. J. Syst. Architect. 122, 102338 (2022)

27. Kára, J.: Ext4 filesystem scaling. https://events.static.linuxfound.org/sites/events/files/slides/ext4-scaling.pdf

28. Méndez Orero, A.: Analysis and mitigation of writeback cache lock-ups in Linux (Bachelor's thesis, Universitat Politècnica de Catalunya) (2020)

29. Lee, C.G., Byun, H., Noh, S., Kang, H., Kim, Y.: Write optimization of log-structured flash file system for parallel I/O on manycore servers. In: Proceedings of the 12th ACM International Conference on Systems and Storage, pp. 21–32 (2019)

30. Bang, J., et al.: Finer-lru: A scalable page management scheme for HPC manycore architectures. In: 2021 IEEE International Parallel and Distributed Processing Symposium (IPDPS), pp. 567–576. IEEE (2021)

31. Lee, C.G., Noh, S., Kang, H., Hwang, S., Kim, Y.: Concurrent file metadata structure using readers-writer lock. In: Proceedings of the 36th Annual ACM Symposium on Applied Computing, pp. 1172–1181 (2021)

32. Zhu, J., Wang, L., Xiao, L., Liu, L., Qin, G.: CFIO: A conflict-free I/O mechanism to fully exploit internal parallelism for Open-Channel SSDs. J. Syst. Architect. 135, 102803 (2023)

33. Borge, M.F., Dinu, F., Zwaenepoel, W.: On the application level impact of SSD performance anomalies. In: 2020 IEEE International Symposium on Performance Analysis of Systems and Software (ISPASS), pp. 170–179. IEEE (2020)
34. Lu, R., et al.: NVMe SSD failures in the field: the fail-stop and the fail-slow. In: 2022 USENIX Annual Technical Conference (USENIX ATC 22), pp. 1005–1020 (2022)
35. Ren, Z., Trivedi, A.: Performance characterization of modern storage stacks: POSIX I/O, libaio, SPDK, and io_uring. In: Proceedings of the 3rd Workshop on Challenges and Opportunities of Efficient and Performant Storage Systems, pp. 35–45 (2023)
36. Lee, C., et al.: Towards enhanced I/O performance of a highly integrated many-core processor by empirical analysis. Clust. Comput. **26**(5), 2643–2655 (2023)

Recognition Method of Tobacco Disease Based on Deep Learning

Jingjing Li, Yi Xu[(✉)], Yapeng Li, Lianying Jv, Xiaohui Xie, and Shuangyan Li

Henan Tobacco Company Jiyuan Company, Jiyuan, China
lijj0202@126.com

Abstract. The inability to accurately and efficiently identify tobacco diseases can significantly impact both the yield and quality of tobacco crops. Based on Deep learning, this study focuses on enhancing the precision, efficiency, and accessibility of tobacco disease identification while minimizing associated costs. The research explores the application of deep learning techniques for this purpose. Initially, samples of 19 prevalent tobacco diseases were gathered from tobacco cultivation regions in Henan Province. These samples were categorized based on expert diagnoses. Following data augmentation, a comprehensive dataset was compiled. Next, the YOLOv5 network model was examined. To facilitate real-time, user-friendly identification suitable for mobile deployment, the model underwent pruning and optimization to reduce its complexity without compromising accuracy. The model was then trained using the prepared dataset. After training, the model was adapted for the Android platform, and a dedicated application was developed. This application not only identifies diseases but also offers insights into their causes and prevention strategies. The final phase involved experimental validation, which demonstrated that the optimized model operates effectively on Android devices, achieving a recall rate exceeding 90% for the majority of the diseases studied. This advancement represents a significant step forward in the practical application of AI for agricultural disease management.

Keywords: Deep Learning · YOLO · Android · Tobacco disease · intelligence recognition

1 Introduction

According to statistics, in 2022, the tobacco industry achieved a total industrial and commercial tax and profit of 1.4413 trillion yuan, a year-on-year increase of 6.12%, and a total fiscal contribution of 1.4416 trillion yuan, a year-on-year increase of 15.86%. Therefore, tobacco industry is one of the main sources of fiscal revenue in China. Henan Province, as one of the largest tobacco-growing provinces north of the Yangtze River, has an annual tobacco planting area of approximately 800,000 mu. However, in recent years, with the intensification of global climate change, the layout and cultivation systems of tobacco crops have changed, and the incidence of tobacco diseases has been rising, causing direct production losses. Therefore, accurate and rapid identification of tobacco diseases is of great significance for tobacco farmers to produce high-quality tobacco leaves.

© The Author(s) 2026
P. Siarry et al. (Eds.): WCNA 2024, LNEE 1550, pp. 103–112, 2026.
https://doi.org/10.1007/978-981-95-6946-5_11

Traditional methods for identifying tobacco diseases mostly rely on manual diagnosis. Due to the wide variety of diseases, manual diagnosis is challenging, inefficient, and costly, and its accuracy depends on the expertise of the technical personnel. Therefore, the tobacco cultivation industry urgently needs an intelligent method for identifying tobacco diseases that is efficient, highly accurate, and low-cost. In recent years, many researchers have begun studying computer-aided identification methods for tobacco diseases, which can be divided into identification methods using traditional image processing techniques and those based on deep learning.

With the advancement of image processing technology, some scholars have done some research work to identify and classify the diseases based on images parameters such as size, shape, color, texture, or a combination of these parameters. For instance Wu [1] extracted image features of tobacco diseases on an Android client, uploaded them to a server, and searched for disease types in a feature database. Yu [2] segmented disease images in the H channel, achieving satisfactory experimental results. Guoet al. [3] established a nonlinear model based on meteorological factors to predict tobacco common mosaic virus. Chen [4] designed a crop pest diagnosis system based on traditional image processing, which provides relevant popular science knowledge and control methods for pests. Zhang [5] proposed a tobacco early disease identification method based on firefly algorithm-optimized support vector machine technology, achieving recognition of two common diseases.

With the in-depth research on deep learning technology in the fields of object detection and image processing, some scholars at home and abroad have conducted research on the application of deep visual recognition in agricultural production. Li [6] used a 6-layer convolutional neural network model to identify 8 types of tobacco diseases, including viral diseases, angular spot diseases, and climate spot diseases. Istiyadi [7] used the VGG16 network model to identify diseases caused by tobacco leaf pests. Sun [8] utilized a pre-trained model to obtain initial feature maps and designed a new multi-attention module, demonstrating good early disease recognition capabilities. Chen [9] used the multi-layer feature extraction method of convolutional neural networks to extract features of buckwheat diseases and classify them based on these features. Zhang [10] used the InceptionV3 network based on transfer learning to construct a tobacco disease recognition model. Liu [11] constructed a tobacco disease detection model based on YOLOv3, achieving recognition of 5 common diseases including common mosaic virus, cucumber mosaic virus, bacterial wilt, tobacco wildfire disease, and climate spot disease.

In summary, while traditional image feature extraction methods are efficient, they struggle to adapt to the diversity of disease images. Because AI-based methods utilize complex networks, they require high-performance computers and often target a limited range of diseases. To achieve low-cost, efficient, and accurate identification, a tobacco disease recognition solution based on the YOLOv5 model is proposed. The deep learning model is deployed on the Android platform, and a corresponding app is designed, enabling offline disease recognition and providing synchronous treatment recommendations.

2   Building Datasets

2.1   Disease Image Acquisition

Common tobacco diseases in Henan Province are captured which are Albinism, Spot disease, Frog eye disease, Potassium depletion, Tobacco mosaic virus disease, Black blight, Cucumber mosaic virus, Etching virus disease, Root-knot disease, Potato y-mosaic, Vascular bundle necrosis, Angular leaf spot, Hollow stalk, Bacterial wilt, Black leg, Fusarium root rot, Black shank disease. A total of 10,334 original images were collected. Additionally, ordinary smartphones were used as data collection devices, and the images were captured using the smartphone camera under outdoor lighting conditions.

2.2   Datasets Building

To adapt to outdoor factors such as varying lighting conditions and shooting angles, this study employed linear and non-linear image enhancement algorithms, multi-scale Retinex color restoration algorithms, brightness non-linear correction, and other methods to extend the brightness and color of the collected images. After processing the brightness and color of the images, operations such as cropping and rotation were performed on the preprocessed original image data to increase the sample size and enhance the network's generalization ability. The augmented dataset was expanded from 10,334 images to 51,670 images, with 50,235 images used for training and 1,435 images used for testing.

The expanded dataset was annotated according to expert diagnostic results using Labelimg software, and the image sizes were standardized to 640*480 pixels. Some examples of disease images are shown in Fig. 1.

(a)Powdery mildew   (b)Albino   (c)cucumber mosaic (d) Etching virus

(e) Spot disease   (f) Root-knot   (g) Potato y-mosaico   (h) Black blight

Fig 1. Image of some common diseases

3   Detection Model of Tobacco Diseases based on YOLOv5

As shown in Fig. 2, the structure of YOLOv5 consists of four parts: Input, backbone, head, and prediction. The input can be an image, number, character, etc. In the Input part, Mosaic data augmentation, adaptive anchor box calculation, and adaptive image scaling are implemented. The backbone part is mainly composed of the Focus structure and CSP structure. The Neck part, similar to YOLOv4, adopts the FPN+PAN structure and utilizes the CSP structure inspired by CSPnet design to enhance the network's feature fusion ability. At the output end, the CIOU_Loss is used as the bounding box loss function [12].

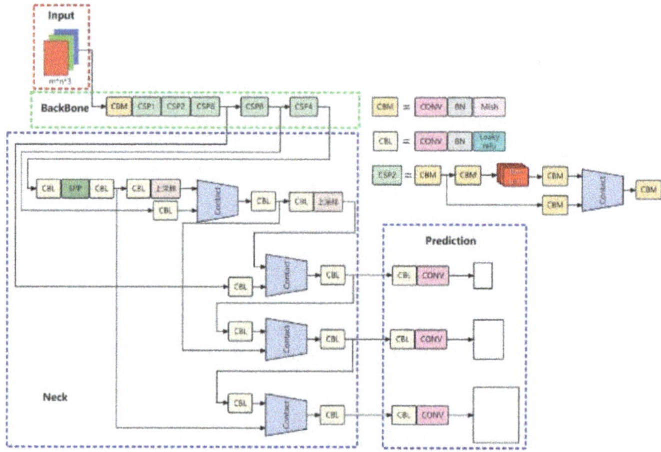

Fig. 2. Framework of YOLOv5

3.1   Networks Pruning

To optimize the YOLOv5 network model for efficient execution on the Android platform, we conducted a pruning process using the NCNN library. The pruning process is outlined as follows:

1) Identify the indices corresponding to the Batch Normalization (BN) layers to be pruned.
2) Before each backward propagation, incorporate L1 regularization into the gradients produced by the BN layer.
3) Set the pruning rate:

   Extract the absolute values of the γ parameters of the BN layers to be pruned into a list and sort them in ascending order. For a pruning rate of 0.8, the value at the 80th percentile of the list serves as the pruning threshold.

   Extract the channels with values below the pruning threshold. If all channels' γ values in a layer are below the threshold, to prevent pruning of the entire layer, retain the channels with the highest γ values (based on the layer_keep parameter, with a minimum of 1, in the form of 2n) in that layer.

4) Validate the mean Average Precision (mAP) of the pruned model.
5) Actually prune the model parameters, and merge β into the running_mean calculation of the BN in the subsequent convolutional layer. Validate mAP, compare model parameter quantity, and inference speed.
6) Generate a new model file.

3.2 Training of the Modle

This paper adopts the ReLU function as the activation function and CIOU_Loss as the loss function. The training parameters are set as follows: the learning rate is set to 0.01, the batch size for iteration is 32, the number of epochs is 1000, and the model iterates for a total of 10,000 iterations. The training loss function, mAP, precision, and recall are shown in Fig. 3.

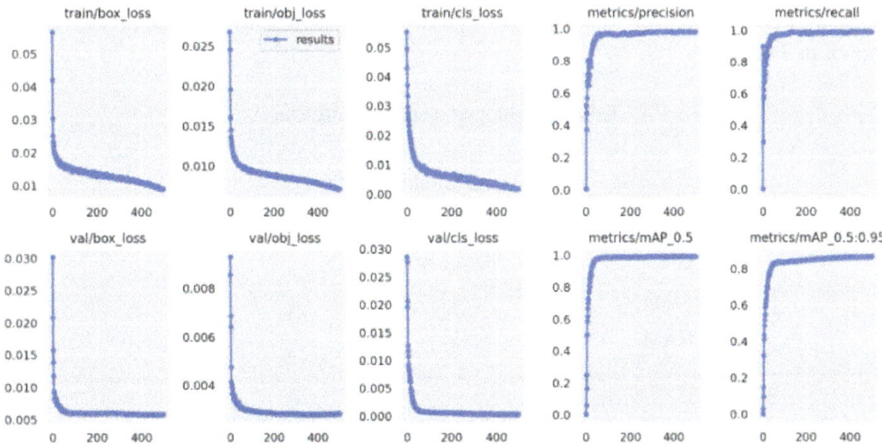

Fig. 3. Curves of model training

3.3 Model Evaluation Indexes

In this paper, Precision, Recall, Recall-Confidence relationship curve and Precision-Recall relationship curve are used as the evaluation indexes of the model. Where Precision indicates the proportion of the number of actual positive samples in the predicted samples to the number of all positive samples, as shown in Eq. (1). Recall indicates the proportion of the number of actual positive samples in the predicted samples to the number of all predicted samples, as shown in Eq. (2).

$$\text{Precision} = \frac{TP}{TP + FP} \tag{1}$$

$$\text{Recall} = \frac{TP}{TP + FN} \tag{2}$$

In which, TP represents the number of positive samples predicted as positive, FP represents the number of negative samples predicted as positive, and FN represents the number of positive samples predicted as negative.

The Recall-Confidence curve is a graph that depicts the relationship between recall and confidence, used to illustrate the relationship between recall rate and confidence. Similarly, the Precision-Recall curve is a graph that depicts the relationship between precision and recall, used to illustrate the relationship between recall rate and precision.

4   Experiments and results

4.1   Training results of the model

To validate the performance of the proposed model, experiments were conducted on platforms with parameters as shown in the table below. A selection of 19 common diseases were chosen for identification. The experimental platform configurations are shown in Table 1.

Table 1. The parameters of hardware

	Android Platform
CPU	64bit
GPU	Dimensity1300
ROM	128G
RAM	12G

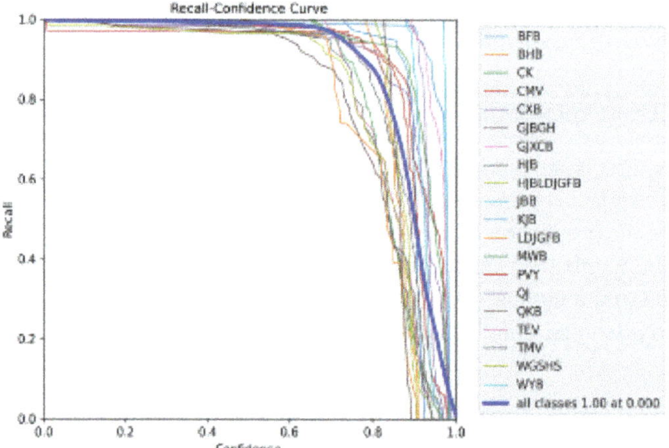

Fig. 4. Recall-confidence curves

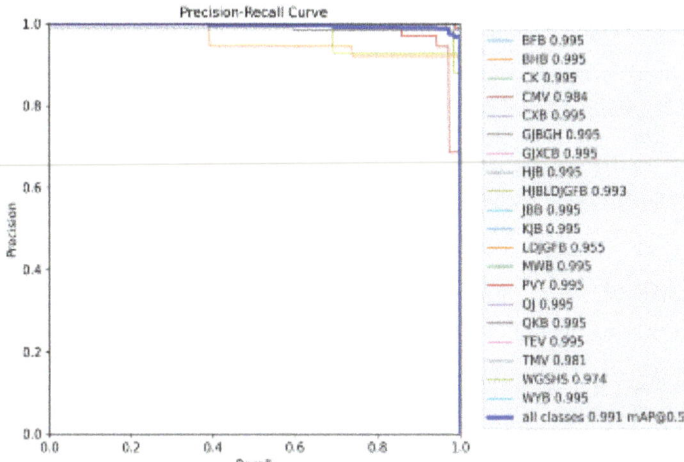

Fig. 5. Precision-recall curves

The test curves for the 19 diseases are shown in Figs. 4 and 5, and the statistical data is presented in Table 2. From the data in the table, it can be observed that, except for Angular Leaf Spot, the recognition rates of the other 18 diseases are all above 90%, with recognition time around 200ms, which can well meet the application requirements of this paper. Some recognition results for certain diseases are illustrated in Fig. 6.

Fig. 6. Recognition results

Table 2. Statistical data of 19 diseases identification

Disease	Recall	Time consume by one image
Powdery mildew	0.97	71ms
Albinism	0.90	75ms
Spot disease	0.96	154ms
Frog eye disease	0.95	144ms
Potassium depletion	0.93	131ms
Tobacco mosaic virus disease	0.94	205ms
Black blight	0.95	69ms
Cucumber mosaic virus	0.96	122ms
Etching virus disease	0.92	104ms
Root-knot disease	0.96	117ms
Potato y-mosaic	0.94	122ms
Vascular bundle necrosis	0.96	116ms
Angular leaf spot	0.89	84ms
Hollow stalk	0.94	104ms
Bacterial wilt	0.92	185ms
Black leg	0.94	90ms
Fusarium root rot	0.95	118ms
Black shank or fusarium root rot disease	0.91	112ms

4.2 Android Platform Deployment and Operational Performance

To promptly and conveniently identify tobacco diseases, this paper deployed the neural network to the Android platform and designed an APP using the Java language. The NCNN library for Android version was utilized to invoke the model for tobacco disease recognition. The specific steps are as follows:

1) Create an Android Studio project, import the OpenCV function module, and install the NCNN Android library.
2) Modify the corresponding build.gradle file.
3) Configure the corresponding CMakelist.txt file.
4) Utilize NCNN to call the YOLOv5 model.

This paper conducted tobacco disease type recognition using the improved YOLOv5 model optimized for Android platform. The results are presented in Table 2. It can be observed that the improved neural network was successfully deployed to the Android platform.

An example of recognition results in APP developed in this research is shown in Fig. 7.

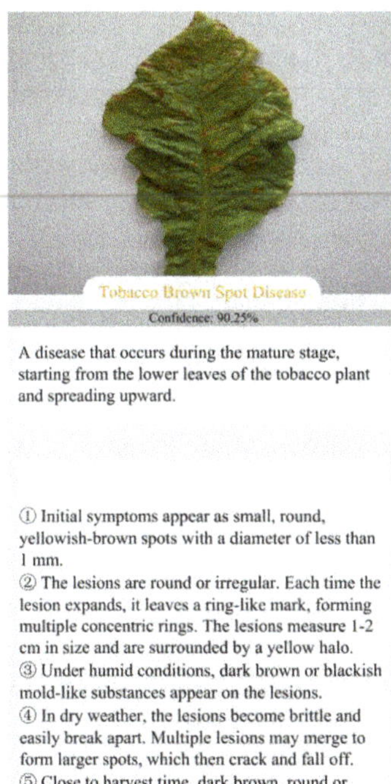

Fig. 7. Example of recognition results in APP [13]. http://36.133.42.215:9094/jybchyst/

5 Conclusion

In response to the low accuracy and efficiency of manual identification of tobacco diseases, this paper proposes a reliable intelligent identification method for tobacco diseases. This method is based on the Android platform and offers the advantages of convenience, low cost, high accuracy, and offline capability. Firstly, images of common tobacco diseases in Henan Province were collected to establish an image dataset. Secondly, the YOLOv5 network was pruned and trained based on the image features of the diseases to be identified. Then, the model was deployed to the Android platform using the NCNN library, and an APP interface was developed. Finally, experimental tests were conducted, and the results showed that this method achieved effective identification of 19 tobacco diseases.

Acknowledgment. This work has been supported by the Science and Technology Plan Project of Henan Provincial Tobacco Company Jiyuan Company (2021410881240101).

References

1. Wu, Z.L.: Research on intelligent recognition system of tobacco pests and diseases image based on android mobile terminal. M.S. thesis, Dept. Fdn. and Info. Eng., Yunnan Ag. Univ., Kunming, Yunnan (2015)

2. Yu, Y., Zhang, Y.W., Wang, J.: Research on recognition of tobacco leaf disease based on computer vision. Comput. Eng. Appl. **51**(20), 167–168 (2015)
3. Guo, S., Wu, L.: A nonlinear prediction model of tobacco mosaic virus based on meteorological factors. Hunan Agricult. Sci. **6**, 101–105 (2019)
4. Chen, T.J., Zeng, J., Xie, C.J., et al.: Intelligent identification system of disease and insect pests based on deep learning. China Plant Protection **39**(4), 26–34 (2019)
5. Zhang, J.T., Zhu, Y., Tan, L., et al.: The recognition of early tobacco disease based on FA-SVM technology. J. Henan Agricult. Sci. **49**(8), 156–161 (2020)
6. Li, J.: Research on automatic identification of tobacco diseases based on convolutional neural network, M.S. thesis. Shandong Ag. Univ. (2016)
7. Istiyadi, S.D., Handayani, T., Chastine, F.: Classification of tobacco leaf pests using VGG16 transfer learning. In: Proc. of 2019 International Conference on Information and Communication Technology and Systems, ICTS, pp. 176–181. Surabaya, Indonesia, (2019)
8. Sun, Y., Wang, H.Q., Xia, Z.Y., et al.: Tobacco-disease image recognition via multiple-attention classification network. In: 4th International Conference on Data Mining, Communications and Information Technology, pp. 1–6. DMCIT, Xi'an, China (2020)
9. Chen, S.X., Wu, S., Yu, X.P., et al.: Buckwheat disease recognition using convolution neural network combined with image processing. Trans. Chinese Soc. Agricult. Eng. **3**, 155–163 (2021)
10. Zhang, W.J., Sun, X.P., Qiao, Y.L., et al.: Tobacco Disease Identification Based on InceptionV3. Acta Tabacaria Sinica **27**(5), 61–70 (2021)
11. Liu, Y.X., Wang, J.F., Du, C.Y., et al.: Detection of various tobacco leaf diseases based on YOLOv3. Chinese Tobacco Science **43**(2), 94–100 (2022)
12. Zheng, Z.H., Wang, P., Liu, W., et al.: Distance-IoU loss: Faster and better learning for bounding box regression. In: Proc. of the AAAI Conference on Artificial Intelligence, pp. 12993–13000. New York, United states (2020)
13. http://36.133.42.215:9094/jybchyst/

Guidance Generation Algorithm Based on Two-Step Sliding Window Least Squares Fusion Processing

Yaodong Yang, Yanqiu Zhang$^{(\boxtimes)}$, Jiachen Shen, Kang Su, Chenglei Liu, and Xianyu Qi

Northwest Institute of Nuclear Technology, Xi'an, China
yangYaodong88@21cn.com, ZhangYanqiuvip@email.cn

Abstract. The significance of guidance data for telemetry and control systems cannot be overstated. Fully utilizing limited input data to generate high-quality guidance data plays a critical role in the efficiency of the entire telemetry and control system and, to some extent, acts as protection for the receiving equipment. This paper addresses the issue of varying input data rates from different sources by designing a guidance generation algorithm for low data rate sources. It is based on the sliding window least squares method, performing two rounds of fitting. The first fitting keeps the data rate unchanged, while the second fitting increases the data rate to meet the guidance requirements. Simulation tests have proven that the two rounds of fitting can achieve more accurate and smoother guidance data under limited window length conditions.

Keywords: Least squares fitting · Guidance · Root mean square error · Average rate of change

1 Introduction

In the field of external measurement data processing, how to fully utilize multi-source heterogeneous external measurement data to generate accurate and smooth real-time guidance data is an important research direction. The sliding window least squares method, due to its simple mathematical principles, low computational complexity, and good smoothing treatment, is widely applied in this field. However, when faced with low data rate sources, external measurement data processing software can only extrapolate based on the existing window data before receiving new data, and after receiving new data, it then completes window updating and data fitting. Generally speaking, the longer the waiting time, the greater the difference between the two fitting functions before and after, which can lead to step changes in guidance data at the window update positions. This not only directly affects the continuity of the guidance data but also places a burden on the operation of the mechanical servo mechanisms of the guided devices. This problem is particularly evident for low data rate sources [1].

This paper addresses this practical problem by proposing a guidance generation algorithm based on dual sliding window least squares fusion processing. This algorithm

P. Siarry et al. (Eds.): WCNA 2024, LNEE 1550, pp. 113–122, 2026.
https://doi.org/10.1007/978-981-95-6946-5_12

adds an additional least squares smoothing process on top of the traditional sliding window least squares method. During the first processing, the data rate of the output data is kept consistent with the input, and in the second processing, higher data rate guidance data is generated according to actual needs. Through simulation verification, it has been confirmed that this algorithm can utilize a limited window length to obtain more accurate and smoother guidance data under certain conditions [2].

2 Basic Principles of the Sliding Window Least Squares Algorithm

The sliding window least squares method is a commonly used real-time mathematical optimization algorithm. It is based on the original measurement data within the sliding window, finding the function that best matches the measurement data by minimizing the sum of the squares of the errors, thereby obtaining predictions that are closer to the true values. With the input of new data, the window data is updated in real time, thus continuously generating new predictions [3] (See Fig. 1).

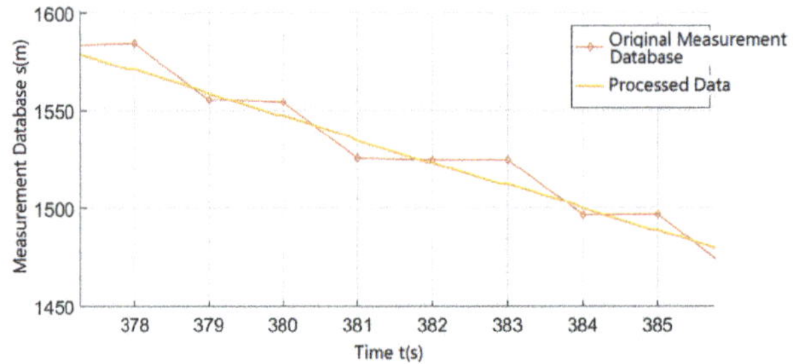

Fig. 1. Least Squares Method Processing of Original Measurement Data

Let the original measurement data be denoted by (t_i, s_i), where $i \in (1, 2, \ldots, m_1)$ represents a set of m_1 measurement data points, with time unit s, and measurement data unit m; the window length is set to l, and the fitting order is $n(l \geq n)$. The fitting function is as follows:

$$h(t_i) = \theta_0 + \theta_1 \cdot t_i + \theta_2 \cdot t_i^2 + \ldots + \theta_n \cdot t_i^n \tag{1}$$

where $\theta_0, \theta_1, \theta_2, \ldots, \theta_n$ are the polynomial coefficients, and $i \in (1, 2, \ldots, l)$. The sum of squared errors between $h(t)$ and the measurement data within the window is taken to obtain the objective function as:

$$E = \sum_{i=1}^{l} (h_\theta(t_i) - s_i)^2 \tag{2}$$

Hence, by taking the partial derivatives of the polynomial coefficients E in equation θ_i, and setting them to zero, we obtain the corresponding values of θ, leading to the following set of equations:

Transforming the above into matrix form:

$$
\begin{cases}
2 \cdot \sum_{i=1}^{l} \left(\theta_0 + \theta_1 \cdot t_i + \theta_2 \cdot t_i^2 + \ldots + \theta_n \cdot t_i^n - s_i \right) = 0 \\
2 \cdot \sum_{i=1}^{l} \left[\left(\theta_0 + \theta_1 \cdot t_i + \theta_2 \cdot t_i^2 + \ldots + \theta_n \cdot t_i^n - s_i \right) \cdot t_i \right] = 0 \\
2 \cdot \sum_{i=1}^{l} \left[\left(\theta_0 + \theta_1 \cdot t_i + \theta_2 \cdot t_i^2 + \ldots + \theta_n \cdot t_i^n - s_i \right) \cdot t_i^2 \right] = 0 \\
\quad\quad\quad\quad \ldots \\
2 \cdot \sum_{i=1}^{l} \left[\left(\theta_0 + \theta_1 \cdot t_i + \theta_2 \cdot t_i^2 + \ldots + \theta_n \cdot t_i^n - s_i \right) \cdot t_i^n \right] = 0
\end{cases}
\tag{3}
$$

Transform the above equation into matrix form.

$$
\begin{cases}
X^T X \theta = X^T Y \\
X = \begin{bmatrix} 1 & t_1 & t_1^2 & \cdots & t_1^n \\ 1 & t_2 & t_2^2 & \cdots & t_2^n \\ 1 & t_3 & t_3^2 & \cdots & t_3^n \\ \cdots & \cdots & \cdots & \cdots & \cdots \\ 1 & t_l & t_l^2 & \cdots & t_l^n \end{bmatrix} \quad
Y = \begin{bmatrix} s_1 \\ s_2 \\ s_3 \\ \cdots \\ s_l \end{bmatrix} \quad
\theta = \begin{bmatrix} \theta_0 \\ \theta_1 \\ \theta_2 \\ \cdots \\ \theta_n \end{bmatrix}
\end{cases}
\tag{4}
$$

Continuing to simplify:

$$
\theta = \left(X^T X \right)^{-1} X^T Y = X^{-1} Y
\tag{5}
$$

After obtaining the θ value, it is possible to solve for the predictive values of subsequent times using the least squares formula.

3 Mathematical Modeling and Analysis

The guidance generation algorithm based on dual sliding window least squares fusion processing is a real-time optimization algorithm. Its core objective is to obtain smoother data within a limited time, making the selection of window length an important evaluation criterion [4, 5]. The first least squares smoothing process does not change the data rate of the original measurement data, while the second process adjusts the data rate to that required for guidance, as illustrated in Fig. 2.

Here, $l_0 = l_1 + l_2$, l_0 represent the total window length, indicating the total amount of data used for fitting; l_1 and l_2 are referred to as the sub-window lengths, indicating the amount of fitting data required during each of the two fitting processes. k_1 and k_2 represent the data rates of the original measurement data and the required data rate for guidance data, respectively, with unit Hz. The guidance data is denoted by r_j, where $j \in (1, 2, \ldots, m_2)$, indicates the generation of m_2 sets of guidance data.

Before conducting experiments, historical measurement data and theoretical data from a particular data source can be used as templates. By using three window length

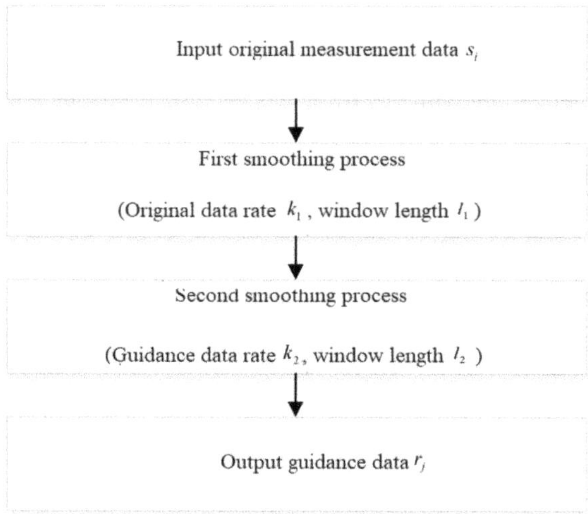

Fig. 2. Mathematical Calculation Process

combinations as variables, the accuracy of guidance data can be compared by calculating the root mean square error [6].

$$MRSE = \sqrt{\sum_{j=1}^{m_2} (q_j - r_j)^2 / m_2}$$ (6)

where the root mean square error is denoted by $MRSE$, with unit m; the theoretical data is denoted by q_j.

The smoothness of different parameter combinations can be obtained by calculating the average rate of change.

$$MR = \sum_{j=1}^{m_2-1} \left[(r_{j+1} - r_j)/(t_{j+1} - t_j) \right]/(m_2 - 1)$$ (7)

where the average rate of change is denoted by MR, with unit m.

Since the dual fitting requires calling the sliding window least squares algorithm twice, by calculating and comparing the average computation times of different fitting methods, it is possible to determine whether they affect the timeliness of the guidance data. The statistical process should exclude the window accumulation time [7].

$$\begin{cases} t_{c1} = T_{c1}/(m_1 - l_0) \\ t_{c2} = T_{c2}/(m_1 - l_1) \end{cases}$$ (8)

where t_{c1} represents the average time using single fitting computation, t_{c2} represents the average time using dual fitting computation, excluding the window length of the total length l_0; T_{c1} represents the total time consumed using single fitting computation, T_{c2} represents the total time consumed using dual fitting computation, excluding the window length of the first fitting l_1.

4 Simulation Verification

Given the historical measurement data of a data source denoted as (t_i, s_i), with $i \in (1, 2, \ldots, m_1)$; the benchmark data as (t_j, q_j), with $j \in (1, 2, \ldots, m_2)$; the measurement data rate as $k_1 = 1Hz$, it is now required to generate guidance data at rate $k_2 = 10Hz$. By analyzing different combinations of window lengths l_0, l_1, and l_2, and comparing with the results of a single fitting, the optimal window combination under limited time window conditions is determined (See Fig. 3).

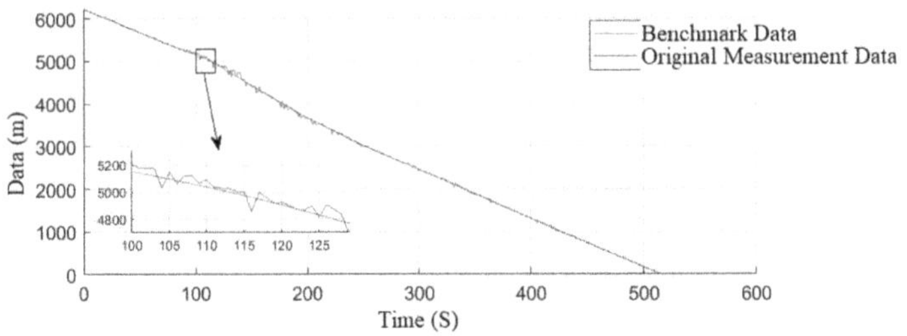

Fig. 3. Benchmark Data vs. Measurement Data

Set l_0 to start accumulating from 60, increasing by 10 each time, to analyze the improvement in guidance data quality with each increase; set $l_1 \in [5, 6, \cdots, l_0 - 5]$, and $l_2 = l_0 - l_1$, by iterating through combinations, to analyze the impact of different sub-window combinations on guidance data quality improvement under a fixed total window length [8].

4.1 Accuracy

Based on preset conditions, calculate the root mean square error between the guidance data obtained from both single and dual fittings and theoretical data, to analyze the accuracy of the guidance data (See Fig. 4).

Integrate the above simulation results to obtain the following data summary chart (See Table 1 and Fig. 5).

Analysis of simulation results and their trends preliminarily yields the following findings:

Beyond a certain window length, the root mean square error of data after dual fitting is better than that of a single fitting, with the dividing point at a total window length of 70;

When $l_0 = 80, l_1 = 70$, $l_2 = 10$, the minimum root mean square error after dual fitting is $MRSE_{\min} = 7.7372$m, an improvement of 3.786% compared to a single fitting;

Whether in single or dual fitting, the pattern of change in root mean square error is to first decrease and then increase, indicating that too long or too short a fitting window will degrade data accuracy [9, 10].

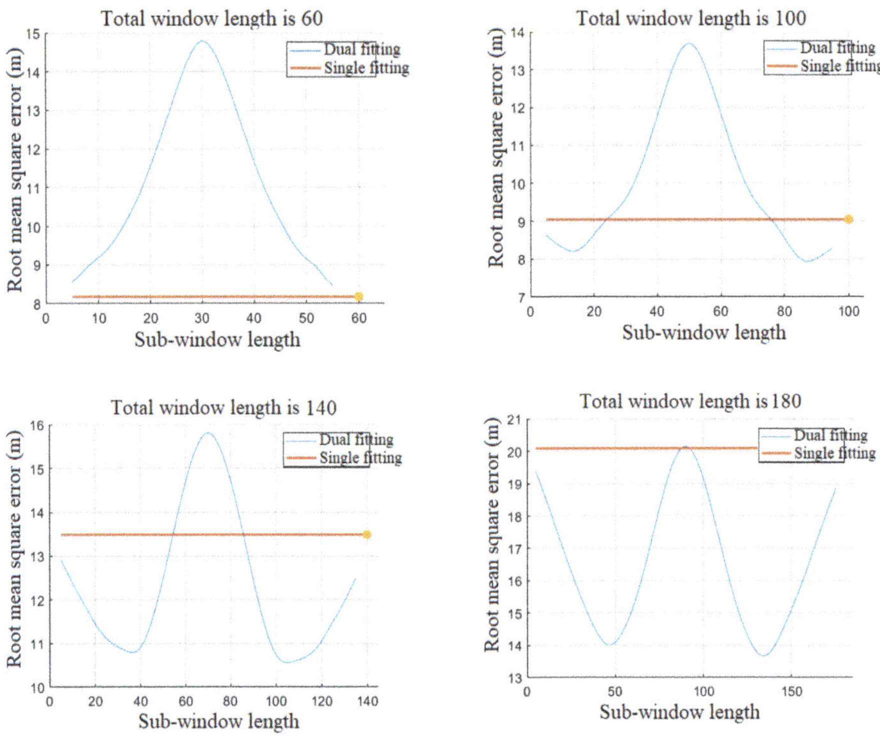

Fig. 4. Comparison Results of Root Mean Square Error by Different Window Lengths

4.2 Smoothness

Based on preset conditions, calculate the average rate of change for both single and dual fittings, to analyze the smoothness of the guidance data (See Fig. 6).

Integrate the above simulation results to obtain the following data summary chart (See Table 2 and Fig. 7).

Analysis of simulation results and their trends preliminarily yields the following findings:

Beyond a certain window length, the smoothness of data after dual fitting is better than that of a single fitting, with the dividing point at a total window length of 76;

When $l_0= 120$, $l_1=100$, $l_2= 20$, the minimum average rate of change after dual fitting is $MR_{min}= 0.4508$m, an improvement of 8.074% compared to a single fitting;

Whether in single or dual fitting, the pattern of change in average rate of change is to first decrease and then increase, indicating that too long or too short a fitting window will degrade data quality [11];

As the fitting window length increases, the growth rate of the average rate of change for dual fitting compared to single fitting is generally on an upward trend, but the rate of increase is unstable.

Table 1. Summary of Root Mean Square Error Comparison

Window Length	Root Mean Square Error for Dual Fitting (m)	Root Mean Square Error for Single Fitting (m)	Improvement Rate
60	8.4583	8.1692	-0.03418
70	7.7555	7.7718	0.00210
80	7.7372	8.0301	0.03786
90	7.8259	8.2354	0.05233
100	7.9310	9.0466	0.14066
110	8.3862	10.2122	0.21774
120	8.8599	11.2609	0.27100
130	9.5822	12.3149	0.28519
140	10.5628	13.4829	0.27645
150	11.4483	14.9497	0.30584
160	12.0819	16.6227	0.37583
170	12.7906	18.3880	0.43762
180	13.6679	20.0841	0.46943

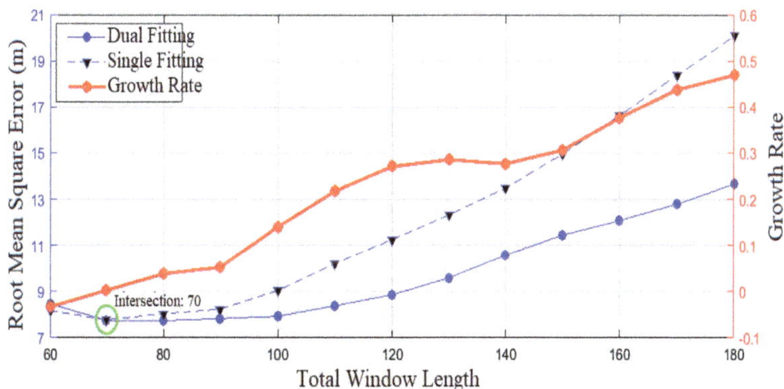

Fig. 5. Summary of Root Mean Square Error Comparison Results

4.3 Timeliness

Select the conditions when the results for accuracy and smoothness are optimal to calculate the time, with the input measurement data quantity as $m_2 = 515$, as shown in Table 3.

It is evident from the table that even when using the dual fitting algorithm, the average computation time is less than 1ms, making the impact on timeliness negligible.

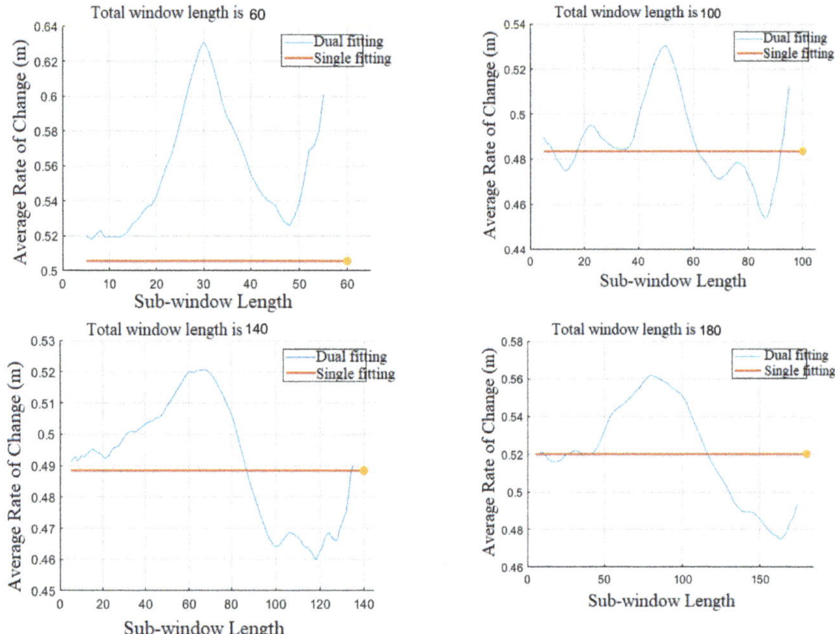

Fig. 6. Comparison Results of Average Rate of Change by Different Window Lengths

Table 2. Summary of Average Rate of Change Comparison

Window Length	Average Rate of Change for Dual Fitting (m)	Average Rate of Change for Single Fitting (m)	Improvement Rate
60	0.5182	0.5054	-0.02469
70	0.4908	0.4870	-0.00765
80	0.4745	0.4916	0.03620
90	0.4619	0.4835	0.04670
100	0.4508	0.4835	0.06533
110	0.4518	0.4888	0.08073
120	0.4508	0.4872	0.08074
130	0.4550	0.4902	0.07733
140	0.4597	0.4884	0.06232
150	0.4628	0.4958	0.07127
160	0.4644	0.5014	0.07970
170	0.4707	0.5102	0.08396
180	0.4747	0.5199	0.09531

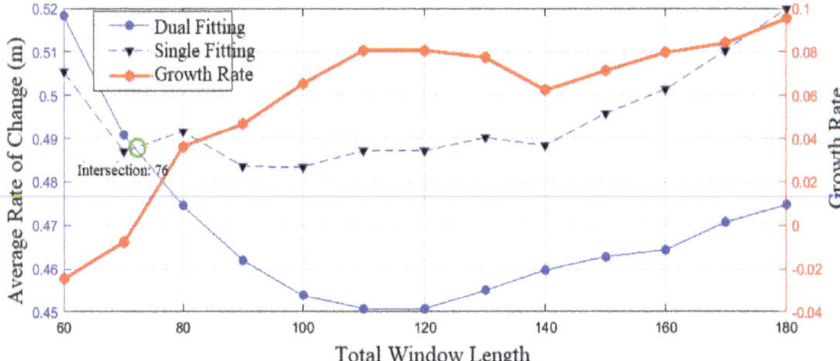

Fig. 7. Summary of Average Rate of Change Comparison Results

Table 3. Time Statistics Table

	Single Fitting		Dual Fitting	
Window Length	80	120	70	100
Total Time (ms)	140	76	364	294
Average Time (ms)	0.3	0.2	0.8	0.7

5   Conclusion

This paper proposes a guidance generation algorithm based on dual sliding window least squares fusion processing. Through simulation testing, it has been proven that dual fitting can utilize a limited window length to obtain more accurate and smoother guidance data, to some extent alleviating the contradiction problem caused when the input data rate is lower than the output data rate.

It was also discovered during the simulation process that choosing a total window length that is too long or too short will increase the output data's root mean square error and average rate of change, and the optimal window combination for both is not consistent. In this set of data, when, the root mean square error is minimized, at this point, the average rate of change improves by 3.620%, which is also effective. Regardless of the method used, the average time is less than 1ms, which essentially does not affect timeliness. Therefore, for different data rates, accuracy, and other characteristics of input data sources, it is necessary to use historical data in advance for simulation verification to obtain the optimal total window length and sub-window combination results, providing necessary support for generating high-quality guidance data.

References

1. Jia, X., Xu, C., Bai, X.: The establishment of the least squares method and its ideological method. J. Northw. Univ. **36**(3), 507–511 (2006)
2. He, J., Ning, G., Wang, X., et al.: Secondary filtering trajectory tracking and prediction of high maneuvering targets in near space. Aerospace Control **41**(2), 24–32 (2023)

3. Jun, L.: Research on moving least squares scatter curve surface fitting and interpolation. Zhejiang University (2011)
4. Cao, J., Yu, L.: Analysis method of flowmeter uncertainty based on least squares fitting. China Measurement & Test **48**(1), 122–128 (2020)
5. Guo, C.: Fast Algorithm of Moving Least Squares and Its Application. Chongqing University (2016)
6. Torres, K., Sacchi, M.: Least-squares reverse time migration via deep learning-based updating operators. Geophysics **87**(6), S315–S333 (2022)
7. Yang, D., Li, M., Guo, J.E., et al.: An attention-based multi-input LSTM with sliding window-based two-stage decomposition for wind speed forecasting. Appl. Energy **375**, 124057 (2024)
8. Rucci, M.A., Hardie, R.C., Martin, R.K., et al.: Atmospheric optical turbulence mitigation using iterative image registration and least squares lucky look fusion. Appl. Opt. **61**(28), 8233–8247 (2022)
9. Ogar, A.O., Bagiwa, M.A., Abdullahi, M.: Image denoising based on an improved least-squares generative adversarial networks. Sci. World J. **17**(1), 155–161 (2022)
10. Wang, H., Xiong, M., Chen, H., et al.: Multi-step ahead wind speed prediction based on a two-step decomposition technique and prediction model parameter optimization. Energy Rep. **8**, 6086–6100 (2022)
11. Farhadmanesh, M., Rashidi, A., Marković, N.: General aviation aircraft identification at non-towered airports using a two-step computer vision-based approach. IEEE Access **10**, 48778–48791 (2022)

Development of Interactive Learning Platform for Industry-University-Research Cooperation Based on Augmented Reality Technology

Feifei Su[✉]

Tianjin University of Finance and Economics Pearl River College, Tianjin, China
sufeifei66@21cn.com

Abstract. The development of interactive learning platform for industry-university-research cooperation based on augmented reality technology aims to integrate the advantages of information exchange and resource sharing among universities, enterprises and researchers, with the fundamental goal of improving the quality of talent training. This paper uses object-oriented system engineering method to carry out elaborate design, the content mainly covers the following aspects. The introduction introduces the background and significance of this topic in detail. Based on the existing literature at home and abroad, the relevant work part makes an in-depth analysis of the status quo of independent innovation research and development, summarizes the preparatory work required for the development of industry-university-research cooperation platform based on augmented reality technology, possible problems, and future development direction. After that, this paper designs an interactive learning platform for industry-university-research cooperation based on augmented reality technology, and tests and analyzes the platform. The results are as follows: Platform A has achieved significant advantages in user authentication and data encryption, reaching 95% and 90% respectively. Platform B, on the other hand, performs slightly less well in these two areas, at 85% and 80%. This means that Platform A has a 10% higher level of user authentication and data encryption compared to Platform B. In terms of secure transmission, although Platform A is slightly lower, reaching 85%, Platform B has a higher level, reaching 90%. Platform B's high level of user authentication and data encryption indicates that user identity information and data are more reliably protected on the platform. This gives users a greater sense of trust and security. Despite the slight shortcomings in secure transmission, Platform B is still committed to improving its own security level to ensure that users' information is protected to the maximum extent possible during transmission.

Keywords: Augmented Reality Technology · Industry-University-Research · Cooperative Interactive · Learning Platform

1 Introduction

With the continuous progress and development of science and technology, the construction of interactive learning platform for industry-university-research cooperation has been paid more and more attention. In China, universities and research institutions

© The Author(s) 2026
P. Siarry et al. (Eds.): WCNA 2024, LNEE 1550, pp. 123–132, 2026.
https://doi.org/10.1007/978-981-95-6946-5_13

are exploring their own unique models of industry-university-research cooperation. The innovative e-learning platform based on virtual reality technology has the characteristics of high autonomy, flexibility and openness. The platform can not only provide enterprises with functions such as information release, real-time monitoring feedback and learning resource management, but also combine traditional teaching with modern information technology to help improve students' knowledge mastery.

According to the platform requirements, this paper designs and develops a set of interactive learning assistance system for industry-university-research cooperation, which aims to effectively enhance the independent innovation and collaboration ability of college students. This paper constructs a multi-dimensional learning environment through the "point-axis" mode, and builds an interactive platform to promote communication and understanding among students. This paper uses virtual network technology to establish a perfect communication feedback mechanism and real-time online test system to realize the remote teaching quality monitoring and management function. The ultimate goal of this paper is to improve the level of college personnel training, enhance the comprehensive competitiveness, and train excellent and high-quality compound science and technology workers for the national economic construction.

The innovations of this paper are as follows:

1. Taking "independent learning" as the core concept, cooperate with college students and enterprises to build a new educational technology exchange mode. This new model makes full use of existing resources and sets up a dedicated agency within the school to develop the platform and provide relevant training and support services to students.
2. In this paper, the relationship of innovation community is built in the virtual network environment: the interactive learning platform system of industry-university-research cooperation is built through virtual reality technology, and the cooperation and communication mode of information sharing and mutual trust is formed to promote the close cooperation between college students and enterprises.

2   Related Work

The study of virtual learning in foreign countries has started early, and has formed a set of relatively complete and mature technology system. This method is through the network connection to achieve remote teaching and courseware production and other ways to assist teachers to complete the classroom teaching, at the same time, it can also provide personalized learning guidance and feedback information according to the needs of students in different learning stages. The study by Hsiu-Yuan Wang et al. aims to explore the use of augmented reality to work with Pokmon -Go robots to increase hospitality intentions in hotels. They identified the potential of augmented reality to enhance the attractiveness of the hospitality industry and investigated the impact of this technology on customer willingness [1]. Carsten Rudolph and others explored the integration of humans and technology, utilizing the possibilities of augmented reality technology. They examined the frontier of interactions between technology and humans, and demonstrated the potential applications of augmented reality technology in human life [2]. Yoori Hwang and others focused their research on the impact of augmented reality and privacy initiation on the technology acceptance model [3]. Christos Papakostas and others studied the behavioral intentions of users in the educational field to adopt mobile augmented reality technology, through an extended technology acceptance model [4]. Alejandro Álvarez-Marín and others investigated the acceptance level of augmented reality technology in engineering education, discussing the impact of technology on the adoption of augmented reality technology in engineering education [5]. Xiao-Ming Wang et al. used augmented reality technology to explore ways for digital technology to assist empathic learning. They used augmented reality to enhance students' fluid experience, motivation, and achievement in biology courses [6]. Kazufumi Suzuki's research explored the use of augmented reality to non-contact measure the angle of the puncture needle in biopsies guided by computed tomography. He designed the stereotaxic coordinates and evaluated their accuracy [7]. Fengyu Na et al. studied the application of augmented reality medical technology combined with contrastenhanced ultrasound in the treatment of rabbit liver cancer with high-energy focused knife. They analyzed the effects and advantages of this technique in the intraoperative assisted operation [8]. Research by Rabia Meryem Yilmaz et al. examined the impact of the use of augmented reality in English learning on vocabulary learning and retention in pre-school children. They explored the potential advantages of this technology in children's English learning [9]. Yevgenia A. Daineko et al. developed an interactive mobile platform based on augmented reality technology for learning radio engineering disciplines. The platform aims to provide an interactive and innovative learning tool to facilitate students' learning of radio engineering disciplines [10]. Through the construction of cooperative learning platform, this paper realizes the close communication and common development between enterprises and students.

3   Method

3.1   Augmented Reality Technology

In augmented reality technology, projection transformation is an important process in mapping the coordinates of virtual objects to the image space captured by the camera. The projection transformation can be expressed by matrix operation, specifically, the perspective projection matrix can describe the two-dimensional coordinates of three-dimensional coordinates after the projection transformation. The matrix usually contains a combination of camera internal parameter matrix, camera external parameter matrix and projection matrix, expressed as follows:

$$P = K[R|t] \tag{1}$$

In augmented reality applications, the projection transformation matrix P, the camera internal parameter matrix K, and the camera external parameter matrix [R|t] including the rotation matrix R and the translation vector t play important roles. Through complex matrix multiplication operations, these matrices realize the transformation process from three-dimensional coordinates to the camera coordinate system and finally to the image coordinate system. Stereo matching technology plays a key role in measuring depth information of objects under different camera viewpoints [11, 12]. Among them, cost aggregation is a core step in the process of stereo matching, which determines the best matching position by calculating the cost function. Assuming the coordinates of the original graph are (x, y) and the coordinates of the distorted graph are (x', y'), the following equation can be used to describe the relationship between them:

$$x' = h_1(x, y)$$
$$y' = h_2(x, y) \tag{2}$$

If g (x, y) is used to represent the grayscale of the original image at point (x, y), and $f(x', y')$ is used to represent the grayscale of the distorted image at point (x', y'), then there should be:

$$g(x, y) = f(x', y') \tag{3}$$

The technology can effectively simulate the real world and be transmitted to students through digital signals to help them understand knowledge content and information processing methods. Through the use of various means such as image display technology, schools can obtain colorful, lively and interesting teaching materials that are easy to operate and control. In order to achieve better results, the existing system can be improved and updated [13, 14]. In the network teaching process embedded in the real scene, teachers can track whether students and teachers encounter problems at any time, and pass real-time information to students or teachers through multimedia courseware, and timely feedback relevant data for other personnel reference.

3.2 Composition of Interactive Learning Platform for Industry-University-Research Cooperation

In order to better meet the needs of industry-university-research cooperation interactive learning, this study designed a learning platform development framework based on augmented reality technology. The platform structure (as shown in Fig. 1) consists of four main parts:

Analysis of functional requirements of the system: it meets application requirements such as communication, interaction and data exchange between different users by processing relevant information. These functional design requirements are fulfilled through embedded micronetworks and mobile terminal devices [15, 16]. At the same time, it provides a scalable, open and high sharing virtual reality technology environment as one of the basic conditions to support the interactive interface built based on augmented reality technology. Through data analysis, text classification and other ways to group users, different types of people are divided into several groups, in each group, members can choose their interests and representative characteristics or have a common interest in the group, establish communication and cooperation, and regularly update the existing contact network information. (2) Auxiliary content analysis and display tools/data interface technology: it realizes the management of basic equipment and resources by targeting different kinds of collaborative tasks. In addition, the module also provides students with relevant communication and testing tools, which is convenient for users to complete corresponding functions according to their own needs. Virtual reality technology is used to simulate the operation scenario in the real environment, while providing real-time online monitoring and auxiliary content, such as updating system parameter information [17, 18]. The platform updates interactive interface data in real time, generates learning tasks and sends task notifications to meet the practical application needs of students. According to the requirements of talent quality and technical conditions of different types of enterprises, integrating and managing existing resources, and realize the basic hardware resources such as required equipment and environmental information. In addition, it also involves software program code writing based on augmented reality technology, development process control, platform architecture design and later maintenance and other related content.

(3) Basic database: it is used to store, manage and retrieve relevant literature and information. The module supports a variety of functions, such as text document library, video conferencing and other application software, and has embedded server and network server applications.

(4) Building a related system based on augmented reality technology and interactive learning platform: creating a platform system supporting learning platform, combining augmented reality technology and interactive functions. The system provides an interactive interface between the real world and the virtual world, enabling students to conduct practical operations in the virtual environment and receive real-time feedback and guidance [19, 20].

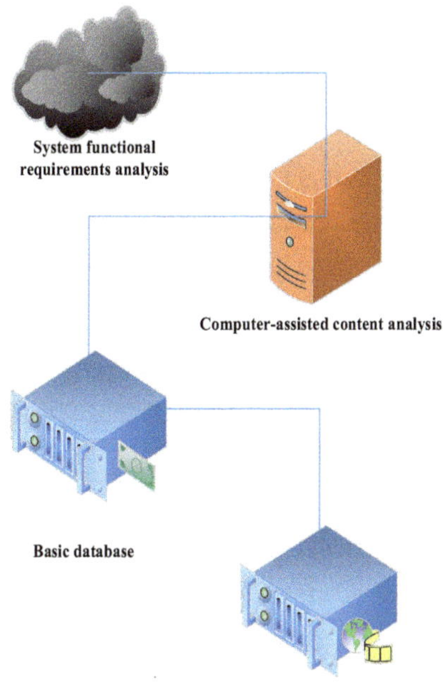

Fig. 1. Interactive learning platform for industry-university-research cooperation

4 Results and Discussion

4.1 The Effectiveness Testing Process of the Interactive Learning Platform for Industry-University-Research Cooperation

This design is mainly based on virtualization and embedded features to build models and test cases, based on development. The platform needs to effectively and accurately respond to the real information of different types of users, collect relevant parameters through front-end devices and store them in database tables to support the operation of functional modules. After entering the test interface, selecting the key technologies of the system, input data and information to generate target documents for matching, and then improve the platform function modules and databases. After the completion of the requirement analysis phase, starting software development and run the program until it is successfully completed. After confirming the completion of the work, it will be put into practical application testing. If there is any problem, it will return to the initial state and continue debugging until the system meets the requirements. After the completion of the system development, a comprehensive test should be carried out to ensure the smooth operation of the platform. Table 1 is a sample of interactive learning data.

Table 1. Interactive learning of the data samples

Platform		Interaction learning duration(h)	Sample data amount	Number of users
Traditional platform	Platform A	8	145	500
Industry-university-research cooperative learning platform	Platform B	8	214	500

Entering each function module through the relevant interface, view the learning content, and test whether the platform is normal use, stable operation, connection and data exchange, including performance testing. Through network connection technology and computer interface protocol, completing website interaction, confirm that the application meets the performance requirements proposed by the research, including friendly interface display and fast response, and verify the performance parameters and running status to ensure the normal operation of the platform.

4.2 Effect Test Analysis

The development goal of this system is to realize the information sharing among enterprises, teachers and students based on the multi-dimensional communication and interaction of industry-university-research cooperation interactive learning platform. In addition, the platform also provides a rich variety of resources combined with virtual reality technology, including virtual environments, multimedia teaching equipment and software technology and other innovative development space. Improving teaching efficiency and enhance learning resource utilization by purchasing or updating computer hardware systems. As shown in Fig. 2, on Platform A, Resource 1 provides 80 learning resources, Resource 2 provides 120 learning resources, and Resource 3 provides 100 learning resources. On Platform B, Resource 1 has 100 learning resources, Resource 2 has 90 learning resources, and Resource 3 has 110 learning resources. Through the comparison in Fig. 2, we can see the difference in the quantity of different learning resources on the two platforms. In Platform A, Resource 2 has the largest number of learning resources, reaching 120. In Platform B, Resource 1 has the most abundant learning resources, with a total of 100. In addition, the number of learning resources distributed on the two platforms is also different. These data changes indicate the strengths and weaknesses of the two platforms in terms of different learning resources. When Platform A is more resource-rich in Resource 2, it may mean that the platform has more resources for learning in a particular topic or area. Platform B, which has more resources in Resource 1, may have other advantages. It is necessary to evaluate the applicability and advantages of the two platforms in different learning resources according to user needs and learning goals.

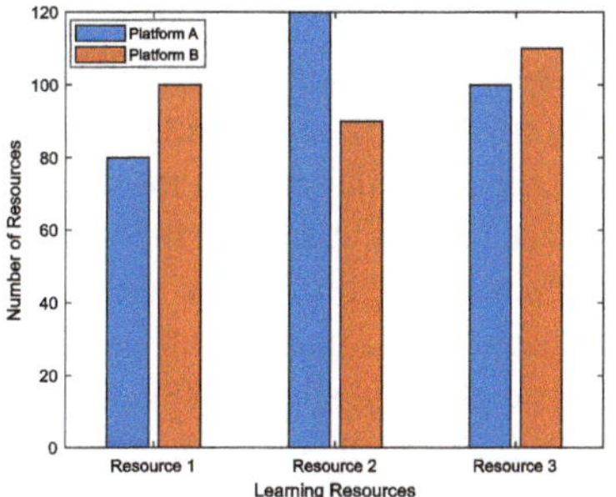

Fig. 2. Learning resources are rich

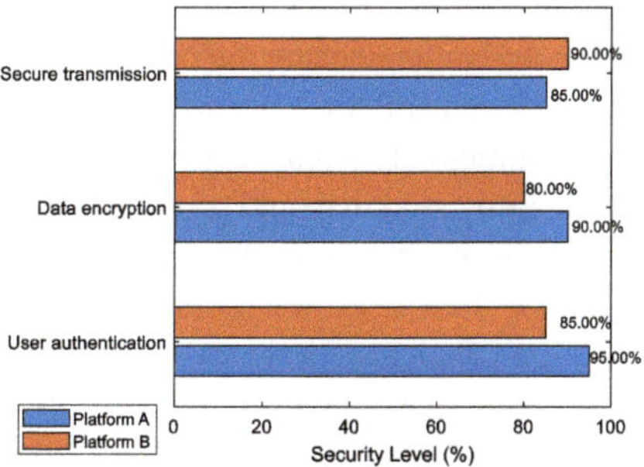

Fig. 3. Platform security

In terms of platform security, the designed features of the platform mainly cover user login, teacher registration, and learning material downloads, among others. The current technology used is based on existing system architecture and databases, which includes data backup functionality. To verify the reliability of the research results, the system will undergo rigorous testing. During the software testing phase, the program will be subjected to strict scrutiny to ensure that the platform can operate normally on a secure foundation. On the other hand, the platform also needs to securely and stably meet user needs. From Fig. 3, we can see that in terms of security level data, Platform A's user authentication reaches 95%, data encryption is at 90%, and secure transmission at 85%. Meanwhile, the data for Platform B are 85%, 80%, and 90%, respectively. By comparing the data of both, it can be seen that Platform A is 10% higher than Platform B in user authentication and data encryption, but slightly lower in secure transmission by 5%. The high security levels in user authentication and data encryption indicate that Platform A is more reliable in these aspects. The higher security level in secure transmission for Platform B may suggest that this platform places more emphasis on the security of the data transmission process, ensuring the confidentiality of data transmission.

5 Conclusion

The industry-academia-research cooperative interactive learning platform based on augmented reality technology represents a new and efficient teaching innovation model. This platform relies on a virtualized network environment, integrates various information resources, and provides an excellent collaboration mechanism, dedicated to creating an open and diverse learning space. With the development of the augmented reality-based industry-academia-research cooperative interactive learning platform, it has been recognized that the system has multiple shortcomings during operation, especially in terms of the interface not being user-friendly and simple. When testing the development model based on augmented reality technology, it is also necessary to effectively address issues that arise during system operation.

For the improvement of system functions, the following two aspects should be paid attention to in the future:

1. Enriching the platform functions, support independent learning, collaborative learning, and provide opportunities for communication and sharing among students.
2. In the system design stage, fully considering the development based on virtual reality technology, and use 3D modeling software and simulation equipment to complete the model production. The advantage of virtual reality technology is to simulate the real world, allow students to be immersed in it, and produce a strong sense of engagement. However, there are still many problems in the current industry-university-research cooperation platform, such as how to build an innovative talent training mechanism, improve the communication efficiency between teachers and students, schools and enterprises, and adapt to the increasing requirements of professional skills talents in Chinese universities.

Acknowledgements. Topic: 2022 Tianjin Municipal Education Science Planning Project. Project Title: Research on the Optimization of Innovation Talent Cultivation Mechanisms at Private Applied Universities in Tianjin under the Background of Industry-Education-Research Collaboration.

References

1. Wang, H.-Y., Wang, J.-H., Zhang, J., Tai, H.-W.: The collaborative interaction with pokémon-go robot uses augmented reality technology for increasing the intentions of patronizing hospitality. Inf. Syst. Frontiers **26**(1), 107–119 (2024)
2. Rudolph, C., Brunnett, G., Bretschneider, M., Meyer, B.: Frank Asbrock:TechnoSapiens: merging humans with technology in augmented reality. Vis. Comput. **40**(2), 1021–1036 (2024)
3. Hwang, Y., Shin, H., Kim, K., Jeong, S.-H.: The effect of augmented reality and privacy priming in a fashion-related app: an application of technology acceptance model. Cyberpsychology Behav. Soc. Netw. **26**(3), 214–220 (2023)
4. Papakostas, C., Troussas, C., Krouska, A., Sgouropoulou, C.: Exploring users' behavioral intention to adopt mobile augmented reality in education through an extended technology acceptance model. Int. J. Hum. Comput. Interact. **39**(6), 1294–1302 (2023)
5. Alejandro Álvarez-Marín, J.: Ángel Velázquez-Iturbide, Mauricio Castillo-Vergara:The acceptance of augmented reality in engineering education: the role of technology optimism and technology innovativeness. Interact. Learn. Environ. **31**(6), 3409–3421 (2023)
6. Wang, X.-M., Hu, Q.-J., Hwang, G.-J., Yu, X.-H.: Learning with digital technology-facilitated empathy: an augmented reality approach to enhancing students' flow experience, motivation, and achievement in a biology program. Interact. Learn. Environ. **31**(10), 6988–7004 (2023)
7. Suzuki, K., Morita, S., Endo, K., Yamamoto, T., Sakai, S.: Noncontact measurement of puncture needle angle using augmented reality technology in computed tomography-guided biopsy: stereotactic coordinate design and accuracy evaluation. Int. J. Comput. Assist. Radiol. Surg. **17**(4), 745–750 (2022)
8. Na, F., Wang, Li., Cuicui, Wu.: Yan Ding:Contrast-enhanced ultrasound combined with augmented reality medical technology in the treatment of rabbit liver cancer with high-energy focused knife. Comput. Intell. **38**(1), 121–138 (2022)
9. Yilmaz, R.M.: Topu, F.B., Tulgar, A.T.: An examination of vocabulary learning and retention levels of pre-school children using augmented reality technology in English language learning. Educ. Inf. Technol. **27**(5), 6989–7017 (2022)
10. Daineko, Y.A., Tsoy, D., Ipalakova, M.T., Kozhakhmetova, B., Aitmagambetov, A., Kulakayeva, A.: Development of an interactive mobile platform for studying radio engineering disciplines using augmented reality technology. Int. J. Interact. Mob. Technol. **16**(19), 147–162 (2022)
11. Calza, F., Carayannis, E.G., Panetti, E., Parmentola, A.: The role of university in the smart specialization strategy: exploring how university-industry interactions change in different technological domains. IEEE Trans. Eng. Manage. **69**(6), 2649–2657 (2022)
12. Gerdsri, N., Manotungvorapun, N.: Systemizing the management of university-industry collaboration: assessment and roadmapping. IEEE Trans. Eng. Manage. **69**(1), 245–261 (2022)
13. Zhu, Z., Tang, H., Zhu, Z.: Realization path of the stability of university-industry coupling symbiotic network. J. Comput. Methods Sci. Eng. **21**(6), 1663–1675 (2021)
14. Andrews, K., MacIntosh, R., Sitko, R.: Commercializing university innovations: a sense-making perspective to communicate between academics and industry. IEEE Trans. Eng. Manage. **71**, 614–625 (2024)
15. Lu, W., Zhao, L., Xu, R.: Remote sensing image processing technology based on mobile augmented reality technology in surveying and mapping engineering. Soft Comput. **27**(1), 423–433 (2023)
16. Kisno, B.W.: Khaerudin: digital storytelling for early childhood creativity: diffusion of innovation "3-D coloring quiver application based on augmented reality technology in children's creativity development." Int. J. Online Biomed. Eng. **18**(10), 26–42 (2022)

17. Uiphanit, T., et al.: Application of augmented reality technology to access facial sunscreen product label information. Int. J. Interact. Mob. Technol. **16**(2), 171–178 (2022)
18. Jajic, I., Spremic, M., Miloloza, I.: Behavioural intention determinants of augmented reality technology adoption in supermarkets/hypermarkets. Int. J. E Serv. Mob. Appl. **14**(1), 1–22 (2022)
19. Elmira, O., Rauan, B., Dinara, B.: Blerta prevalla etemi: the effect of augmented reality technology on the performance of university students. Int. J. Emerg. Technol. Learn. **17**(19), 33–45 (2022)
20. Majeed, B.H., Salim Alrikabi, H.Th.: Effect of augmented reality technology on spatial intelligence among high school students. Int. J. Emerg. Technol. Learn. **17**(24), 131–143 (2022)

Design of Performance Evaluation System of Computer Network Platform based on Error Reversal Algorithm

Cuiyuan Yu[✉], Xiaolin Song, Yuye Zhu, and Xuexia Duan

College of Computer Science, BINZHOU POLYTECHNIC, Binzhou, China
`ycyjianqiang2002@163.com`

Abstract. The government agencies and various enterprises and institutions in China have increasingly high requirements for performance appraisal systems. In order to better protect the rights and interests of employees, this article introduces the principle of the error deduction algorithm and analyzes its advantages in performance appraisal applications. Based on the error deduction algorithm, a computer network platform performance appraisal system is established. The system is developed using JSP language, with MyEclipse and Tomcat as the development platforms and backend server platforms, respectively. After online testing, the system can effectively obtain real democratic evaluation data, becoming a reliable basis for personnel selection.

Keywords: Computer network · Error backpropagation · Performance evaluation

1 Introduction

The implementation of performance appraisal system is a necessary measure for employees and employers to achieve "win-win". A reasonable performance appraisal system will greatly improve employees' sense of responsibility and mission, and they will devote themselves to their work more attentively; Quantification and feedback results will make employees clearly understand the strengths and weaknesses of their work, better realize their personal status and help achieve their career development goals; The assessment results can also be used as an important basis for personnel deployment, job promotion and salary division of the employer.

In recent years, government agencies and enterprises at all levels have regularly conducted various forms of performance evaluation mechanisms through computer networks. Despite continuous improvements during use, these evaluation mechanisms still have varying degrees of flaws, failing to ensure fairness and impartiality in the assessment process and unable to guarantee the authenticity of democratic evaluation data. This system is based on an error correction algorithm; the greater the discrepancy between the scores given by evaluators and those from democratic evaluations, the lower the score assigned by the evaluator. The system has effectively enhanced the authenticity of the evaluation results, with multiple pilot units reporting positive feedback.

© The Author(s) 2026
P. Siarry et al. (Eds.): WCNA 2024, LNEE 1550, pp. 133–140, 2026.
https://doi.org/10.1007/978-981-95-6946-5_14

2 The System Requirements Analysis

In order to better understand the needs of users and reach a consensus with customers, we must stand in the user's perspective through the process of demand analysis.Starting with the function and risk of the system, a specific development process is formed, which lays a good foundation for the later system construction. To achieve a complete performance appraisal function, this system needs to be composed of three modules: system administrator, unit administrator and employee. The details are as follows (Fig. 1):

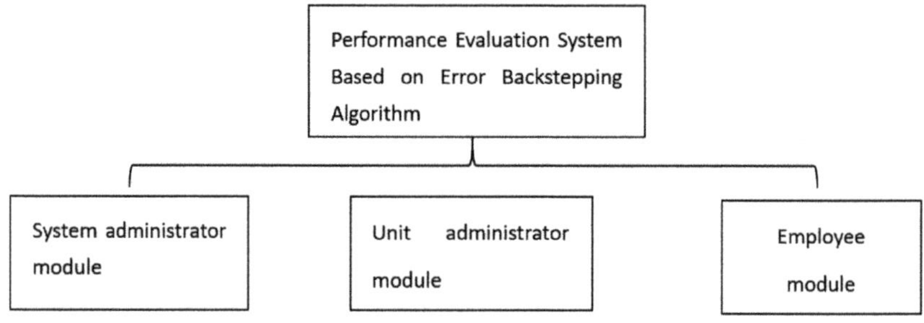

Fig. 1. System module composition

This system uses ASP and Java as programming languages, which are commonly used as similar systems and software development languages, with clear syntax, easy to understand and simple architecture design. In order to improve compatibility and reduce the cost of crossplatform development, the system runs with the help of virtual machine (JVM), which gets rid of the restrictions of the operating system and is feasible. The application of DAO encapsulation database design pattern makes the system code simplified, and the program readability and portability are greatly improved.

3 Detailed System Design

3.1 Overall Design Concept

In order to simplify the development process and facilitate later maintenance, the system does not choose to place all the processing logic in the JSP page, but defines the properties of the system administrator table as java classes, and embeds Java code in the JSP page. With the help of the model layer, view layer, and controller layer of Java, various services of the system are logically processed. The full functionality of the evaluation system is encapsulated by the classes of the DAO design pattern. The following diagram illustrates the JSP page request process (Fig. 2).

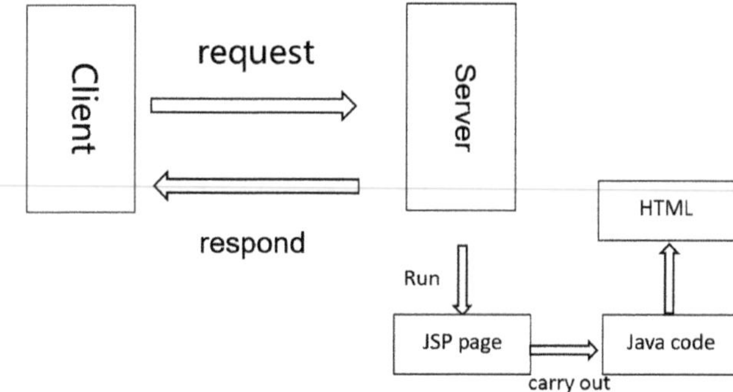

Fig. 2. JSP page request process

3.2 System Development Technology

a. Error Deduction Algorithm

The basic idea of the error deduction algorithm is that the final score is obtained by subtracting the score of the error term. It contains four key factors: firstly, the deduction standard, which is the prerequisite for the deduction algorithm. The deduction standard can be determined based on specific factors such as error occurrence rate and severity caused by the situation; Next is the initial score, which can be determined based on the requirements of the testing or evaluation system; Once again, it is error identification, which is the key to the error deduction algorithm. Whether the error can be accurately identified directly determines the fairness and impartiality of the evaluation results; Finally, there is the score calculation. After determining the deduction criteria and identifying errors, the final score is obtained by subtracting the corresponding score based on the deduction criteria.

b. Simulated Server

This system uses Tomcat server, an open-source Servlet container developed by Apache Corporation, which can run independently as a separate process and is widely used in small and medium-sized systems as well as large systems. This server supports JSP and Servlet, and has specific features such as security domain management, Tomcat management, and Tomcat custom valves. The Tomcat server can also be treated as a standalone web server, as there is an HTTP server included within Tomcat. Tomcat itself also includes a configuration management tool. When the XML format file is edited and configured correctly, Tomcat will run Servlet and JSP pages, while Apache provides support and assurance for JSP pages.

c. MyEclipse Development Platform

The development tool used in this system is MyEclipse, an enterprise-level integrated tool development package widely used for developing Java, Java EE, and mobile applications. Essentially, it builds upon Eclipse by adding plugins, but compared to Eclipse, MyEclipse offers more powerful features and broader applications, especially excelling in supporting the development of various open-source products. MyEclipse is also commonly used for developing Web Services, Java Persistence, UML, J2EE, and various application servers and extended databases. It fully supports mainstream third-party frameworks and multiple plugins based on the Eclipse platform, making it an efficient development tool that covers all major Java open-source products. Additionally, users can customize and extend functionalities through plugins.

d. SQL Server Database

The system will face a large amount of data during use, and these data are uniformly stored in the database. Choosing a cost-effective database management is conducive to evaluating the system's function implementation, and finally SQL Server 2019 is determined to be used. This database management system is developed by Microsoft integrates a variety of development and management tools, and on the basis of inheriting the characteristics of seamless combination and strong system compatibility of the previous version, it has also improved in terms of compatibility with related software, scalability, availability and security functions, and can be applied to server platforms of various types of operating systems. At the same, SQL Server 2019 supports Microsoft's engine service, providing a friendly interface for users to query and search.

3.3 System Development Pattern Design

This system adopts a B/S (Browser/Server) architecture design, with the server in the form of a virtual server to store all information and data. Customers no longer need to install client programs separately; they only need to enter the server URL into their browser to connect to the server and perform operations on system data as needed. The architecture is shown in the following figure (Fig. 3):

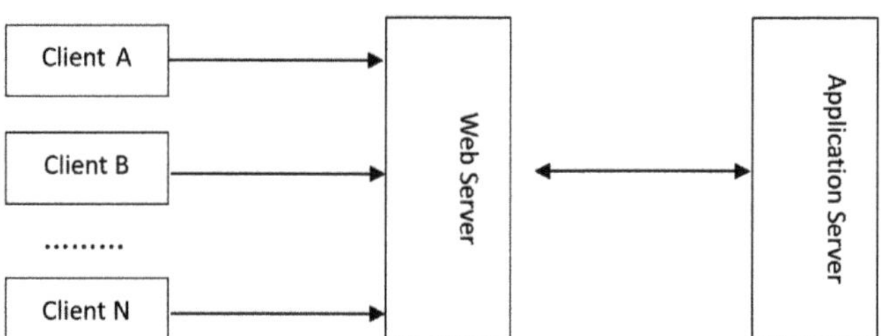

Fig. 3. System architecture

4 System Implementation

4.1 System Administrator Function Module

Admin login process (Fig. 4):

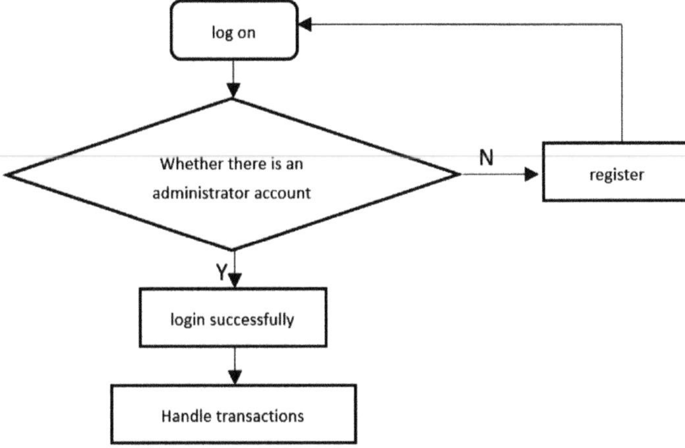

Fig. 4. The administrator login process

The administrator module can achieve the functions of entering unit administrators, adding units, etc., and the specific business process is shown in the figure below (Fig. 5):

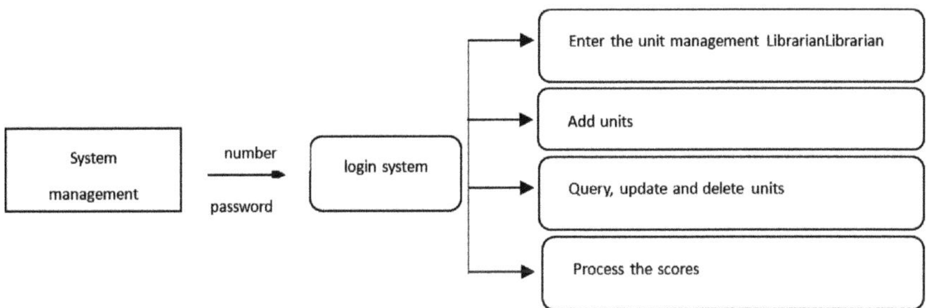

Fig. 5. Business flow of the administrator module

4.2   Unit Administrator Function Module

The unit administrator module can realize the functions of entering employees and searching for employees, as shown in the following figure (Fig. 6):

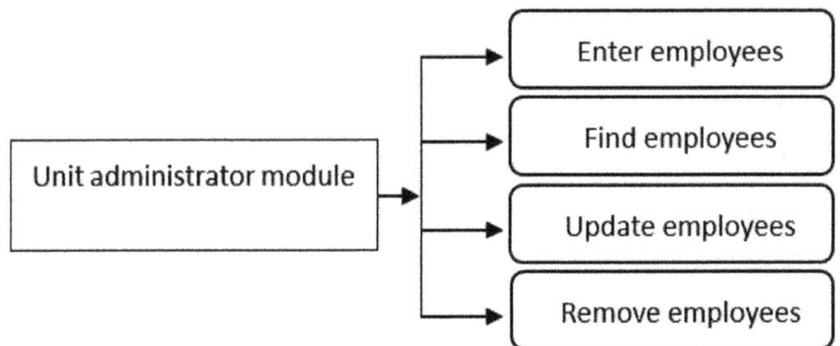

Fig. 6. Unit Administrator Module Business Flow

4.3 Employee Function Module

The employee module can realize the functions of personal information filling, score evaluation and so on. The specific business flow is shown in the following figure (Fig. 7):

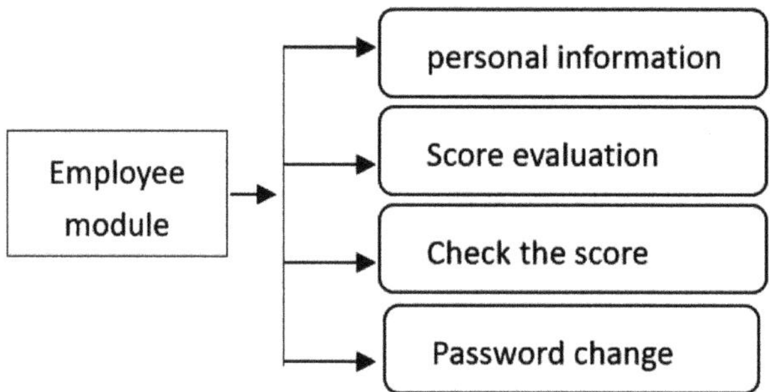

Fig. 7. Employee module business flow

5 System Testing

In order to ensure that the functional modules of the performance evaluation system can work normally, it is necessary to test the main functional modules of the system in a computer network environment.

5.1 Test Environment Configuration

Hardware: Intel Core 2 Duo CPU 1G Memory
  Software: MyEclipse, Tomcat and SQL Server 2019 are installed under the Windows 7 operating system
  network:Internet

5.2 Test Case

Enter different usernames and passwords into this system to test if the expected results are achieved. When an incorrect username or password is entered, the username or password is empty, or the selected account type does not match the username and password, it is impossible to log in correctly; Otherwise, it is possible to successfully implement the functions of each module (Table 1).

Table 1. Test case table

Test Account Category	User Name	Password	Test Result
Admin account	*Admin*	*Admin*	*Administrator login successful*
Standard account	*abc*	*abc*	*User login successful*

(continued)

Table 1. (*continued*)

Test Account Category	User Name	Password	Test Result
Incorrect account	*123*	*123*	*Incorrect user name or password*
Empty account	*empty*	*empty*	*Please enter your user name or password*

6 Summary

In the global economic society, all major enterprises are facing great challenges. In order to maintain a good sustainable development trend of enterprises, it is necessary to ensure that employees have core competitiveness. The performance appraisal system is an effective means to manage employees, improve their business level and promote the rapid progress of enterprises.

The performance evaluation system for computer network platforms designed in this paper, based on the error inversion algorithm, provides users with an excellent user interface. After testing, the assessment results are fair and practical, serving as a valuable supplement to current corporate performance evaluation methods. There are also areas that need improvement in the system, such as certain parameter settings requiring further refinement, how to introduce technical control mechanisms during use to effectively prevent malicious data manipulation, and so on. These issues should be addressed with targeted enhancements during use.

References

1. Li, F., Wang, R.: Research on network teaching management based on performance management theory. China Education Informatization (10) (2010)
2. Phillips, L.A.: Cleverly Learning and Using HTML4. Electronic Industry Press (8) (2004)
3. Li, Z., Wang, X., Ma, S.: Java Language Programming. Tsinghua University Press (2009)
4. Ma, Y., Yan, D.: Performance Evaluation of Foreign Digital Resources ERS for Academic Achievements: A Case Study of Foreign Electronic Journals in Suzhou University Library. Jintu Journal (05) (2019)
5. Chen, Y.: All average, Jiang Yanxia COUNTER 5: Key Changes and Application Prospects. Library Forum (01) (2020)
6. Yang, H., Tan, H., Ma, Y.:ntao Research on the coordination mechanism of the performance evaluation subject of "Internet plus+government service" -- based on the perspective of stakeholders. J. Party School of the CPC Tianjin Municipal Committee (03) (2020)
7. Qin, C., Zhang, Y.: Improvement of association classification algorithm based on classification pruning. Computer System Application **28**(4) (2019)
8. Chen, X.: Probe into the Construction of Hospital Internal Control System under the New Performance Appraisal System. Quality and Market (12) (2022)
9. Shi, Y.: Algorithm design of frequent pattern mining for distributed multidimensional data streams. J. Jilin Univ. (Information Science Edition) (01) (2023)
10. Qi, W., et al.: Exploration and implementation of the assessment system for party branches in public hospitals. China Digital Medicine (10) (2020)

Enterprise Skill Training Performance Evaluation Method Integrating Big Data and BP Neural Network

Changsheng Bao[1] and Qing Zhu[2]($\boxtimes$)

[1] School of Economics and Management, Shanghai University of Political Science and Law, Shanghai, China
[2] Shanghai Technical Institute of Electronics and Information, Shanghai, China
`ZhuQing_STIEI@21cn.com`

Abstract. In order to solve the problems of low efficiency and insufficient accuracy in the performance evaluation of traditional enterprise skills training, the author proposes a performance evaluation system for enterprise skills training based on big data technology. This system integrates big data collection, storage, and processing technologies, combined with BP neural network algorithms and data visualization tools, to achieve deep analysis and dynamic evaluation of employee training data. The system leverages the self-learning capability of the BP neural network and integrates nonlinear fitting abilities, enabling it to effectively process complex training datasets, reduce evaluation bias, and improve decision-making accuracy. The system can accurately identify key performances during the training process and effectively predict the potential for employee skill development. The research results indicate that the system significantly improves the scientificity and real-time performance evaluation of training, providing strong support for enterprises to optimize training resource allocation and management decisions. Meanwhile, the application of this system has accelerated the digital transformation of enterprise human resource management. In the future, to enhance the adaptability of the BP neural network, research will focus on optimization through the integration of deep learning techniques, thereby further improving the system's predictive capabilities and enhancing its scalability across different industries. With the further development of intelligent algorithms and dynamic data analysis technology, this system will play a greater role in applicability and flexibility, helping enterprises enhance their core competitiveness.

Keywords: Big data technology · Skill training · BP neural network · performance evaluation

1 Introduction

As the world's economic continues to grow and the IT develops rapidly, Large Data Technique has permeated all parts of our society, which has brought about great changes in all fields. For enterprises, big data is not just a concept, it has become an important

© The Author(s) 2026
P. Siarry et al. (Eds.): WCNA 2024, LNEE 1550, pp. 141–151, 2026.
https://doi.org/10.1007/978-981-95-6946-5_15

tool for management decision-making, greatly enhancing organizational competitiveness. Especially in modern enterprise operations, employee skills training is one of the key means to improve organizational competitiveness, and its importance is self-evident [1]. However, the reality is that many companies still rely on traditional training performance evaluation methods, which often lack data support and scientific analysis, making it difficult to comprehensively and accurately reflect the evaluation results of training effectiveness. Currently, a performance evaluation system for enterprise skill training based on big data technology has emerged. This innovative technology can break the limitations of traditional methods by building a multi-dimensional data source integration system [2]. The system integrates employee training data, job performance data, and feedback data, and uses advanced algorithms for analysis to establish a data-driven evaluation model. This not only enables comprehensive, real-time, and dynamic analysis of training effectiveness, but also effectively predicts the potential for employee skill development. This comprehensive evaluation model provides enterprises with a comprehensive and accurate insight into the performance of skills training, helping them gain a deeper understanding of the actual effectiveness of training and make more accurate training planning and resource optimization decisions [3].

2 Literature Review

In today's fiercely competitive business environment, enterprise skills training is no longer just a tool for employee development. It has evolved into a strategic resource, enhancing the overall market competitiveness of the company by improving employee quality, promoting the construction of corporate culture [4]. As the importance of such training becomes increasingly prominent, organizational management is also placing greater emphasis on incorporating it into the core aspects of daily management. The rise of big data technology has brought revolutionary changes to training management [5]. Big data not only provides massive and multi-dimensional data analysis capabilities, but also enables prediction of future trends, identification of patterns, and discovery of new opportunities. Therefore, many researchers and practitioners have begun to explore in depth how to effectively utilize big data technology to optimize training processes, including designing more personalized learning paths, developing accurate evaluation indicator systems, and improving feedback mechanisms for training effectiveness. Hachim Almrshed, S. K. and others mainly developed a big data knowledge management model for small and medium-sized enterprises by examining several business situations. The writer adopts the method of qualitative data analysis. In this paper, we have gathered the sample of SME's large data, and applied it to the KM model. Collecting, encoding, and analyzing the data are the three major parts of qualitative data analysis. Collecting large data is a commercial endeavor for SMEs, who will have to explain their large data plans and deal with any possible organisational problems [6]. Gabdullin, M.T., and others researchers, as well as a number of useful insights into optimization of large data applications. A comparison was made between the text files and graphs that were used in the Word Count and the PageRank task, and a performance test was carried out on a variety of data sets. It is shown that Spark performs better than Hadoop in many applications, especially in iterative computing and real time data processing [7]. Xu, Z. et al. have adopted IT mature mode, PDCA management period and so on, and have researched the optimal project of IT management. The assessment model is applied to evaluate the efficiency of MIS reformation [8].

3 Research Methods

3.1 Big Data Technology

Big Data is a mass of data that can't be gathered and handled effectively by conventional software. These data have many features such as big amount of data, high demand for processing, variety of data types, and small ratio. Along with the unceasing development of Large Data, a number of techniques are being developed to deal with large scale data. In response to the actual needs of enterprise skills training, a performance evaluation system based on big data technology can effectively collect, store, and analyze data from multiple sources, including employee training records, work performance, and feedback information. Table 1 shows the classification and application methods of big data technology, which can provide accurate training effectiveness evaluation, optimize training programs, and improve employee skill development efficiency for enterprises, thereby better supporting talent cultivation and performance management [9].

From the Table 1, it can be seen that the performance evaluation system for enterprise skill training based on big data technology mainly involves five core technology areas: Basic models, data collection, storage systems, data processing, and data communication. There are various technical means and tools that can be applied to the construction of performance evaluation systems in every field. In terms of basic models, cloud computing, distributed systems, virtualization technology, and monitoring systems provide strong support capabilities for enterprise skill training performance evaluation systems, ensuring the stability and flexibility of the system. The data collection process utilizes information collection and data transmission technology to achieve real-time data collection and transmission of employee training process and performance. The storage system ensures the secure storage and efficient management of large-scale training data through technologies such as databases, cloud storage, storage, data fusion, and memory. In terms of data processing, visualization data processing, predictive analysis techniques, data graph conversion, and intelligent processing methods can help the system extract valuable insights from massive data to support performance evaluation and decision-making. Finally, data communication achieves timely display of training data and rapid transmission of information through screen display, wireless communication, and communication lines, enabling training managers to obtain and monitor training performance in real time and further optimize training programs [10].

3.2 Data Visualization Technology

The author uses data visualization technology to visualize the data in the enterprise skill training performance evaluation system, enabling various departments of the enterprise to have a clear understanding of training performance information, thereby facilitating management decision-making and optimizing training programs. Data visualization

Table 1. Classification of Big Data Technologies

Big data classification	Means and tools
base model	cloud computing
	distributed system
	Virtualization technology
	monitoring system
data acquisition	information acquisition
	data transmission
Storage system	database
	Cloud data
	storage
	Data Fusion
	Memory
data processing	Visualization processing
	Prediction and processing
	Graphic Conversion
	Intelligent processing
data communication	Screen display
	telecommunications
	Communication line

technology can also provide support for enterprises in handling complex training data through various display methods, such as screen display, real-time presentation of training data and performance evaluation results. Based on the various departments and training content of enterprise skills training, design corresponding data visualization structures, as shown in Fig. 1, to help managers comprehensively grasp the effectiveness of employee training and improve the efficiency and accuracy of training management [11].

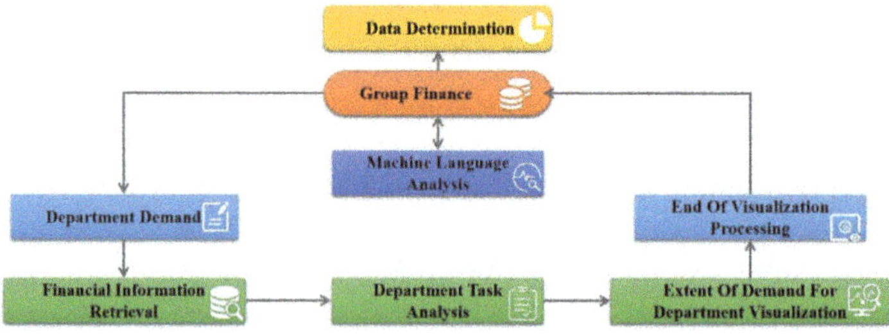

Fig. 1. Data visualization structure diagram

The performance evaluation system for enterprise skills training first collects employee training related data through the training department, and then inputs this data into the requirements of each department, which applies according to their own training needs. Next, conduct training data retrieval to analyze whether there are sufficient resources and conditions to meet the training needs of each department. By retrieving and analyzing training data, and combining it with specific departmental tasks, the visualization level of skill training needs and performance evaluation in different departments is determined, and ultimately the visualization processing of training performance data is completed. Finally, the visualized training performance evaluation results are output to the enterprise management. The entire process forms a cycle, and with the support of big data technology, the training performance evaluation system can complete tasks more efficiently and accurately, improving the overall training management effectiveness of the enterprise [12].

3.3 BP Neural Network Algorithm

Since a lot of training data are stored in the Cloud Data Platform, the safety of the training data has become a key problem. To solve the problem of data safety, this paper presents an early-warning system for the safety of training data by BP Neural Network. BP Neural Network is a kind of multi-dimension neural network which has the ability to classify and recognize models. By inputting and outputting data, a BP neural network model is constructed to achieve secure prediction, processing, and storage of training data, ensuring data sharing and security of the enterprise skill training performance evaluation system. Firstly, assume that the input vector set of the BP neural network is:

$$X = \{x_1, x_2, \cdots, x_n\}^T \tag{1}$$

The output vector set of BP neural network is:

$$Y = \{y_1, y_2, \cdots, y_n\}^T \tag{2}$$

From this, it can be inferred that the data source input for the BP neural network's dark layer is:

$$K_j = \sum_{j=1}^{n} w_{ij} - \chi_j, j = 1, 2, \cdots, k \tag{3}$$

Wij represents the weight of training data, which is the input amount from training indicators to intermediate data elements in each department. It indicates the quantity of training work contained in every section, i. e., the quantity of middle data elements. Moreover, xj is a threshold that can be accepted by the middle data unit.

Finance messages are sent across all sectors, and the Common Centre produces a forecast model that has the following functionality:

$$f(x) = \frac{1}{1 + e^{-x}} \tag{4}$$

Based on the above-mentioned function and the middle data element inputted, we get the middle data element's output function, which is:

$$\lambda_i = \frac{1}{1 + \exp\left(-\sum_{i=1}^{n} w_{ij} - \chi_j\right)} \tag{5}$$

The data element follows an exponential trend, where λi serves as the output function of the intermediate data element. This establishes a direct exponential relationship between the input and the output function. Furthermore, the output of the intermediate data element is inversely proportional to the exponential of the input. This means that as the input increases, the persistence of the intermediate data element's output diminishes progressively [13, 14].

According to the input-output model function of middle data unit, we can carry out the fuzzy arithmetic to get the data index. The input for later storage is obtained as follows:

$$\zeta_t = \sum_{j=1}^{n} v_{jt}\lambda_j - \xi_t \tag{6}$$

In the formula, v_{jt} represents the weight of intermediate data elements to subsequent storage units; λ_j represents the output function of intermediate data elements; ξ_t represents the maximum value that intermediate data elements can output.

On the basis of the input weighting function which is derived from the fuzzy computation and the sum of the data elements, the output of the succeeding memory can be derived as follows:

$$\eta_t = \frac{1}{1 + \exp\left(-\sum_{j=1}^{n} v_{jt}\lambda_j - \xi_t\right)} \tag{7}$$

It can be found from the above equation that the output function of the following assessment has the same relation with that of middle data element. Likewise, the output function has a linear relation with its input, and its later assessment is the inverse of its input [15]. What this means is that as the amount of input goes up, the downstream output is going to fall. The two have numerical differences, but their functional relationships are similar. According to this formula, it is possible to recognize the input and output of each data element in the system, thereby calculating the level of training effectiveness risk in the enterprise skill training performance evaluation system [16].

4   Results Analysis

The author selected the designer training work of Company A, and through observation, investigation, and analysis of the designer training work of Company A, identified performance gaps, analyzed the reasons for performance gaps, designed problem solutions based on the reasons for performance gaps, and followed up on the implementation and effectiveness of improvement plans [17].

4.1 Training Types and Content

The training content for company designers can be divided into two types based on the training time: Pre job training and on-the-job training.

(1) Pre job training

Pre-employment training is designed to give new employees a general understanding of the company and division. Pre-employment training is the beginning of a professional development of an enterprise. It implies that staff should abandon some ideas, values and behaviour models, conform to the demands and objectives of the organization, and study the new rules and practices. In this phase, it is necessary for a new employee to build a good rapport with co-workers and a group of people, setting up a realistic expectation and a positive attitude.

(2) On the job training

The on-the-job training for the design department is mainly divided into two parts: The first part is brand concept and feature training, fabric composition and feature training, process training, and trend training, which are four fixed training programs. The training programs in the second part are not fixed, and employees can independently propose their needs, which will be investigated and analyzed by the training department to determine the content of the training. For the non fixed training programs for employees' autonomous needs in the second part, they will be implemented according to the process in Table 2:

Table 2. Employee Self training Demand Declaration Process

Step	Key points of content
Step	Key points of content
Step 1	Employees declare training needs
Step 2	The human resources training department compiles statistics on employees with similar training needs and their training requirements. Analyze the necessity of training, and if it is indeed necessary to provide similar training, the next step of training arrangements will be made after approval by relevant management personnel
Step 3	Select suitable training instructors, organize the writing of training materials, and have the relevant leaders review and supplement the training materials
Step 4	Determine the specific training time and participants, organize and coordinate the implementation of the training
Step 5	Collect feedback from training participants

A company has provided various trainings for designers in order to improve their abilities and work results. During the assessment after the training, it was found that the results were very good. However, the author obtained contradictory conclusions through interviews with leaders of the design department and the operations department. Through searching and analyzing the archived materials of the training department, the author found that the assessment scores of the knowledge-based training projects in the design department's training work are relatively high, as shown in Table 3:

Table 3. Average scores of the first 5 knowledge training project assessments before improvement

Training Program	Average assessment score
Enterprise related knowledge training	91
Departmental knowledge training	92
Training on Fabric Composition and Characteristics	94
Craft training	86

From Table 3, it can be seen that the results of the knowledge-based training are good. However, through in-depth interviews conducted by the author, it was found that the head of the design department has a poor evaluation of the designer's understanding of the process, and mentioned that designers often do not consider the cost of the process, resulting in excellent design styles. However, due to the complexity of the production process and high costs, it is not recognized by customers. The leader in charge of operations reported that the designer's introduction of the work at the ordering meeting was too simplistic and failed to capture the attention of customers. Overall, it was impossible to market their work. The human resources department has reported that designers often have serious overtime situations in the early stages of ordering meetings, and their work status is relatively relaxed, which inevitably affects the quality of their work [18].

4.2 Training Analysis Based on Performance Technology

In this case, the designer team is the value pillar and core competitiveness of the company, and is the backbone of A company's capital appreciation. Designers possess knowledge capital, which is hidden in their minds and transfers value to the enterprise through the development of products. The enterprise sets up a series of training programs for designers, with the aim of enabling employees to transfer their knowledge capital to tangible products faster and more. So what is the target performance status of designer training work? The author will sort out the target performance status of the enterprise for designer training work through interviews and surveys with relevant personnel. Create the following survey questionnaire to investigate and analyze the current performance status of designer training and the gap with the target performance. The expected goals in the questionnaire are jointly discussed and determined by the training department experts, product development department leaders, and senior management of the company.

Create the following survey questionnaire to investigate and analyze the current performance status of designer training and the gap with the target performance. The expected goals in the questionnaire are jointly discussed and determined by the training department experts, product development department leaders, and senior management of the company. Distribute the above survey questionnaire to 17 senior designers at the level of training department managers, supervisors, and product development department managers. Evaluate the attitudes, knowledge, and skills of designers, junior designers, and design assistants (who have participated in all designer training programs) in five design teams. The evaluation results are shown in Fig. 2:

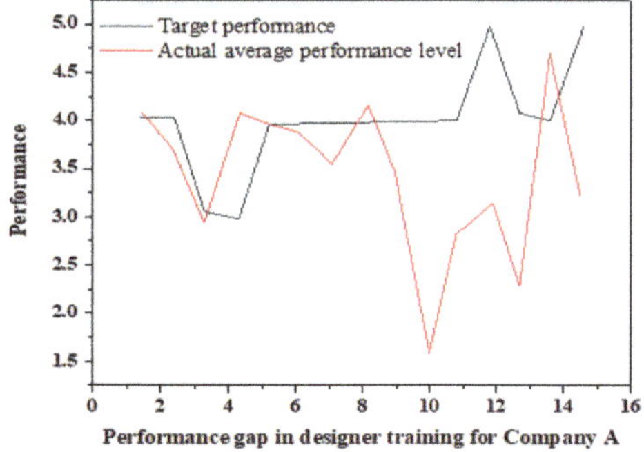

Fig. 2. Performance Gap in Designer Training of Company A

Figure 3 compares the expected goals of designer training with the actual overall average level. The blue solid line represents the target performance level, while the red dashed line represents the actual average performance level. The distance between the two lines reflects the gap between expected and actual performance. It is evident that only the indicators numbered 1, 4, 5, 6, 8, and 14 have reached the expected level, while others have not, especially the indicators numbered 10, 11, 12, 13, and 15, which have a significant gap from the expected level.

5   Conclusion

The enterprise skill training performance evaluation system based on big data technology has brought profound changes to traditional training performance evaluation methods. The author proposes a comprehensive and scientific training performance evaluation system by deeply analyzing the characteristics of big data technology and combining data collection, storage, processing, and visualization techniques. In this system, the author uses BP neural network algorithm to intelligently analyze training data, improving the accuracy of evaluation and data security. In addition, through data visualization technology, enterprise managers can grasp training performance in real time, optimize resource allocation and decision-making efficiency. Research has shown that evaluation systems based on big data technology can comprehensively and dynamically capture the performance of knowledge-based employees during training, accurately predict the potential for employee skill development, and provide data support for enterprises to develop more targeted training programs. This innovative model not only improves the effectiveness of enterprise training, but also provides important reference for the digital transformation of enterprises in the field of human resource management.

References

1. Nasrullah, Ahmad, N., Sharma, N., Niazy, M.S.: A study of performance evaluation and comparison of nosql databases choosing for big data: hbase and cassandra using ycsb. IOSR J. Comp. Eng. (3), 25 (2023)
2. Chen, X., Zhang, D., Wang, B., Ahmad, K.: Application-based big data development framework for health sciences libraries. Health Info. Libr. J. **41**(3) (2024)
3. Omar, H.K., Frikha, M., Jumaa, A.K.: Pytorch and tensorflow performance evaluation in big data recommendation system. Ingénierie des Systèmes d'Information **29**(4) (2024)
4. Gao, D., Zhang, Y., Wei, G.: Athlete-focused student physique test and evaluation system utilizing body test big data: enhancing performance and health monitoring. International Journal of Medicine & Science of Physical Activity & Sport / Revista Internacional de Medicina y Ciencias de la Actividad Física y del Deporte **24**(97) (2024)
5. Kumar, R.: Simulative analysis and performance evaluation for data variety aware power optimization technique using big data. Wireless Personal Communications: an Internaional Journal **133**(3), 1987–2002 (2023)
6. Hachim Almrshed, S.K., Abdullah Al_Shaalan, A.A., Flayyih, M.R.: Evaluation of the implications of big data analytics with organizational performance in small and medium enterprises and its associated role of knowledge management. South Asian J. Soc. Sci. Human. **5**(5) (2024)
7. Gabdullin, M.T., Suinullayev, Y., Kabi, Y., Kang, J.W., Mukasheva, A.: Comparative analysis of hadoop and spark performance for real-time big data smart platforms utilizing iot technology in electrical facilities. J. Electr. Eng. Technol. **19**(7), 4595–4606 (2024)
8. Xu, Z., Wang, J.: Exploration of enterprise informatization management innovation strategy model driven by big data technology. Appl. Math. Nonlin. Sci. **9**(1) (2024)
9. Song, C., Liang, Y.: Exploring the teaching mode of secondary english education based on big data technology. Appl. Math. Nonlin. Sci. **9**(1) (2024)
10. Yi, G., et al.: Innovation of classroom teaching of chemical principles based on big data in boppps teaching model. Appl. Math. Nonlin. Sci. **9**(1) (2024)
11. Yan, H.: Research on innovative model of piano informatization teaching under the background of big data and soft computing. J. computat. Methods Sci. Eng. **23**(5), 2425–2435 (2023)
12. Zhao, J., Wang, Z., Zhang, J.: Construction on precise-personalized-learning evaluation system based on cipp evaluation model and integrated fce-ahp method. J. Intel. Fuzzy Sys. Appl. Eng. Technol. **45**(3), 3951–3963 (2023)
13. Nasir, F., Ahmed, A.A., Kiraz, M.S., Yevseyeva, I., Saif, M.: Data-driven decision-making for bank target marketing using supervised learning classifiers on imbalanced big data. Comp. Mat. Continua **81**(1) (2024)
14. Hu, Y., Zhang, C., Cui, Y., Wei, L., Ni, Z.: Construction and performance evaluation of big data prediction model based on fuzzy clustering algorithm in cloud computing environment. J. Electr. Sys. **19**(4) (2023)
15. Li, J.: Optimization and upgrading of big data processing techniques in high performance computing environments. Appl. Math. Nonlin. Sci. **9**(1) (2024)
16. Li, C., Ju, W., Chu, S.: Design and application of a quality evaluation index system for sustainable cultivation of big data professionals in universities based on bloom cognitive domain classification method. Discover Sustainability **5**(1), 1–16 (2024)
17. Wen, Z., Wang, F., Yang, N.: Optimized evaluation of the quality of sensor video internet of things (viot) by the integration of big data and artificial intelligence. Discover Computing **27**(1), 1–16 (2024)
18. Su, X., Yin, W.: Retracted article: application of internet big data analysis technology based on deep learning in tourism marketing performance evaluation. Soft. Comput. **28**(Suppl 2), 595 (2024)

Dynamic Time Warping-Based Similarity Analysis for Track and Field Athlete Movements

Mi Chen and Chuan Huang[✉]

Chongqing College of Architecture And Technology, Chongqing, China
HuangChuanCQ@email.cn

Abstract. In track and field sports, athletes' movements are highly personalized and random, which makes it difficult to directly compare and analyze the similarities between movements made by different athletes or the same athlete in different situations. Aiming at the problem of similarity analysis of track and field athletes' movements, this study designs and implements an analysis system based on the Dynamic Time Warping (DTW) algorithm. The overall architecture of the system includes four modules: data preprocessing, time series alignment, distance calculation, and result output and analysis. The data preprocessing module extracts key features and converts them into time series format. The time series alignment module uses the DTW algorithm to achieve accurate alignment of time series. The distance calculation module quantitatively evaluates the similarity of actions by calculating the Euclidean distance between the aligned time series. The result output and analysis module helps users understand the differences and similarities between athletes' movements. Compared to traditional template matching methods, the accuracy of motion sequence alignment and computational efficiency have been significantly improved through the optimization of the DTW-based system. Experimental results show that the system reduces alignment error, enhances similarity measurement accuracy (with the correlation coefficient increasing from 0.68 to 0.76), and significantly optimizes computational performance while reducing processing time and memory consumption. Additionally, the system enables dynamic tracking of athletes' movement variations across different scenarios, providing real-time technical optimization suggestions and offering better support for personalized training strategies.

Keywords: DTW · template matching algorithm · similarity measurement · alignment error rate · processing time

1 Introduction

In the field of competitive sports, athlete motion analysis has always been an important means to improve athletic performance, prevent sports injuries and optimize training strategies. As one of the basic events in competitive sports, the accuracy and consistency of track and field technical movements have a decisive impact on the athletes' competitive performance.

© The Author(s) 2026

P. Siarry et al. (Eds.): WCNA 2024, LNEE 1550, pp. 152–162, 2026.
https://doi.org/10.1007/978-981-95-6946-5_16

This paper applies the DTW algorithm to the similarity analysis of track and field athletes' movements and constructs a DTW-based similarity analysis system for track and field athletes' movements. The system achieves efficient and accurate analysis of track and field athletes' movement sequences. This paper not only explains the system construction process, algorithm principles and implementation details but also verifies the effectiveness and practicality of the system through experiments.

The paper first introduces the research background, contributions of this paper and the structure of the paper; then, it reviews the research progress in related fields, providing a theoretical basis for the research of this paper; then, it explains the construction process of the system, including four major modules: data preprocessing, time series alignment, distance calculation, and result output and analysis; then, the performance of the system is verified through the experimental part, and the experimental results are discussed; finally, the research results of this paper are summarized, and prospects for future research directions are given.

2 Related Work

Researchers are constantly exploring new methods and technologies to understand athletes' physical functions, sports performance, and the prevention of sports injuries. Ren et al. [1] selected 15 college basketball athletes to analyze the lower limb biomechanics of vertical jumps before and after muscle fatigue. They used a 10-camera three-dimensional motion capture system and a three-dimensional force platform to collect the kinematics and ground reaction parameters of the subjects' lower limb hip, knee, and ankle joints when they landed from a vertical jump. The lower limb kinematic and kinetic data were calculated using Visual 3D software, and the SPSS paired sample T test was used to compare and analyze the lower limb biomechanical parameters of basketball athletes performing vertical jumps before and after fatigue. Wu and Wang [2] used Noraxon wireless surface electromyography to collect surface electromyographic signals of the main muscle groups of athletes' bilateral rotation movements, and used non-negative matrix decomposition method to extract muscle synergy data to analyze the differences in control modes between the dominant and non-dominant side rotation movements of athletes. Chen Helin [3] used testing, mathematical statistics, and logical analysis methods to conduct functional movement screening (FMS) on young male and female volleyball players in Zhangzhou City, and comprehensively understood the physical movement function characteristics and weak links of young male and female volleyball players in Zhangzhou City. The results showed that the FMS level of young male and female volleyball players in Zhangzhou City was not high, and nearly half (46.67%) of the athletes were at risk of sports injuries. Most of the young male and female volleyball players in Zhangzhou City had insufficient flexibility in the hips, knees, and ankles, and their basic core strength needed to be further consolidated. Wang [4] improved the technical performance of tennis players by individualized biomechanical optimization method. Yu and Yang [5] used hierarchical clustering algorithm to identify the stress sources of track and field athletes, and recorded the characteristics of various sports anxiety states through hidden unit centers, function widths and connection weights.

Belkhouja et al. [6] proposed a framework called Dynamic Time Warping Adversarial Robustness (DTW-AR) that uses the dynamic time warping metric to create adversarial samples and proves that the DTW metric is more effective than the Euclidean distance metric commonly used in the image field. Landmesser [7] used the DTW method to analyze COVID-19 data in Poland, dealing with the scaling and deformation in time series data and more accurately capturing the changes in the epidemic dynamics. Kumar et al. [8] introduced DTW as an alternative method to seismological applications to improve the matching accuracy of seismic signals. Herrmann and Webb [9] proposed an early abandonment and pruning strategy to optimize the calculation of elastic distances such as DTW. Remya Krishnan and Arun Raj Kumar [10] used DTW to analyze the time series characteristics of network traffic, identify abnormal patterns, and thus detect attack behaviors. As a powerful time series analysis tool, DTW has demonstrated its unique value in many fields. The application of this technology is constantly expanding, providing new ideas and methods for research in related fields. However, in the field of track and field sports, although motion analysis is important for optimizing athletes' techniques and improving their performance, there is still a lack of an efficient, accurate and applicable analysis system for complex motion sequences. Therefore, this paper will focus on the research of the DTW-based track and field athlete motion similarity analysis system, aiming to develop a system that can effectively evaluate the motion similarity of track and field athletes, optimize technical performance and assist in training by introducing DTW technology, providing a scientific basis for track and field training, and further expanding the scope of application of DTW technology in the field of sports and promoting the development of sports science.

3 Methods

3.1 Overall System Architecture

The overall architecture of the track and field athlete movement similarity analysis system based on the dynamic time warping (DTW) algorithm covers four core modules, each of which carries specific functions and responsibilities [11]. The data preprocessing module collects the motion data of track and field athletes and performs preprocessing work including data cleaning, denoising, feature extraction and time series conversion to lay the foundation for subsequent analysis. The time series alignment module uses the DTW algorithm to accurately align the preprocessed time series and adjust the time axis to eliminate the differences in length and rhythm of the action time series. The distance calculation module calculates the Euclidean distance between the two time series based on the aligned time series to quantitatively evaluate the similarity between the actions. Finally, the result output and analysis module is responsible for outputting and visualizing the calculated action similarity measurement results, while providing data analysis functions to help users understand the differences and similarities between athletes' actions. Figure 1 shows the system architecture:

These four modules work together to form the complete architecture of the system, ensuring efficient operation and accurate analysis of the system.

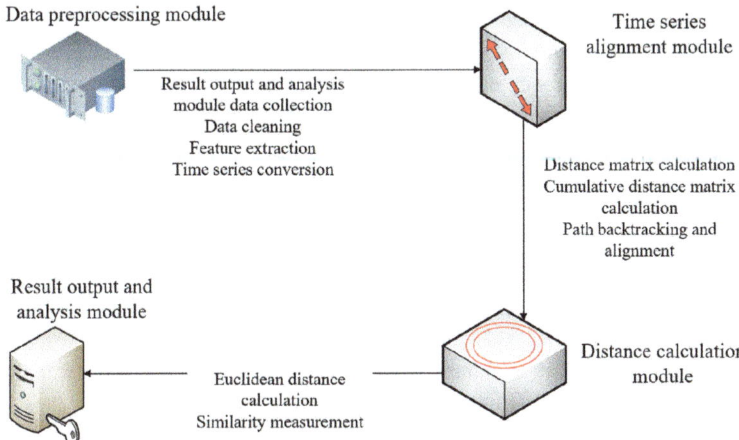

Fig. 1. System architecture

3.2 Data Preprocessing Module

In the DTW-based track and field athlete motion similarity analysis system, data collection in the data preprocessing module depends on multiple sources such as motion capture systems, video records, and sensor data [12]. The motion capture system records the athlete's movement trajectory and posture with high precision; the video record provides an intuitive display of the athlete's movements, which is convenient for the subsequent extraction of key movement data; and the sensor data such as the readings of the accelerometer and gyroscope reveal the real-time dynamics of the athlete's movement. As shown in Table 1, the data collected for the athlete's movement:

Table 1. Movement data

Athlete ID	Reaction Time (s)	Hip Joint Angle (°)	Knee Joint Angle (°)	Ankle Joint Angle (°)	Accelerometer Reading (m/s^2)	Gyroscope Reading (°/s)
001	0.152	45.3	82.1	23.4	2.15	120.5
002	0.148	46.7	80.9	22.8	2.20	118.3
003	0.160	44.2	83.5	24.1	2.05	122.0
004	0.155	47.1	81.2	23.0	2.18	119.8
005	0.158	45.9	82.8	23.6	2.12	121.2
006	0.149	46.5	81.5	22.9	2.25	117.6
007	0.153	45.6	83.0	23.3	2.10	120.9
008	0.157	44.8	82.3	24.0	2.08	121.5

After data collection, since the original data may be mixed with noise, outliers or missing data, filtering, smoothing and other technologies are used to pre-process the data to remove high-frequency noise and make the data smoother and more continuous. Missing values are filled by interpolation methods to ensure the integrity and accuracy of the data.

Feature extraction carefully selects information such as joint angles, velocities, accelerations and displacements from the raw data that can represent the core characteristics of the athlete's movements.

Finally, time series transformation converts the extracted feature data into a time series format to accommodate subsequent time series alignment and similarity calculations, including converting the data into a format suitable for time series analysis, such as CSV, and performing standardization to eliminate differences in dimensions and numerical ranges between features.

3.3   Time Series Alignment Module

In the similarity analysis system, the time series alignment module uses the DTW algorithm to accurately align the preprocessed time series to solve the problem of difficulty in direct comparison due to differences in action time length and speed.

The core idea of the DTW algorithm is to find the best alignment path between two time series through dynamic programming, so that the cumulative distance between the aligned sequences is minimized. This paper first initializes a distance matrix D, which contains the distances between all possible alignment points in the two time series:

$$D(i,j) = d(x_i, y_j) \tag{1}$$

$d(x_i, y_j)$ is the Euclidean distance between point x_i and point y_j.

Next, the cumulative distance matrix C is calculated by dynamic programming, where each element $C(i,j)$ refers to the minimum cumulative distance from the starting point $(1,1)$ to the point (i,j):

$$C(i, j) = D(i, j) + \min\{C(i - 1, j), C(i, j - 1), C(i - 1, j - 1)\} \tag{2}$$

Then, backtrack from the end point of the accumulated distance matrix to find the best alignment path P:

$$P = \arg \min_{P} \sum_{k} D(i_k, j_k) \tag{3}$$

$D(i_k, j_k)$ is an element in the distance matrix, which represents the distance between the i_k-th point of time series X and the j_k-th point of time series Y.

Finally, the two time series are aligned according to the alignment path to obtain the aligned sequence.

In the alignment process, the stretching and compression of the time axis is achieved by adjusting the alignment path. In order to limit the excessive stretching or compression of the time axis, the Sakoe-Chiba Band local path constraint is introduced.

3.4 Distance Calculation Module

The distance calculation module is responsible for quantitatively evaluating the similarity of the aligned time series. The module calculates the Euclidean distance and combines the point reuse strategy in the DTW algorithm to accurately measure the similarity of the athletes' movements [13].

In the DTW algorithm, point reuse allows a point in one sequence to correspond to multiple points in another sequence, or multiple points to correspond to one point during the alignment process. This flexibility allows the DTW algorithm to find the best alignment path between two time series. Through the point reuse strategy, the DTW algorithm minimizes the overall distance - the cumulative distance between the aligned sequences [14, 15]. This cumulative distance comprehensively considers the similarity of all points in the sequence and provides a comprehensive evaluation of the similarity of the athletes' movements. As shown in Table 2, the measurement results between different athlete movement sequences are:

Table 2. Measurement results

Sequence Pair	Sequence Length A	Sequence Length B	Cumulative Euclidean Distance (DTW)	Number of Point Reuses
1	50	55	120.5	7
2	45	50	135.2	5
3	60	65	110.8	9
4	55	60	140.3	8
5	40	45	105.6	6

In the distance calculation module, the system combines the Euclidean distance calculation and the point reuse strategy in the DTW algorithm to accurately and quantitatively evaluate the similarity of the aligned time series. This similarity measure not only takes into account the feature similarity at each time point in the sequence but also reflects the impact of differences in time length and speed on the similarity by searching for the best alignment path and minimizing the overall distance.

3.5 Result Output and Analysis Module

The result output and analysis module helps users fully understand the differences and similarities between athletes' movements through intuitive visualization and data analysis functions.

The module provides a variety of result presentation methods, including chart display, animation simulation and interactive interface. Users can intuitively see the similarity measurement results between different athletes' movements, understand the alignment of movement sequences on the timeline and the changing trend of Euclidean distance. The animation simulation dynamically displays the alignment results of the action sequence, allowing users to more vividly experience the differences and similarities between actions. The interactive interface allows users to adjust the visualization parameters according to their needs to achieve personalized result presentation.

The module also provides in-depth analysis of differences and similarities. The system uses cluster analysis to identify groups of athletes with similar movement characteristics; trend analysis reveals the changing trends of athlete movement sequences over time; and feature comparison helps users compare the differences in key movement characteristics of different athletes. The system also has functions such as correlation analysis, anomaly detection and predictive analysis to further explore the intrinsic connections between movement features, identify abnormal movement sequences and predict the development trend of athletes' future movements.

4 Results and Discussion

4.1 Experimental Design

In the study, 15 athletes from different track and field events are selected as experimental subjects to cover the diversity and representativeness of the movements. The collected data include key features such as joint angles, speeds, accelerations, etc. of athletes when performing key actions. These data are cleaned and feature extracted and converted into time series format, providing a basis for subsequent alignment and similarity calculations.

In order to verify the advantages of the DTW algorithm, the experiment set up two groups of comparison algorithms: the DTW algorithm as the experimental group, responsible for processing the preprocessed time series and calculating the similarity between actions; and the template matching algorithm as the control group, used to evaluate the performance of the DTW algorithm on the same task. The experiment uses multiple quantitative indicators to comprehensively evaluate the performance of the two algorithms, including alignment accuracy (evaluated by calculating the alignment error rate), similarity metric accuracy (comparing the correlation between the algorithm calculation results and the actual competition results), processing time (recording the time required for the algorithm to process data) and memory usage (monitoring resource consumption during the algorithm operation).

4.2 System Comparison Results

Based on the above experimental settings, this paper conducts comparative experiments and analyzes the experimental results. The alignment error rate comparison results are shown in Fig. 2.

As shown in Fig. 2, the alignment error rate of the experimental group is generally lower than that of the control group, showing the advantage of the DTW algorithm in aligning the time series of track and field athletes' movements. The alignment error rate of the experimental group ranges from 0.21% to 0.79%, while the alignment error rate of the control group fluctuates between 0.96% and 2.5%. This difference shows that the DTW algorithm can handle the alignment problem of action time series more accurately and reduce the errors caused by differences in time length and rhythm. The DTW algorithm effectively solves the problems of different time series lengths and rhythm differences through dynamic programming methods, thereby achieving more accurate action alignment. This advantage is crucial for the similarity analysis of track and field athletes' movements, because it can ensure the accuracy of similarity measurement, thereby providing a more reliable basis for athletes' technical improvement and training optimization.

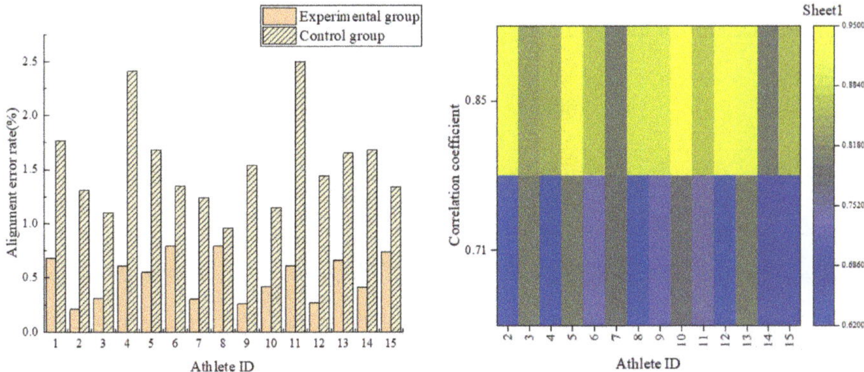

<table>
<tr><td>Fig. 2. Alignment error rate</td><td>Fig. 3. Correlation</td></tr>
</table>

Analyzing the data in Fig. 3, this paper finds that the correlation coefficient of athlete 1 in the experimental group is 0.85, but only 0.71 in the control group; and the correlation coefficient of athlete 2 in the experimental group is 0.92, but only 0.64 in the control group. This trend remains consistent in all data, and the correlation coefficient of the experimental group is always higher than that of the control group. Therefore, it can be concluded that when quantifying the similarity of track and field athletes' movements, the system based on the DTW algorithm can provide more accurate and stable similarity measurement results, providing a more reliable basis for athletes' technical analysis and training optimization.

4.3 Results Analysis

This paper uses a DTW-based system to quantify the similarity of the movements of different track and field athletes. The results show that the system can accurately distinguish the differences between the movements of different athletes and give corresponding similarity scores. The movement similarity scores of sprinters and long jumpers are generally low, which reflects the differences in movement patterns, muscle force methods and movement trajectories between the two. On the contrary, despite individual differences among different athletes in the same event, their movement similarity scores are generally high, especially in the key links of technical movements, showing high consistency.

By analyzing these quantitative results, it is found that the system can accurately capture subtle differences in athletes' movements, such as changes in key indicators such as step frequency, stride length, and take-off angle, thereby providing support for athletes' technical improvement and personalized training. In addition, the system can also perform cluster analysis on the movements of different athletes, revealing the commonalities and differences in movement styles, and providing a scientific basis for the coaching team to formulate targeted training plans.

In order to fully evaluate the performance of the system, we also quantitatively analyzed the similarity of the movements of the same athlete in different situations. These situations include different game stages, training intensity, environmental conditions, etc. The results are shown in Table 3:

Table 3. Action similarity scores

Athlete ID	Critical Race Stage	High-Intensity Training	Adverse Weather Conditions (Rainy)	Low-Intensity Recovery Training	Pre-Race Preparation
1	93	79	71	89	87
2	95	83	76	92	90
3	91	77	70	86	85
4	94	81	74	91	89
5	90	75	68	84	83
6	92	80	73	88	86
7	96	84	77	93	91
8	91	78	72	87	85

Athletes tend to focus more on the details of their technical movements during critical stages of the competition, striving to perform at their best, and their movement similarity scores are usually higher at this time. However, under high-intensity training and harsh environmental conditions, athletes' movements become deformed due to fatigue and external interference, resulting in a decrease in movement similarity scores. These changes reflect the impact of factors such as the athlete's psychological state, physical condition and adaptability in different situations on movement consistency.

5 Conclusion

This study successfully constructs a track and field athlete motion similarity analysis system based on the DTW algorithm. By introducing the DTW algorithm, the system solves the difficulties of traditional motion analysis methods in time series alignment and quantitative evaluation of motion similarity, providing a scientific basis for track and field athletes' technical optimization, training improvement and sports injury prevention. However, data collection in this study is limited by experimental conditions and equipment accuracy, which may affect the accuracy of the analysis results; the system currently focuses on the movement similarity analysis of track and field athletes, and its applicability to other sports needs further verification. Looking into the future, this study will continue to deepen the application of DTW algorithm in the motion analysis of track and field athletes and explore more efficient and accurate analysis methods and techniques. At the same time, the scope of data collection will be expanded, data quality will be improved, and the analysis performance and applicability of the system will be further enhanced. In addition, the system's automation and intelligence will be strengthened, and a more convenient and user-friendly interface will be developed to better serve track and field athletes and coaching teams.

References

1. Size, R., Yiwen, Ma., Wanke, T.: Biomechanical analysis of lower limbs of male basketball players' vertical jump due to muscle fatigue. Zhejiang Sports Science **47**(1), 97–105 (2025)
2. Ying, W., Xin, W.: Difference analysis of muscle coordination patterns in bilateral rotation movements of freestyle snowboard athletes. Med. Biomechan. **39**(S01), 498 (2024)
3. Lukai, C., Jing, L.: Analysis of the functional movement screening (FMS) characteristics of young volleyball players in Zhangzhou City - Taking the Zhangzhou Youth Sports School preparing for the volleyball competition of the 17th Fujian Provincial Games as an example. Bulletin of Sports Sci. Technol. Literat. **32**(5), 48–51 (2024)
4. Xing, W.: Biomechanical analysis of technical movements of high-level tennis players. Contemp. Sports Sci. Technol. **14**(8), 1–3 (2024)
5. Yu, Z., Yang, G.: Analysis of anxiety and stress sources of track and field athletes based on RBF neural network model. J. Jilin Ins. Phys. Edu. **40**(4), 98–108 (2024)
6. Belkhouja, T., Yan, Y., Doppa, J.R.: Dynamic time warping based adversarial framework for time-series domain. IEEE Trans. Pattern Anal. Mach. Intell. **45**(6), 7353–7366 (2022)
7. Landmesser, J.M.: The use of the dynamic time warping (DTW) method to describe the COVID-19 dynamics in Poland. Oeconomia Copernicana **12**(3), 539–556 (2021)
8. Kumar, U., Legendre, C.P., Zhao, L., et al.: Dynamic time warping as an alternative to windowed cross correlation in seismological applications. Seismological Soc. America **93**(3), 1909–1921 (2022)
9. Herrmann, M., Webb, G.I.: Early abandoning and pruning for elastic distances including dynamic time warping. Data Min. Knowl. Disc. **35**(6), 2577–2601 (2021)
10. Remya Krishnan, P., Arun Raj Kumar, P.: Detection and mitigation of smart blackhole and gray hole attacks in VANET using dynamic time warping. Wirel. Pers. Comm. **124**(1), 931–966 (2022)
11. Mohammadzade, H., Hosseini, S., Rezaei-Dastjerdehei, M.R., et al.: Dynamic time warping-based features with class-specific joint importance maps for action recognition using Kinect depth sensor. IEEE Sens. J. **21**(7), 9300–9313 (2021)
12. Gassouma, M.S., Benhamed, A., El Montasser, G.: Investigating similarities between Islamic and conventional banks in GCC countries: a dynamic time warping approach. Int. J. Islam. Middle East. Financ. Manag. **16**(1), 103–129 (2023)
13. Deriso, D., Boyd, S.: A general optimization framework for dynamic time warping. Optim. Eng. **24**(2), 1411–1432 (2023)

14. Hollerbach, A.L., Conant, C.R., Nagy, G., et al.: Dynamic time-warping correction for shifts in ultrahigh resolving power ion mobility spectrometry and structures for lossless ion manipulations. J. Am. Soc. Mass Spectrom. **32**(4), 996–1007 (2021)
15. Langfu, C.U.I., Zhang, Q., Yan, S.H.I., et al.: A method for satellite time series anomaly detection based on fast-DTW and improved-KNN. Chin. J. Aeronaut. **36**(2), 149–159 (2023)

AI-Driven Speech Recognition and Automatic Evaluation for English Oral Error Correction

Pingying Hou[✉]

Beijing University of Civil Engineering and Architecture, Beijing, China
`houpy.bucea@email.cn`

Abstract. Standardized pronunciation assessment is crucial for English learners, especially given the inefficiencies of traditional methods. To address these challenges, this study develops an AI-based speech recognition and automatic feedback system aimed at improving the accuracy of oral error correction. This system is composed of two parts, namely, the voice recognition sensor and the neural network. Using Dynamic Time Warping (DTW) and Hidden Markov Model (HMM) to analyse the time sequence of voice signals, it can precisely detect voice mistakes and offer real time error-correcting feedback. The author used a standard speech corpus and scoring mapping model to construct an efficient scoring mechanism, significantly improving the reliability and intelligence level of speech processing. The experimental results show that when using this system to recognize selected natural speech segments, the recognition rate is higher than 94%, and the average video rate reaches over 96.5%; The recognition time fluctuates between 0.6 and 1.5 s, with an average recognition time of approximately 1.2 s. This system has significant advantages in error detection accuracy, feedback rationality, and user experience optimization. This study not only provides intelligent solutions for English oral teaching, but also opens up new directions for the application of speech recognition and artificial intelligence technology in the field of education. By integrating AI-driven speech recognition technology with a real-time feedback mechanism, users can enhance their English-speaking proficiency, providing valuable insights for the broader application of artificial intelligence in language education.

Keywords: Natural speech processing · Speech correction · artificial intelligence · Deep learning · English oral teaching

1 Introduction

With the rapid development of the world and the acceleration of globalization, English has become particularly important as an international lingua franca. Among many English learners, improving their oral English proficiency has become an important goal they pursue. Unfortunately, in traditional teaching models, oral error correction often relies on the manpower input of teachers, which is not only inefficient, but also subjective and has obvious limitations [1]. In order to overcome these challenges, the author introduces speech recognition sensors and artificial intelligence technology as effective

© The Author(s) 2026
P. Siarry et al. (Eds.): WCNA 2024, LNEE 1550, pp. 163–172, 2026.
https://doi.org/10.1007/978-981-95-6946-5_17

ways to solve the problem of English oral error correction. Firstly, speech recognition sensors can accurately capture learners' speech input. Unlike traditional microphones, it utilizes sound sensing technology to capture speech signals and convert them into digital form for subsequent speech analysis and evaluation. Secondly, the application of artificial intelligence technology, especially deep learning based automatic evaluation systems, can automatically recognize speech errors from speech recognition data, analyze pronunciation differences, and provide learners with more personalized feedback [2].

At the forefront of current technology, English oral correction systems rely on advanced English speech recognition technology and artificial intelligence algorithms, which have not only made breakthrough progress in the field of technical research, but also demonstrated their unique advantages in actual language teaching and communication. Along with the development of the Sensor Hardware Technique, it has offered a more accurate and more sensitive way to the voice capture system, which makes it more accurate and reliable [3]. Meanwhile, the AI assessment system can enhance the precision and the precision of rectification by optimizing the algorithm, enabling the system to more accurately identify and correct common oral errors, reducing the need for human intervention and making the error correction process more efficient and natural. In response to this deficiency, the application of artificial intelligence technology in the design of natural speech error correction and feedback systems is studied to accurately correct errors in natural speech and provide feedback to users on the corrected errors, thereby improving the accuracy of natural speech spelling [4].

2 Literature Review

Along with the fast developing of the computer identification technique, the study of the Computer Voice Recognition System has reached an advanced stage. Of these, the English pronunciation proof-reading system is the most popular one. Voice Checking System is used to assess and compute sound quality and to check for pronunciation mistakes. It is relatively easy to evaluate and calculate the quality of pronunciation based on many algorithms at home and abroad, and the level and accuracy of calculation are also relatively high. Jing, W. et al. applied the theory of voice recognition and voice synthesis. By means of voice recognition, this system is able to translate spoken words into words and assess their veracity. Through the use of voice synthesis technique, this system is able to produce the corrected sound according to the correction rule, and help the students to learn the pronunciation. Through evaluation and correction of actual learners' spoken language, the system can accurately estimate their pronunciation and produce appropriate correction [5]. Garneau, N. et al. investigated the possibility of using an automated voice recognition (ASR) system to help record the transcript of a trial. Three ASR models (with both commercial and open source alternatives) were compared with a well-designed data set [6]. Tac et al. introduced a novel depression detection model that extracts multi-level mixed features from speech and audio signals. They used the MODMA dataset, which includes audio signals from individuals with and without major depression. Advanced techniques such as hybrid manual feature generation and multi-level discrete wavelet transform are used for effective feature extraction [7]. Ha,

S. et al. utilized artificial intelligence technology to extract and analyze features of speech signals to improve auditory rehabilitation outcomes. Wav2vec 2.0 is used to recognize phonemes in 2000 sound files recorded by hearing-impaired patients. The average speech intelligibility score calculated by comparing the difference between the reference sentence and the sentence obtained through speech to text (STT) exceeds 0.92%. It is possible to accurately distinguish individual phonemes from speech files recorded by hearing-impaired individuals and measure the comprehensibility of their speech by evaluating their speech clarity [8].

Based on this research, the author proposes a speech error correction method based on deep learning and dynamic time warping (DTW) algorithm, which can accurately detect pronunciation and provide real-time feedback, effectively improving learners' oral expression ability. He developed a scientific pronunciation quality evaluation system using a standard speech corpus and rating mapping model, achieving quantitative analysis of pronunciation errors and providing precise improvement directions for oral learning. By combining Hidden Markov Models (HMM) and deep learning algorithms, the author significantly improved the accuracy and real-time performance of speech signal processing, providing a reference example for the application of speech recognition technology in the field of education.

3 Research Methods

3.1 Overall System Structure Design

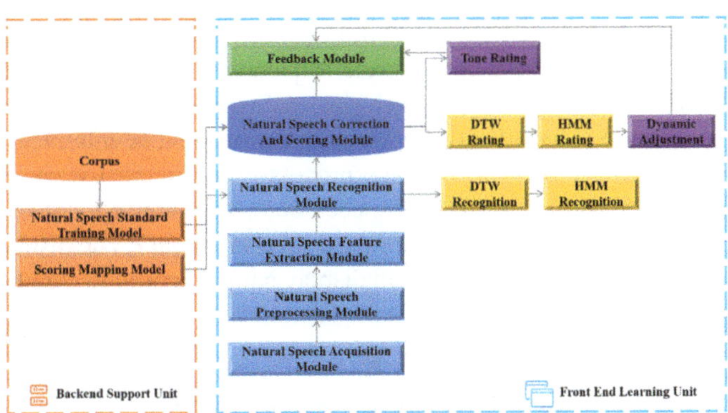

Fig. 1. Overall System Structure Diagram

The NMI is composed of the front end and the back end. The front end of the study is composed of the natural voice collection module, the characteristic extracting module, the nature voice identification module, the error-correcting and grading module, and the back end of the system is composed of the data base language, the training model of the nature language, and the evaluation map model [9]. Figure 1 illustrates the general architecture of the SDI.

In this system, the natural voice is acquired in real time, and the related parameters are extracted; Using Dynamic Time Warping (DTW) and Natural Voice Recognition with Hidden Markov Model (HMM), and completing natural speech error correction according to the set natural speech specification model. Based on the correspondence of the evaluation model with voice parameters, we can get the natural voice score, and then give the user a score. The natural speech error correction and feedback system requires strong computing power, so a DSP chip with built-in DSP operating system, chain bridge technology, and start-up tools is used to design the front-end support unit. The main advantages of DSP chips are small size, strong digital signal processing capability, and low cost. Figure 2 shows the structure diagram of the DSP chip.

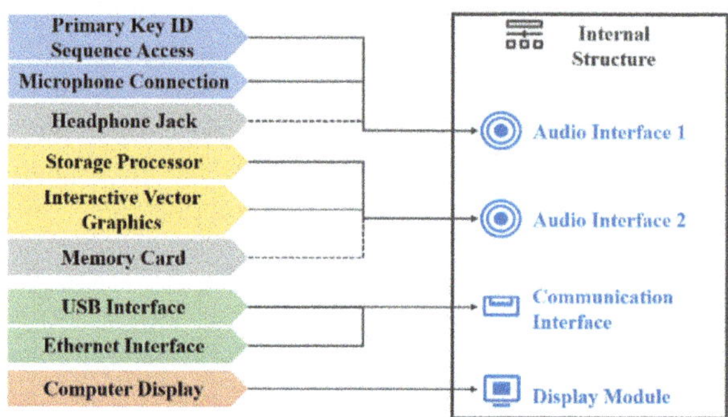

Fig. 2. DSP chip structure diagram

The selected DSP chip includes two audio interfaces, one communication interface, and one display module. The chip connects sequence data and microphone through the first audio interface for natural speech data recognition and error correction, and can also provide voice feedback to users through the headphone interface; The second audio interface is connected to the storage processor and interactive vector graphics, and its main function is to cache natural speech data. The interface connected to USB and Ethernet is mainly used for transmitting external input data; The display module is connected to the computer monitor and can display natural speech recognition and error correction results to the user [10, 11].

3.2   Design of Natural Speech Feature Extraction Module

After the system collects natural speech, it performs preprocessing such as filtering and digitization, and divides the initial phonemes and final phonemes in natural speech according to the phonological segmentation algorithm based on the distance between segments, distinguishing the different feature vectors of natural speech frames. The system determines the distance between natural speech segments through the measurement rules and methods of segment spacing, and then describes the overall differences between different natural speech segments based on the distance.

3.3 Design of Natural Speech Recognition Module

The HMM model is a common method for voice identification, and it can express the stochastic process of an unknown condition with a given probability. Not only is it able to show short term natural voice stability, but also it can be used to express the steady change relation of the natural voice signal. The HMM model consists of a variety of conditions that are transformed with time. The observation of the vector produces the respective output probabilities for the various states. In the course of HMM recognition, if the mode is not able to move forward, it can only be fixed at present or moved to the following status [12].

Taking a 4-state HMM model as an example, with O = (o1, o2,... or) and a; J (i = 1,2,3,4; j = 1,2,3,4) represents the transition probability between the input natural speech signal sequence and the i state to the j state, respectively. The input sequence is also an observation sequence. In the case of continuously moving time, we can express the various states in the output statistic model as S1, S2.., Sn, in which y is the number of states. The uninterrupted mixture Gaussian HMM model can be represented as σ = {Y, A, φ, B}, where Y and A = {aij} respectively represent the number of states and state transition probability matrix within the HMM model, and φ = {φ i} and B = {bi (o)} respectively represent the starting probability distribution and output probability density function of different states.

The determination of output probability is one of the most important problems in natural voice recognition. Therefore, it is possible to calculate the output probability by multiplication of the positive and negative probabilities, and to deduce the positive and negative probabilities by recursion. In a given situation where 0, σ, and y are known, a positive and negative probability at time t = 1,2 is used to determine an output probability of 0 in T:

$$P(O|\sigma) = \sum_{i=1}^{Y} \alpha_t(i)\beta_t(i) = \sum_{i=1}^{Y} \alpha_t(i),\ 1 \leq t \leq T-1 \tag{1}$$

In formula (1), α t (i) and β t (i) are the forward probability and backward probability, respectively, and their recursive formulas are:

$$\begin{cases} \alpha_1(i) = \phi_i b_i(o_1),\ 1 \leq i \leq Y \\ a_{t+1}(j) = \left(\sum_{i=1}^{Y} \alpha_t(i) \right) b_i(o) \\ 1 \leq t \leq T-1,\ 1 \leq j \leq Y \end{cases} \tag{2}$$

$$\begin{cases} \beta_t(i) = 1,\ 1 \leq i \leq Y \\ \beta_t(j) = \sum_{i=1}^{Y} a_{ij} b_{i_j}(o_{t+1})\beta_{t+1}(j) \\ 1 \leq t \leq T-1,\ 1 \leq j \leq Y \end{cases} \tag{3}$$

According to the formula, the author uses HMM model to represent the time-varying and statistical features of natural speech signals, and uses Baum Welch algorithm to determine the training problem of natural speech signal model parameters. After several iterations, the model parameters are obtained, and the average output probability of all learning samples reaches the upper limit [13, 14].

3.4 Design of Natural Speech Error Correction and Scoring Module

Natural voice correcting and grading module uses the AI technique of Dynamic Time Normalization, Based on B2 standard language, it performs natural voice correcting and grading by comparing the two kinds of voice characteristics. The dynamic time normalization algorithm, based on dynamic regularization technology, transforms complex global optimization problems into a large number of local optimization problems. Its core idea is: Using $R = \{r1, r2,..., ri\}$ and $K = \{k1, k2,..., kj\}$ as the comparison of natural speech feature vector sequences and the recognition results of the input natural speech signal to be corrected, where the i value is inconsistent with the j value. The main purpose of the dynamic time normalization algorithm is to determine an optimal time regularization function, with the goal of minimizing overall distortion, so that the time axis j (the natural speech recognition result to be corrected) is non linearly mapped to the time axis i (compared to the natural speech feature vector sequence). Use formula (4) to represent the time regularization function:

$$H = \{h(1), h(2), \cdots, h(N)\} \tag{4}$$

In formula (4), N represents the length of the matching path, and the distortion value of N can be represented by d (ri (n), kj (n)), which can describe the local matching distance. The dynamic time integration algorithm utilizes the process of local optimization to achieve the goal of minimizing the weighted distance as a whole. The formula is described as follows:

$$D = \min_{h} \frac{\sum_{n=1}^{N} \left[d\left(r_i(n), k_j(n)\right) \times Q_n \right]}{\sum_{n=1}^{N} Q_n} \tag{5}$$

In formula (5), Qn represents the weight of the path.

J (n) represents the i-th matching point composed of the i-th (n) th feature vector of the comparison sequence and the i-th (n) th natural speech signal recognition result to be corrected, which can be represented by d (ri (n), kj (n)).

3.5 Matching Filtering Processing of Speech Signals

The input voice is analysed, digitized, and transformed into digital signals. A low pass filter is adopted to restrict the frequency of the signal so that the frequency is less than half of the sampling rate. A high-frequency filter is used to inhibit 50 Hz power disturbances. Generally, a band pass filter is employed to process voice signals and gather the processed voice signals. Based on the above-mentioned method, the voice signal of English spoken language examination system is simulated by means of the above-mentioned sensor array, and then the voice signal is matched and filtered to get rid of the interfering noise. The author designs an audio collector for the English oral automatic pronunciation verification system, which uses a high-frequency meson amplitude collector. The collected data is compared and recorded with data encoding for verification and evaluation. The architecture is shown in Fig. 3.

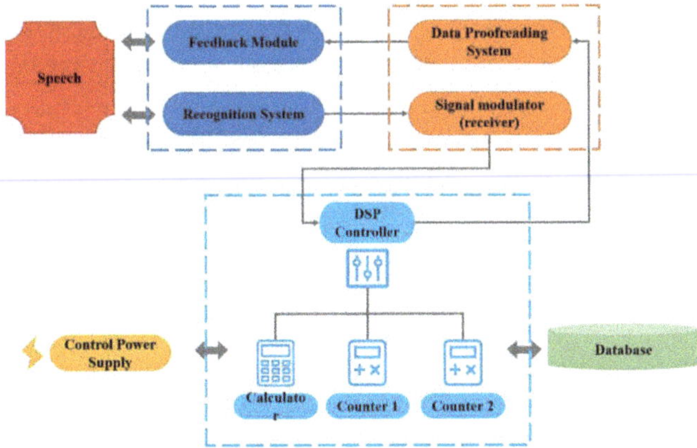

Fig. 3. Hardware System Architecture Diagram

4 Results Analysis

In order to comprehensively evaluate the practical application effectiveness of this research method, a rigorous experiment was conducted in a language research laboratory at a renowned university. The laboratory constructed a natural speech database through manual recording, which contains a large number of finely processed natural speech segments recorded by professionals. Based on this resource, we selected 600 randomly distributed speech segments as the test set. The research mainly focuses on three different methods: Method 1 focuses on correcting errors in speech through manual intervention; Method 2 attempts to automatically recognize and correct problems in speech through natural language processing technology; The method designed by the author combines the advantages of the above two methods, aiming to improve the overall error correction effect by simultaneously performing error correction and feedback [15].

4.1 Natural Speech Recognition Results

In the testing environment, we carefully selected 13 different types of natural speech segments as test samples, covering topics ranging from everyday conversations to more complex speech content. In order to evaluate the performance of the system designed by the author, we compared three different methods: Method 1, Method 2, and the new method proposed by the author. Each fragment is randomly assigned to these three algorithms for recognition, and the results are summarized to present two key indicators: recognition rate and recognition time. These data are clearly presented in the form of charts in Fig. 4. Through in-depth analysis of Fig. 4, it can be seen that when using the author's design method to process these natural speech segments, we observed a very high recognition rate. Specifically, the recognition rate of all test segments exceeded 94%, with the average recognition rate of video segments reaching over 96.5%. This achievement is significantly higher than the other two methods, demonstrating the system's outstanding performance in natural speech recognition. At the same time, we also noticed the distribution of recognition time. Although the recognition time of each segment varies, overall, their fluctuation range is between 0.6 and 1.5 s, with an average recognition time of approximately 1.2 s. This time range is significantly shorter compared to methods 1 and 2, indicating that the system is more efficient and fast in executing tasks [16].

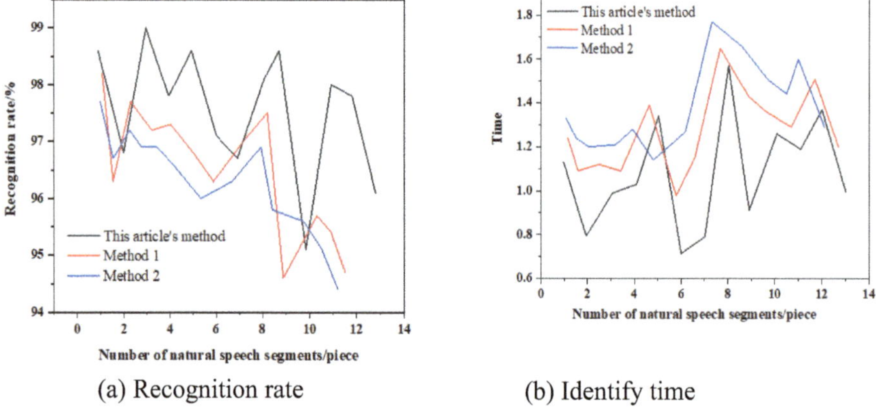

(a) Recognition rate   (b) Identify time

Fig. 4. Accurate Recognition Rate of Natural Speech

4.2 Application Performance Test Results

Table 1 shows the accuracy of easily mispronounced pronunciations during the spelling process using Method 1, Method 2, and the author's design method.

Table 1. Comparison Results of Pronunciation and Spelling Accuracy for Pronunciation with Easy Errors

Pronunciation that is prone to errors	Accuracy/%		
	Method 1	Method 2	Author's Design Method
T	83.23	83.61	90.63
D	82.54	83.47	90.36
K	74.36	70.74	87.81
S	70.74	80.17	84.30
CH	71.26	70.00	83.18
R	86.17	82.74	92.30

According to Table 2 analysis, it can be seen that using this system has a higher spelling accuracy than methods 1 and 2, and shows a significant upward trend, indicating that the application of this system can improve the accuracy of natural speech spelling.

4.3 System Error Correction Test Results

In the course of the test, both the conventional and the new one are recorded, as illustrated in Table 2. Based on the results of Table 2, this paper shows that the English Language

Automatic Sound Correcting System can be used to verify the accuracy of the text. These three values show that the author's designed English oral automatic pronunciation proofreading system can effectively proofread English oral.

Table 2. Test Results/%

Proofreading error types	accuracy	recall	F value
traditional method	60.42	61.32	60.87
Author's Method	95.12	93.2	94.15

According to the analysis of Table 2, the robustness of the English oral automatic pronunciation proofreading system designed by the author is extremely high, indicating that the system is very stable.

5 Conclusion

The author conducts research on the design and implementation of a natural speech error correction and feedback system based on artificial intelligence technology, and proposes a complete solution from system structure design, natural speech feature extraction, speech recognition model to error correction and scoring module. By introducing the Dynamic Time Warping (DTW) algorithm and Hidden Markov Model (HMM), accurate recognition and error correction of natural speech signals have been achieved, effectively improving the robustness and reliability of the system. Meanwhile, utilizing DSP chips to design front-end support units significantly enhances the efficiency and real-time performance of speech processing. Research has shown that the application of AI based speech correction systems in English oral teaching has significant advantages, providing learners with efficient, accurate, and personalized feedback, overcoming the shortcomings of traditional teaching models such as strong subjectivity and low efficiency. Through endpoint detection, feature extraction, and multi model fusion analysis of natural speech signals, this system achieves precise localization and correction of pronunciation errors, effectively improving the efficiency of English oral learning and learners' language expression ability.

Acknowledgment. This paper is supported by the Beijing Higher Education "Undergraduate Teaching Reform and Innovation Project."

References

1. Xu, C.: Speech feature extraction in broadcast hosting based on fluctuating equation inversion. J. Adv. Comput. Intell. Intell. Inform. **28**(4), 39 (2024)
2. Mori, D., Ohta, K., Nishimura, R., Ogawa, A., Kitaoka, N.: Recognition of target domain japanese speech using language model replacement. EURASIP J. Audio Speech Music Process. **2024**(1), 1–14 (2024)

3. Chandra, S., Jayamukesh, Singh, G.K., Verma, A.R.: Size and inference time optimized automatic speech recognition model. In: International Conference on Applied Soft Computing and Communication Networks. Springer, Singapore (2024)
4. Mishra, S., Bhatnagar, N., Prakasam, P., Sureshkumar, T.R.: Speech emotion recognition and classification using hybrid deep CNN and bilstm model. Multimed. Tools Appl. **83**(13), 37603 (2024)
5. Jing, W.: Speech recognition sensors and artificial intelligence automatic evaluation application in English oral correction system. Meas. Sens. **32**, 101070 (2024)
6. Garneau, N., Bolduc, O.: The state of commercial automatic french legal speech recognition systems and their impact on court reporters et al (2024)
7. Tac, B.: Multilevel hybrid handcrafted feature extraction based depression recognition method using speech. J. Affect. Disord. **364**, 9–19 (2024)
8. Ha, S., Lee, S.: Feature extraction and analysis of ai-based speech signals for auditory rehabilitation. Asia-pacific J. Converg. Res. Interchange (2023)
9. Anmella, G., et al.: Automated speech analysis in bipolar disorder: the caliber study protocol and preliminary results. J. Clin. Med. **13**(17), 4997 (2024)
10. Jo, J., Kim, S.K., Yoon, Y.C.: Text and sound-based feature extraction and speech emotion classification for Korean. Int. J. Adv. Sci. Eng. Inf. Technol. **14**(3), 873 (2024)
11. Ahmed, I., Irfan, M.A., Iqbal, A., Khalil, A., Siddiqui, S.I.: Efficient feature extraction and classification for the development of Pashto speech recognition system. Multimed. Tools Appl. **83**(18), 54081 (2024)
12. Yu, L., Xu, F., Zhou, Q.K.: Speech emotion recognition based on multi-dimensional feature extraction and multi-scale feature fusion. Appl. Acoust. **216**(Jan), 109752.1-109752.10 (2024)
13. Boulal, H., Hamidi, M., Abarkan, M., Barkani, J.: Amazigh CNN speech recognition system based on Mel spectrogram feature extraction method. Int. J. Speech Technol. **27**(1), 287 (2024)
14. Kar, P., Debbarma, S.: Multilingual hate speech detection sentimental analysis on social media platforms using optimal feature extraction and hybrid diagonal gated recurrent neural network. J. Supercomput. **79**(17), 19515 (2023)
15. Ghorpade, T., Shinde, S.: Itts model: speech generation for image captioning using feature extraction for end-to-end synthesis. Int. J. Intell. Syst. Technol. Appl. **21**, 176–198 (2023)
16. Zhou, X., Zhang, Y., Wang, Y., Tian, J., Xu, S.: Pyramid feature attention network for speech resampling detection. Appl. Sci. **14**(11), 2076–3417 (2024)

Deep Learning-Based Motion Correction and Training Assistance for Track and Field Athletes

Chuan Huang and Mi Chen[✉]

Chongqing College of Architecture And Technology, Chongqing, China
chenmiwin@21cn.com

Abstract. In track and field training, precise motion analysis is a crucial learning process that plays a vital role. Traditional training methods lag behind in capturing motion features, leading to inconsistencies in performance evaluation. To enhance the accuracy of motion correction while improving detection efficiency, this study proposes a deep learning-based system designed to help track and field athletes optimize their technical movements. The system optimizes the CNN network structure and combines batch normalization technology to accurately extract and identify erroneous movements during training, providing athletes with immediate feedback and targeted correction suggestions. The experimental findings indicate that all four TPR parameters outperform the other two methods, with the detection accuracy of this approach exceeding 97%, thereby confirming the exceptional effectiveness of deep convolutional neural networks. This system can achieve high recognition accuracy in different sports events, effectively reducing the problem of human error in traditional teaching methods. The system functions include real-time image acquisition, deep learning analysis, intelligent scoring, and online guidance, providing more efficient and intelligent technical support for physical education teaching and training.

Keywords: Deep learning · Convolutional neural network · Physical education teaching · Error action detection

1 Introduction

The technical movements of track and field athletes are not just simple repetitions of body movements, but also a display of technique that includes strength, speed, and coordination. These technical details are crucial in every leap and stride on the field. In the field of competitive sports, especially in track and field competitions, athletes' performance often depends on the precision and efficiency of their movements. For example, whether athletes can quickly and accurately complete a series of movements such as starting, accelerating, and sprinting directly affects their final results and position on the track [1]. Therefore, sports trainers must conduct in-depth research and master scientific training methods, as well as use advanced technical means to help athletes correct movements and improve skills. This includes but is not limited to targeted training of

© The Author(s) 2026

P. Siarry et al. (Eds.): WCNA 2024, LNEE 1550, pp. 173–183, 2026.

https://doi.org/10.1007/978-981-95-6946-5_18

muscle groups, fine adjustments to exercise rhythm and breathing control, and regulation of athletes' psychological states. By implementing such a comprehensive training strategy, athletes can effectively improve their motor skills and achieve better results in competitions [2].

Recently, with the fast progress of AI, particularly in the area of Deep Learning, Computer Vision and Movement Analysis have been paid more and more attention. Not only do they have the ability to identify the player's actions, but they can also analyze and evaluate their performance in real time. In particular, Convolutional Neural Networks (CNN) and Long Short Term Memory Networks (LSTM) have been used in the field of sports training. CNN is widely praised for its excellent action recognition ability, which can extract key information from video clips, such as the position of the ball, the body posture of athletes, etc. [3]. LSTM is good at processing time series data and can capture the dynamic changes and skill development trends of athletes during competitions. These advanced algorithms enable coaches to more accurately identify deficiencies in athletes during training and provide timely and effective feedback [4].

2   Literature Review

In recent years, deep convolutional neural networks have shown great promise in many applications such as image classification. Unfortunately, because of the bad hierarchy of the model and the lack of ability to deal with motion data, it is not suitable for PE training. Lianju et al. presented a new approach for detecting badminton leaping by means of an enhanced deep learning algorithm. Make a collection of badminton video, follow the motion object of badminton match, get the player's active zone. Based on the 3D convolution neural network, they extracted the motion characteristics of badminton players from their active regions [5]. Babazaki et al. introduced a new approach to detect spatial and temporal interactions, which improves the performance of object recognition. Zero-Sample spatiotemporal action detection refers to identifying the behavior of individuals in a video and determining the time and place of those activities without having previously been trained for these particular activities. Deep convolutional neural networks such as CLIP have shown remarkable performance on a wide range of computer vision tasks, but have difficulty finding local features and relations [6]. Hongwei, W. et al. studied a new approach to correct the position of the swimming pose in real time by using a new method of vision motion recognition. First of all, it is used to digitalize the body image. First of all, the text is classified into a frame, and then CNN is used to get the annotation image. Then, the labeled frontal posture data is pre-processed by translation and normalization [7]. Based on this research, the author designed an intelligent sports teaching and training system using deep convolutional neural networks (CNN), which can automatically identify erroneous movements during the training process, provide real-time feedback to athletes, and improve the scientific and intelligent level of training. In response to the complexity and temporal characteristics of movements in sports training, the author optimized the CNN structure and combined batch normalization technology to improve the accuracy and generalization ability of error action detection.

3 Research Methods

3.1 Software Composition and Function Introduction

3.1.1 Software Composition

.The purpose of this teaching platform is to provide correct movement guidance for fitness beginners. The composition and functional process of the system are shown in Fig. 1. Teaching starts with the most basic action correction, and the system provides feedback information to provide suggestions for users. At the same time, it can also interact with practice to provide deeper guidance for users. This requires the implementation of action collection on the client side, communication with the server side, and display on the screen; Implement analysis and processing of received data on the server side, and provide feedback on the results to the client. Image recognition processing technology is the core technology of software, which uses electronic products with cameras to collect image information and uses neural network algorithms to dynamically analyze images. And feedback the results to the user's handheld electronic device, helping the user adjust their posture and achieve the goal of error correction [8, 9].

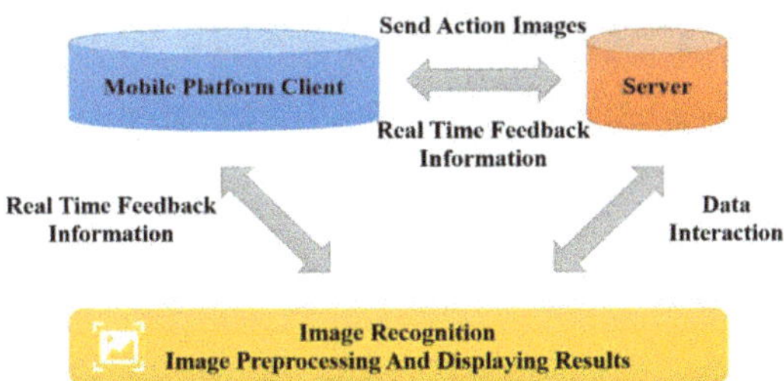

Fig. 1. Schematic diagram of system composition and functional flow

3.1.2 Function Introduction

This system has real-time action monitoring and action error correction functions. When the similarity between the user's action and the system's action library is less than 80%, the system will display alarm prompts and voice reminders, achieving the function of timely correcting the user's incorrect posture.

(1) Coach online teaching mode. When users imitate actions in videos for learning, the system will also rate each action of the user, and users can judge their exercise effectiveness based on the high or low score. Coaches can watch the learning progress of each student through the shadow coaching platform. When there are too many errors in the user's actions, the system will alert and prompt, and the coach can provide online guidance.

(2) Live streaming mode Coaches can demonstrate their teaching skills through live streaming and also use this platform to showcase their abilities Users can share their videos or live stream on the platform to become internet celebrities.

(3) Multiplayer mode. Users can work out with friends, play games together, and enjoy the joy of fitness together.

(4) Ranking mode. Users can understand their level of strength and stimulate their competitiveness through qualifying matches. In order to reach the next level, it takes some time to make users more obsessed with fitness and enhance their stickiness [10].

3.2 Software Design Proposal

The communication process of the system is shown in Fig. 2. The client is an app developed based on Android 6.0, mainly responsible for collecting human body images and sending pictures. Use the built-in camera on electronic mobile devices to capture motion images, and then send them to the server via wireless network. Call the system camera and screen capture function to obtain the current preview image, use the Android system timer to determine the frequency of image acquisition, compress the image matrix, and send data using TCP/P communication protocol. When the service IP and port number are obtained, data transmission can be carried out. The client is in standby mode after successful sending, waiting for a response from the server [11].

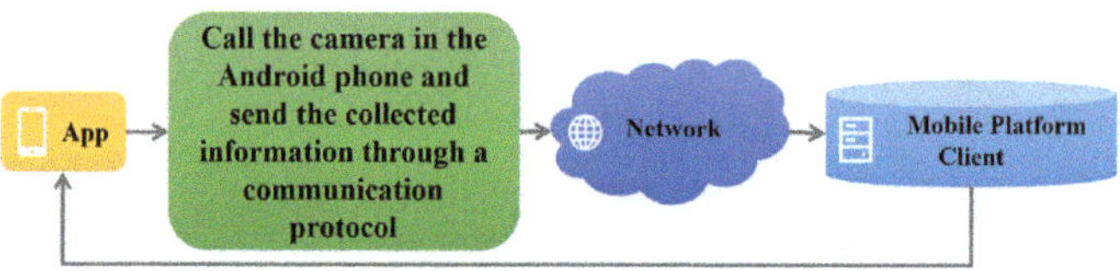

Fig. 2. System Communication Process

The server is responsible for receiving and processing data and providing feedback. The server is divided into two parts, one responsible for communication between both parties, and the other for performing image recognition processing. The communication part simultaneously listens to two ports and temporarily saves the image data locally after receiving it, waiting for the image recognition result. After obtaining the information, it is displayed on the electronic device screen, and a voice broadcast function is also added to make the platform more intelligent;

3.3 Deep Convolutional Neural Networks

Deep neural networks have been proposed as an approach to solving the problem of multiple layers in order to solve the problem. Since the weights of the convolutional neural networks are shared, it is easy to compute them. The results show that they can be applied to analyzing time sequence data, which can alter the capability of the PE Learning Learning Error Detecting Mode by means of the deep and wide network architecture. Convolutional neural networks have been widely used in the field of computer vision, especially in the field of computer vision. CNN is an efficient way to extract the key characteristics of the input data by means of a special layer architecture, such as convolution, pooling, and full connection. The kernel of this approach is to utilize convolution to extract the local features from the original data, and to improve the nonlinear properties of the neural network [12]. The neural network structure diagram is shown in Fig. 3.

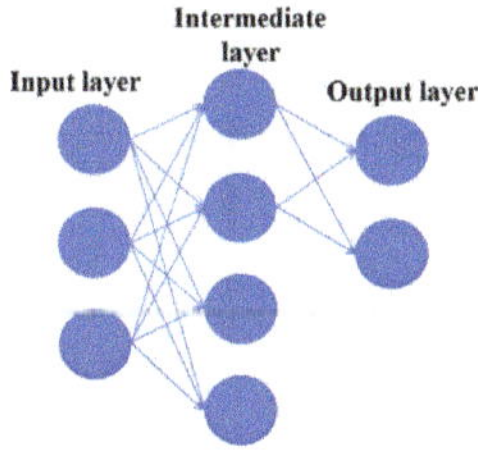

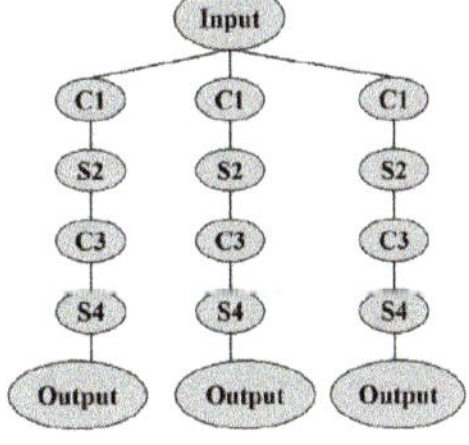

Fig. 3. Neural Network Structure Diagram   **Fig. 4.** Structure of Deep Convolutional Neural Network

3.3.1 Structure of Deep Convolutional Neural Networks

Convolutional neural networks consist of three levels: the first, the second, the second, and the second, the third. The hidden layer consists of multiple convolution and pooling layers. The architecture of this model is illustrated in Fig. 4.

First, the original data is inputted into the input layer, and then, in the convolution (C1), the input data is processed by convolution of the convolution of the convolution to get the respective convolution. Then, by means of the pooling layer (S2), a combined characteristic map is acquired. Then, a simulation of C1 and S2 is carried out in the hidden layer (C3, S4). Through the establishment of convolutions and pooling, it is possible to extract the wrong behavior characteristics efficiently and to increase the tolerance of the distorted invariant characteristic in PE teaching and training. Meanwhile, we can get lots of characteristic data by improving the resolution of the picture. Finally, we can send the test results by full connection [13, 14].

3.3.2 Convolutional Layer

To make sure the input and output dimensions are identical, pre-processed acceleration sensor data x, y and z (depth 3) need to be added as input data. The conversion of the same convolutions has no effect on their weights in the convolution, which is common in the x axis data. It is shown that this method is effective in decreasing the number of parameters in deep convolutional neural networks. Convolutional neural networks are widely used to extract features automatically. Based on the convolution kernel, we can get all kinds of information about the PE training picture by means of the convolution kernel.

If the depth and width of the convolution core are set as fh and fw, a two-dimensional convolution is obtained (1):

Using ReLU function, we derive the input and output of the entire convolution layer (2):

3.3.3 Maximum Pooling Layer

Pooling is used for selection and reduction of output dimensions. During the detection of improper motion in PE, we adopt the maximal pooling policy, which takes 2×2 and s as the step size. Pooling core height and width are Ph Pw, Get the largest pooling (2):

$$
y_{n,m} = A \left\{
\begin{array}{cccc}
x_{n,m} & x_{n+1,m} & \cdots & x_{n+f,m} \\
x_{n,m+1} & x_{n+1,m+1} & \cdots & x_{n+f,m1} \\
\vdots & \vdots & \vdots & \vdots \\
x_{n,m+f_h} & x_{n+1,m+f_h} & \cdots & x_{n+f_w,m+f_h}
\end{array}
\right\}
\tag{1}
$$

$$
y_{i,j} = \max \left\{
\begin{array}{cccc}
x_{is,js} & x_{is,js+1} & \cdots & x_{is,js+p} \\
x_{is+1,js} & x_{is+1,js+1} & \cdots & x_{is+1,js+p} \\
\vdots & \vdots & \vdots & \vdots \\
x_{is+p_h,js} & x_{is+p_h,js+1} & \cdots & x_{is+p_h,js+p_w}
\end{array}
\right\}
\tag{2}
$$

$$
Y = \mathrm{relu}(\sigma(WB + b))
\tag{3}
$$

3.4 Feature Extraction and Detection Result Output

The precision of the detection of wrong motion in athletic training is dependent on the depth of NN, which is positively correlated with the performance. Deep neural networks can compute the characteristics of all wrong motion data, and the more accurate the results are, the better the performance is. In the course of deepening the network, it is easy for the gradient to disappear, which results in the degradation of the network performance. To overcome these difficulties, a new approach is proposed to extract the fine characteristics from the PE training sample. The algorithm is called Resnet101. In order to improve the efficiency of the network, ResNet was adopted to batch increase the normalization level and the remaining block in the convolution layer and the pooling layer. In this paper, a batch normalization method is used to combine the data from the network level into the PE teaching and training. Batch normalization is represented as (4):

Among them, x represents the vector input to a certain layer within the deep convolutional neural network; X represents the sample set, and an input group of the overall training set can be described by X = {x1, x2,…, xN}.

Batch normalization needs a lot of calculation to deal with each level's input, and it also needs a lot of time to get the covariance matrix. For this, two simplified ways of improving:

(1) Replacing the joint normalization of each dimension of data with independent batch normalization yields the following formula (5):

$$
\widehat{X} = norm(x, X)
\tag{4}
$$

$$
\hat{X}^{(k)} = \frac{x_i^{(k)} - \mathrm{E}\left[x^{(k)}\right]}{\sqrt{\mathrm{var}\left[x^{(k)}\right]}}
\tag{5}
$$

The k-th dimension of the input specimen is expressed as x (k), the expected value is expressed as E (x (k)), and the variance is expressed as var (x (k)). Independent batch normalization is an effective method to speed up the learning rate of the network, but it is not able to ensure the stability of the original description. To keep the batch normalization procedure constant, we add parameter λ (k) and β (k) into the th dimension of every input sample:

$$y^{(k)} = \lambda^{(k)} \hat{X}^{(k)} + \beta^{(k)} \tag{6}$$

In this case, $\lambda(k)$ represents the variance of the k-th dimension of the input sample following a scale transformation, while $\beta(k)$ corresponds to the expected value of the k-th dimension after a translation transformation. Both $\lambda(k)$ and $\beta(k)$ serve as key parameters in adjusting the input sample. By incorporating these parameters into the model during network training, it becomes possible to significantly minimize the output error of deep neural networks.

(2) Random gradient training of deep convolutional neural networks is performed using micro batch samples. By computing the estimated mean and variance of each level on each sample, gradient propagation in the opposite direction can be achieved [15].

Suppose that the mini-lot is B, the sample size is given by M, and the input dimension at some level is expressed as x. Normalization by dimension is represented as:

$$BN_{\lambda,\beta} : x_1, \cdots, x_m \rightarrow y_1, \cdots, y_m \tag{7}$$

Using trained deep convolutional neural networks for error detection in physical education training, accurate error detection results can be obtained.

4   Results Analysis

To validate the effectiveness of the method based on deep convolutional neural networks for detecting incorrect movements during physical education training, a group of students from a specific physical education college was chosen as the experimental sample. The parameter configurations for the deep convolutional neural network are outlined in Table 1.

Table 1. Parameter Settings for Deep Convolutional Neural Networks

Number of layers	Layer function	Number of layers	Function
1	Input layer	10	Maximum pooling layer
2	Convolutional layer 3-64	11	Convolutional layer 3-512
3	Convolutional layer 3-64	12	Convolutional layer 3-512
4	Maximum pooling layer	13	Maximum pooling layer
5	Convolutional layer 3-128	14	Convolutional layer 3-512
6	Convolutional layer 3-128	15	Convolutional layer 3-512
7	Maximum pooling layer	16	Fully connected layer 3072
8	Convolutional layer 3-256	17	Fully connected layer 1024
9	Convolutional layer 3-256	18	Output layer

The study employed three methods—the deep convolutional neural network-based detection method, Method 1 (two-dimensional wavelet packet action detection), and Method 2 (spatial clustering-based action detection)—to identify incorrect actions during physical education instruction and training. The effectiveness of these methods in detecting erroneous movements was evaluated as the number of experiments increased. A non-public dataset was used for the experiments, with machine vision technology capturing the physical movements during sports teaching and training. These captured movements were then denoised and enhanced to boost the accuracy of the results. The three methods were applied to detect errors in the experimental samples, and the error detection rates for each method were compared. The overall performance of the methods in detecting errors during physical education teaching and training was assessed based on these comparative results, as shown in Table 2.

Table 2. Comparison of Detection Error Rates of Different Methods

Number of experiments/time	Error rate/%		
	Author's Method	Method 1	Method 2
15	0.020	0.077	0.14
25	0.018	0.088	0.142
35	0.044	0.152	0.248
45	0.012	0.123	0.198
55	0.031	0.085	0.131
65	0.074	0.087	0.132

As shown in Table 2, the detecting error of PE in PE teaching and training is still very low with the increase of experimental quantity. In PE instruction, the mean error ratio of PE instruction and practice was about 0.033%, but Method 1 was about 0. 102%, and Method 2 was about 0. 167%, which showed that it was more accurate. The application of this approach to the detection of wrong motion in PE instruction can make it possible to control the error in a rational scope.

To validate the validity and reliability of the proposed approach in the detection of wrong motion during athletic practice, we made a quantitative comparison between ACC (precision), TPR (sensitivity), FPR (specificity) and PPV (positive prediction ratio). The concrete calculation equation is described as follows:

$$ACC = \frac{TP + TN}{TP + TN + FP + FN} \tag{8}$$

$$TPR = \frac{TP}{TP + FN} \tag{9}$$

$$FPR = \frac{TP}{TN + FP} \tag{10}$$

$$PPV = \frac{TP}{TP + FP} \tag{11}$$

Of these, TP indicates the number of specimens determined to be positive, that is, the actual count of the positive specimens; FP indicates the count of the specimens determined to be positive, but in reality it is a negative sample count; TN indicates the count of the negative specimens determined to be negative, and FN indicates that the number of specimens determined to be negative, but is in reality a positive number. The more ACC and TPR are, the less FPR and PPV are.

The test results for the detection of improper motion during PE instruction are given in Table 3.

Table 3. Detection Results of Incorrect Actions in Physical Education Teaching and Training

Test indicators	Author's Method	Method 1	Method 2
ACC	0.973	0.842	0.825
TPR	0.943	0.843	0.842
PPV	0.081	0.243	0.225
FPR	0.006	0.031	0.106

TPR technology outperforms two solutions adopted by other competitors in all four parameter indicators, demonstrating its outstanding performance. It is worth mentioning that the approach presented in this paper has made significant progress in detecting precision, with an accuracy of more than 97 percent. Not only does this prove that the performance of deep learning models is very high, but it also shows that there is a great deal of progress in the field of vision.

5 Conclusion

The author developed an advanced system for detecting errors in sports teaching and training actions using deep convolutional neural networks (CNN). By optimizing the network architecture, the system significantly enhanced both the accuracy and efficiency of action recognition. The research results indicate that using CNN models combined with batch normalization techniques can effectively extract erroneous motion features in sports training and achieve accurate detection, providing athletes with more targeted correction solutions. The author collaborates with the client and server to use real-time image acquisition and deep learning algorithms to analyze athletes' training movements. This not only provides real-time feedback, but also has various functions such as intelligent scoring, online guidance, and live teaching, further enhancing the intelligence level of physical education teaching. The experimental results show that the system can achieve high recognition accuracy in different sports events and effectively reduce the misjudgment problem caused by human factors in traditional teaching methods, providing scientific and accurate technical support for sports training.

References

1. Agarwal, H., Mishra, D., Kumar, A.: A deep-learning approach for turbulence correction in free space optical communication with laguerre–gaussian modes. Optics Commun. **556**, 130249 (2024)

2. Ding, F., Zhang, D.: Spatial correction and deblurring fusion algorithm for vehicle license plate images based on deep learning. Int. J. Model. Simul. Sci. Comput. **15**(05), 2430003 (2024)

3. Hewlett, M., Petrov, I., Johnson, P.M., Drangova, M.: Deep-learning-based motion correction using multichannel MRI data: a study using simulated artifacts in the fastmri dataset. NMR in Biomed. **37**(10), E5179 (2024)

4. Kashtanova, V., Pop, M., Ayed, P.S.M.: Simultaneous data assimilation and cardiac electro-physiology model correction using differentiable physics and deep learning. Interface Focus **13**(6), 20230043 (2023)

5. Lianju, L., Haiying, Z.: Correction to: research on badminton take-off recognition method based on improved deep learning. J. Ambient. Intell. Humaniz. Comput. **15**(12), 4101 (2024)

6. Babazaki, Y., Shibata, T., Takahashi, T.: Zero-shot spatio-temporal action detection by enhancing context-relation capability of vision-language models. In: International Conference on Pattern Recognition. Springer, Cham (2025)

7. Hongwei, W.: Real-time swimming posture image correction framework based on novel visual action recognition algorithm. In: 2023 8th International Conference on Communication and Electronics Systems (ICCES), pp. 1714–1718 (2023)

8. Bolzonello, L., Bruschi, M.: Nonlinear optical spectroscopy of molecular abemblies: what is gained and lost in action detection? J. Phys. Chem. Lett. **14**(50), 11438–11446 (2023)

9. Shang, Z., Liu, B.: Facial action unit detection based on multi-task learning strategy for unlabeled facial images in the wild. Expert Syst. Appl. **253**, 124285 (2024)

10. Rodrigues, N.R.P., et al.: Fusion object detection and action recognition to predict violent action. Sensors **23**(12), 14248220 (2023)

11. Thrall, S.F., et al.: A comparison of wavelet-based action potential detection from the neuroamp and the iowa bioengineering nerve traffic analysis system. J. Neurophysiol. **131**(6), 1168 (2024)

12. Su, C., Wei, J., Guan, K.Y.L.: A novel model for fall detection and action recognition combined lightweight 3d-cnn and convolutional LSTM networks. Pattern Anal. Appl. **27**(1), 31–316 (2024)

13. Tallec, G., Dapogny, A., Bailly, K.: Multi-order networks for action unit detection. IEEE Trans. Affect. Comput. **14**(4), 2876–2888 (2023)

14. Wang, P., Zeng, F., Qian, Y.: A survey on deep learning-based spatio-temporal action detection. Int. J. Wavelets Multiresolut. Inf. Process. **22**(4), 2350066 (2024)

15. Jung, S., Jeoung, J., Hong, D.-E.: Visual–auditory learning network for construction equipment action detection. Comput. Aided Civ. Infrastruct. Eng. **38**(14), 1916–1934 (2023)

Multimodal Data-Driven Intelligent Decision Support for Traditional Chinese Medicine (TCM) Acupuncture and Moxibustion

Xiaojun Li[✉]

Shandong University of Traditional Chinese Medicine, Jinan, Shandong Province, China
bybj131419@163.com

Abstract. In view of the low efficiency of manual analysis when dealing with massive medical records in clinical practice, this study constructs an AI-powered intelligent analysis and decision support system for TCM acupuncture based on multimodal data mining. The system integrates multiple functional modules to enhance diagnostic accuracy and treatment efficiency. The multimodal data fusion module leverages an autoencoder model to integrate tongue images, facial images, and pulse information into a comprehensive feature vector, combining key textual descriptions to generate a holistic patient health profile. The feature extraction and selection module applies decision trees (DT) to identify the most critical features for acupuncture treatment decision-making. The intelligent analysis and decision support module employs a support vector machine (SVM)-based prediction model to process the extracted features and generate personalized acupuncture treatment plans. The user interface and interactive module enable seamless data entry, treatment recommendation visualization, and historical query access, ensuring ease of use for physicians. Experimental results demonstrate a significant reduction in diagnosis and treatment time compared to traditional methods, while achieving a treatment accuracy exceeding 98%. The system also outperforms conventional approaches in symptom improvement rates and quality of life scores, with symptom improvement exceeding 98% in numerous cases and the highest recorded quality of life score reaching 10. By integrating advanced data mining techniques and intelligent decision support, this system not only enhances the efficiency and accuracy of TCM acupuncture therapy but also provides a scalable framework for future AI-driven clinical applications.

Keywords: multimodal data mining · acupuncture treatment · diagnosis and treatment time · symptom improvement rate · quality of life

1 Introduction

Although traditional Chinese medicine acupuncture diagnosis and treatment methods have a long history, they rely on classical Chinese medicine theories and the rich experience of clinicians, and are highly subjective and inefficient. Multimodal data mining technology has made progress in many fields, providing new ideas and methods for the intelligent development of Chinese medicine acupuncture therapy.

© The Author(s) 2026

P. Siarry et al. (Eds.): WCNA 2024, LNEE 1550, pp. 184–193, 2026.

https://doi.org/10.1007/978-981-95-6946-5_19

This paper constructs an intelligent analysis and decision support system with multiple modules. The autoencoder model is used to achieve deep fusion of data from different channels and in different forms, improving the quality and representativeness of the data. Through DT, the most critical features for acupuncture treatment decision-making are screened and optimized to improve the representativeness and discrimination of the features. A prediction model is built based on support vector machine (SVM) to generate personalized acupuncture treatment plans, assisting doctors in formulating more scientific and effective treatment plans. Finally, through comparative experiments, the system's advantages in diagnosis and treatment efficiency, accuracy and improvement of patient treatment effects are verified.

2 Related Work

The application and research of multimodal data are increasingly becoming a key force in promoting innovation in various fields. Wu et al. [1] constructed a data set based on the visual, physiological and other multimodal data of individual learners in a synchronous classroom, extracted action and emotional features from it, and formed a concentration evaluation model. Cui et al. [2] proposed a multimodal data modeling technology for cloud-edge-end collaboration based on the data types of the three layers of cloud-edge-end data. They gave a tuple-based multimodal data model definition and designed six base classes to solve the problem of difficulty in unified representation of multimodal data. Finally, they gave a demonstration application of a multimodal data model for cloud-edge-end collaboration. Xue et al. [3] designed a research framework for online learning cognitive load assessment based on multimodal data, which includes three parts: multimodal data collection, multimodal feature extraction, and evaluation model construction. Jiang et al. [4] reviewed existing research on smart classroom behavior analysis, and constructed a three-dimensional analysis framework using sound, image, and text data, capturing changes in verbal activities, body movement, and technology use. Hu et al. [5] determined the target state vector by utilizing the state update function and the formed candidate image template; an additional weighting stage was added to the particle filter (PF) so that the PF could adaptively synchronize multimodal data streams to achieve robust target tracking.

Bae et al. [6] used meta-analysis to evaluate the effects of acupuncture and moxibustion on intestinal microbiota. Yan et al. [7] analyzed clinical research literature on acupuncture, focusing on methods, efficacy and safety. Masuda et al. [8] discussed traditional Japanese acupuncture, especially the global spread of Hokushin-kai style. Lee et al. [9] analyzed patient visits to acupuncture departments in South Korea. Kim et al. [10] reported a case of oculomotor nerve palsy treated with acupuncture. Based on these studies, this paper combines multimodal data mining with acupuncture knowledge to build an intelligent decision support system. The system analyzes physiological indicators, imaging data, and clinical manifestations to offer personalized treatment suggestions and assist in formulating effective plans.

3 Methods

3.1 Overall System Architecture

The overall architecture of the intelligent analysis and decision support system for traditional Chinese medicine acupuncture therapy based on multimodal data mining consists of multiple key modules, each of which has a specific function and works together to achieve the overall goal of the system [11, 12]. The data acquisition module collects patients' medical data from a variety of sources, including integrating electronic medical record system interfaces to obtain medical history records, using high-definition cameras and image processing technology to capture tongue and facial images, and measuring pulse information through pulse sensors. The data preprocessing module uses noise filtering algorithms to clean up raw data, unify data formats through standardized processes, and perform preprocessing operations on text data. The multimodal data fusion module uses the autoencoder model to fuse multimodal data such as tongue image, facial image, pulse information, etc. to form a comprehensive health portrait of the patient. The feature extraction and selection module uses the decision tree algorithm to select the most critical features for acupuncture treatment decision-making from the comprehensive feature vector [13]. The intelligent analysis and decision support module builds a prediction model based on the support vector machine algorithm and outputs a personalized acupuncture treatment plan. Finally, the user interface and interaction module designs a simple and intuitive operation interface, providing functions such as data entry, treatment recommendation viewing, and history query to ensure that physicians can easily get started and use the system efficiently [14]. The system architecture is shown in Fig. 1:

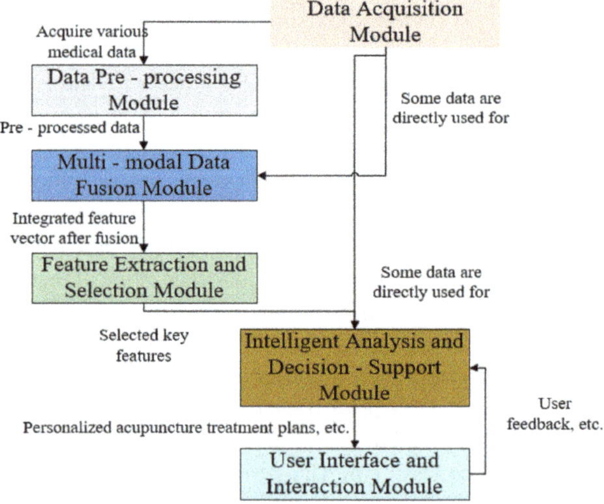

Fig. 1. System Architecture

These modules work together to provide intelligent analysis and decision support for traditional Chinese medicine acupuncture therapy.

3.2 Data Acquisition Module

The data acquisition module constitutes the starting point and core of the data processing of the system proposed in this paper. This module is carefully designed to integrate with the interface of the electronic medical record system, and realizes seamless connection with the electronic medical record system used in modern medical institutions. Through this connection, the system accesses and integrates the patient's physical data, detailed medical history, past diagnosis, treatment plan and any relevant medical records in real time, ensuring the timeliness and integrity of the data. In addition, the data collection module also integrates cutting-edge tongue image and facial diagnosis image collection technology. Using a high-definition camera, the system accurately captures the patient's tongue color, texture, shape and other tongue characteristics, as well as facial skin color, luster, facial features and other details, which are important bases for judging physical constitution and condition in the TCM theory system. At the same time, the data acquisition module also integrates pulse information collection technology, which is another essence of TCM diagnosis. Through pulse sensors such as precision pressure sensors and photoelectric sensors, the system monitors the patient's pulse in real time and accurately, capturing subtle changes in pulse fluctuations, rhythms, and strength. These pulse characteristics are important for the formulation of TCM acupuncture therapy.

3.3 Data Preprocessing Module

The data preprocessing module first performs the noise filtering task, using the noise filtering algorithm to identify and eliminate invalid, abnormal, and redundant information in the data such as outliers caused by equipment failures and human input errors, thereby improving the purity of the data.

Next, data standardization is performed to convert the data into a unified format and range, and data from different sources are encoded into a unified format, numerical data are scaled to a specific range, and missing values are handled using filling methods.

For text data, the module uses word segmentation technology to segment continuous text into independent vocabulary units. The word segmentation process not only considers the boundaries of vocabulary but also combines professional terms and context information in the field of traditional Chinese medicine to ensure the accuracy and professionalism of word segmentation.

Finally, feature information useful for subsequent analysis is mined from the text data - extracting key words, analyzing sentiment, identifying topic categories, and providing rich structured data support for intelligent analysis and decision-making of traditional Chinese medicine acupuncture therapy. Through feature extraction, the system understands the text content and mines valuable information hidden behind the data.

3.4 Multimodal Data Fusion Module

The multimodal data fusion module uses the autoencoder model to achieve deep fusion of data from different channels and in different forms. Based on its unique encoder and decoder structure, the autoencoder model not only learns the intrinsic representation of the data but also retains the key information of the data while reducing the dimension, removes noise and redundancy, and improves the quality of the data:

$$z = f(Wx + b) \tag{1}$$

x is the input data, W is the weight matrix, b is the bias term, f is the activation function, and z is the encoded feature representation.

In the process of multimodal data fusion, the model first preprocesses the data of each modality, and then extracts deep features from the data of each modality. These features can reflect the intrinsic structure and key information of the data. In the feature fusion stage, the module organically integrates the features of each modality, such as tongue image, facial image and pulse information, to generate a comprehensive feature vector. This vector not only contains the unique information of each modality data but also reveals the inherent connection between them, providing comprehensive and accurate data support for the construction of the patient's health portrait:

$$h = g(z_1, z_2, ..., z_n) \tag{2}$$

$z_1, z_2, \ldots, z_n$ are the feature representation of each modal data, g is the fusion function, and h is the comprehensive feature vector.

Through the application of autoencoder models and the generation of comprehensive feature vectors, the multimodal data fusion module successfully transforms complex and multi-dimensional data into concise and powerful feature representations. This process not only improves the accuracy and efficiency of data analysis but also provides a scientific basis for the personalized treatment of TCM acupuncture and moxibustion, and promotes the intelligent development of TCM diagnosis and treatment technology. Table 1 shows the feature screening and optimization data:

Table 1. Feature screening and optimization data:

Feature Name	Information Gain (IG)	Gini Coefficient	Feature Importance (%)	Optimized Feature Weight
Tongue Color	0.85	0.12	25	0.92
Pulse Strength	0.78	0.15	22	0.88
Facial Color	0.72	0.18	20	0.85
Coating Texture	0.68	0.20	18	0.80
Pulse Frequency	0.65	0.22	15	0.75

3.5 Feature Extraction and Selection Module

The feature extraction and selection module uses the decision tree algorithm (DT) to analyze the comprehensive feature vector generated after multimodal data fusion to screen out the most critical features for acupuncture treatment decisions and optimize them.

DT builds the model by recursively partitioning the feature space. In the feature extraction and selection module, DT analyzes the importance of each feature in the comprehensive feature vector and their association with acupuncture treatment decision. Specifically, the algorithm evaluates the contribution of each feature in the classification or regression task, and measures the importance of the feature by calculating indicators such as information gain and Gini coefficient:

$$IG(T, A) = H(T) - \sum_{v \in A} \frac{|T_v|}{|T|} H(T_v) \tag{3}$$

IG(T,A) is the information gain of feature A, H(T) is the entropy of data set T, T_v is the data subset when feature A takes the value v, $|T_v|$ is the number of samples in the subset, and |T| is the total number of samples.

$$\text{Gini(T)} = 1 - \sum_{i=1}^{C} p_i^2 \tag{4}$$

p_i is the proportion of category i in the dataset T, and C is the total number of categories.

Based on these evaluation results, DT automatically screens out key features that have a significant impact on acupuncture treatment decisions.

After selecting key features, the module will also optimize these features. The purpose of optimization is to further improve the representativeness and discrimination of features and reduce redundant information. The optimization process reduces the dimension of features through principal component analysis (PCA) to remove the correlation between features.

Through the application of DT and the screening and optimization of key features, the feature extraction and selection module can effectively extract the most critical information for acupuncture treatment decision-making from the comprehensive feature vector.

3.6 User Interface and Interaction Module

The user interface and interactive module design focuses on simplicity and intuitiveness, avoiding complex elements and redundant information, aiming to enable users to quickly get started and use the system efficiently. The interface layout is reasonable, and important functions and information are clear at a glance, which is convenient for users to quickly locate and operate. At the same time, the module covers functions such as data entry, treatment recommendation viewing, history query, and user personalized settings to meet the diverse needs of users in different scenarios.

To improve the user experience, the user interface and interaction module adopt an interactive design, which supports users to interact with the system through natural actions such as clicking, dragging, and sliding, making the operation process smoother and more intuitive. In addition, the system also has responsive design capabilities, which can automatically adapt to different devices and screen sizes, ensuring that users can get a consistent and good user experience in different environments [15].

In terms of user experience optimization, the module provides real-time feedback mechanisms such as loading progress bar and operation prompt information, allowing users to grasp the system's operating status and operation results at any time, enhancing user trust and satisfaction. At the same time, the system also allows users to adjust the interface theme, font size, etc. according to personal preferences, further improving the comfort and satisfaction of use. In addition, in order to reduce the user's learning cost and usage threshold, the system also provides user education and guidance functions such as operation guides and video tutorials to help users quickly master the system's usage methods and skills.

4 Results and Discussion

4.1 System Performance

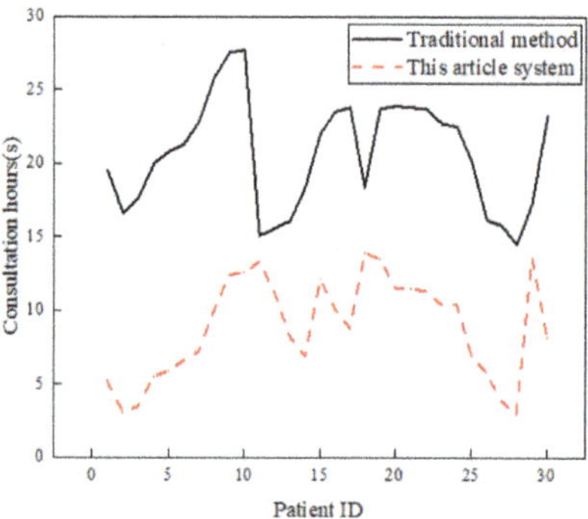

Fig. 2. Treatment time   **Fig. 3.** Diagnosis and treatment accuracy

Traditional Chinese medicine acupuncture treatment has a long history and has long relied on classical Chinese medicine theories and the rich experience of clinicians. In order to comprehensively evaluate the advantages and improvements of the new system, this paper designs a comparative experiment to directly compare the performance of this system with traditional treatment methods, especially in terms of diagnostic accuracy and diagnostic time. Through this comparison, we can see how the system can improve the efficiency and accuracy of diagnosis and treatment through technological innovation while maintaining the essence of traditional Chinese medicine. The results are shown in Figs. 2 and 3, respectively:

As shown in Fig. 2, there are differences in the diagnosis and treatment time between the traditional Chinese medicine acupuncture diagnosis and treatment method and the intelligent analysis and decision support system proposed in this paper. The diagnosis and treatment time of the traditional method ranges from 14.5 to 27.7 s, with an average of about 20.7 s, with large fluctuations, and some cases (such as cases 9 and 10) take a long time. The diagnosis and treatment time of the intelligent system ranges from 3 to 13.9 s, with an average of 8.9 s, which takes less time than the traditional method, and has less fluctuations, showing higher stability. The average diagnosis and treatment time of the intelligent system is about 57% shorter than that of traditional methods, and it can still maintain high efficiency when dealing with complex cases. For example, in Case 18, the traditional method took 18.4 s, while the intelligent system only took 13.9 s. This data comparison directly highlights the advantages of the intelligent analysis and decision support system of Traditional Chinese Medicine Acupuncture Therapy based on multimodal data mining in improving diagnosis and treatment efficiency. Not only does the intelligent system speed up the diagnosis and treatment process by reducing the average diagnosis and treatment time, but its smaller fluctuation in diagnosis and treatment time also indicates that it has higher stability and consistency in handling different cases.

Based on the data in Fig. 3, this paper divides the accuracy range into different ranges (below 90%, 90%–94.9%, 95%–97.9% and 98% and above), and counts the number of cases of the two methods in each range. The number of cases involved in the intelligent system in the accuracy range below 90% is 0, and in the accuracy range of 90%–94.9%, although the intelligent system also has a certain number of cases, the number of cases is relatively small, reflecting that the intelligent system is more inclined to provide a higher accuracy diagnosis. In the accuracy range of 95%–97.9%, the number of cases of the intelligent system increased, which proves that the system can provide very accurate diagnosis and treatment services in most cases. More impressively, in the accuracy range of 98% and above, the number of cases of the intelligent system far exceeds that of the traditional method, showing its ability and advantage in pursuing extreme accuracy. The accuracy of traditional methods is all below 90%. This comparison result deeply reveals the advantage of the system in the accuracy of Chinese medicine acupuncture diagnosis and treatment, and fully demonstrates the potential and application value of the system in improving the accuracy of diagnosis and treatment.

4.2 Effect of Multimodal Data Fusion

The effect of multimodal data fusion is undoubtedly a key indicator for achieving system effectiveness. By effectively integrating multimodal data from different channels and in different forms, the system can more comprehensively capture the patient's health status and disease characteristics, laying a solid foundation for subsequent intelligent analysis and accurate decision-making. This paper compares the system performance before and after multimodal data fusion, especially in key indicators such as disease recognition accuracy, personalized treatment plan generation rate, patient satisfaction, overall system efficiency, and overall data processing speed efficiency, as shown in Table 2:

Table 2. System performance

Metric	Before Multimodal Data Fusion	After Multimodal Data Fusion	Percentage Improvement
Personalized Treatment Plan Generation Rate	60%	90%	+50%

(continued)

Table 2. (continued)

Metric	Before Multimodal Data Fusion	After Multimodal Data Fusion	Percentage Improvement
Patient Satisfaction	8.2/10	9.5/10	+15.85%
Overall System Effectiveness	Moderate	High	Significant Improvement
Data Processing Speed	1,000 Cases/Hour	3,000 Cases/Hour	+200%

These data fully demonstrate the role of multimodal data fusion in improving the performance of intelligent analysis and decision support systems for traditional Chinese medicine acupuncture therapy.

5 Conclusion

This paper improves the efficiency and accuracy of diagnosis and treatment of traditional Chinese medicine acupuncture by integrating multimodal data such as tongue images. Through system design and experimental verification, this paper successfully solves the problems of improving diagnosis and treatment efficiency, improving diagnosis and treatment accuracy, and improving patient treatment effects. Although the system performs well in many aspects, its performance is highly dependent on the quality and quantity of multimodal data. Incomplete or inaccurate data may affect the analysis and decision-making of the system. The multimodal data fusion and feature extraction process involves complex models and algorithms, which may lead to high demand for computing resources and increased training time. In the future, the system will be further optimized to address existing limitations and expand its application scope. This includes introducing more data sources and improving data collection technology to improve the quality and integrity of multimodal data; reducing computing resource requirements to improve the real-time and response speed of the system; conducting clinical verification in more medical institutions and patient groups to evaluate the actual application effect and adaptability of the system; and introducing more modal data such as genetic data and metabolic data to further enrich the system's data foundation and improve the accuracy of analysis and decision-making. Based on this, we hope to further improve the performance and practicality of the system, promote the intelligent development of traditional Chinese medicine acupuncture therapy, and contribute to the modernization and internationalization of traditional Chinese medicine diagnosis and treatment technology.

References

1. Wu, F., Gao, S., Lai, S.: Analysis of concentration based on multimodal data in synchronous classroom. Mod. Distance Educ. Res. **37**(1), 93–101 (2025)
2. Cui, S., Wu, X., Wang, H., Wu, H.: Multimodal data modeling technology and its application for cloud-edge-device collaboration. J. Softw. **35**(3), 1154–1172 (2024)

3. Xue, Y., Wang, K., Qiu, Y., Zhu, F.: Cognitive load assessment of online learning based on multimodal data. Mod. Educ. Technol. **34**(3), 79–88 (2024)

4. Jiang, J., Yu, W., Wang, H.: Research on student learning behavior in smart classroom based on multimodal data. China Educ. Inf. **30**(4), 107–117 (2024)

5. Hu, G., Zhao, J., Hao, Y.: Mobile target tracking based on state observability and multimodal data PF. Radio Eng. **54**(6), 1504–1511 (2024)

6. Bae, S.J., Jang, Y., Kim, Y., et al.: Gut microbiota regulation by acupuncture and moxibustion: a systematic review and meta-analysis. Am. J. Chin. Med. **52**(05), 1245–1273 (2024)

7. Yan, S.Y., Xiong, Z.Y., Liu, X.Y., et al.: Review of clinical research in acupuncture and moxibustion from 2010 to 2020 and future prospects. Chin. Acupunct. Moxib. **42**(1), 116–118 (2022)

8. Masuda, T., Takeshita, Y., Okumura, Y., et al.: Rethinking traditional Japanese acupuncture and moxibustion: spread the Hokushin-kai style of acupuncture to the world. Med. Acupunct. **36**(2), 61–62 (2024)

9. Lee, Y.R., Cha, H.J., Choi, H.K., et al.: Statistical analysis of patients visiting department of acupuncture and moxibustion in Korean medicine hospital before and after COVID-19-focusing on a Korean medicine hospital in Daejeon. J. Korean Med. **42**(2), 31–49 (2021)

10. Kim, S.H., Lee, J.H., Kyung, D.H., et al.: A case report of oculomotor nerve palsy treated with acupuncture and moxibustion. J. Korean Med. Ophthalmol. Otolaryngol. Dermatol. **35**(2), 61–71 (2022)

11. Amalia, F.S., Alita, D.: Application of SAW method in decision support system for determination of exemplary students. J. Inf. Technol. Softw. Eng. Comput. Sci. **1**(1), 14–21 (2023)

12. Deveci, M., Mishra, A.R., Gokasar, I., et al.: A decision support system for assessing and prioritizing sustainable urban transportation in metaverse. IEEE Trans. Fuzzy Syst. **31**(2), 475–484 (2022)

13. Tutun, S., Johnson, M.E., Ahmed, A., et al.: An AI-based decision support system for predicting mental health disorders. Inf. Syst. Front. **25**(3), 1261–1276 (2023)

14. Megawaty, D.A., Silitonga, A.: Decision support system feasibility for promotion using the profile matching method. J. Data Sci. Inf. Syst. **1**(2), 50–56 (2023)

15. Rani, P., Kumar, R., Ahmed, N.M.O.S., et al.: A decision support system for heart disease prediction based upon machine learning. J. Reliable Int. Environ. **7**(3), 263–275 (2021)

Joint Decoding-Based Dynamic Motion Analysis for Track and Field Athletes

Haili Meng[(✉)]

Chongqing Vocational Institute of Engineering, Chongqing, China
`haili_meng@21cn.com`

Abstract. The existing system for dynamic motion analysis of track and field athletes faces the limitation of a single data source, which makes it impossible to fully and accurately capture the dynamic performance and details of athletes, affecting the accurate evaluation of training and competition. Traditional methods rely too heavily on a single modality of data, making it difficult to capture a holistic view of an athlete's movements, which in turn leads to suboptimal training adjustments. This paper introduces a multimodal neural network (MMNN) and an attention mechanism to effectively integrate and jointly decode the dynamic motion analysis system of track and field athletes based on a joint decoding strategy. This can make up for the shortcomings of various data sources, reduce information loss, solve the problem of inconsistent information between different data sources, and improve the accuracy and robustness of the system, thereby providing decision support and personalized training plans for athletes. The experimental results show that the introduction of the attention mechanism has a lower loss value and is more effective than the comparative graph convolutional network strategy system. The error value fluctuates between 0.5 and 1, which can better reduce the MSE and achieve better accuracy, thereby formulating a good training plan for athletes, strengthening the training status of athletes, and improving their competitive level.

Keywords: Joint Decoding · Track and Field Athletes · Motion Analysis · Multimodal Neural Network · Attention Mechanism

1 Introduction

With the continuous development of science and technology, the accuracy and personalization requirements of athletes' training and competition are increasing day by day. Among them, motion analysis technology has gradually been widely used in the field of track and field sports. Motion analysis is not limited to traditional visual evaluation and manual recording. More and more high-tech equipment and systems are being used in athletes' training and competition [1]. Especially in track and field sports, the dynamic performance and details of athletes are crucial. The core elements of track and field sports include multi-dimensional motion states such as running posture, cadence, stride, speed, etc., and changes in these states often have a profound impact on competition results. Therefore, accurately monitoring and analyzing athletes' movement changes has

© The Author(s) 2026
P. Siarry et al. (Eds.): WCNA 2024, LNEE 1550, pp. 194–204, 2026.
https://doi.org/10.1007/978-981-95-6946-5_20

become the key to improving training effects and competition results. However, existing analysis systems often rely on a single data source for analysis, which leads to the problem that data cannot be fully integrated and accurately decoded [2]. A single data source cannot accurately and comprehensively capture the dynamic state of athletes and the subtle differences in the movement process, resulting in deviations in the evaluation of training effects and competition status, which in turn affects the training effects and performance of athletes. Therefore, integrating multiple data sources to improve the accuracy and robustness of the analysis system has become a key issue in current sports analysis technology [3].

This paper proposes a dynamic motion analysis system for track and field athletes based on a joint decoding strategy, which innovatively combines a multimodal neural network model to fuse sensor data, video data, and physiological data. Through the joint decoding strategy, the information from multiple data sources is effectively integrated to solve the differences in time synchronization, feature extraction, and data scale between different data sources, thereby comprehensively and accurately analyzing the dynamic state of athletes. Compared with the traditional single data source method, the advantage of this system is that it uses MMNN to conduct deep learning on the multi-dimensional data of athletes, extract and integrate various types of information, and provide a more comprehensive and accurate sports state assessment. The system can not only monitor the dynamic characteristics of athletes such as step frequency and stride in real time, but also evaluate important indicators such as athletes' fatigue status and movement accuracy, providing a scientific basis for athletes to formulate personalized training plans and competition strategies. In order to verify the effectiveness of this method, this paper conducted multiple comparative experiments. The experimental results show that the multimodal neural network model based on the joint decoding strategy has shown significant advantages in the dynamic analysis of athletes. At the same time, the real-time and feedback capabilities of the system have also been verified, which can help coaches and athletes adjust strategies in time during training to improve training effects and competition results.

2   Related Works

In the field of dynamic motion analysis of track and field athletes, many studies have attempted to solve problems such as athlete movement accuracy and dynamic state monitoring. In recent years, researchers such as Rana M [4] have focused on using sensor data to analyze athletes' athletic performance, using accelerometers and gyroscopes to monitor gait and speed in real time, but such research has failed to effectively solve the problem of integrating information from multiple data sources. Researchers such as Deng L [5] have used visual recognition technology to track and analyze athletes' movements. However, existing visual methods are often limited by environmental factors, such as lighting changes and background clutter, which results in reduced accuracy in complex scenes. Based on physiological signal analysis, researchers such as Ishaque S [6] monitored the physical condition of athletes through indicators such as heart rate and electromyography. However, this method was not directly combined with the dynamic movements of athletes, and the connection between physical fitness evaluation and sports performance was weak. In order to enhance the real-time performance and accuracy of sports analysis, scholars such as Geng X [7] combined video analysis with sensor data and proposed a fusion method. Although this method improved the analysis effect to a certain extent, it still faced the difficulties of data fusion and feature selection. In recent years, Ren H [8] has attempted to introduce deep learning algorithms to analyze athletes' movements, but most of them focus on specific data sources and have failed to make significant progress in multimodal data fusion. A single deep learning model relies on a large amount of labeled data, but the lack and high cost of labeled data limit the universality of these methods. Most existing systems have problems such as incomplete data fusion, insufficient feature extraction, and poor model robustness. There is an urgent need to propose more effective solutions in data fusion and multimodal learning.

3 Methods

3.1 Data Preprocessing and Synchronization

In the dynamic motion analysis system for track and field athletes, this paper adopts the method of multimodal data fusion [9], which integrates different types of data sources (sensor data, video data, and physiological data, etc.) and uses deep learning models to extract and fuse their features, thereby providing more comprehensive and accurate athlete dynamic analysis results. In sports analysis, data sources may include: (1) sensor data (accelerometers, gyroscopes, pressure sensors, etc.); (2) video data (recording the athlete's movement process through a camera to provide spatial characteristics); (3) physiological data (athlete's heart rate, breathing rate, electromyography, etc., reflecting the athlete's physiological state). This paper combines these different types of data sources to make up for the limitations of a single data source, enhance the model's discriminative ability, and improve the accuracy and robustness of predictions.

In multimodal data such as sensor data and video data, the presence of noise often affects the accuracy of analysis. Noise may be caused by sensor errors, environmental factors, or transmission problems. In the denoising process, this paper adopts a moving average filter [10]:

$$\hat{x}_t = \frac{1}{N} \sum_{i=t-N+1}^{t} x_i \tag{1}$$

N is the size of the sliding window.

Data from different sources have different scales and units, and they are standardized or normalized to ensure that they participate in fusion at the same scale.

$$x\prime = \frac{x - \mu}{\sigma} \tag{2}$$

$$x\prime = \frac{x - \min(x)}{\max(x) - \min(x)} \tag{3}$$

Missing value filling:

$$x_{\text{missing}} = x_1 + \frac{(x_2 - x_1)(t - t_1)}{t_2 - t_1} \tag{4}$$

t represents the time point to be interpolated.

Feature extraction:

Different methods of feature extraction are used for data of different modalities (sensors, videos, physiological data, etc.). Video data usually contains spatiotemporal information, so when processing video data, it is necessary to extract spatial and temporal features. Using convolutional neural networks to extract from video data can effectively extract spatial features in images.

$$F_{\text{spatial}}(I_t) = \text{CNN}(I_t) \tag{5}$$

A 3D convolutional neural network is used to extract temporal information from video data [11].

$$F_{\text{temporal}} = 3D - \text{CNN}(I_1, I_2, \ldots, I_t) \tag{6}$$

For the time series data of the sensor, the Fast Fourier Transform is used to convert the time series data from the time domain to the frequency domain, so as to extract the frequency domain features.

$$X(f) = \sum_{n=0}^{N-1} x_n e^{-i2\pi fn/N} \tag{7}$$

Physiological characteristics:

$$\mu_{\text{HR}} = \frac{1}{n} \sum_{i=1}^{n} hr_i \tag{8}$$

Among them, μ_{HR} represents the mean of heart rate.

For data synchronization, data at different time steps are aligned to the same time axis to better integrate multiple different data.

$$A_{\text{aligned}}(t_i) = A(t_1) + \frac{(t_i - t_1)}{(t_n - t_1)}(A(t_n) - A(t_1)) \tag{9}$$

And the down-sampling method is used to unify the different sampling rates of different data into a standard frequency.

$$x_{\text{downsampled}}(t_k) = x(t_{k \cdot \Delta t}) \tag{10}$$

3.2 Integration of Multimodal Data Sources

Through the above data preprocessing steps, different data are feature extracted and information is effectively integrated in the same model, thereby improving the accuracy, stability and comprehensiveness of the analysis. Data fusion of multiple data sources can be performed at different levels: ① Concatenation fusion concatenates the feature vectors extracted from each modality. ② Weighted sum assigns different weights to the features of different modalities and then performs weighted sum. ③ Late fusion fuses the independent model outputs of each modality, which is usually used in the decision-making layer. This paper focuses on introducing the attention mechanism fusion [12] for data fusion. The core idea of the attention mechanism is to assign different weights to input features so that the model can focus on the most important information when processing data.

$$e_{ij} = \mathrm{score}(h_i, h_j) = v^T \tanh(W_1 h_i + W_2 h_j) \tag{11}$$

Among them, $\mathbf{f}_{\mathrm{fusion}}$ represents the concatenated vector of all modal features.

The attention mechanism can dynamically assign weights to features of different modalities according to the importance of the input data, so that the model can focus on the most useful features and reduce the impact of redundant information on model performance. In multimodal deep learning, data from different modalities have different importance. The attention mechanism can adaptively select and strengthen information that is more important to the target task and ignore those less relevant parts. It also overcomes the limitations of simple weighting in traditional methods, making the fusion of information from different modalities more refined and more flexible.

3.3 Joint Decoding Model Design

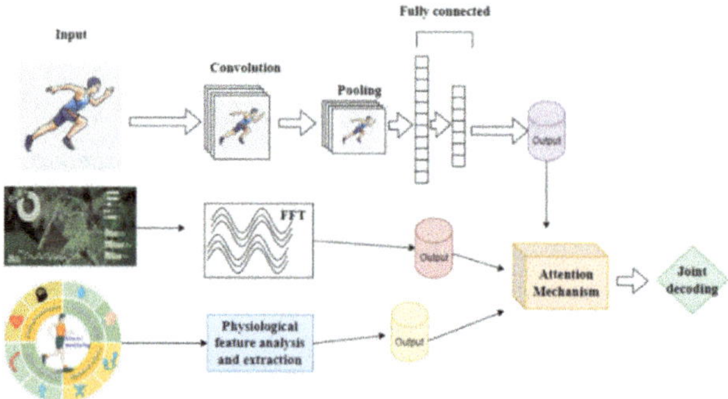

Fig. 1. Multimodal joint strategy structure diagram

Figure 1 shows the multimodal joint structure diagram. The input data is extracted according to different categories, and the image and video data are recognized through the CNN (convolution neural network) [13] convolution layer. The feature space data is filtered, the information is obtained from the sensor, and the vital signs are extracted through the change function. After feature extraction, the attention mechanism fusion strategy is combined to fuse multimodal data, and finally the joint strategy is used for optimization analysis to obtain the corresponding optimization solution results.

3.4 System Visualization

Figure 2 is the visualization page of the system. The page contains multiple key elements. It uses the chart.js chart library to generate corresponding data monitoring charts based on the form data information entered by the user, and conducts real-time analysis. It is more intuitive to formulate better training plans for coaches and athletes based on the charts to enhance the quality of athletes' competitions.

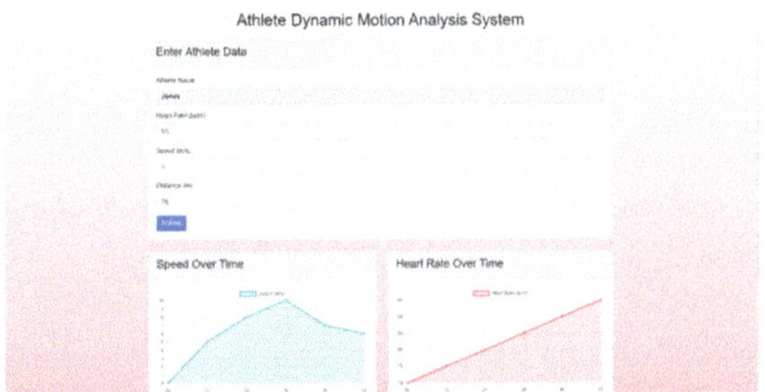

Fig. 2. Visualization page

4 Experimental Design

4.1 Dataset Collection

This experiment focuses on the strategy analysis of hurdlers in the competition, as shown in Table 1, including athlete ID, training time, training heart rate, step frequency, stride length, starting reaction time, hurdle technical score and other information.

Table 1. Hurdles sports dataset

ID	Gender	Training duration	Heart rate	Cadence	Stride	Reaction time	Score
1	Male	14	145	180	2.4	0.16	85
2	Female	12	140	170	2.2	0.14	88
3	Male	16	150	185	2.3	0.18	82
4	Female	15	148	175	2.1	0.17	80
5	Male	18	155	190	2.5	0.15	90
6	Female	13	142	172	2.3	0.16	87
...	...	...	...	...	...	...	...

4.2　Training Optimization

For the optimization of the system training process, this paper uses the Adam (Adaptive Moment Estimation) optimizer [14] for optimization. Adam combines the advantages of Momentum and can adapt to changes in learning rate and effectively handle sparse gradient problems. It can automatically adjust the learning rate of each parameter, avoiding the complexity of manually adjusting the learning rate.

$$m_t = \beta_1 m_{t-1} + (1 - \beta_1)g_t \tag{12}$$

$$v_t = \beta_2 v_{t-1} + (1 - \beta_2)g_t^2 \tag{13}$$

$$\theta_{t+1} = \theta_t - \frac{\alpha}{\sqrt{\hat{v}_t} + \epsilon}\hat{m}_t \tag{14}$$

Among them, β_1 is the parameter for controlling momentum, and β_2 is the parameter for controlling square momentum.

In order to prevent the model from overfitting during training, this paper uses regularization to prevent it.

$$\mathcal{L}_{reg} = \mathcal{L} + \lambda \sum_i \theta_i^2 \tag{15}$$

λ is the regularization coefficient.

4.3　Experimental Plan

A: Comparison of prediction accuracy between joint decoding strategy and graph convolutional network strategy: This paper verifies that the prediction accuracy of the system based on joint decoding strategy is improved in athlete dynamic motion analysis compared with other optimized data fusion strategies.

B: Comparison between attention-based systems and other optimization fusion mechanisms: Experimental verification shows that after the introduction of the attention mechanism, the joint decoding strategy improves the performance of the athlete dynamic motion analysis system.

C: Experimental application of the joint decoding strategy in dynamic real-time motion evaluation: This paper focuses on the performance of the system in real-time data processing and feedback delay under different load factors.

5　Results

5.1　Accuracy Comparison and Verification

This paper compares other optimized data fusion strategies (graph convolutional network strategy [15]) to verify whether the joint decoding strategy can effectively improve the training load prediction of athletes. The two strategies are tested using the same data set.

Evaluation indicator: Mean Square Error (MSE) is used to average the square of the error between the actual value and the predicted value.

Figure 3 shows the system's prediction of the athlete's training load indicators under two different strategies to see whether it can accurately predict their training load. The red line is the joint decoding strategy designed in this paper, which can process multiple modal data and reduce information loss. The error of its prediction result is also stably controlled in the range of 0.5–1, and the error value is small. For the relatively excellent graph convolutional network strategy, the MSE curve of the joint decoding strategy can be lower than that of the graph convolutional network strategy, and the error of the prediction result of the graph convolutional network strategy can be improved, and the error value can fluctuate between 0.6 and 1.2. The joint decoding strategy can achieve lower errors, reduce MSE and achieve better accuracy in the training load prediction task, formulate good training plans for athletes, strengthen athletes' training status and improve their competitive level.

5.2 Ablation Attention Mechanism

In order to verify the effect of the innovative use of attention mechanism for multi-source data fusion, and compare the fusion methods such as splicing fusion, weighted summation, and late fusion to see whether they can improve the system performance. The results are shown in Fig. 4.

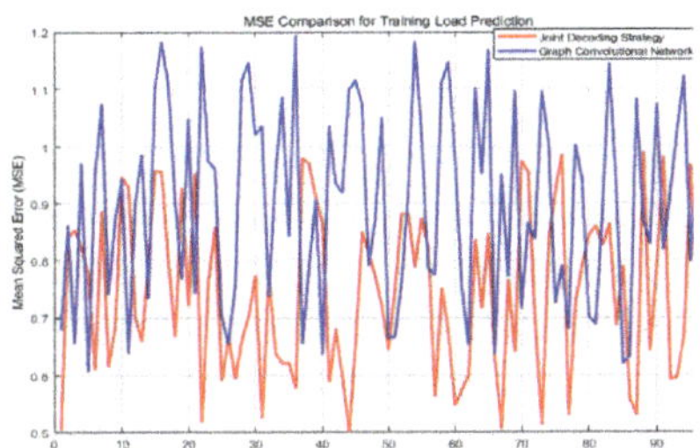

Fig. 3. Comparison results

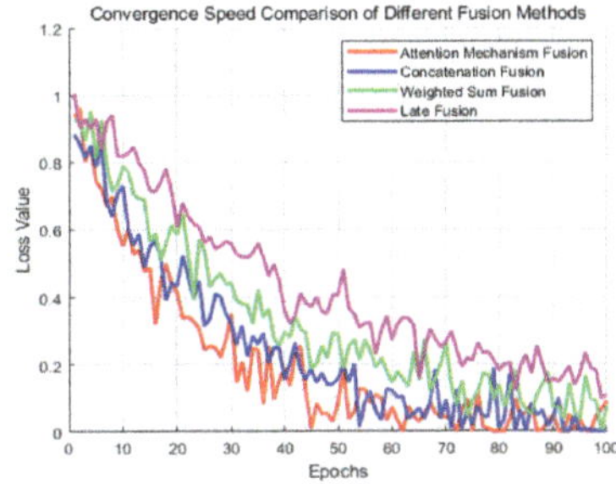

Fig. 4. Comparison of fusion methods

As can be seen from Fig. 4, after the introduction of the self-attention mechanism, the convergence speed of the attention mechanism is faster, and the loss value has a clear downward trend. As the training progresses, the loss value decreases rapidly, and compared with other fusion methods, the loss value is at the bottom. As the iterative training increases, the average loss value can reach 0.20; the loss value of the splicing fusion method decreases slowly. Its average loss value can reach 0.24; the average loss value of weighted sum fusion is 0.33, which is still not as good as the performance of the splicing fusion method. The loss of late fusion decreases the slowest, and the loss value fluctuates greatly, with an average loss value of 0.43. It can be clearly seen from these data that the introduction of the attention mechanism greatly accelerates the convergence speed of the training process, improves the training efficiency of the model, and shows higher accuracy and stability in terms of loss value, far exceeding other fusion methods.

5.3 System Performance Test

The performance of the system is simulated and verified by adding different loads, which are the number of athletes involved in the system analysis process or the complexity of the sports scene. This change in load can directly affect the performance of the system.

Table 2. System indicators with different loads

Load conditions	Frame rate (FPS)	Feedback delay (ms)
No load	130	20
Low load	125	20
Medium load	120	21
High load	120	23
Ultra-high load	120	24

Table 2 shows the data analysis results of the frame rate and feedback delay of the system under different loads. It can be seen that under no-load conditions, the frame rate of the joint decoding strategy system can reach 130 frames per second. The system can capture more details of the athletes every second and can quickly process action information. As the load intensity increases, the system is less disturbed. Under ultra-high load conditions, it can still maintain an ultra-high frame rate of 120 frames per second, providing strong support for athlete motion analysis. For the feedback delay indicator, the time required from the system receiving the input signal to the output result, low feedback delay is an important factor in improving the training effect and accuracy of athletes. The system takes 20ms to provide timely feedback when there is no load. When the load increases, the delay time can reach 24ms, and it can still provide good feedback. The above data results show that the dynamic motion analysis of track and field athletes based on the joint decoding strategy can provide good feedback under different loads, provide dynamic analysis for athletes, and improve their competitive state.

6 Conclusions

This experiment designed and verified a dynamic motion analysis system based on a joint decoding strategy. By improving multiple key indicators, it proved the effectiveness of the system in athlete training and competition strategy optimization. By adopting a joint decoding strategy, the system has achieved improvements in multiple key indicators. The system can reach a maximum frame rate of 130 FPS under no-load conditions and maintain an ultra-high frame rate of 120 FPS under high-load conditions, ensuring real-time capture of athlete movement details and effective processing. At the same time, the system's feedback delay is maintained between 20ms and 24ms, which can quickly respond to athlete movement changes and provide timely feedback to athletes. This experiment also includes a number of different verification experiments to ensure the full optimization effect of the system. The experimental results show that the joint decoding strategy shows a lower error in MSE, indicating its advantage in training load prediction. This paper compares traditional fusion methods such as splicing fusion and weighted summation. The results show that the attention mechanism significantly improves the convergence speed of the model during training and effectively reduces the loss value; Under different load conditions, the system can still maintain a high frame rate and low feedback delay, ensuring reliable data processing and real-time feedback in complex and high-load scenarios, further improving the training effect and competitive state of athletes.

However, this experiment still has certain shortcomings. (1) The system performs stably under low and high load conditions, but under extreme ultra-high load conditions, the system's performance still degrades significantly, especially when processing complex motion scenes, where slight delays may occur. Therefore, future work can further optimize the system's performance under ultra-high load conditions. (2) The dataset used in this paper is limited to specific hurdlers and lacks extensive validation in other sports or different groups of athletes, which may limit the universality and scope of application of the system. (3) The effect of multi-source data fusion is improved by adding the attention mechanism, but the fusion strategy still needs to be further optimized to enhance the system's adaptability to more complex motion patterns, especially in terms of the accuracy of motion analysis.

References

1. Bortolini, M., Faccio, M., Gamberi, M., et al.: Motion Analysis System (MAS) for production and ergonomics assessment in the manufacturing processes. Comput. Ind. Eng. **139**, 105485 (2020)
2. Abbaspour, S., Lindén, M., Gholamhosseini, H., et al.: Evaluation of surface EMG-based recognition algorithms for decoding hand movements. Med. Biol. Eng. Comput. **58**, 83–100 (2020)
3. Li, D., Wang, X.: On monitoring and detecting abnormal physiological state of athletes from internet of bodies. Internet Technol. Lett. **4**(3), e282 (2021)
4. Rana, M., Mittal, V.: Wearable sensors for real-time kinematics analysis in sports: a review. IEEE Sens. J. **21**(2), 1187–1207 (2020)
5. Deng, L., Deng, Y., Bi, Z.: Simulation of athletes' motion detection and recovery technology based on monocular vision and biomechanics. J. Intell. Fuzzy Syst. **40**(2), 2241–2252 (2021)
6. Ishaque, S., Khan, N., Krishnan, S.: Trends in heart-rate variability signal analysis. Front. Digi. Health **3**, 639444 (2021)
7. Geng, X.: Research on athlete's action recognition based on acceleration sensor and deep learning. J. Intell. Fuzzy Syst. **40**(2), 2229–2240 (2021)
8. Ren, H.: Sports video athlete detection based on deep learning. Neural Comput. Appl. **35**(6), 4201–4210 (2023)
9. Gao, J., Li, P., Chen, Z., et al.: A survey on deep learning for multimodal data fusion. Neural Comput. **32**(5), 829–864 (2020)
10. Luo, W., Wei, D.: A frequency-adaptive improved moving-average-filter-based quasi-type-1 PLL for adverse grid conditions. IEEE Access **8**, 54145–54153 (2020)
11. Chen, J., Bi, S., Zhang, G., et al.: High-density surface EMG-based gesture recognition using a 3D convolutional neural network. Sensors **20**(4), 1201 (2020)
12. Wang, G., Gan, X., Cao, Q., et al.: MFANet: multi-scale feature fusion network with attention mechanism. Vis. Comput. **39**(7), 2969–2980 (2023)

13. Bhatt, D., Patel, C., Talsania, H., et al.: CNN variants for computer vision: history, architecture, application, challenges and future scope. Electronics **10**(20), 2470 (2021)
14. Yang, J., Long, Q.: A modification of adaptive moment estimation (adam) for machine learning. J. Ind. Manag. Optim. **20**(7), 2516–2540 (2024)
15. Hodson, T.O.: Root mean square error (RMSE) or mean absolute error (MAE): When to use them or not. Geosci. Model Dev. Discuss. **2022**, 1–10 (2022)

Augmented Reality-Based Virtual Model Room for Interactive Interior Design

Hongbo Xia[(⊠)] and Chuang Peng

Chongqing College of Architecture and Technology, Chongqing, China
xhbcq@21cn.com

Abstract. Traditional interior design methods lack immersive user interaction and are often limited by insufficient visualization capabilities, making it difficult for clients to fully comprehend design concepts. In order to address the limitations of traditional interior design solutions in terms of display and user interaction, the author proposes an augmented reality based virtual model room display and interaction system for interior design. The system combines real-time rendering, speech recognition, and multimodal interaction technology, allowing users to intuitively feel the spatial layout, material selection, and decorative style, and adjust and optimize them through voice or gestures. The research results indicate that in augmented reality environments, prolonged hovering and swinging of arms for interaction significantly increases user fatigue. In addition, in terms of comfort, the average scores of the two are 2 and 4 points respectively, which may be due to the fact that augmented reality technology allows users to interact in more natural postures without the need for significant gestures. In terms of acceptability score, the average score of augmented reality technology is higher than that of traditional gesture interaction methods, at 4 and 3 points respectively. This system can effectively reduce design communication costs, improve user participation and decision-making efficiency. In the future, the application of AR technology in the field of interior design will further develop towards intelligence and personalization, promoting the digital transformation of the industry.

Keywords: Augmented Reality (AR) · Virtual model room · Interior design · interactive system

1 Introduction

Augmented reality (AR) is a way of observing the world that combines virtualization technology. Based on the technique of the computer, it can be superimposed on the actual picture or the space, creating a 3D picture that can interact with the real world. Because of its constant function in realistic and virtual environment, Augmented Reality offers a new approach to human and machine interaction, which can meet the needs of the user [1]. Augmented reality technology is a technology that embeds virtual data information into real space in the form of images, text, videos, etc., to enhance visual and interactive experience effects. Therefore, the study of Augmented Reality is one of the most

© The Author(s) 2026
P. Siarry et al. (Eds.): WCNA 2024, LNEE 1550, pp. 205–214, 2026.
https://doi.org/10.1007/978-981-95-6946-5_21

significant areas in man-machine interaction. The types of augmented reality devices that people can use are also increasing, including Microsoft Hololens, Rokid Glass, Ned + GlassX2, and so on. However, how to enhance people's interactive experience when using AR devices has become a cutting-edge topic in the field of human-computer interaction [2]. Compared with it, virtual showrooms based on augmented reality technology can overlay virtual models of design schemes onto real spaces through mobile devices or smart glasses, helping users intuitively feel the design effect and provide modification suggestions for the design scheme through real-time interaction, thereby improving user experience and design efficiency [3]. The author aims to construct an augmented reality based virtual model room display and interaction system for interior design, and explore the potential application of AR technology in virtual model room display based on the actual needs of interior design [4].

2 Literature Review

With the increasing development of virtual reality technology, its application fields are becoming more and more extensive. "VR+" continues to expand, and "VR + real estate" is also a very important application direction of virtual reality worldwide. According to the survey, the earliest virtual reality model rooms were simply 360 degree panoramic views of the house. Mohammad Rezaei et al. have explored the use of AR and Virtual Reality (VR) techniques, using Unity's light-sensor to create complicated illumination simulations and improve indoor design practices. Emphasis is placed on strategically placed and dynamically analyzing optical sensors in a virtual environment for evaluation and improvement of illumination performance [5]. Xu, W. et al. studied the interactive forms and applications of AR technology in RV interior design and display, providing new ideas and solutions for RV interior design, optimizing customer user experience, and improving work efficiency in the process of RV customized interior design [6]. Xue, Z. explores the application characteristics and implementation approaches of language models in virtual education research laboratories. By analyzing the basic principles of language models and the characteristics of virtual education research laboratories, the relationship between the two was systematically studied, and corresponding practical paths were proposed [7].

Based on this research, the author combines real-time rendering and multimodal interaction to achieve a more immersive and intuitive virtual showroom experience, enhancing users' understanding and perception of the design scheme. Through natural interaction, users can independently adjust the spatial layout, materials, and decorative styles, improve personalized customization capabilities, and optimize user experience. Through practical verification of the feasibility of AR technology in design display and interaction, it provides technical support for future intelligent and personalized interior design, and promotes the digital transformation of the industry.

3   Research Methods

3.1   Virtual Reality and Augmented Reality Technology

Virtual Reality (VR) technology, also known as Lingjing technology or artificial environment, is a system simulation of interactive 3D dynamic scenery and physical behavior that integrates multiple sources of information. This technology is a collection of various technologies such as computer graphics, human-computer interaction, multimedia technology, network technology, sensing technology, etc. It is a challenging interdisciplinary and research field. Its main characteristics are immersion, imagination, interactivity, and multisensory nature. The principle of implementing the application of virtual reality technology is shown in Fig. 1:

The AR technique consists of three dimensional modeling technique, multi-media technique, real time visual display and control technique, multisensor fusion, real time tracing and registering, and so on. AR offers a variety of messages that are not normally perceived by people [8]. The key technologies mainly include three parts: Virtual real object registration, display technology, and realistic rendering technology. The primary goal of achieving AR effects is to seamlessly overlay and synthesize real scene images with computer-generated virtual objects. In fact, it is a course of alignment, which is to confirm the relation of imaginary target and observer, and then it can be projected into the space of time by correctly projecting it [9]. Need accuracy (no lag), velocity (accurate, non-shaking), and robust (unaffected by illumination, shading and movement of the subject) (35). The implementation principle of AR technology is shown in Fig. 2:

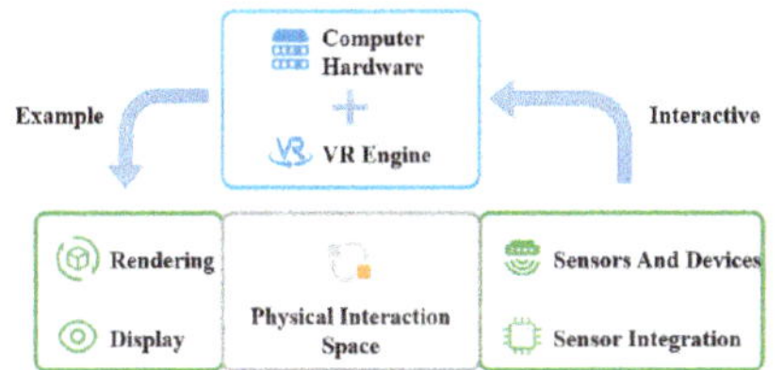

Fig. 1. Schematic diagram of virtual reality technology implementation

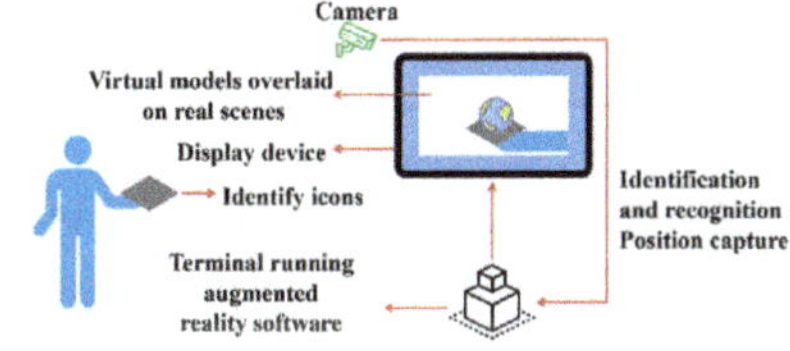

Fig. 2. Schematic diagram of augmented reality

3.2   Design of Voice Interaction Algorithm for Virtual Template Room Based on Virtual Reality

Simple voice recognition techniques are not surprising anymore, and they are widely used in many areas. Nowadays, voice recognition techniques have become more mature. The technique of voice identification has been widely used in the fields of voice call, speech guide, room device control, audio file search and text input. But so far, it has been used as a kind of interaction in VR. However, as the VR technique is being developed, voice communication will be the main stream. This system has conducted research on voice interaction based on virtual reality. Through the integration of Pocket phinx library and

UE4 engine, the speech recognition process is encapsulated in C++ encoding to embed functional functions into the UE4 project. The visual encoding function of UE4 is directly called to achieve voice control function in virtual reality. The specific implementation process is as follows [9].

3.2.1 Speech Preprocessing

Firstly, the input raw speech signal is preprocessed, mainly including filtering out unimportant information and background noise in the speech signal, and performing endpoint detection, speech framing, pre emphasis and other related operations on the speech signal. Among them, pre emphasizing the speech signal is to emphasize the high-frequency part of the speech signal, remove the influence of lip radiation, and increase the high-frequency resolution of the speech. We implement pre emphasis using a first-order FIR high pass digital filter with a transfer function of $H(z) = 1 - az^{-1}$, where a is the coefficient of the pre emphasis filter, typically ranging from 0.9 to 1.0, and the author takes 0.96. Assuming the speech sampling value at time n is x (n), the pre emphasized signal is:

$$y(n) = x(n) - a * x(n - 1) \tag{1}$$

3.2.2 Feature Extraction

The author uses the Mel Frequency Cepstral Coefficients (MFCC) method to extract features. MFCC parameters are based on human auditory characteristics. They utilize the critical band effect of human hearing and use MEL cepstral analysis technology to process speech signals to obtain MEL cepstral coefficient vector sequences, which represent the frequency spectrum of input speech using MEL cepstral coefficients. Set up several bandpass filters with triangular or sinusoidal filtering characteristics within the range of the speech spectrum, then pass the speech energy spectrum through the filter bank, calculate the output of each filter, take its logarithm, and perform discrete cosine transform (DCT) to obtain the MFCC coefficients. The solution formula is as follows:

$$C_i \sum_{k=1}^{p} \log F(k) \cos\left[\pi (k - 0.5)i/p\right], \ i = 1, 2, \cdots L \tag{2}$$

Among them, Ci is the feature parameter; K is a variable, $0 \leq k < P$; P is the number of triangular filters, with P set to 26; F (k) is the output data of each filter; I is the data length, which is the number of MFCC coefficients, and L is set to 12. Using the characteristics of the training voice database, we train the sound model parameters. In the process of identification, we can get the identification result by matching the characteristic parameters of the voice with the acoustic model. In this paper, we adopt Hidden Markov Model (GMM-HMM), which is a hybrid Gauss model. The form of the joint probability density function of the mixture Gaussian model is as follows:

$$P(x) = \sum_{m=1}^{M} \frac{C_m}{(2\pi)^{D/2}\left|\sum m\right|^{1/2}} \exp\left[-\frac{1}{2}(x - u_m)^T \sum_{m}^{-1} (x - u_m)\right]$$

$$= \sum_{m=1}^{M} C_m N\left(x; u_m, \sum m\right) \tag{3}$$

Among them, M represents the number of Gaussians in the mixture Gaussian model, Cm represents the weight, um represents the mean, $\sum_m$ represents the covariance matrix, and D is the dimension of the observation vector [10].

3.3 Design of Dynamic Gesture Interaction Algorithm for Virtual Template Room Based on Augmented Reality

The augmented reality module of the virtual prototype room system is implemented using the AR development toolkit ARToolKit. ARToolKit is an open-source augmented reality SDK, which is a C/C++ language library that provides fast and accurate tag tracking, making it easy for us to write AR applications. We combined ARToolKit with the UE4 engine to add dynamic gesture detection, recognition, and control functions, completing the corresponding functions of this module [11].

By following the above steps, the overlay effect of virtual 3D graphics and the real environment can be achieved. The AR effect we see. The principle is shown in Fig. 3:

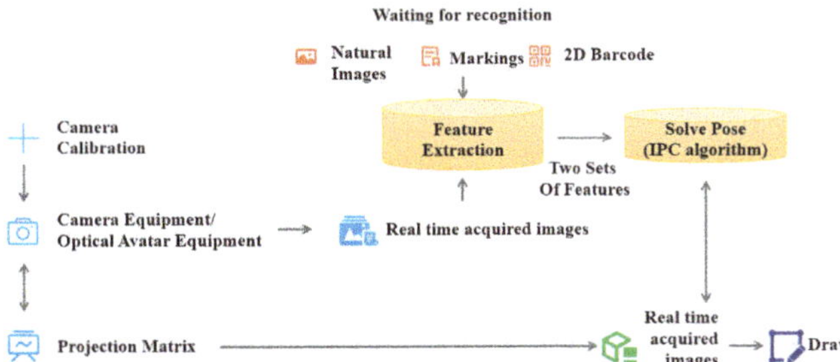

Fig. 3. Schematic diagram of ARToolKit

It should be noted that the implementation of augmented reality relies heavily on the training and production of its labels. The author uses NFT (Natural Feature Tracking) as an example to implement AR technology for tag tracking. The tag tracking of ARToolkit mainly includes benchmark tag tracking, NFT tag tracking, and two-dimensional barcode matrix tracking [12].

3.3.1 Dynamic Gesture Recognition in Augmented Reality

This system adds dynamic gesture recognition control in the augmented reality function module. Through dynamic gestures, virtual models in augmented reality scenes can be controlled. Based on the movement of the hand, the model simulates the feedback of the force received and can also follow dynamic gesture movements to achieve multi-dimensional observation of the model. After turning on the camera, add OpenCV video information processing and initialize the video stream. First, perform motion detection. If dynamic gestures are detected, perform image processing, preprocess the gesture image sequence, convert it to HSV mode, perform morphological processing, draw contour lines, compare calibration information and contour information, set direction vectors, and finally simulate the force on the model to achieve interaction between dynamic gestures and virtual models. Specifically, this dynamic gesture interaction achieves gesture recognition and control in three-dimensional space. The dynamic hand obtained from the video stream is two-dimensional information. Here, through matrix operations, the transformation matrix of the camera relative to the detected identifier is compared with the calculated matrix to obtain a three-dimensional direction vector of the force exerted by the gesture on the model, thereby achieving force simulation of the model in different directions in three-dimensional space.

After a simple motion detection step, if motion information is detected, processing of video frame images containing motion gestures begins. Firstly, the image is subjected to median filtering using OpenCV's medianBlur function to remove salt and pepper noise. Then, the image is subjected to color space conversion using the cvtClor function to obtain its HSV space data, and the brightness v value in HSV space is reset to a relatively small brightness value (reducing the interference of static skin color). Reset the v of HSV space as shown in Eq. (4):

$$\begin{cases} v_{\text{temp}} = \begin{cases} r/g, & g \neq 0 \\ 4, & g = 0 \end{cases} \\ v = \begin{cases} 10r/g, & v_{temp} \leq 2 \\ 255, & v_{temp} > 2 \end{cases} \end{cases} \tag{4}$$

Among them, r and g are the red and green pixels of the skin color region of interest, and r > g. Then, by comparing the obtained motion binary image with the binary image obtained through backprojection, and performing some image morphology closure operations, a relatively complete motion skin gesture binary image can be obtained [13].

After preliminary morphological operations to remove noise and make the boundaries of the hand clearer, the gesture contour is obtained through OpenCv's findContours function, and then the pseudo contour is removed. Due to the constant movement of our hands, the contours we obtain are also constantly changing. We compare the contours obtained from each frame and set comparison conditions. Assign values to directional indicator variables through comparison. Status comparison and analysis are shown in Table 1:

Table 1. Status Analysis

Dynamic Gesture	Contour variation	Direction markers	Model dynamics
Palm from far to near	The shape changes from small to large, and the center point of the contour position remains unchanged	X-axis direction vector	Move forward

(continued)

Table 1. (*continued*)

Dynamic Gesture	Contour variation	Direction markers	Model dynamics
Palm from bottom to top	The shape changes little, but the center point of the contour position changes	Z-axis vector	Move upwards
Twist the palm in different directions	Irregular shape changes, but the center point of the contour position remains unchanged	The center point of the three-dimensional space torsion direction vector remains unchanged	Make corresponding movements based on the twisting direction vector

After the dynamic gesture is judged by the contour, the virtual model is subjected to force simulation operations based on different judgment results. Based on the values of direction markers during the contour judgment process, the coordinate values of the model in three-dimensional space will be multiplied on the x, y, and z axes. By changing the coordinate values, the position of the model can be changed to simulate the force.

4 Results Analysis

4.1 System Function Implementation

Build the model of the prototype room using 2D drawings and 3DMax modeling tools. Using the UE4 engine to edit the model, including creating textures, materials, adding physical collisions, adding lighting, illumination, and special effects to the overall environment, and baking and rendering to make the visual effect of the overall prototype closer to the real prototype. At the same time, in the UE4 engine, set up the up, down, left, and right keys on the keyboard, bind the Up, Down, Right, and Left direction control functions, and bind the Turnaround control function for the mouse to achieve scene roaming between the entire virtual template. This allows for a 720 degree panoramic view of the model room and indoor roaming display using a mouse and keyboard [14]. A single PC display cannot meet the experience of multiple users. The present invention generates a QR code through the scene and scans the QR code to achieve the display of multiple users on the mobile phone. Through the QR code connection, the mobile phone jumps to the panoramic display page of the virtual template room. On the mobile end, users can enable the gyroscope of their phone, switch to VR split screen mode, set corresponding phone parameters, and use simple VR glasses to experience each scene in the virtual prototype room, achieving a 720 degree perspective display and giving users a real-time immersive experience of the scene.

4.2 System Testing

(1) Run the program and enter the VR mode interface of the virtual model room display system. Click on the AR editor in the upper right corner to switch to the home design function module in augmented reality mode. Click the exit button to exit the system.

(2) Sample room roaming function test. By using the up, down, left, right directional keys (or W, A, S, D keys) on the keyboard and the mouse, you can move and roam around the virtual model room at any position and angle. However, before roaming to the closet in the room, the closet material selection menu is displayed on the left side of the interface. Clicking on the material can change the material style of the closet in real time. Similarly, when roaming to the front of the single chair in the bedroom, the single chair style selection menu will be displayed, and click any model in the menu to display the real-time model replacement in the virtual sample room.

(3) Functional testing of changing the style of the TV wall in the virtual model room. Roam to the front of the TV wall, press the M key to change the texture style of the TV wall. This system has four styles set, and use the M key to control the cycle of replacement. Each press changes one style.

(4) Functional testing of TV video playback and stop in the virtual model room. The TV has added multimedia objects and bound the keyboard X key to control playback for the effect of real TV playback in 3D space. Press the X key on the keyboard to start playing the video, and press the X key again to stop playing the video.

(5) Random movement of the model and testing of the automatic sensing door function. Click on the white single sofa in the model room with the mouse, select the model, and the model will follow the mouse movement. Move the mouse to any position in the room and right-click to place the model in that position. When we wander to a certain area in front of the bedroom door, the door that enters the bedroom will automatically open, achieving the simulation effect of an automatic sensing door, and we can enter the bedroom. Especially, due to the physical collision of the automatic sensing door, we can only enter the bedroom through the opened door, which is also a simulation operation of the real model room. When my door is not in the designated area in front of the bedroom door, it is closed and only opens automatically when we wander to a specific area in front of the bedroom door.

4.3 User Subjective Evaluation and Discussion

During and after the experiment, evaluations and feedback from all participants were collected and ranked according to the 5-point Likert scale, as shown in Fig. 4. The traditional gesture based selection method and the author's proposed augmented reality technology have average scores of 5 and 3 in terms of user fatigue, respectively. In augmented reality environments, prolonged hovering and swinging of arms for interaction can significantly increase user fatigue. In addition, in terms of comfort, the average scores of the two are 2 and 4 points respectively, which may be due to the fact that augmented reality technology allows users to interact in more natural postures without the need for significant gestures. In terms of acceptability score, the average score of augmented reality technology is higher than that of traditional gesture interaction methods, at 4 and 3 points respectively. Due to certain limitations of both interaction methods, the overall highest score did not exceed 4 points. The experimental results show that this system is relatively complete, fully demonstrating the authenticity, interactivity, and immersion characteristics of virtual reality technology, and effectively solving the problems in previous model room display systems proposed by the author in the research background.

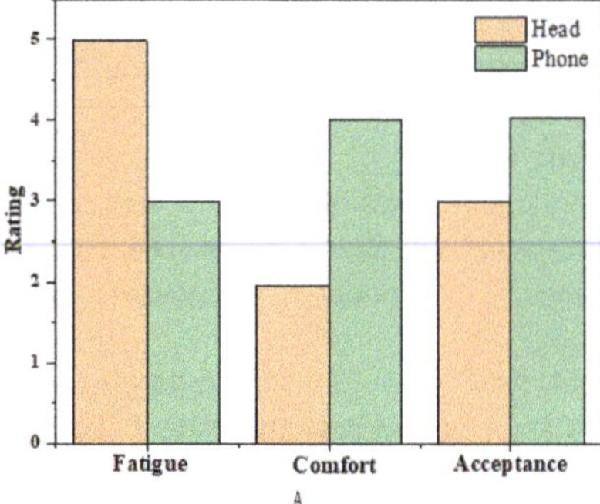

Fig. 4. Subjective evaluation result chart

5   Conclusion

In this paper, we have built an interactive and visual model for indoor design project, which is based on Augmented Reality (AR), to improve the visual representation and interactive experience of the indoor design project. By analyzing the VR and AR techniques deeply, we have probed into its possible use in architectural and indoor areas, and have integrated speech, real time rendering and multisensor integration to create a more immersive and interactive experience. Research has shown that compared to traditional floor plans and physical model rooms, AR based virtual model rooms can significantly reduce design communication costs, improve user understanding and participation in design schemes, and enhance users' intuitive perception of spatial layout, material selection, and decorative styles. In addition, by combining multimodal interaction methods such as speech recognition and gesture interaction, users can more conveniently adjust and optimize the design scheme, improving the flexibility and efficiency of the design. In the future, it will be possible to make the system more efficient, integrate it with reality, and use AI to strengthen its smart interactive ability, so as to satisfy more individual and diversified customer requirements. As AR continues to develop, it will provide more creative opportunities for architectural and indoor design, which will propel the sector toward smarter and more digitalized ways.

References

1. Sahria, Y., Santoso, B., Kuat, T.: Innovative learning media for interior design using augmented reality at vocational high schools. Internet Things Artif. Intell. J. (IOTA) **4**(2), 199 (2024)
2. Maulana, F.I., Azis, B., Primadani, T.I.W., Hasibuan, P.R.A.: Funar-furniture augmented reality application to support practical laboratory experiments in interior design education. Indones. J. Electr. Eng. Comput. Sci. **31**(2), 845 (2023)
3. Almuzaini, T., et al.: Virtual reality cleanroom suite introduction video on students' knowledge and confidence during USP 797 simulation. Am. J. Pharm. Educ. **87**(8), 100541 (2023)
4. De, S., Jackson, C.D., Jones, D.B.: Intelligent virtual operating room for enhancing nontechnical skills. JAMA Surg. **158**(6), 662 (2023)

5. Mohammadrezaei, E., Sarshartehrani, F., Behravan, M., Graanin, D.: Exploring the effectiveness of augmented and virtual reality for visualizing complex lighting simulation data structures and evaluating user experience. In: International Conference on Human-Computer Interaction. Springer, Cham (2025)
6. Xu, W.: Research on user demands and functional design of an ar-based interior design and display platform for recreational vehicles. Appl. Sci. **14**, 10568 (2024)
7. Xue, Z.: Research of language big model + virtual teaching research room. Open Access Libr. J. **11**(1), 8 (2024)
8. Ricci, E., Hulme-Beaman, A., Ressel, L.: The virtual post-mortem room: immersive gross pathology experience in the metaverse. J. Comp. Pathol. **210**, 52 (2024)
9. Krishnan, S., Ademi, Z., Malone, D., Abebe, T.B., Lim, A.: Substitution of a traditional face-to-face workshop with virtual escape room in higher education. Simul. Healthc. J. Soc. Simul. Healthc. **20**(1), 61–69 (2024)
10. Feng, B., Wang, Y., Mo, L., Yang, J., Luo, H., Wei, Z.: Three-dimensional model construction of power distribution room and its digital twin physical structure research. Appl. Math. Nonlinear Sci. **9**(1), 1 (2024)
11. Ringgold, V., Kaiser, P., Eskofier, B.M., Rohleder, N.: #68 - - the impact of a deceptive virtual agent on stress markers in the virtual reality stroop room. Psychoneuroendocrinology **160**, 106852 (2024)
12. Daniel, K., Benjamin, H., Garrett, H.-y: Utilization of virtual reality for operating room fire safety training: a randomized trial. Virtual Real. **27**(4), 3211–3219 (2023)
13. Wrottesley, C.: The physical and virtual space of the consulting room: room-object spaces. Br. J. Psychother. **39**(2), 228 (2023)
14. Steiner, R.A.: Introduction to the virtual thematic issue on room-temperature biological crystallography. IUCrJ **10**, 248–250 (2023)

Optimizing Sustainable Supply Chain Management with Blockchain Technology

Qi Zhang and Wubo Zhang(✉)

Dongbei University of Finance & Economics, Dalian, China
`wubo_zhang@126.com`

Abstract. Aiming at the problems of difficulty in tracking the flow path in the product source of the existing Supply Chain Management System (SCM) and low transparency and traceability, this paper combines blockchain technology to optimize the design and performance of sustainable SCM, aiming to improve the transparency and traceability of system management. First, based on demand analysis, the system architecture is designed from four levels: user interface layer, application service layer, blockchain core layer, and data management layer. Then, the information of each product or raw material is recorded as a block, and data is recorded and tracked based on the blockchain. The Merkle tree is introduced to recursively calculate the blocks, and the block generation interval is automatically adjusted according to the current transaction volume of the network. Finally, in user queries, an index table is created for each important data point to record the location and key attributes of the block where it is located. Experimental results show that compared to SCM under a centralized database, the proposed system significantly enhances supply chain transparency and further improves traceability. The blockchain-based approach reduces information asymmetry among supply chain participants, ensuring the reliability of product provenance and transaction records while providing data protection. By leveraging decentralized ledger technology, the system effectively minimizes the risk of unauthorized modifications, offering enterprises a more secure supply chain management framework. These findings indicate that blockchain-driven SCM holds great potential in enhancing trust, operational efficiency, and sustainability, supporting the further optimization of modern supply chain networks.

Keywords: Sustainable Supply Chain Management · Blockchain Technology · Transparency and Traceability · Systems Design · Performance Optimization

1   Introduction

In today's globalized economy and increasingly fierce market competition, consumers are paying more and more attention to the sustainable development and social responsibility awareness of enterprises, prompting enterprises to pay more attention to SCM design [1, 2]. Traditional SCM has great limitations in terms of transparency, information sharing, and traceability, making it difficult to achieve transparency in the entire process from raw materials to products. There are many deficiencies in tracing the flow of

© The Author(s) 2026
P. Siarry et al. (Eds.): WCNA 2024, LNEE 1550, pp. 215–225, 2026.
https://doi.org/10.1007/978-981-95-6946-5_22

components or materials, which directly affects the operating efficiency and cost of the supply chain, and even affects the company's performance and customer expectations [3, 4]. With the continuous development of digital technology, blockchain technology has made great progress. It has the characteristics of decentralization, immutability and transparency. In the SCM design, this paper uses blockchain technology to establish an open and trusted system that can monitor in real time, share data and effectively collaborate. It is of great significance to improve the rapid response capability and reliability of the supply chain and promote the sustainable development of the entire supply chain.

In order to improve the transparency and traceability of SCM and realize efficient information sharing, this paper combines blockchain technology to study the sustainable SCM design and performance optimization. Taking a manufacturing enterprise as the object, this paper compares the system with the management system based on a centralized database. In terms of query response time, compared with the control system, the average query response time of the proposed system under each test case is reduced by 37.8%; in terms of transparency, the average transparency of the proposed system in the test case is 30.0% higher than that of the control system; in terms of traceability, the average traceability rate of the SCM based on blockchain in this paper is 22.4% higher. In practical applications, the blockchain-based system in this paper can effectively enhance the transparency and traceability of the supply chain.

2 Related Works

Existing research on sustainable SCM mainly focuses on improving the performance and environmental friendliness of the system [5–7]. To achieve sustainability and improve profitability, Ehtesham Rasi Reza improved SCM based on an integer linear programming model and used a multi-objective genetic algorithm and particle swarm algorithm to solve it. The results showed that the proposed method provided stronger robustness [8]. In order to improve delivery quality and achieve a sustainable supply chain, Salcuk Kafiye proposed a closed-loop supply chain network model, which was developed using goal programming techniques in the supply chain process. He combined the developed model with decision variables regarding the number of workers and verified the effectiveness of the method through management practice [9]. Existing research has promoted enterprises to actively manage their environmental and social impacts and has enhanced the overall sustainability of the supply chain to a certain extent, but there are still deficiencies in terms of transparency and traceability.

Blockchain provides more possibilities for building a more open and transparent SCM [10, 11]. Oriekhoe Osato Itohan identified the significant impact of blockchain on improving supply chain transparency and efficiency. Through the analysis of its implementation status in various industries, challenges and solutions in adoption, and comparative performance analysis with non-blockchain systems, he emphasized the advantages of blockchain in reducing costs, improving operational efficiency and environmental sustainability in the supply chain [12]. Chowdhury Rakibul Hasan provided a ledger built on an operational wall based on blockchain technology, which significantly improved the transparency and traceability of the supply chain process. He demonstrated the advantages of the proposed method in terms of supply chain clarity and functionality through application practice [13]. Jan Amin used blockchain technology to realize the transmission of green environment-related information, and achieved greener supply chain governance practices by storing data in an immutable ledger accessible to network members and sharing compliance with multiple stakeholders in real time [14]. Existing research uses blockchain to provide strong technical guarantees for the authenticity and compliance of product sources, but there are still limitations in the balance between the efficiency and transparency of the system.

3 Methods

3.1 Requirements Analysis

3.1.1 Functional Requirements

Functional requirements are considered from three levels:

(1) Transparency

In SCM, transparency is a critical issue. Traditional SCM suffers from information asymmetry and decentralized data storage, which makes the information disclosure of enterprises less transparent, making it difficult for consumers and regulators to certify whether it is sustainable and meets ethical standards. High transparency of information can ensure that all participants obtain authentic and reliable data. This requires the system to not only comprehensively record all links from raw material purchase to final sales, but also establish a mechanism that can be updated and shared with authorized users in a timely manner.

(2) Traceability

To track the entire product cycle, it is necessary to be able to keep detailed records of the logistics trajectory of each product and quickly find specific batches. This requires the system to have strong data processing capabilities, to be able to efficiently manage and extract massive transactions, and to add accurate time tags to each transaction. Ensuring the accuracy of the event sequence is the key to tracking the historical development trajectory of commodities.

(3) Integration and Automation

Since the supply chain is a multi-agent collaborative process, the system must have the ability to integrate and automatically control in order to effectively implement various agreements and rules on the supply chain. For example, when goods are delivered, if certain conditions (such as quality inspection, etc.) are met, smart contracts can be used to achieve efficient payment while reducing manual intervention. Through integration and automation, the behaviors of both parties in the transaction are standardized, the trust in each link of the supply chain is improved, and the risk of fraud in the transaction is reduced.

3.1.2 Performance Requirements

In addition to meeting functional requirements, the performance of the system is not only related to its practicality, but also has a direct impact on the user experience and the scalability of the system. This paper also considers its performance requirements at three levels:

(1) Efficient processing capability

Since SCM involves a large number of transactions and data needs to be updated in real time, the system needs to have a high processing speed to ensure the rapid execution of various businesses. Especially during peak periods such as promotional seasons and emergencies, it is required to support high concurrent processing capabilities while ensuring the normal operation of the business.

(2) Scalability

As the scale of business operations continues to expand, the market environment is also constantly changing, and the number of participants and the scale of transactions can continue to increase. This requires the system to have good lateral expansion capabilities, that is, to increase the capacity of the network by adding nodes on the original basis. In addition, the modular design makes the system very flexible, and new functional modules can be added or the existing structure can be adjusted according to actual conditions to meet market requirements.

(3) Security

In sustainable SCM, multiple security mechanisms need to be adopted to prevent unauthorized access to sensitive information of the enterprise. When sharing sensitive information, the authenticity of the information must be authenticated without leaking specific data. In system design, an effective access control mechanism needs to be built to ensure that relevant data resources are only used by authorized users to avoid information leakage and abuse.

3.2 Overall Design

3.2.1 System Architecture

Based on the demand analysis, the overall design of the sustainable SCM architecture is carried out. The core of the system is to give full play to the characteristics of blockchain and improve the transparency and traceability of the system. The essence of blockchain is a distributed ciphertext database, which aims to ensure the reliable storage and exchange of data through decentralization and trusted auditing [15]. The mechanism consists of multiple nodes, each of which holds the same data and stores it as the hash value of the previous block, thus forming an unalterable chain structure. When new data is to be added to a new block, the consistency mechanism must be used to make all nodes agree, and then add the data to the new block. Such a data structure ensures the high security and reliability of blockchain technology. Since any modification of data can cause the hash value to change, tampering can be quickly detected. The overall architecture of this system is shown in Fig. 1:

As shown in Fig. 1, the system architecture of this paper consists of the user interface layer, application service layer, blockchain core layer and data management layer.

The first layer is the user interaction layer, where users can complete tasks such as placing orders and checking product status through the interface. Through the application program interface, it can interact with other programs to achieve data sharing and service collaboration. Common query results can be preprocessed and cached, thereby shortening the system's response speed and improving the system's operating efficiency.

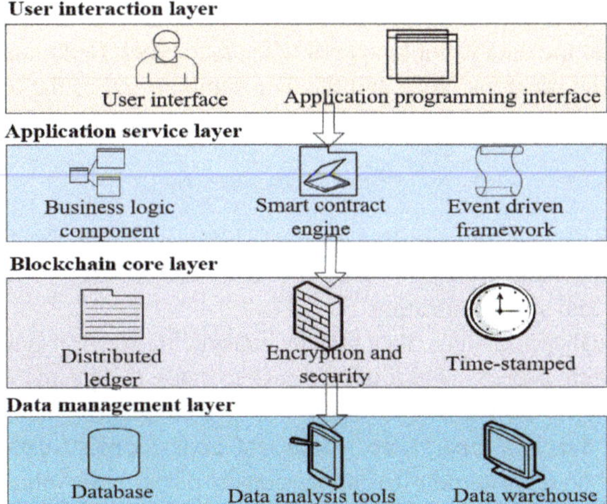

Fig. 1. Overall design of SCM architecture

The second layer is to manage complex business logic to ensure that various businesses can be carried out according to predetermined rules. The smart contract engine can automatically complete qualified transaction processing, making the cooperation process between enterprises smoother. The event-driven framework enables the system to respond to changes in the external environment in a timely manner.

At the third layer, a distributed ledger is used to store each business data in a new data block, which is jointly maintained by each node. The encryption and security module can ensure the security of information during transmission and prevent unauthorized access. At the same time, the timestamp service can also provide users with an accurate event sequence and improve the traceability of the system.

The last layer is the data management layer, which mainly stores user data and enterprise data, including user data and enterprise information. It also provides corresponding data analysis tools. By deeply analyzing a large amount of data, valuable information can be obtained, thus providing a reference for enterprise decision-making. The data warehouse is used to centrally store historical data for long-term analysis.

3.2.2 Blockchain Mechanism

(1) Data Recording and Tracking

This paper uses blockchain as a basis to establish a sustainable SCM to ensure transparency from production of raw materials to final consumption. The information of each product or raw material is recorded in the form of a module and connected to the previous module to form a chain structure. It contains both the basic information of the product (such as name, batch number, etc.) and all links from production to processing, transportation to sales.

Assuming that a product has a total of n events in its life cycle, the overall form of the chain can be expressed as:

$$C = \{B_1, B_2, \cdots, B_n\} \tag{1}$$

Here, B_i represents the block corresponding to the i th transaction or operation. To ensure that the data is not tampered with, each block B_i contains the hash value $H(B_{i-1})$ of the previous block and the data summary D_i of the current block, that is:

$$B_i = H(B_{i-1}) + D_i \tag{2}$$

In this way, any change to historical data can cause the hash value of all subsequent blocks to change, so it can be easily found.

(2) Data update and synchronization

In practical applications, data update and synchronization is a very tedious and time-consuming process, and it is very easy to cause delays and errors. This paper introduces the concept of Merkle tree to improve the security and reliability of data. The Merkle tree is a binary tree whose leaf nodes consist of the hash values of its two child nodes, and non-leaf nodes consist of the hash values of its two child nodes. For any block B_i, its Merkle root can be obtained by recursion:

$$R_i = Hash(Hash(L_{i-1})Hash(R_{i-1})) \tag{3}$$

Here, L_{i-1} and R_{i-1} represent the root node hash values of the left and right subtrees, respectively. The Merkle tree method can quickly confirm a specific data, and in most cases, only the hash value of the corresponding branch needs to be passed to other nodes, thereby reducing the amount of data transmission.

(3) User query

The index mechanism can be introduced into the user's query process, an index table is established for each key data point, and the index table is recorded in the location and key attributes of each node. After the user makes a query request, the system can directly find the corresponding block for retrieval, greatly improving the system's response speed. Assuming that the purpose of a certain query is to obtain the specific circulation route of a batch of products, the retrieval index table can be used to quickly find the entire block number set $S = \{ID_1, ID_2, \cdots, ID_m\}$ corresponding to the batch, and read out the contents of these blocks in turn to complete the query.

4   Results and Discussion

In order to verify the effect of sustainable SCM design and performance optimization based on blockchain technology, this paper conducts a systematic test on it.

4.1   Experimental Data and Settings

This paper takes a manufacturing enterprise as the object and uses the real transaction records between the enterprise and its upstream and downstream enterprises as test data. From raw materials to product sales, the entire supply chain process is complete and representative, which can well reflect the role of the system in practice. The test data covers the entire supply chain process and is anonymized for confidentiality purposes. The specific information is shown in Table 1:

Table 1. Test data information

Data type	Classification	Number of samples	Time span
Transaction records	Purchase information	21772	From January 2024 to December 2024
	Transportation information	32819	
	Warehousing information	29455	
Logistics tracking	Product Location	50308	
	State change	33982	
Inventory management	Inventory level	10353	
	Inventory changes	17769	
Quality inspection	Raw material testing	18344	
	Finished product testing	10592	

The centralized database management system currently used by the company is used as a comparison. This system relies on a single server to store and process supply chain-related data. In order to objectively evaluate the performance of the two systems, a comparison is made from three indicators:

(1) Query response time

Query response time refers to the time from when a user sends a query request to when a result is received. It is calculated as:

$$Response_{time} = T_{response} - T_{request} \qquad (4)$$

(2) Transparency

Transparency refers to comparing the ratio of published data to total data to determine the transparency of the system in data management:

$$TI = \frac{D_{public}}{D_{total}} \times 100\% \qquad (5)$$

(3) Traceability

Traceability evaluates the clarity of a product's origin based on the degree to which each product can be traced back to its original supplier:

$$TS = \frac{N_{traced}}{N_{total}} \times 100\% \qquad (6)$$

4.2 Experimental Results

(1) Query response time results

According to the test data classification, 9 types of data were randomly selected as test cases, and the system response time was compared, as shown in Fig. 2.

As can be seen in Fig. 2, the query response time of the proposed system is shorter than that of the control system. The average query response time of the centralized database-based system under each test case is 0.82 s, while the average query response time of the proposed system under each test case is 0.51 s, which is 37.8% lower. From the specific comparison results, the traditional centralized database relies on one server to complete all tasks, which is very easy to have bottlenecks in a high-concurrency environment, thereby increasing the query response time. The blockchain-based system in this paper introduces the concept of Merkle tree, and has a distributed ledger and smart contract mechanism, which reduces unnecessary intermediate steps and adopts efficient consistency and caching mechanisms to improve data processing efficiency.

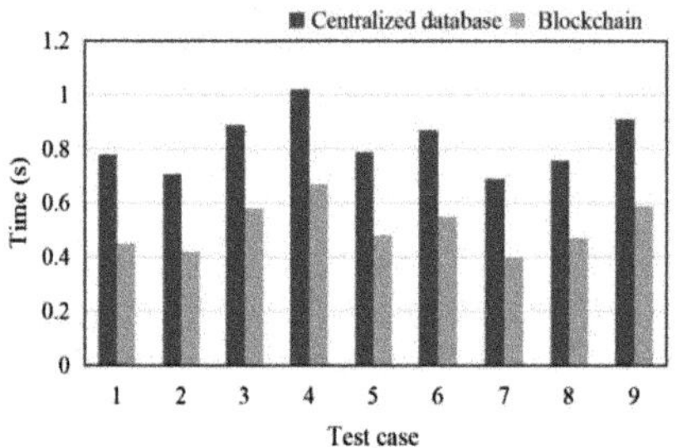

Fig. 2. Comparison of query response time

Fig. 3. Transparency comparison

(2) Transparency results

In the transparency index comparison, the 9 types of data extracted were also used as test cases to compare the degree of openness of the two types of systems in data association, as shown in Fig. 3.

In Fig. 3, the transparency of the two types of systems under test cases shows a significant difference. The average transparency of the system based on the centralized database under each test case is 59.4%, and the average transparency of the system in this paper under each test case is 89.4%. Compared with the control system, the average transparency of the blockchain-based SCM in this paper is 30.0% higher. Traditional centralized database systems, due to their reliance on a single data storage center, often encounter the problem of "information islands" and are unable to achieve all-round information sharing and transparent management, resulting in a low overall level of transparency. This paper adopts a distributed approach based on blockchain to ensure that multiple nodes collaborate to save transaction records and implement pre-set rules through smart contracts, thereby improving the authenticity and immutability of data.

(3) Traceability results

Comparing the traceability of each system under the test case, the specific results are shown in Fig. 4.

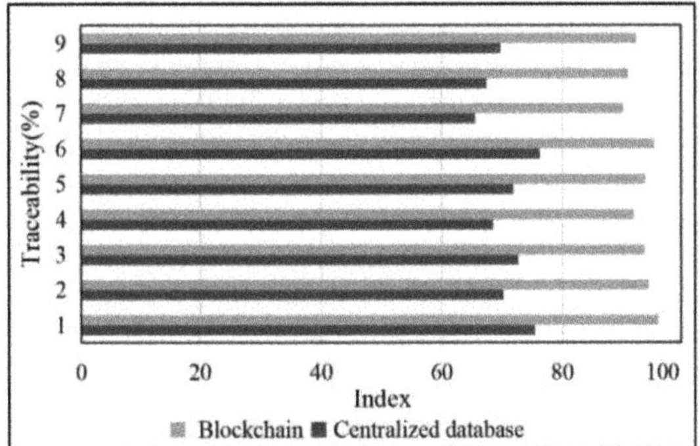

Fig. 4. Traceability comparison

As can be seen from Fig. 4, the system in this paper has a more significant advantage in the traceability comparison. In the specific comparison, the average traceability rate of each test case in the system in this paper reached 93.3%, and the average traceability rate of the system based on the centralized database was 70.9%. In comparison, the average traceability rate of the blockchain-based SCM in this paper was 22.4% higher. The results show that blockchain technology can effectively ensure that all transaction records are accurately and completely saved, and can be accurately traced back to each link through timestamp services, thereby improving the traceability and reliability of the supply chain.

4.3 Discussions

This paper verifies the significant advantages of sustainable SCM based on blockchain technology in terms of query response time, transparency and traceability through comparative tests. The test shows that compared with the centralized database, the average query response time of the proposed system is reduced by 37.8%; the average transparency is 30.0% higher than the control system and the average traceability rate is increased by 22.4%. The results show that the use of blockchain technology can effectively improve the real-time and responsiveness of the system and improve the transparency and traceability of the supply chain. This paper takes blockchain technology as the core, and on the premise of ensuring the efficiency of supply chain operation, it meets the stringent requirements of modern supply chain for information transparency and credibility, and provides certain guidance for building a more reliable and effective SCM.

5 Conclusions

This paper combines blockchain technology to study the design and performance optimization of sustainable SCM. By introducing Merkle tree, dynamic adjustment strategy and index mechanism, an efficient, transparent and traceable sustainable SCM is established. It can be applied to sustainable SCM, providing some support for solving the problem of full process transparency in all links of the supply chain and building a more credible and efficient SCM. However, this paper also has some shortcomings. During the system testing process, no further analysis was done on the challenges of network latency and scalability. Future research can focus on how to optimize the system under high concurrency, and can introduce technologies such as the Internet of Things and organically integrate them with blockchain technology to improve the efficiency of the entire SCM, promote the development of enterprise management, and further enhance the competitive advantage of the enterprise.

Acknowledgment. Fund Number: Liaoning Provincial Basic Research Program for Young Scholars, JYTQN2023174.

References

1. Fu, Q., Rahman, A.A.A., Hui, J., Jawad, A., Ubaldo, C.: Sustainable supply chain and business performance: the impact of strategy, network design, information systems, and organizational structure. Sustainability **14**(3), 1080–1094 (2022)
2. Sheng, X., et al.: Green supply chain management for a more sustainable manufacturing industry in China: a critical review. Environ. Dev. Sustain. **25**(2), 1151–1183 (2023)
3. Xiao, Q., Chen, L., Xie, M., Wang, C.: Optimal contract design in sustainable supply chain: interactive impacts of fairness concern and overconfidence. J. Oper. Res. Soc. **72**(7), 1505–1524 (2021)
4. Zimon, D., Tyan, J., Sroufe, R.: Drivers of sustainable supply chain management: practices to alignment with un sustainable development goals. Int. J. Qual. Res. **14**(1), 219–236 (2020)
5. Shekarian, E., Behrang, I., Amirreza, Z., Jukka, M.: Sustainable supply chain management: a comprehensive systematic review of industrial practices. Sustainability **14**(13), 7892–7921 (2022)
6. Madani, B., Saihi, A., Abdelfatah, A.: A systematic review of sustainable supply chain network design: optimization approaches and research trends. Sustainability **16**(8), 3226–3258 (2024)
7. Bernal, G., Paola, J., Castro, J.A.O., Mantilla, C.E.M.: The sustainable supply chain: concepts, optimization and simulation models, and trends. Ingenieria **25**(3), 355–377 (2020)
8. Ehtesham, R., Reza, R., Mehdi, S.: A multi-objective optimization model for sustainable supply chain network with using genetic algorithm. J. Model. Manage. **16**(2), 714–727 (2021)
9. Salcuk, K., Sahin, C.: A novel multi-objective optimization model for sustainable supply chain network design problem in closed-loop supply chains. Neural Comput. Appl. **34**(24), 22157–22175 (2022)
10. Van, N., Truong, H., Cong, P., Minh, N.N., Li, . Mohammadreza, A.: Data-driven review of blockchain applications in supply chain management: key research themes and future directions. Int. J. Prod. Res. **61**(23), 8213–8235 (2023)
11. Sahoo, S., et al.: Blockchain for sustainable supply chain management: trends and ways forward. Electron. Commer. Res. **24**(3), 1563–1618 (2024)

12. Oriekhoe, O.I., Omotoye, G.B., Oyeyemi, O.P., Tula, S.T., Daraojimba, A.I., Adefemi, A.: Blockchain in supply chain management: a systematic review: evaluating the implementation, challenges, and future prospects of blockchain technology in supply chains. Eng. Sci. Technol. J. **5**(1), 128–151 (2024)
13. Chowdhury, R.H.: Automating supply chain management with blockchain technology. World J. Adv. Res. Rev. **22**(3), 1568–1574 (2024)
14. Jan, A., Salameh, A.A., Rahman, H.U., Alasiri, M.M.: Can blockchain technologies enhance environmental sustainable development goals performance in manufacturing firms? Potential mediation of green supply chain management practices. Bus. Strategy Environ. **33**(3), 2004–2019 (2024)
15. Han, Y., Fang, X.: Systematic review of adopting blockchain in supply chain management: bibliometric analysis and theme discussion. Int. J. Prod. Res. **62**(3), 991–1016 (2024)

Quantitative Analysis of the Influence of Color Feature Extraction Based on Computer Vision and K-means Clustering Algorithm on Consumers' Purchase Intention

Bing Xu[✉]

School of Business Administration, Shanghai Lixin University of Accounting and Finance, Shanghai, China
xubingsh@yeah.net

Abstract. This study proposes a color feature extraction method combining computer vision and the K-means clustering algorithm to quantitatively analyze the impact of lotus pattern colors on consumers' purchase intentions. Based on the HSV color space, the primary color features of different lotus patterns were extracted using the K-means clustering algorithm. The SOR (Stimulus-Organism-Response) model was applied to analyze the relationship between these color features and consumers' emotions (arousal, pleasure) and purchase intentions. The experiment was conducted using the E-prime visual stimulation environment, involving 40 participants in a psychological experiment. The data collected were analyzed using SPSS and AMOS software for reliability and structural equation model validation. The results indicate a significant correlation between the saturation and brightness of the colors in the patterns and consumers' emotional responses and purchase intentions. Lotus patterns with colors closer to natural states (high brightness, low saturation) were more likely to stimulate consumer interest. This study provides technical support for optimizing the color design of traditional patterns based on computer vision and offers valuable insights into enhancing consumer perception in product design.

Keywords: Computer Vision · K-means Algorithm · HSV Color Space · Consumer Purchase Intention · Quantitative Analysis

1 Instruction

Lotus pattern, as one of the traditional auspicious patterns in China, is widely used as decorative patterns of various textiles and garments because of its diverse shapes and rich colors. Since color is the most significant influence in visual stimulation [1] and closely related to people's sensation, therefore, the study chooses the color composition of lotus pattern as the main research subject. As for traditional lotus patterns, former research have discussed the application of lotus patterns in clothing and textile design from theoretical perspectives based on their design elements such as their shapes and colors [2–4]. The results help to enhance consumers' esthetic experience when purchasing

© The Author(s) 2026
P. Siarry et al. (Eds.): WCNA 2024, LNEE 1550, pp. 226–236, 2026.
https://doi.org/10.1007/978-981-95-6946-5_23

relevant products. Although some studies have explored the influence of the shapes and colors of Chinese traditional patterns, especially lotus patterns, on consumers' subjective aesthetic feelings and their purchase intentions, most of the studies remain theoretical. Few studies focused on establishing the quantitative relationship based on computer image feature extraction algorithm between patterns' design elements and consumers' subjective feelings. Therefore, further research should be done on the relevant area.

2 Theoretical Basis

Firstly, the Stimulus-Organic-Response (SOR) theory was originally proposed from the field of environmental psychology to explore the influence of environmental cues on individuals' emotions and their further behavioral responses [5]. Relevant research showed the significant correlation between clothing patterns (visual stimuli) and consumers' emotional feelings (organism's pleasure and arousal emotions) [6, 7]. Therefore, SOR theory is applicable to this study.

Secondly, in the field of computer image processing, because of the easy operation, K-means clustering algorithm is often used in image color feature extraction in order to quantify image color information [8]. Therefore, this paper chose K-means clustering algorithm for color extraction in HSV color space and further analysis.

3 Research Hypothesis

3.1 Hypothesis on the Correlation Between Color Stimulation and Consumer Emotion

Some studies have shown that visual stimuli can have important effects on consumers' mood [9, 10]. Meanwhile, in all elements of clothing design, patterns can be seen as the most significant visual stimulation for consumers [7, 11].

Therefore, the following hypotheses are proposed:

H1a: The color composition of lotus patterns positively affects consumers' arousal emotion.
H1b: The color composition of lotus patterns positively affects consumers' pleasure emotion.
H1c: Consumers' arousal emotion positively affects consumers' pleasure emotion.

3.2 Hypothesis on the Correlation Between Color Stimulation and Consumer Patronage Intention

Mehrabian and Russell believed that, in the SOR model, emotions can be divided into pleasure dimension for happiness and arousal dimension for excitement, which have positive impacts on people's consumption behavior [5]. Studies have shown that clothing patterns influence consumers' purchase intentions positively [6, 7, 10], among which, colors of patterns played a major role [11–13].

Therefore, the following hypotheses are proposed:

H2a: Consumers' arousal emotion positively affects their patronage intention.

H2b: Consumers' pleasure emotion positively affects their patronage intention.
H2c: The color composition of lotus patterns positively affects consumers' patronage intention.

4   Research Method

4.1   Sample Design

10 lotus samples (P1-P10) with similar structures and representative color schemes are shown in Fig. 1.

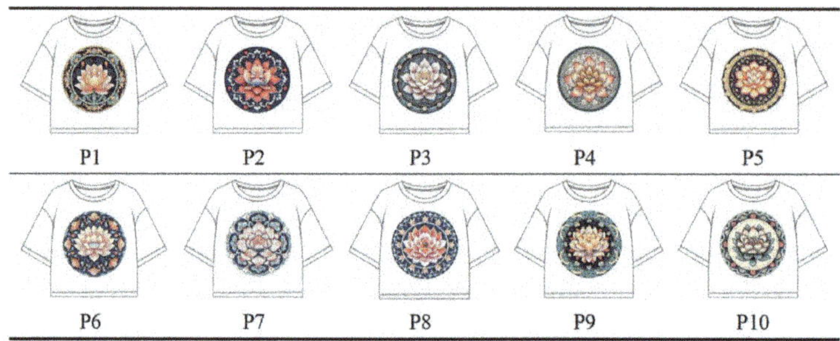

Fig. 1. The 10 Research Samples

4.2   Measurement Scale Design

The scale is a seven-point Likert scale, and its each item is designed based on the existing references. The meanings of 1 to 7 differed according to questions. The designed scale and references are shown in Table 1.

Table 1. Designed Scale and References

Variable	Question	References
Pleasure	This T-shirt makes me feel: annoyed—pleasant	Mehrabian & Russell [5]
	This T-shirt makes me feel: unease—ease	
	This T-shirt makes me feel: anxious—relaxed	
Arousal	This T-shirt makes me feel: striking	Mehrabian & Russell [5]
	This T-shirt makes me feel: exciting	
	This T-shirt makes me feel: exaggerated	

(continued)

Table 1. (*continued*)

Variable	Question	References
Patterns Color Visual Effect	I think the design of this T-shirt is: unsuccessful—successful	Baker et al. [14]
	I think the design of this T-shirt is: dissatisfied—satisfied	
	I think the design of this T-shirt is: boring—desirable	
Patronage Intention	When I see this T-shirt, I: have interest to buy it	Baker et al. [14]
	When I see this T-shirt, I: am willing to learn more about it	
	When I see this T-shirt, I: intend to buy it	
	When I see this T-shirt, I: am willing to recommend it to my friends	

4.3 Research Method and Participants

Study was conducted by E-prime experimental environment measured by designed scale. Individually and randomly, 10 research samples were presented in the screen center with a white background.

The study recruited 40 college students (20 male and 20 female) for the psychological experiment. The age range of all the participants was 18–24 years old, with an average age of 21.07 years old and a standard deviation of 2.142 years old.

5 Scale Data Analysis and Hypothesis Testing

SPSS 25.0 and AMOS 25.0 software were used to analyze the research data and to test the hypothesis.

5.1 Reliability and Validity Testing

By calculating the reliability of each dimension and the overall reliability in the scale, the reliability coefficients of all dimensions are higher than 0.6, and the overall Cronbach Alpha is greater than 0.8. Therefore, the results are high in reliability and the research data can be used for further analysis. By confirmatory factor analysis using AMOS 25.0, it was shown that all the absolute values of the Std. Estimate for 13 questions in the scale were greater than 0.6, which means the scale has good validity.

5.2 Model Fitting Testing

The model fitting index χ^2 is 104.247, degrees of freedom is 59. The p value is 0.000, indicating that the model fitting effect is significant. The ratio of χ^2 to degrees of freedom is 1.767. The CFI is 0.989, the GFI is 0.979, the TLI is 0.985, which are all greater than 0.900. The RMSEA is 0.032 and the SRMR is 0.024, which are all less than 0.080. Therefore, the model fitting effect is good.

5.3 Model Verification

All path coefficients of the model were further tested, and the results were shown in Fig. 2.

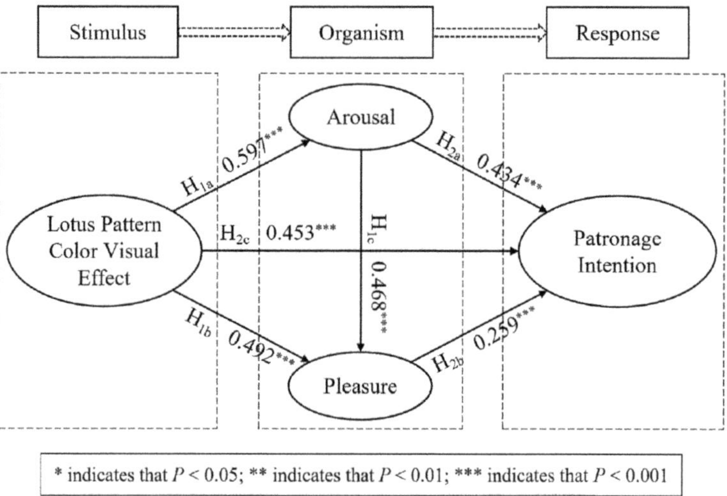

Fig. 2. Model and Its Influence Path

It can be seen from Fig. 2 that all the hypotheses of H1a, H1b, H1c, H2a, H2b, and H2c are proved to be true.

6 Pattern Color Information and Consumer Subjective Experience

For the choice of color representation, both RGB color space and CMYK color space are mainly used in computer software, while HSV color space is a model proposed according to the features of color observation by humans. HSV color space uses H (Hue), S (Saturation) and V (Value) to describe the color features, which can reflect the relationship between colors directly. In order to analyze the relationship between consumers' subjective experience and different color information visually, HSV color space was chosen for further analysis.

6.1 Color Space Conversion

The default image format is in BGR mode, so the BGR mode need to be converted into the RGB mode firstly. Then the RGB mode can be converted into HSV mode finally. The formula of converting RGB color space into HSV color space is shown below. Set max and min as the maximum value and minimum value of r, g and b respectively, that is, (h,s,v) in HSV color space is:

$$
h = \begin{cases}
0° \ \textit{if } \max = \min, \\
60° \times \frac{g-b}{\max - \min} + 0° \ \textit{if } \max = r \textit{ and } g \geq b, \\
60° \times \frac{g-b}{\max - \min} + 360° \ \textit{if } \max = r \textit{ and } g < b, \\
60° \times \frac{b-r}{\max - \min} + 120° \ \textit{if } \max = g, \\
60° \times \frac{r-g}{\max - \min} + 240° \ \textit{if } \max = b;
\end{cases}
$$

$$
s = \begin{cases}
0 \ \textit{if } \max = 0, \\
\frac{\max - \min}{\max} = 1 - \frac{\min}{\max} \ \textit{others};
\end{cases}
$$

$$
v = \max .
$$

6.2 Color Extraction Method

In order to quantify the color information of research patterns, this paper selected K-means algorithm with simple operation and strong applicability to extract the color features of research samples.

The basic thought of K-means clustering algorithm is to minimizes the sum of squared distances between each pixel within the color classifications and its corresponding cluster center. The formula is,

$$
J = \sum_{n=1}^{N} \sum_{k=1}^{K} r_{nk} \| C(n) - \mu_k \|^2
$$

In the formula, "n" represents a pixel coordinate in the background, "C(n)" represents the color value of the pixel, "N" represents the color data sample of the background, and "K" represents the number of color classifications. And then, "k" represents the cluster center of each color classification, "r_{nk}" represents whether the current color belongs to the classification of number k. If r_{nk} is 0, the current color does not belong to class k. If r_{nk} is1, the current color belongs to class k. A few K-means cluster algorithm may result in local optima rather than the global optima. To overcome the defect, several times of K-means analysis on the background image should be done for the optimal main colors.

The K-means cluster algorithm consists of six steps.

(1) The first step is to determine the value of K. Specify a possible maximum cluster randomly at first, then increase the cluster number from 1. When the set number of the cluster is approaching the real number, "J" will decrease rapidly. By plotting a

graph of k, the inflection point while "J" decreasing can be found, thus, the value of k can be better determined.

(2) The second step is to determine the center of the initial classification by selecting K data points randomly or in a certain way.

(3) The third step is to calculate the distance from each point to each center separately, then each point can be divided into the classification closest to the center.

(4) The fourth step is to select the new center by taking the mean of each classification when each center has a number of points.

(5) The fifth step is to compare the new center with the previous center. If the distance between the new center and the previous center is less than a certain threshold, or the number of iterations exceeds certain threshold, then the clustering is considered to be converged, and the processes stop.

(6) The sixth step is to continue the iterative execution if step 5 is not complete until the processes stop.

6.3 Lotus Pattern Color Feature Extraction

Separate the Patterns Off. Since lotus patterns are on the T-shirts, the first step of the color extraction is to separate the patterns off from the T-shirts. Therefore, colored images need to be converted to the grayscale images, then the image binarization can be done. Thus, the patterns can be detected and separated off. Then the K-means algorithm can be used to extract the main colors of the lotus patterns.

The specific processes are shown in Fig. 3.

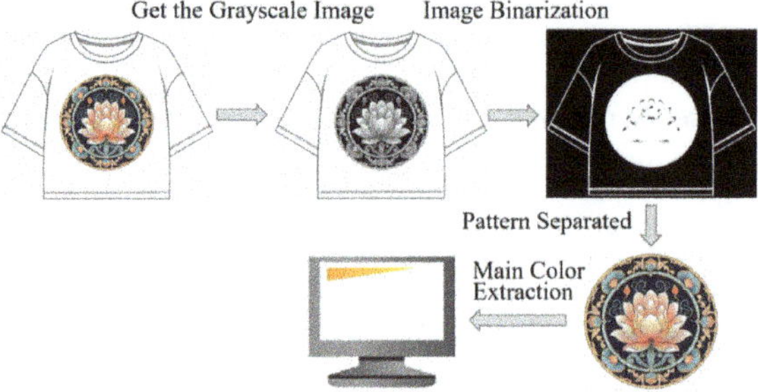

Fig. 3. Processes of Patterns' Color Extraction

Patterns' Color Feature Extraction. Since the color composition of each pattern is different, the numbers of main colors for each lotus pattern, which are the K values, need to be determined firstly. Take sample P1 and sample P2 as an example, it can be seen from Fig. 4 that when k is 7 in graph (a) and 6 in graph (b), both curves show the obvious inflection points, therefore, the k values for P1 and P2 are 7 and 6 respectively.

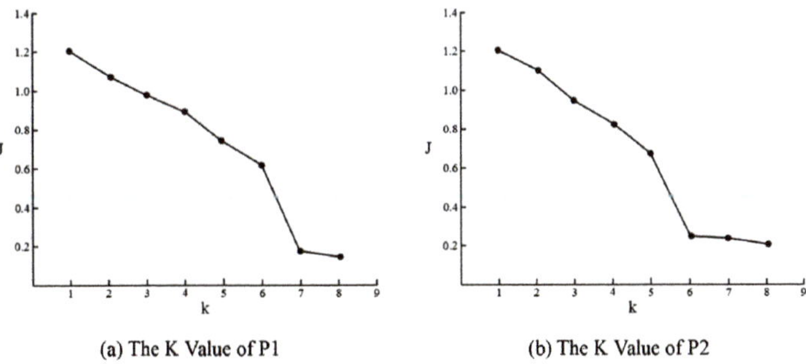

Fig. 4. The Determination of K Value

Then, after several rounds of iterative clustering, the main color schemes of 10 lotus patterns and the proportion of each color were obtained. The main colors and their proportion of 10 research samples are shown in Fig. 5.

Sample No.	Main Colors	Proportion (%)	H	S	V
P1		36.02	179	100	17
		23.48	21	48	100
		15.70	29	40	86
		11.25	179	20	86
		8.03	179	23	69
		3.37	33	31	99
		2.15	44	50	99
P3		30.43	180	51	35
		20.08	239	49	33
		16.95	359	20	86
		12.74	181	33	52
		9.38	60	11	99
		6.29	29	34	84
		4.13	209	49	68
P5		36.58	44	49	99
		30.17	239	51	35
		15.29	34	30	99
		10.53	22	48	99
		7.43	2	78	91
P7		23.17	0	0	86
		21.42	3	14	99
		19.16	209	100	34
		17.39	179	25	69
		12.28	183	20	87
		6.58	357	59	84
P9		31.46	204	87	18
		23.57	184	36	64
		19.69	34	31	99
		15.33	44	30	98
		6.79	5	83	75
		3.16	359	22	87

Sample No.	Main Colors	Proportion (%)	H	S	V
P2		42.17	180	100	17
		21.49	210	100	35
		13.75	33	31	100
		9.53	0	60	86
		7.69	0	75	69
		5.37	0	48	100
P4		35.27	2	13	99
		21.18	179	50	35
		18.14	179	25	69
		13.45	34	31	99
		7.63	15	65	100
		4.33	56	13	100
P6		37.90	239	51	35
		18.25	0	14	100
		16.17	20	60	86
		12.68	34	31	99
		9.43	208	37	83
		5.57	29	34	84
P8		36.40	213	81	40
		19.82	0	14	100
		16.45	21	48	100
		12.18	34	29	99
		9.42	0	75	68
		5.73	62	14	100
P10		30.17	3	14	99
		24.74	179	50	35
		17.36	179	25	69
		10.65	59	20	87
		7.28	40	14	98
		5.31	0	39	85
		4.49	181	32	52

Fig. 5. The Main Colors and the Proportion of 10 Lotus Patterns

6.4 Patterns' Color Information and Consumers' Subjective Experience

Since a seven-point Likert scale was used, it can be considered that when the measurement result is 5 or greater, consumers' overall perception of the sample was positive. In contrast, when the measurement result is less than or equal to 3, consumers' overall perception of the sample was negative. The research samples with evaluation scores of both emotional feelings and purchase intentions greater than or equal to 5 were picked out to be the favorable samples, namely P1, P4, P7, P9 and P10. On the contrary, samples with both evaluation scores less than or equal to 3 were picked out to be the unfavorable samples, namely P3 and P5. The classification results and their color composition are shown in Fig. 6.

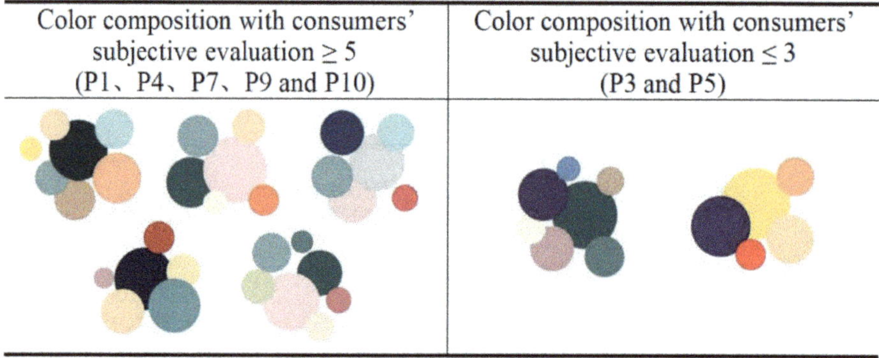

Fig. 6. Samples Classification Results and Their Color Composition

By comprehensive analysis, lotus patterns with higher scores from consumers are tend to use colors with higher value (V ranges from 67% to 100%) and moderate to lower saturation (S ranges from 0% to 66%). In terms of color choice and matching, lotus patterns ranked higher are tend to use light yellow, pink and different shades of green as their main colors, and it can be found that these color schemes are closely related to the colors of real lotus and their leaves in nature. Therefore, it can be inferred that consumers are tend to prefer the colors of patterns that are closer to the colors of things observed in daily lives subconsciously due to the influence of living habits.

7 Conclusion

This study systematically analyzed the impact of lotus pattern color features on consumers' purchase intentions by combining computer vision technology and the K-means clustering algorithm. In this study, the K-means algorithm based on the HSV color space was employed for color feature extraction, and through multiple rounds of clustering analysis, the primary color features of the lotus patterns were successfully extracted and their relationship with consumers' subjective perceptions was quantified. This research not only provides technical support for the modern application of traditional pattern design but also validates the application value of computer vision and quantitative algorithms in visual marketing. The results indicate that the saturation and brightness of colors have a significant positive effect on consumers' emotional responses (pleasure and arousal) and purchase intentions. This finding suggests that the selection and optimization of pattern colors play a crucial role in enhancing product market competitiveness.

Future research can further incorporate multimodal data to explore the impact of traditional pattern design on consumer behavior more comprehensively. Additionally, the analytical methods in this study can be extended to other visual design fields to provide scientific support for design decisions.

References

1. Wang, Q.B.: Enterprise color marketing strategy. J. Jiangnan Univ. **05**, 126–128 (2004)
2. Chen, H.Y.: The application of Ming Dynasty porcelain wrapped lotus grain in modern clothing design. Design **36**(21), 18–21 (2023)
3. Li, K.J., Yu, Z.R.: Research on the application of traditional lotus pattern in silk bedding design. J. Nanning Vocat. Techn. Coll. **18**(05), 15–17 (2013)
4. Meng, F.D.: The lotus pattern in modern clothing design. Text. Guide **10**, 132–134 (2016)
5. Mehrabian, A., Russell, J.A.: An Approach to Environmental Psychology. The MIT Press, Cambridge (1974)
6. Chen, Y.F., Liu, C.: The kansei engineering application of Chinese painting patterns in cheongsam design in Song Dynasty. Wool Spinning Technol. **49**(03), 47–53 (2021)
7. Wei, Y.H., Xia, M., Su, Z.W., Sun, L., Wu, K.M.: Evaluation of influencing factors of clothing pattern design effect based on PAD model. Apparel J. **5**(06), 507–512 (2020)
8. Muller, K.R., Mika, S., Ratsch, G.: An introduction to kernel-based learning algorithms. IEEE Trans. Neural Netw. **12**(2), 181–201 (2001)
9. Wang, T.X., Ma, Z.Q., Yang, L.: Creativity and sustainable design of wickerwork handicraft patterns based on artificial intelligence. Sustainability **15**, 1574–1580 (2023)
10. Yang, C., Won, L.S.: Study on the influence of clothing design elements on purchase intention. Des. Res. **6**, 197–209 (2021)
11. Cheng, L.: Research on modern fashion design based on emotion. Chem. Fiber Text. Technol. **51**(01), 163–165 (2022)
12. Su, Z.W., Wei, Y.H.: Analysis of dose-effect relationship between clothing pattern design effect and consumer preference. J. Wuhan Text. Univ. **34**(01), 50–54 (2021)
13. Hao, X.Y., Liu, K.X.: Research on perceptual evaluation of socks decorative pattern design. Wool Spinning Technol. **49**(03), 54–59 (2021)
14. Baker, J., Parasuraman, A., Grewal, D.: The influence of multiple store environment cues on perceived merchandise value and patronage intentions. J. Mark. **66**, 120–141 (2002)

Challenges and Prospects of Unmanned Long-Distance Non-Contact High-Altitude Strong Sound Directional Non-Lethal Dispersal System

Xiaofeng Zhang[✉]

Engineering College of Information, Engineering University of PAP, Xi'an, China
zxf801031@126.com

Abstract. The unmanned, long-range, non-contact, high-altitude, directed strong sound non-lethal dispersal system represents an emerging technological means in the fields of public safety and military security. Leveraging advanced directional acoustic technology, this system can precisely emit high-intensity sound waves from a high altitude at a long distance, achieving non-lethal dispersal of target crowds or areas. It faces numerous challenges, such as susceptibility to interference in complex environments affecting dispersal effectiveness, the need for improved long-range positioning accuracy to ensure action on specific targets, system stability under prolonged operation and extreme conditions, and strict legal and ethical scrutiny. However, with continuous technological advancements, its prospects are broad. It will play a significant role in maintaining order at large events, border control, counter-terrorism response, and more, potentially becoming an effective alternative to traditional dispersal methods, driving innovative developments in non-lethal equipment technology, and providing new technical support and strategic options for security safeguards.

Keywords: unmanned control · non-contact · dispersal system · high-altitude directed strong sound non-lethal · public safety · military security

Public safety and military security fields require, in multiple scenarios, an unmanned, long-range, non-contact, high-altitude, directed strong sound non-lethal dispersal system. Such a system should integrate dispersal, broadcasting, communication, and music playback functions, enabling flexible deployment and reducing interference through directional sound technology, while intelligent remote control enhances response speed and system flexibility. This system is a comprehensive non-lethal weapon system designed to achieve long-range dispersal, broadcasting, and communication functions through high-intensity directional sound waves. It adopts advanced directional sound technology, capable of operating effectively in different environments and suitable for various complex scenarios such as counter-terrorism, public security and stability maintenance, and crowd incident control and management.

© The Author(s) 2026
P. Siarry et al. (Eds.): WCNA 2024, LNEE 1550, pp. 237–244, 2026.
https://doi.org/10.1007/978-981-95-6946-5_24

1 Challenge Analysis

1.1 Technical Challenges and Innovation Difficulties

1) *High-Precision Positioning and Navigation*

Achieving high-precision positioning and navigation in complex environments poses a significant technical challenge for non-lethal UAV (Unmanned Aerial Vehicle) strong sound systems. Due to factors such as dense urban buildings, complex and varied terrain, and electromagnetic interference, traditional GPS positioning technology often fails to meet requirements. Therefore, there is a need to develop more advanced positioning and navigation technologies, such as vision-based positioning methods and positioning methods combining inertial measurement units (IMUs) with GPS, to improve UAV positioning accuracy and stability, ensuring reliable operation in complex environments.

2) *Autonomous Decision-Making and Intelligent Obstacle Avoidance*

During mission execution, UAVs need to possess high levels of autonomous decision-making and intelligent obstacle avoidance capabilities to cope with various emergencies and obstacles. However, the autonomous decision-making capabilities of most UAVs are still relatively weak, mainly relying on preset flight paths and rules for operation, lacking flexibility and adaptability. Therefore, it is necessary to strengthen the research and application of artificial intelligence technology to enhance UAVs' autonomous decision-making and intelligent obstacle avoidance capabilities, enabling them to make optimal decision-making schemes based on actual situations and timely avoid obstacles to ensure successful mission completion.

3) *Energy Management and Endurance*

Energy management and endurance are key factors restricting UAV development. Most UAVs currently use lithium batteries as their power source, which have limited endurance and long charging times, long-term continuous operation requirements. Therefore, there is a need to develop more efficient energy management systems and new battery technologies to improve UAV endurance and charging speed, meeting the demands of long-term continuous operation. At the same time, exploring the utilization of renewable energy sources such as solar energy can provide UAVs with continuous and stable energy supply.

1.2 Safety Risks and Privacy Protection

1) *Information Security Risks*

With the rapid development of information technology, network security issues are increasingly prominent, and non-lethal UAV strong sound systems also face challenges from information security risks. Malicious actions such as hacker attacks and virus intrusions can lead to UAV loss of control and data breaches, causing severe consequences. Thus, it requires enhanced protective measures and well-rounded protective system on information security for UAV. Such a system shall include functions with encrypted communication, identity authentication, access control, etc. in order to forestall cyber hackings and data breaches.

2) *Infringement of Personal Privacy*

UAV's operation conducting process might engage the collecting and processing of personal privacy information, such as recording surveillance video and audio footages and the like. An abuse or a leakage of these constitutes a grave infringement upon individual privacy rights, and social discontent and legal disputes would ensue. Therefore, to ensure a legitimate and safe usage of information, relevant laws and regulations shall be strictly complied with, and the protective management on personal privacy information ought to be strengthened as well. Meanwhile, the training and management towards UAV's operating personnel also needs reinforcement. Doing so promotes personnel's legal awareness and professional ethics, therefore wards off abuse or leakage of individual privacy information.

3) *Prevention of Accidental Injury Risks*

In UAV's operation conducting process, there might occur casualties or property losses caused by operational errors or malfunctions. Hence, we need to enhance the training and management towards UAV's operating personnel to improve their operational skills and safety awareness. We should also affix value to the equipment maintenance to ensure its normal and good performance. Besides, comprehensive emergency response plans and rescue mechanisms should be set up, so that once there are collateral damages by accident, we can resort to them to conduct effective rescue response in hope of less losses or negative impacts.

2 Development Prospects of Non-Lethal UAV Strong Sound Systems

2.1 Technological Innovation and Intelligent Development

1) *Application of Artificial Intelligence Technology*

Self-navigation and target identification: As AI technology continues to evolve, non-lethal UAV-deployed acoustic hailing device (AHD) system will live up to a higher autonomous navigation capability in both accuracy and efficiency. Via the in-depth learning of algorithms, UAV can analyze environmental data in an autonomous manner, identify potential threats, and implement appropriate response measures. Furthermore, the application of image recognition technology would enable UAV to accurately identify the type and location information of targets, serving as a solid backbone for subsequent operations. Intelligent decision-making algorithms could present optimal solutions based on real-time data, thereby promoting the quality and efficiency in conducting tasks.

Big data analysis and cloud computing: vast amount of data generated by non-lethal UAV-deployed AHD system could be processed with big data analytics technology and restored on the cloud computing platform. We can extract valuable information and rules via the analysis of these data, which can further serve as scientific evidences and supports. Moreover, data instant transmission and data pooling are also feasible on the platform. This facilitates collaboration and communication across departments, thereby improving work efficiency and magnifying synergistic effects.

Integration of IoT Technology: Non-lethal UAV strong sound systems can connect and communicate with other IoT devices, achieving information interconnection and sharing. For example, UAVs can communicate in real-time with ground monitoring systems, transmitting monitoring data and image information; simultaneously, UAVs can also interact with other sensor devices to achieve multi-dimensional environmental monitoring and data collection, providing comprehensive data support and technical guarantees for environmental protection and resource management. Furthermore, the integration of IoT technology can promote cooperation and exchanges between different fields, driving the innovative development and application expansion of related technologies.

2) *High-Precision Positioning and Navigation*

Advanced Positioning Technology: Achieving high-precision positioning and navigation in complex environments poses a significant technical challenge for non-lethal UAV strong sound systems. Traditional GPS positioning technology often fails to meet requirements, necessitating the development of more advanced positioning technologies, such as vision-based positioning methods and positioning methods combining IMUs with GPS. These technologies can advance positioning accuracy and stability for UAV, ensuring a fail-safe operation in complex environment.

Intelligent obstacle avoidance system: Superior performance in intelligent obstacle avoidance enables the non-lethal UAV-deployed AHD system to respond to various emergencies and obstacles. By integrating advanced sensors and algorithms, UAV can timely perceive the surrounding environmental changes, forecast potential collision risks, and adopt corresponding measures to avoid crash. This boosts UAV's safety metrics while minimizing collision-induced damages and impacts.

Optimized energy management: Flaws in energy management and defects in endurance significantly impede the development of UAV. To build up range and accelerate charging rate, we need to hold researches on and develop energy management system with higher efficiency and new battery technologies. Additionally, further researches can be conducted into exploring solar and other renewable energy sources to provide sustainable and stable power support for UAV.

2.2 Application Scenario Expansion

1) *Urban Safety and Public Security Management*

Patrol and monitoring: non-lethal UAV-deployed AHD system also exert significant role in the field of urban security and public order management. Mounted with high-definition cameras, infrared thermal imagers and other sensor devices, UAV can realize real-time surveillance and patrol within key urban areas and major traffic routes. Once it detects abnormal situations or potential threats, UAV would sound prompt warning alarms, deter and turn away threats with its acoustic hailing dispersal function. This can effectively safeguard social stability and public order with reduced labor costs and increased work efficiency.

Emergency Response: In response to emergencies, non-lethal UAV strong sound systems can quickly respond and arrive at the scene for assessment and disposal. For example, in cases of natural disasters such as fires and earthquakes, UAVs can carry fire-extinguishing equipment or rescue supplies for rapid deployment and rescue work; in terrorist attacks and other emergencies, UAVs can deter and stop terrorists through the strong sound dispersal function.

Counter-Terrorism and Anti-Riot: In the field of counter-terrorism and anti-riot, non-lethal UAV strong sound systems have unique advantages and roles. By equipping UAVs withdevices such as loudsound generators, they can emit loud warning sounds to deter and stop terrorists. At the same time, UAVs can also carry non-lethal weapons such as tear gas and stun grenades for dispersal and restraint. This can not only effectively protect the lives and property safety of the people but also reduce casualties and property damage.

2) *Border Patrol and Control*

Remote Monitoring: In border areas, due to complex terrain and harsh climate conditions, traditional manual patrol methods face many difficulties and challenges. However, using non-lethal UAV strong sound systems for remote monitoring can effectively solve these problems. Equipped with sensors such as high-definition cameras and infrared thermographs, UAVs can provide real-time monitoring and patrol of border areas. Upon discovering illegal border crossings, smuggling, or other illegal activities, UAVs can immediately issue warning sounds and use the strong sound dispersal function to deter and disperse individuals.

Intelligence Gathering: Intelligence gathering is a crucial aspect of border patrol. Non-lethal UAV strong sound systems can monitor and analyze border areas' terrain, meteorological conditions, and other factors in real-time by equipping various sensors. At the same time, UAVs can track and monitor suspicious individuals and vehicles to promptly obtain their dynamic information and take corresponding measures.

Emergency Response: In the event of emergencies in border areas, non-lethal UAV strong sound systems can quickly respond and arrive at the scene for assessment and disposal. For example, in cases of illegal border crossings, UAVs can deter and stop individuals through the strong sound dispersal function. At the same time, UAVs can carry non-lethal weapons such as tear gas and stun grenades for dispersal and restraint to ensure the security and stability of border areas.

3) *Environmental Protection and Ecological Monitoring*

Pollution Source Monitoring: In the field of environmental protection and ecological monitoring, non-lethal UAV strong sound systems can play an important role. Equipped with various sensors, UAVs can monitor and analyze pollution sources such as industrial emissions and agricultural non-point source pollution in real-time. Upon discovering excessive emissions or violations, UAVs can immediately issue warning sounds and use the strong sound dispersal function to deter and correct behaviors. This can not only effectively protect ecological environment quality but also promote the implementation of sustainable development strategies.

Biodiversity Protection: Biodiversity protection is an essential part of environmental protection. Non-lethal UAV strong sound systems can monitor and analyze wildlife populations, distribution ranges, and other factors in real-time by equipping sensors such as high-definition cameras and infrared thermographs. At the same time, UAVs can track and monitor illegal hunting, logging, and other destructive behaviors to promptly obtain their dynamic information and take corresponding measures to stop and correct behaviors.

Disaster Warning and Response: In terms of natural disaster warning and response, non-lethal UAV strong sound systems also have unique advantages and roles. Equipped with various sensors, UAVs can monitor and analyze meteorological conditions, geological changes, and other factors in real-time. Upon discovering abnormalities or potential threats, UAVs can immediately issue warning sounds and use the strong sound dispersal function to alert and remind relevant departments to take timely measures for response and disposal, thereby reducing disaster losses and impact ranges (Table 1).

4) *Key Technologies for System Implementation*

Table 1. Key Technologies

Key Technology	Technical Challenges	Breakthrough Directions
High-Directivity Sound Wave Emission Technology	Sound Wave Divergence Angle Control ($\leq 5°$); Sound Pressure Attenuation Compensation at 300m Distance;	Parametric Array Technology: Utilizes ultrasonic carriers to self-demodulate and produce highly directional audible sound; Phased Array Control: 128-channel digital beamforming algorithm;
UAV Acoustic Payload Coupling Technology	Rotor Airflow Interference with Sound Wave Propagation;; Vibration-Induced Acoustic Component Failure;	Aeroacoustic Joint Simulation: Coupling optimization of Computational Fluid Dynamics (CFD) and Acoustic Finite Element Analysis (FEA); Active Vibration Damping System: Magnetorheological damper + predictive control algorithm;
Low-Power High-Efficiency Acoustic System	Energy Supply for Instantaneous Sound Wave Emission Power > 5kW; Thermal Management of High Heat Density Components;	Pulse Energy Reuse Technology: Supercapacitor energy storage + pulse modulation circuit; Phase Change Material Cooling: Graphene-based composite phase change module;

(continued)

Table 1. (*continued*)

Key Technology	Technical Challenges	Breakthrough Directions
Human-Machine Cooperative Safety Control Technology	Precise Dose Control for Non-Lethal Effects; Safety Mechanism in Case of System Failure;	Closed-Loop Feedback Control: Real-time sound pressure monitoring + PID dynamic adjustment; Triple Redundancy Safety Architecture: Hardware interlock + software watchdog + mechanical emergency stop;
Complex Environment Penetration Enhancement Technology	Urban Environment Sound Wave Reflection/Diffraction Interference; Sound Energy Attenuation Compensation in Rain/Fog Weather;	Adaptive Waveform Coding: OFDM sound wave modulation technology; nonlinear acoustic focusing; Utilizing Nonlinear Air Effects to Enhance Penetration;

3 Conclusion

The unmanned, long-range, non-contact, high-altitude, directional, non-lethal sound dispersal system has unique application value but also faces a series of challenges that cannot be ignored. At the technical level, the complex and ever-changing external environment can easily interfere with sound wave propagation, making it difficult to guarantee the dispersal effect. Moreover, precise targeting at long distances remains a difficult problem to overcome, and the system's long-term stable operation also requires continuous optimization. From a non-technical perspective, the system must strictly comply with legal and ethical norms to avoid misuse and other issues.

However, its prospects are still promising. With the continuous development of technology, related technical bottlenecks are expected to be gradually broken through. Issues such as sound wave propagation interference can be improved through algorithm optimization and hardware upgrades, and positioning accuracy can also be enhanced. In terms of application fields, it will play a significant role in scenarios such as security for large events, border security maintenance, and responses to terrorist attacks. With its non-lethal advantages, it can effectively ensure personnel safety and maintain social order, becoming an indispensable innovative equipment in future security work and pushing the security industry towards a more efficient and humane direction.

References

1. Chen, M., Liu, J.: Visualization analysis of combat effectiveness evaluation research based on CiteSpace. Sci. Technol. Innovation **13** (2024)
2. Song, S., Liu, Y., Mao, D., Liu, T., Liu, G.: Evaluation of air force ground attack schemes based on fuzzy-AHP. Ordnance Autom. **03** (2024)

3. Tang, J., Liu, S.: Current status and technological development of new special police equipment. Police Technol. **06** (2023)

4. Zhang, S., Li, Q., Zhang, J., Zou L., Bai, W., Zhang, R.: Analysis of internal ballistics and pressure relief diaphragm of dual-chamber high-low pressure variable initial velocity launch. J. Gun Launch Control (2023)

5. Jiang, Q.: Successful Medical Intervention for 3 Cases of Head and Facial Trauma Caused by Sound and Light Grenade Explosions. J. Trauma Surg. (2023)

6. Zhai, H., Cui, X., Cao, Y.: Design of main charge formula for combustion type tear gas grenade based on uniform design method. J. Ordnance Equipment Eng. (2023)

7. Huang, T., et al.: Characterization method of anti-damage capability of multi-axis special vehicle tire system using equal damage line. J. Natl. Univ. Defense Technol. (2023)

8. Lu Y., Long S., Zhao H., Feng G., Zhao X.: Performance evaluation of heterogeneous UAV swarms based on hybrid model. J. Syst. Simul (2024)

9. Li, R., Liu, H.: Research on enterprise financial performance evaluation based on entropy weight TOPSIS-CNN deep learning. Math. Pract. Theor. (2023)

10. Zhang, R., Jiang, D.: Research on development requirements of anti-access/area denial equipment system for internal security corps. China Equipment Eng. (2023)

11. Yang, S., Li, W., Chen, J.: Research on safety of sulfide ore self-combustion based on PCA-RBF network model. Gold Sci. Technol (2022)

12. Guo, J., et al.: Research on combat effectiveness evaluation method of aviation anti-submarine torpedoes based on entropy weight method. Ship Electron. Eng. (2022)

13. Zou, Q., et al.: Design of combat effectiveness evaluation system for attack-type UUV. J. Unmanned Underwater Veh. Syst. (2022)

14. Zhang, R., Pei, Y., Hou, P., Ge, Y.: Research on damage assessment of early warning aircraft by fragmentation warhead. J. Beijing Inst. Technol. (2022)

15. Zhang, Y., Kang, P., Zhu, Z., Zheng, D.: Research on anti-damage degree measurement method of defense system. Mod. Defense Technol. (2022)

16. Mao, T., Zhang, D., Niu, Y., Yu, M., He, M.: Evaluation method of operational effectiveness for linear time-varying combat systems. Fire Control Command Control (2022)

17. Li, P., Zhang, H., Yan, D., Wang, J., Chang, G.: Research on combat effectiveness evaluation of unmanned aerial vehicle missile weapon system. J Projectiles Rockets Missiles Guidance (2022)

18. Yang, J., Dong, Y., Bian, Y., Yao, T., Geng, X.: System effectiveness evaluation method in joint operations context. Sci. Technol. Rev. (2022)

Data Middle Platform Construction and Application of Natural Resource Survey and Monitoring-Taking Chongqing as an Example

Fu Zou[✉], Zhiyue Zhou, Kai Yang, and Weijian Li

Geographic Information InstituteChongqing Institute of Surveying and Monitoring for Planning and Nature Resources, Chongqing, China
615883454@qq.com

Abstract. The current problems in the field of natural resource survey and monitoring, such as unclear data base, information silos, duplicate data storage, low efficiency of data sharing applications, and slow application of big data analysis, are difficult to meet the demands of natural resource survey and monitoring business. Taking Chongqing as an example, this paper proposes the construction ideas, system architecture, and core capacities of the natural resource survey and monitoring data middle platform. Through practical application, a comprehensive management platform integrating data collection, aggregation, storage, governance, management, analysis, application, and service has been constructed, providing basic support for the digital, informational, and intelligent transformation of data-driven natural resource survey and monitoring.

Keywords: data middle platform · investigation and monitoring · data governance · data application

1 Introduction

In the information age, data has become a fundamental strategic resource and a new type of production factor for the nation [1]. General Secretary Xi Jinping has, on multiple occasions, discussed data resources, proposing the use of big data to enhance the modernization of national governance. With the advancement of special projects such as the Second National Land Survey, unified real estate registration, the "single map" of natural resources, the Third National Land Survey, territorial spatial planning, the delineation of "three zones and three lines," comprehensive monitoring and supervision, special farmland monitoring, and routine monitoring, along with the further maturation and application of data acquisition technologies like 3S technology, drones, and oblique photography, the natural resources sector has accumulated a large amount of spatial data in recent years. This data is characterized by its large volume and the diversity of its sources and structures. However, issues such as low data standardization, low spatialization, poor data currency, and poor correlation also exist [2], necessitating urgent

P. Siarry et al. (Eds.): WCNA 2024, LNEE 1550, pp. 245–254, 2026.
https://doi.org/10.1007/978-981-95-6946-5_25

data governance and integrated management of the existing data. Following the establishment of the Ministry of Natural Resources, the continuous deepening of information technology applications in natural resource surveys and monitoring has led to an increase in independent business application systems. The data generated by these systems and businesses has also become increasingly massive, forming numerous information silos. This results in problems such as redundant data storage, low efficiency in data sharing and application, low system collaboration efficiency, and slow big data analysis applications, making it difficult to meet the modern management needs of natural resource surveys and monitoring.

The rapid advancement and application of new-generation information technologies, including the Internet, big data, cloud computing, and artificial intelligence, offer crucial technical support for addressing challenges in natural resource management. It is therefore essential to establish a management system based on a data middle platform. This involves creating a data hub for natural resource survey and monitoring applications, breaking down data silos, and enabling data assetization and empowerment. Through data sharing and reuse, we can support the rapid development of business and application innovation, thereby providing fundamental support for the digital, informational, and intelligent transformation of data-driven natural resource survey and monitoring.

2 Overview of Data Middle Platform

The concept of a data middle platform originated in the internet sector. With the proliferation and advancement of technologies such as the internet and 5G, the volume of data acquired by enterprises and governments has grown exponentially, with data types expanding to include unstructured data like images and audio [3]. To address the challenges of collecting, processing, and analyzing massive amounts of heterogeneous data, and to establish data assets for rapid response to long-term applications, the concept of a data middle platform emerged. Alibaba first proposed this concept in 2015, defining it as a comprehensive, interconnected, and intelligent data processing platform that integrates technology, guides business operations, and establishes standardized definitions. The construction goal was to efficiently meet the data analysis and application needs of the front-end. In 2016, Alibaba completed the initial framework of its data middle platform, establishing a "big middle platform, small front-end" business model [4], which drove business profit growth by accelerating product iteration and reducing costs. Subsequently, other leading internet companies joined the trend of data middle platform construction.

Data middle platforms are currently widely implemented across various sectors, including the internet, smart cities [5], intelligent coal mines [6], healthcare insurance [7], and the power industry [8]. Within the natural resources domain, Qu Xiaobo [2] and colleagues have proposed a data middle platform solution and construction framework for the natural resources sector, leveraging a "Kubernetes + Docker + microservices" cloud-native architecture. Taking the Guangxi Zhuang Autonomous Region as a case study, Li Qian [9] and others have suggested a natural resources data governance approach that preliminarily addresses data source localization, data catalog organization, data flow definition, and data quality control through data middle platform technology. By integrating data acquisition, processing, and service delivery, they have further refined data sharing and exchange methods, offering a valuable reference for effective natural resources data governance.

2.1 Overall Architecture of the Natural Resource Survey and Monitoring Data Middle Platform

The Natural Resource Survey and Monitoring Data Middle Platform leverages internet technologies, big data analytics, and cloud computing. It is built upon the framework of the "14th Five-Year Plan" for natural resource informatization. This platform consolidates vast amounts of heterogeneous data from various survey and monitoring operations, including land change surveys, routine monitoring, and specialized assessments, as well as data related to territorial spatial planning, land use control, and cultivated land protection. It establishes a comprehensive management platform that integrates data acquisition, aggregation, storage, governance, management, analysis, application, and service provision. By implementing unified data standards and specifications, the platform creates a shareable and analyzable data resource pool. This data-driven approach supports business innovation and transformation, rapidly facilitating front-end business applications. Furthermore, it enhances data analysis and mining efficiency through big data analytics and knowledge graphs, offering a one-stop data solution for natural resource survey and monitoring management departments. Ultimately, it provides crucial data support and decision-making insights for natural resource management.

The overall architecture of the natural resource survey and monitoring data platform comprises an infrastructure layer, a data aggregation layer, a data governance layer, a data-specific layer, a platform support layer, and an application service layer, as illustrated in Fig. 1.

(1) Infrastructure Layer. The infrastructure layer provides the platform with hardware resources such as servers, storage devices, and networks, as well as software resources such as GIS professional software and database software. Based on Tsinghua Cloud, the hardware resources are managed in a unified manner, and a virtualized environment and cloud computing resource pool are provided. Hardware and software resources are the basic support for the entire data middle platform, providing infrastructure resources on demand for applications.

(2) Data Aggregation Layer. By standardizing the classification of data sources, multi-source heterogeneous data from various sources, such as district and county submissions, business department collections, other unit sharing, and the Internet, are classified, organized, and aggregated for storage to form an aggregation library. This library is managed according to business dimensions, and the original information of various data sources is preserved for archiving.

(3) Data Governance Layer. Based on GeoScene Pro, GA Plus, ArcMap, and Python scripts, general data governance tools such as format conversion, coordinate transformation, data merging, topological governance, and data quality inspection are provided. According to data governance rules, the data in the aggregation library is structurally organized, processed, cleaned, and reconstructed to form a complete

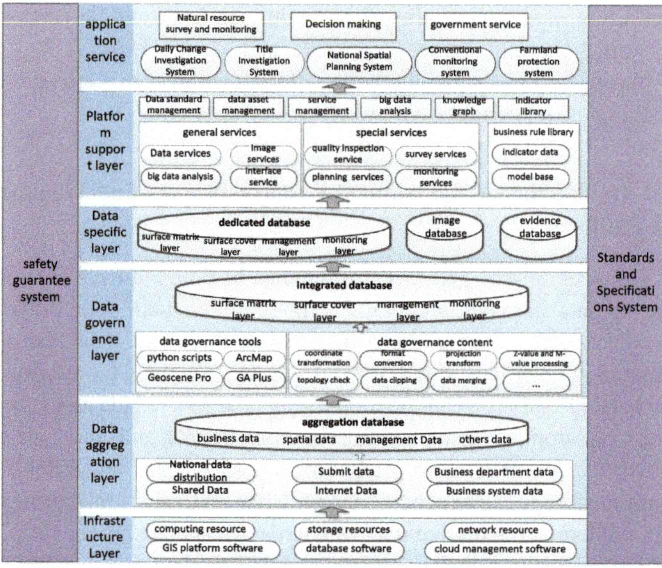

Fig. 1. Overall architecture of the central data platform for natural resource survey and monitoring data.

library. According to the data layering and classification standards of the "Overall Plan for the Construction of Natural Resources Survey and Monitoring System," the complete library is organized and managed according to the surface matrix layer, surface coverage layer, management layer, and monitoring layer.

(4) Dedicated Data Layer. Following data governance procedures, including field adjustments, data from the comprehensive database are stored as spatial data in a PostgreSQL database, creating a dedicated repository. The data storage format is standardized as PG_Geometry. Data are published as services based on business system application requirements, establishing data formats and structures that support business applications and big data analytics.

(5) Platform Support Layer. This layer provides general service functions such as data services, image services, and big data analysis. It also offers specialized services including three-line quality inspection, survey-specific services, planning-specific services, and monitoring-specific services. Furthermore, it provides functionalities for data asset management, service management, data standard management, data display, big data analysis, knowledge graphs, and indicator management.

(6) Application Service Layer. This layer supports natural resource survey monitoring, regulatory decision-making, and natural resource government service applications by providing data and services. Application systems include the land daily change survey system, natural resource ownership survey system, land spatial planning one-map implementation supervision information system, natural resource routine monitoring system, and the government "Yu-Geng-Bao" system. Additionally, interface call instructions are provided for front-end applications, adhering to relevant industry service specifications.

2.2   Core Capabilities of Data Middle Platform

Given the characteristics of the natural resource survey and monitoring sector's data, a data middle platform must possess core capabilities such as unified data standards, data aggregation and integration, unified data management, and data monetization. These capabilities are essential to facilitate users and business applications for more convenient and sustainable data utilization.

The Capacity for Standardized Data Protocols

The capacity for standardized data protocols involves establishing a unified system encompassing database construction, data submission, management, service provision, and application, achieved through the standardization of data naming conventions, storage methods, management practices, and service delivery, as well as data quality control. This approach facilitates data sharing and circulation, thereby resolving inconsistencies in data and service standards. Specifically, this includes: (1) Uniform data storage protocols, encompassing standards for vector data, imagery data, three-dimensional data, evidential data, and unstructured data. (2) Standardized data asset management protocols, including data ledgers, data catalogs, database standards, metadata information, and data dictionaries. (3) Consistent data service protocols, including vector map services, image map services, 3D scene services, data function services, and data interface services. (4) Unified data quality protocols, including data security access controls, data access permissions, and data quality assurance systems.

Unified Data Integration Capabilities

The capacity for unified data integration refers to the ability, within a standardized foundational framework, to consolidate natural resource survey and monitoring data from diverse sources and formats. This process establishes a comprehensive inventory of data resources. It also provides universal data governance tools for data processing, cleansing, and restructuring, culminating in unified storage. This approach facilitates the uniform governance and warehousing of data, alongside the establishment of data update mechanisms. Specifically, this encompasses: (1) A unified foundational environment, including standardized storage and computational infrastructure, as well as uniform hardware planning, system design, environmental setup, and product deployment. (2) Unified data aggregation, which involves the consolidation of natural resource survey and monitoring data from multiple, heterogeneous sources, thereby addressing issues such as decentralized data storage across various operational departments, independent management practices, and unclear data inventories. (3) Unified data governance, which entails a comprehensive review of natural resource survey and monitoring data, the enhancement of metadata, and the provision of universal data governance tools to expedite operational workflows. (4) Unified data storage, which ensures the uniform warehousing of data, thereby mitigating data redundancy and enhancing operational efficiency. Furthermore, it is essential to establish varied data backup strategies and mechanisms, considering factors such as data update frequency, usage frequency, and criticality.

Unified Data Management Capabilities

Unified data management capabilities involve establishing a consolidated foundation for natural resource survey and monitoring data. This entails the centralized administration, presentation, and monitoring of all data, services, tools, metrics, and interfaces, thereby providing unified data support for business applications. This approach enables business users to access, search, and download data efficiently. Specifically, this includes: (1) A unified data warehouse, which constructs a comprehensive natural resource big data system integrating collection, complete, specialized, shared, image, and evidentiary databases to offer unified data support. (2) Standardized data submission and retrieval, providing a unified entry point for data submission and retrieval, facilitating the completion of data submission details, and accessing data metadata information to enhance the efficiency of data exchange and utilization. (3) Unified asset management, which centrally manages all data, services, tools, statistical indicators, and interfaces, along with a unified data asset query service, to address the challenge of locating data through catalogs. (4) Unified data presentation, offering a unified data display window with functionalities such as attribute viewing, data statistics, and overlay analysis, to address the challenge of locating data spatially. (5) Unified asset monitoring, which provides centralized supervision and control over all asset access and operational processes within the various business systems supported by the central platform.

Unifying the Capacity for Data Value Realization

The ability to monetize unified data value refers to the capacity to rapidly enable quick and easy data retrieval and shared application, facilitating data utilization. This involves establishing a sustainable mechanism for transforming data into assets that support business operations, providing data application and management capabilities tailored to specific business scenarios. Specifically, this encompasses: (1) Unified data interaction and sharing, including internal data sharing within the organization, data sharing within the planning and resources system, and data sharing with external entities and the public. (2) Unified data application capabilities, enabling rapid reuse and providing business systems with unified data services, database interfaces, evidentiary data, indicator interfaces, and indicator views, among other diverse data access methods. This also includes the ability to quickly locate data, offering business personnel the capability to find data through directory searches, spatial searches, and knowledge-based searches. (3) Unified big data analysis and computing capabilities, providing a big data platform to meet the multi-data processing needs of business departments, supporting rapid data statistics, business analysis, and application. (4) Efficient spatiotemporal knowledge aggregation capabilities, achieved by clarifying data relationships and establishing data knowledge graphs to enable diverse data searches. The development of a global knowledge graph will link data, services, and system information, thereby improving retrieval efficiency.

3 Applied Practice Scenarios

Given Chongqing's significant topographic variability and diverse geomorphology, coupled with the complex and demanding nature of natural resource monitoring and management, there is a pressing need for advanced technological support. This is crucial to address issues such as information silos, redundant data storage, inefficient data sharing and application, low system collaboration efficiency, and slow big data analysis, ultimately enhancing the management and application of natural resource surveys and monitoring across the city. In June 2020, the nation's first land survey cloud sub-center was established in Chongqing. In 2022, the construction of the land survey cloud Chongqing sub-center application system commenced. This initiative is aligned with Chongqing's "one cloud, one database, one platform, and multiple applications" framework within the planning and natural resources sector. The system integrates technologies such as satellite remote sensing, artificial intelligence, the Internet of Things, and big data to establish an integrated terrestrial, aerial, and space-based monitoring network. Built upon a "one data center" and a "one service center," it supports "multiple business applications" for planning and natural resource surveys and monitoring, providing comprehensive technical support for survey and monitoring operations and empowering the modernization of the natural resource governance system and its capabilities.

3.1 Data Sorting and Aggregation

In terms of data sorting, approximately 220 terabytes of survey and monitoring results across the city were comprehensively reviewed. Based on the requirements of the implementation plan for the establishment of the natural resources survey and monitoring system in Chongqing, a unified, standardized, and layered directory for natural resources survey and monitoring data resources was developed, based on the surface substrate layer (including the underground resource layer), surface cover layer, management layer, and monitoring layer, following the unified standards, spatial references, and classification system.

In terms of data aggregation and integration, we have thoroughly organized the existing data in the field of planning, natural resource surveying, and monitoring, encompassing basic geography, change surveys, territorial spatial planning, business management data, spatial use control, routine and specialized monitoring, and law enforcement oversight. We have integrated and managed the scattered, independent, multi-source heterogeneous data from a five-dimensional perspective, including data source, time, coordinates, format, and business aspects. Based on the data resource directory, we have established a natural resource survey and monitoring data management system, which integrates a collection database, a comprehensive database, a dedicated database, a shared database, a remote sensing image database, and an evidence sharing database. This system has produced unified and authoritative survey and monitoring results, providing spatial data and information technology support for change surveys, planning approvals, use controls, monitoring alerts, law enforcement oversight, decision-making analysis, and more. Currently, our achievements encompass approximately 220TB of data across six categories: basic geography, land survey, territorial spatial planning, natural resource approvals, supervision and law enforcement, and change monitoring. This includes around 170 items of business production data spanning over 500 layers, approximately 100 items of external aggregation data covering more than 200 layers, and over 150 layers of external logic access.

3.2 Construction of Data Center System

The Chongqing Natural Resources Survey and Monitoring Data Center System provides functional modules such as data submission, data requisition, data governance, data visualization, data assets, service assets, big data analysis, data graph, indicator management, notification and announcement, and a management center, supporting the entire data management process, including data aggregation, data governance, data storage, data display, data management, data application, and data sharing. Some key functions are described below.

Data Visualization. As a unified display window for hospital-wide data, the data window is divided into surface cover layer, surface substrate layer, monitoring layer, and management layer according to the classification standard of natural resource survey monitoring data, and is visually displayed. It provides multiple analysis functions, such as basic base map tool, rolling curtain analysis, multi-screen comparison analysis, spatial overlay analysis, iQuery, three-dimensional spatial analysis, etc., to realize the rapid access, accurate statistics, and analysis application of spatial data, so that users can not only view the macro statistical information of the data but also understand the details such as data metadata. Currently, the data window displays about 350 data services.

Data asset management. Data assets record data collection information in a ledger-style format, enabling the tracing of data lineage. Based on different data aggregation processes and usage objectives, it is categorized into three major asset databases: complete, special, and shared databases. The complete database contains thematic data that has undergone data collection, quality inspection, and governance. The special database includes spatial data that is extracted from the complete database and then stored in PostgreSQL according to business requirements. Through an integrated storage and management model, a unified data asset catalog is formed and presented in the form of an asset ledger, offering functionalities such as asset ledger browsing, full-text resource retrieval, asset permission application and approval, asset downloads, CRUD operations, and asset statistics.

Service Asset Management. It provides functions such as service statistics, service list, and service invocation. Service statistics allows for the aggregation and statistics based on the total number of services, total calls, response time, and access system, recording detailed information such as service name, access system, calling system, and service address. The service list presents the service assets under the unified management of the data platform in reverse chronological order as cards based on the time of service release. Users can query and filter according to different metadata information of the service and view the details of the service. Service invocation offers service invocation methods and examples utilizing the localized GeoScene service engine for publishing and proxying. Currently, approximately 500 data services are uniformly registered on the data platform.

Big data analysis. Leveraging the distributed computing framework, the institution has developed a unified big data analysis and calculation platform, GA Plus 4.1, which provides an interactive analysis interface to visually display spatial data resources, analysis operator resources, and business model resources. Users can complete the construction of complex business scenarios through no-code online drag-and-drop, reuse and share models, and address the issues of high thresholds for big data analysis and calculation, difficulty in finding data, and long time consumption in the past. Currently, GA Plus 4.1 provides more than 200 analysis operators, has executed over 3700 tasks, and has developed more than 1200 models, supporting the big data analysis and calculation needs of Chongqing farmland protection, the one-map implementation supervision information system for land spatial planning, conventional monitoring, farmland monitoring, land survey cloud, and other businesses.

Indicator library management. Combined with the big data platform, a basic indicator library based on business requirements has been built, providing management capabilities for indicator metadata, indicator values, and indicator views. It realizes the cross-reference function between basic data, big data models, and indicators, and establishes a linked update mechanism for data, models, and indicators. Currently, 800 indicators have been created, 1 million indicator values have been stored, and 200 indicator views have been created, providing convenient and unified indicator statistics for systems such as Chongqing Farmland Protection, the Supervision Information System for the Implementation of Territorial Spatial Planning on One Map, CSPON, and Comprehensive Monitoring and Supervision Systems.

Knowledge graph. By improving data metadata information and clarifying data relationships and contexts, this paper constructs a knowledge graph from the business perspective, data perspective, and other dimensions. It also provides a thematic knowledge graph centered around the third land survey and a data warehouse, which illustrates the historical development and current status of data. This enables a diversified retrieval method that allows users to find data based on ledgers, spatial locations, and knowledge graphs.

Acknowledgment. With the continuous advancement of new-generation information technology and the expanding application scenarios of natural resource survey and monitoring, the application prospects of data mid-platforms in natural resource management will become even broader. In the future, the Chongqing Natural Resource Survey and Monitoring Data Mid-Platform will continue to optimize its data processing and analytical application capabilities, enhance the intelligence level of data and knowledge services. Meanwhile, it will also strengthen the integrated application with technologies such as the Internet of Things (IoT) and artificial intelligence (AI), promoting the digital transformation and high-quality development of natural resource management.

References

1. The State Council. Opinions of the CPC Central Committee and the State Council on Establishing a Data-Based System to Better Leverage the Role of Data as a Factor of Production. 2 Dec 2022, (No. 1, 2023)
2. Xiaobo, Q., Anhui, C., Xiaowei, Y.: Thoughts on the construction of data mid-platforms in the natural resources industry. Nat. Resour. Informatization **2022**(5), 31–38 (2022)
3. Su, M., Jia, X., Du, X.iaomeng, et al.: Progress and development trends of data mid-platform technology. Front. Data Comput. **1**(05), 120–130 (2019). CNKI:SUN:KYXH.0.2019-05-012

4. Zhonghua, D.: Big Data, Big Innovation: The Way of Alibaba's Cloud-Based Data Mid-Platform, pp. 4–33. Electronics Industry Press, Beijing (2018)
5. Miaofeng, X., Hui, X.: Design and application of data mid-platforms in smart cities. Post Telecommun. Des. **2**, 3 (2020). CNKI:SUN:YDSJ.0.2020-02-019
6. Lichun, S.: Research on the architecture and key technologies of data mid-platforms for intelligent coal mines. Autom. Min. Coal Metall. **47**(6), 5 (2021). https://doi.org/10.13272/j.issn.1671-251x.2020120052
7. Huang, Y., Peng, S., Luo, Y., et al.: Practice and reflection on the construction of data mid-platforms for medical insurance. China Healthcare Insur. [2024-08-09]
8. Cuicui, Z., Min, X., Jiali, S., et al.: Research and application of data mid-platform construction standards for the power industry. Mod. Comput. **17**, 5 (2021). https://doi.org/10.3969/j.issn.1007-1423.2021.17-018
9. Qian, L., Jian, Z., Guangxue, C., et al.: Practice of natural resource data governance based on data mid-platforms: a case study of Guangxi Zhuang autonomous region. Nat. Resour. Informatization **5**, 158–162 (2022)

Automatic Tail Number Recognition for Civil Aviation Remote Apron Control Towers

Aijun Cui[1], Xuan He[2(✉)], Kai Wang[2], Mozhen Tang[2], and Donglin He[2]

[1] Airfield Management Department, Beijing Capital International Airport Co., Ltd., Beijing, China
cuiaj@bcia.com.cn

[2] Research and Development Center, The Second Research Institute of Civil Aviation Administration of China, Chengdu, China
{hexuan,wangkai,tangmozhen,hedonglin}@caacsri.com

Abstract. To enhance the automation level of remote apron control towers, a simple and effective automatic aircraft tail number recognition algorithm is proposed. For aircraft images obtained by remote apron control towers, the tail number area is first quickly located through the Hough transform. Then, YOLOv12 is used to accurately locate the tail number character areas. After that, the tail number is segmented by the threshold method, and finally, the recognition is completed using the hole-based template matching method. Experimental results show that this algorithm has a good effect on aircraft tail number recognition and strong robustness, with broad application prospects.

Keywords: Tail number location · Tail number recognition · Image segmentation · Character recognition

1 Introduction

In recent years, as an innovative control mode, remote apron control towers have been increasingly widely used in large-scale airports [1]. By virtue of video transmission technology, remote apron control towers transmit real-time images of the airport surface to the apron control room, replacing the visual observation method of traditional tower controllers. This change is not simply a visual substitution but significantly improves the controller's surveilling ability of the airport surface by integrating advanced means such as video enhancement technology, infrared technology, and aircraft signage technology.

The aircraft signage technology [2] plays a crucial role in improving the controller's work efficiency. It superimposes key information from aircraft surveillance data and flight dynamic data, such as flight numbers, aircraft types, and parking positions, on the aircraft in the video in an intuitive way, enabling controllers to quickly obtain important information and make accurate decisions. However, when the aircraft's transponder is not turned on, the background system cannot identify the aircraft, resulting in the failure of the signage function and reducing the controller's work efficiency. Therefore, developing a non-cooperative aircraft identification method, especially a technology that

© The Author(s) 2026
P. Siarry et al. (Eds.): WCNA 2024, LNEE 1550, pp. 255–264, 2026.
https://doi.org/10.1007/978-981-95-6946-5_26

can accurately identify aircraft when the transponder fails or is not turned on, has become an urgent problem to be solved in the field of remote apron control.

The aircraft tail number is the unique identifier of an aircraft. By identifying the tail number, not only basic information such as the aircraft type and the affiliated airline can be obtained, but also key data such as the flight number can be further associated. Thus, when the aircraft's transponder is not turned on, it is possible to effectively identify the aircraft and ensure the continuity and uninterrupted operation of aircraft signage.

In recent years, image processing-related technologies have developed rapidly and have become a popular research field at home and abroad. Among them, image detection and recognition technologies, as the core branches of image processing, have been widely applied in many fields such as security monitoring [3], autonomous driving [4–8], traffic surveilling [9], UAV scene analysis and robot vision[10, 11]. In the civil aviation field, many scholars have also carried out relevant research. For example, Fengxi Yan [12] proposed a method for identifying aircraft types based on remote - sensing images, but this method can only identify the aircraft type and cannot obtain more detailed information; Hongying Zhang [13] explored the layout positions of high-definition cameras on the taxiway for identifying aircraft tail numbers; Xi Zhang [14] presented a method for identifying characters on the overhead panel of the aircraft cockpit but did not deeply explore the recognition of tail numbers.

Existing research still has many problems in aircraft tail number recognition, and the location of the tail number is the biggest difficulty. Due to the continuous change of the relative angle between the camera and the aircraft, the position of the tail number in the image is highly uncertain. Traditional methods use dedicated cameras to shoot at a fixed angle from a specific position, which, to a certain extent, solves the problem of tail number location. However, due to its strict usage conditions, it is severely restricted in practical airport environments.

In response to the above problems, this paper proposes an automatic aircraft tail number recognition algorithm based on image processing technology. This algorithm quickly locates the tail number area through the Hough transform, then accurately locates the tail number character areas using YOLOv12, and finally segments the tail number by the threshold method and completes the recognition of the tail number using the hole - based template matching method. Experimental results show that the algorithm proposed in this paper can efficiently and accurately identify aircraft tail numbers in complex airport environments, providing strong support for the operation of remote apron control towers.

2 Tail Number Recognition

The aircraft tail number recognition algorithm mainly consists of four parts: quick tail number location, accurate tail number location, tail number segmentation and tail number recognition. The task of tail number location is to give the position of the tail number in the image. The main task of tail number segmentation is to segment the characters in the located tail number area, and the task of tail number recognition is to recognize the segmented characters. The algorithm flow chart is shown in Fig. 1.

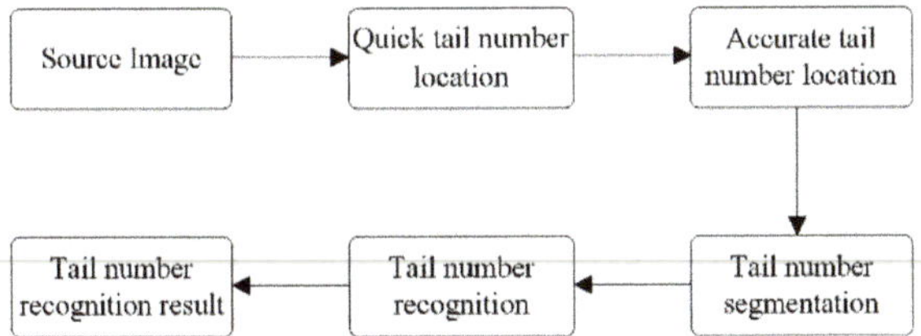

Fig. 1. Algorithm Flow Chart

2.1 Preprocessing

To facilitate computer processing, the source image is converted into a grayscale image P. The image grayscale conversion formula is:

$$Y = W_r \times R + W_g \times G + W_b \times B \tag{1}$$

where R, G, and B are the color components of the image, W_r, W_g, and W_b are the weighting coefficients of the three components, Y is the grayscale value of the converted image. The values of the weighting coefficients are based on the human eye visual model. A larger weight is assigned to the green color, which is more sensitive to the human eye, and a smaller weight is assigned to the blue color, which is less sensitive to the human eye. Generally, the weighting coefficients 0.299, 0.587, and 0.114 are used. The source image and the converted image $P0$ are shown in Figs. 2 and 3, respectively.

Fig. 2. Source Image

Fig. 3. Grayscale Image

Fig. 3.Grayscale Image

2.2 Quick Tail Number Location

According to the regulations of the International Civil Aviation Organization, the tail number is located near the vertical stabilizer of the aircraft. Therefore, the tail number can be quickly and preliminarily located by finding the position of the vertical stabilizer. There are two straight lines with relatively large inclination angles at the edge of the vertical stabilizer, and the Hough transform can be used to detect these two lines.

The Hough transform converts the straight line in the Cartesian coordinate system:

$$y = \left(-\frac{\cos\theta}{\sin\theta}\right)x + \frac{r}{\sin\theta} \tag{2}$$

into a sine curve in the parameter space:

$$r = x\cos\theta + y\sin\theta \tag{3}$$

A point in the Cartesian space will be converted into a sine curve in the parameter space. When N points on a straight line are transformed, N sine curves will be generated in the parameter space, and the intersection point of these sine curves is (r, θ), where r is the distance from the origin in the Cartesian coordinate system to the straight line, and θ is the angle between the perpendicular line from the origin to the straight line and the X-axis. The main steps of the preliminary quick location are shown in Fig. 4.

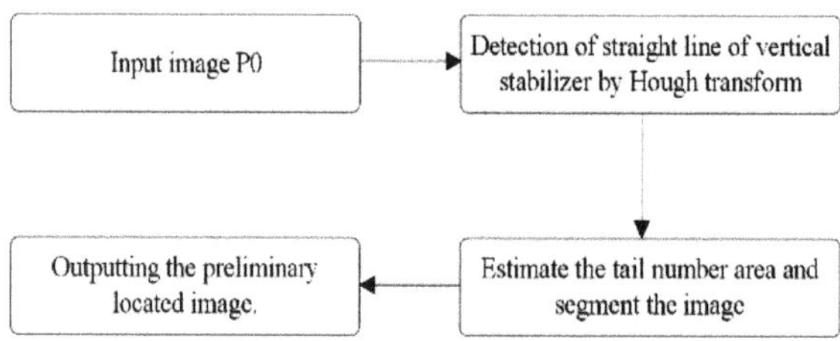

Fig. 4. Preliminary Tail Number Location

Input the image P, use the Hough transform to detect the two straight lines $\overline{P1P2}$ and $\overline{P3P4}$ of the vertical stabilizer, as shown in Fig. 5. Estimate the area of the tail number according to the prior information of the aircraft tail number, and segment the image $P1$ within the rectangle, as shown in Fig. 6.

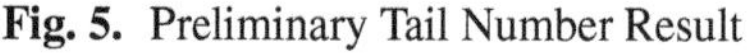

Fig. 5. Preliminary Tail Number Result

Fig. 6. Preliminary Tail Number Location Area

2.3 Accurate Tail Number Location

In character recognition applications, the selection of target detection technology is of great importance. Traditional algorithms rely on manually designed features, such as edge detection and color analysis, but their accuracy is limited in complex environments. Using YOLOv12 for tail number detection has significant advantages.

Network Design

The YOLOv12 tail number detection network adopts a convolutional neural network (CNN) structure, with multiple functional modules working together to achieve image feature extraction and target detection.

Input Layer: The Mosaic data augmentation technology is used. This technology effectively improves the model's recognition ability for small-sized targets by augmenting data of small-sized targets, which is crucial for tail number detection. At the same time, an adaptive anchor box calculation method is adopted to improve the prediction accuracy during the training process.

Backbone Network: The Focus structure is introduced. By slicing and recombining the input image, it optimizes the feature extraction process and improves the feature extraction efficiency. The application of the CSP structure enhances the network's feature extraction and learning ability. Through cross-stage partial connection, it reduces the computational load while improving the network performance and efficiency.

Neck Structure: The FPN and PAN structures are combined to construct a multi‑scale feature extraction network, which can effectively improve the recognition performance for targets of different scales. The FPN structure transmits strong semantic features from top to bottom, and the PAN structure transmits strong positioning features from bottom to top. The combination of the two enables the network to more accurately detect targets at different scales.

Prediction Layer: The CIOU_LOSS is used to optimize the correspondence between the prediction box and the actual target box. This loss function comprehensively considers the overlap rate of the bounding box, the distance between the center points, and the length of the boundary, making the positioning more accurate. At the same time, the non-maximum suppression method is used to simplify the prediction results, effectively ensuring the detection accuracy.

Network Training

The network training process involves two stages: forward propagation and backpropagation.

Forward Propagation: Forward propagation refers to the process in which input data is calculated forward through the network layers to obtain prediction results. Suppose the input data is X, the network parameters are θ, and the output prediction result is Y:

$$\hat{Y} = f(X; \theta) \tag{4}$$

where f is the forward-calculation process of the network.

In YOLOv12, forward propagation performs feature extraction and target detection on the input image through a series of convolutional layers, pooling layers, and fully-connected layers, and finally outputs the predicted bounding boxes and class probabilities.

Backpropagation: Backpropagation adjusts the network parameters by calculating the loss function and gradient descent. The loss function $L(\theta)$ is the difference between the prediction result and the true label:

$$L(\theta) = \frac{1}{N} \sum_{i=1}^{N} l\left(\hat{y}_i, y_i\right) \tag{5}$$

The parameters are updated through gradient descent:

$$\theta_{t+1} = \theta_t - \eta\nabla_\theta L(\theta) \tag{6}$$

where η is the learning rate, which controls the step size of each parameter update. Through multiple iterations, the parameters gradually converge to the values that minimize the loss function.

After detection by YOLOv12, the tail number detection results are marked on the image $P1$, as shown in Fig. 7.

Fig. 7. Accurate Tail Number Location Result

2.4   Tail Number Segmentation

The image $P2$ of the aircraft tail number is segmented from the aircraft image $P1$ according to the accurate tail number location result. The segmentation and recognition of the aircraft tail number are carried out in a binary image. Therefore, the image $P2$ is first binarized to remove the non-registration number area. In this paper, the global threshold segmentation method is used to binarize the image. The process of obtaining the threshold is as follows:

Suppose an image has N pixels, and its gray levels are $\{0,1,2,\ldots,L\text{-}1\}$. The number of pixels with gray level i is n_i, then $N = \sum_{i=0}^{L-1} n_i$. Normalize the image histogram, and the probability of gray level i is:

$$p_i = n_i/N \tag{7}$$

where $\sum_{i=0}^{L-1} p_i = 1$。

The threshold t divides the image pixels into two classes C_0 and C_1, corresponding to pixels with gray levels $\{0,1,2,\ldots,t\}$ and $\{t+1, t+2, \ldots, L-1\}$ respectively.

The probability density and mean value of class C_0 are:

$$\omega_0 = P(C_0) = \sum_{i=0}^{t} p_i = \omega(t) \tag{8}$$

$$\mu_0 = \sum_{i=0}^{t} iP(i|C_0) = \sum_{i=0}^{t} ip_i/\omega_0 = \mu(t)/\omega(t) \tag{9}$$

The probability density and mean value of class C_1 are:

$$\omega_1 = P(C_1) = \sum_{i=t+1}^{L-1} p_i = 1 - \omega(t) \tag{10}$$

$$\mu_1 = \sum_{i=t+1}^{L-1} iP(i|C_1) = \sum_{i=t+1}^{t} ip_i/\omega_1 = \frac{\mu_t - \mu(t)}{1 - \omega(t)} \tag{11}$$

where $\mu(t) = \sum_{i=0}^{t} ip_i$, $\mu_t = \mu(L) = \sum_{i=0}^{L-1} ip_i$.

According to the probability density and mean value of classes C_0 and C_1, the inter-class variance is:

$$\sigma_B^2 = \omega_0(\mu_0 - \mu_T)^2 + \omega_1(\mu_1 - \mu_T)^2 \tag{12}$$

When the inter-class variance σ_B^2 is the largest, the image binarization effect is the best. Then the optimal threshold t^* is:

$$\sigma_B^2(t^*) = \max_{0 \leq t \leq L} \sigma_B^2(t) \tag{13}$$

The image $P2$ is binarized using the threshold t^* to obtain the image $P3$, as shown in Fig. 8.

Due to the shooting angle of the camera, the aircraft registration number may be horizontally inclined. Therefore, horizontal correction is performed before segmenting the characters. The projection method is used for correction, and the horizontal correction result image $P4$ is obtained, as shown in Fig. 9. In this paper, the method in reference [15] is used to segment individual characters.

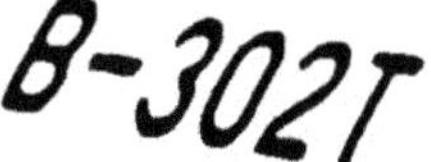

Fig. 8. Binarized Image

Fig. 9. Horizontal Correction Result

2.5 Tail Number Recognition

The segmented single-character images are recognized. There are many mature methods for text recognition, and there are many character recognition algorithms based on different principles, each with its own advantages and disadvantages. Combining the characteristics of aircraft tail numbers, this paper uses an improved hole-based template matching method for character recognition.

Aircraft tail numbers are composed of only 26 uppercase letters and 10 Arabic numerals. These 36 characters can be divided into three categories: 1) Characters with no holes {C, E, F, G, H, I, J, K, L, M, N, S, T, U, V, W, X, Y, Z, 1, 2, 3, 5, 7}; 2) Characters with one hole {A, D, O, P, Q, R, 0, 4, 6, 9}; 3) Characters with two holes {B, 8}. When detecting aircraft tail number characters, the number of holes is found, and then they are matched with the corresponding template library. The XOR operation is performed between the character to be matched and each character in the corresponding template library, and the similarity is evaluated using the Hamming distance. The character with the highest similarity is taken as the matching result.

3 Experimental Results and Analysis

Experimental Results

The hardware environment used in the experiment is as follows: The CPU is an Intel(R) Core(TM) i5-8265U CPU @ 1.60 GHz, with 16 GB of RAM; the GPU is an NVIDIA RTX 5000. The operating system is Windows 10. The algorithm is implemented using the Python programming language and OpenCV library. The original input image size is 1920 × 1080 pixels.

The experimental data set comes from images captured by surveillance cameras at different locations in the airport. A total of 500 images containing aircraft tail numbers were collected, covering different weather conditions, different lighting conditions, and different shooting angles. The data set is divided into a training set, a validation set, and a test set, with proportions of 70%, 15%, and 15% respectively. The training set is used to train the YOLOv12 model, the validation set is used to adjust the model parameters, and the test set is used to evaluate the performance of the algorithm.

The trained model and algorithm are used to detect and recognize the tail numbers in the test set images. The experimental results show that under sunny and well - lit conditions, the algorithm can achieve a high recognition accuracy rate. The tail number detection rate reaches 94%, and the recognition accuracy rate reaches 90%.

Algorithm Comparison

The algorithm proposed in this paper and the algorithm in reference [13] are used to detect and recognize aircraft tail numbers in the test-set images. The experimental results show that the detection and recognition rate of the algorithm in this paper is relatively high, and the recognition time is relatively short. Table 1 presents detailed comparison data (Figs. 10 and 11).

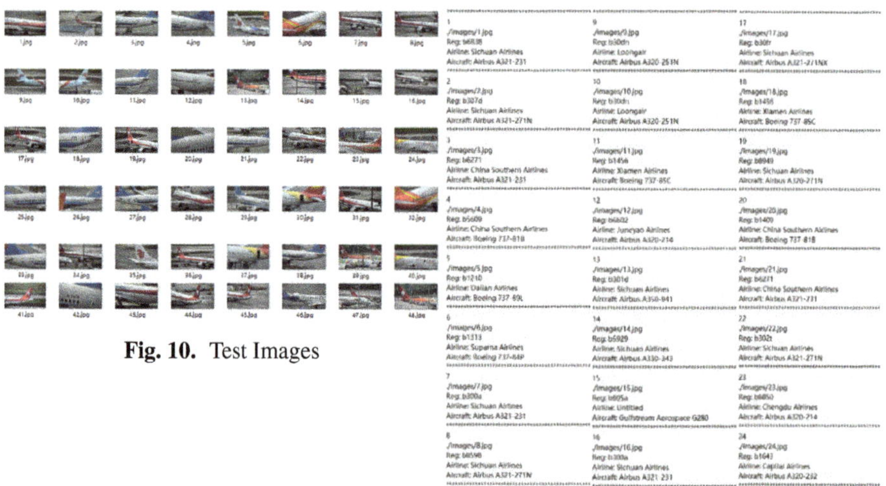

Fig. 10. Test Images

Fig. 11. Recognition Result

Table. 1. Table Type Styles

	Detection Rate	*Recognition Rate*	*Average Time*
Algorithm in this paper	94%	90%	0.79s
Algorithm in reference [13]	86%	78%	1.57s

4 Conclusion

This paper proposes an automatic recognition algorithm for civil aviation aircraft registration numbers based on image processing technology. The area of the aircraft tail number is quickly located through the Hough transform. Then, YOLOv12 is used to accurately detect the position of the tail number. After segmenting the characters, the template matching method based on holes is used to recognize the registration number characters. The experimental results verify the effectiveness of this algorithm. However, currently, the algorithm does not consider the complex meteorological conditions at the airport, resulting in a decrease in the performance of the image-quality-related algorithm in bad weather. Therefore, the next step of research will focus on how to achieve automatic recognition of aircraft tail numbers under complex meteorological conditions, improving the robustness and adaptability of the algorithm to meet the diverse requirements of actual airport operations.

Acknowledgment. This work is supported by National Natural Science Foundation of China, grant No. U2333207.

References

1. Ye, W., Yu, Q., He, N., et al.: Augmented reality for assist apron operation safety supervisor in aircraft inspection processes. 2021 IEEE International Conference on Power Electronics, Computer Applications (ICPECA). IEEE, pp. 273–276 (2021)
2. Tang, Y., Hu, M., Wu, H., et al.: A new method for automatically hanging signs on aircraft in airport videos. J. Jiangsu Univ.: Nat. Sci. Ed. **34**(6), 681–686 (2013)
3. Xie, Y.: Research and training of helmet recognition model based on deep learning. International Conference on Mechatronic Engineering and Artificial Intelligence (MEAI 2023). SPIE, vol. 13071, pp. 684–689 (2024)
4. Li, Y., Zhang, X., Zhang, M., et al.: Object detection in autonomous driving scenarios based on improved efficientdet. J. Comput. Eng. Appl. **58**(6) (2022)
5. Zhao, C., Li, L., Pei, X., et al.: A comparative study of state-of-the-art driving strategies for autonomous vehicles. Accid. Anal. Prev. **150**, 105937 (2021)
6. Xu, X., Ma, Y., Qian, X., et al.: Scale-Aware real-time pedestrian detection in autonomous driving scenarios. J. Image Graph. **26**(1), 93–100 (2021)
7. Chen, X., Wan, M., Ma, C., et al.: Small object detection in remote sensing images using SSD with multi-scale feature fusion. Opt. Precis. Eng. **29**(11), 2672–2682 (2021)
8. Liu, C., Zhang, L., Huang, H.: Visualization of multi-target tracking in traffic intersection surveillance video. Chin. J. Comput. **41**(1), 221–235 (2018)

9. Wang, X., Wang, C.: Dense crowd safety surveilling system based on deep learning. Internet Things Technol. **9**(11), 8–12, 17 (2019)
10. Ma, J., Yao, Z., Xu, C., et al.: Real-time tracking algorithm for multi-UAV based on improved Pp-YOL0 and Deep-SORT. Comput. Appl. **42**(9), 2885–2892 (2022)
11. Zheng, T., Jiang, M., Feng, M.: Research review on target recognition and location of picking robot based on vision. Chin. J. Sci. Instrum. **42**(9), 28–51 (2021)
12. Yan, F., Xu, Y., Cai, S., et al.: Aircraft detection in remote sensing images based on improved YOLOv5 algorithm. Comput. Eng. Des. **44**(9), 2794–2802 (2023)
13. Zhang, H., Hu, Z.: Research on civil aviation aircraft registration number recognition. 2012 IEEE 2nd International Conference on Cloud Computing and Intelligence Systems. IEEE, vol. 1, pp. 47–51 (2012)
14. Zhang, X., Xu, H., Li, Q.: Research on the character recognition method of the overhead panel in the aircraft cockpit based on SVM. Electron. Meas. Technol. (2020)
15. Suzuki, S.: Topological structural analysis of digitized binary images by border following. Comput. Vis. Graph. Image Process. **30**(1), 32–46 (1985)

Untargeted Trigger Reconstruction for Backdoor Defense in Self-Supervised Learning

Jianpeng Li[1], Jiayu Du[2(✉)], and Fan Zhang[2]

[1] Zhengzhou University, Zhengzhou, China
`ljp1747877075@gs.zzu.edu.cn`
[2] Purple Mountain Laboratories, Nanjing, China
`dujiayu@Pmlabs.Com.Cn`

Abstract. Self-supervised learning (SSL) has gained significant attention for its ability to leverage large-scale unlabeled data for training, demonstrating remarkable applications in computer vision and natural language processing. However, recent studies have shown that SSL models are highly vulnerable to backdoor attacks, where adversaries can implant hidden triggers into the pretraining process, leading to compromised downstream tasks. Existing backdoor defense mechanisms, while effective to some extent, often rely on labeled data, incur high computational costs, or involve complex operations, making it difficult to achieve a balance between model performance and defense effectiveness. In this paper, we propose a backdoor defense method called UTRBD, which is based on untargeted trigger reconstruction, combined with a contrastive fine-tuning strategy. It effectively defends against backdoor attacks using only a small amount of unlabeled data and minimal computational resources, while maintaining high model performance. Through extensive experiments on the ImageNet100 and CIFAR10 datasets, our method performs excellently across various self-supervised learning models and different encoder architectures, successfully mitigating multiple types of backdoor attacks.

Keywords: self-supervised learning · backdoor attack · backdoor defense · untargeted trigger reconstruction · contrastive fine-tuning

1 Introduction

Self-Supervised Learning (SSL) [1–4] is a method that trains an encoder using large amounts of unlabeled data. The core idea is to learn general feature representations from unlabeled data by designing pretraining tasks. The trained encoder can then be transferred to various downstream tasks. In recent years, SSL has been widely applied in fields such as computer vision [5] and natural language processing [6], becoming an effective solution to issues like high data annotation costs and the scarcity of labeled data.

© The Author(s) 2026
P. Siarry et al. (Eds.): WCNA 2024, LNEE 1550, pp. 265–275, 2026.
https://doi.org/10.1007/978-981-95-6946-5_27

However, self-supervised learning models are also vulnerable to backdoor attacks [7–10]. Unlike backdoor attacks in supervised learning, backdoor attacks in self-supervised learning primarily occur at the feature level. Attackers implant triggers (such as specific patterns or noise) into the training data, causing the model to incorrectly associate the trigger features with the target class features in the feature space. Attackers can introduce backdoors through various methods, such as contaminating the training dataset or controlling the training process. Since encoders trained via self-supervised learning are typically used for multiple downstream tasks, the impact of the backdoor is further amplified, posing serious security risks to the practical applications of self-supervised learning.

In response to the issue of backdoor attacks, researchers have proposed various backdoor removal methods [12–16], among which the trigger reconstruction-based fine-tuning approach has become a common paradigm. For example, SSL_Cleanse [13] is a backdoor defense method specifically designed for self-supervised learning scenarios. Its core idea is to use a small amount of clean data to generate feature clusters, and then reconstruct the corresponding triggers for each cluster. Finally, the model is fine-tuned to forget these triggers, thereby removing the backdoor. However, SSL_Cleanse has several notable drawbacks: first, self-supervised learning models typically learn from unlabeled data containing a large number of classes, while the defender's data is limited, making it difficult to cover all classes that attackers may target; second, SSL_Cleanse removes the backdoor during the fine-tuning process by reducing the distance between normal samples and trigger samples, but this process does not adequately preserve feature discriminability, leading to a significant drop in model performance.

In this paper, we improve upon the trigger reconstruction and fine-tuning paradigm, with the following key contributions:

- we propose an Untargeted Trigger Synthesis method that requires only a small amount of unlabeled data to reconstruct triggers for any class.
- we design a contrastive fine-tuning method that, while forgetting triggers, maximizes the preservation of feature discriminability, thus better safeguarding model performance.
- we conduct extensive experiments across multiple datasets, self-supervised learning methods, and encoder architectures, validating the effectiveness of our proposed method against various attack strategies.

2 Background and Related Works

2.1 Self-Supervised Learning

Self-Supervised Learning (SSL) [1–4] has emerged as one of the most popular methods for learning representations from complex, unlabeled data. By pretraining an encoder on large-scale unlabeled data and fine-tuning it with a small amount of labeled data, SSL methods have demonstrated performance that rivals supervised learning approaches that rely on large labeled datasets for various downstream tasks. Furthermore, self-supervised learning techniques based on instance discrimination, such as SimCLR [1] and BYOL [2], have gained increasing popularity. A typical self-supervised learning classification task involves pretraining an image encoder, building a classifier, and fine-tuning the model in subsequent steps.

2.2 Backdoor Attacks in SSL

The earliest backdoor attack method for self-supervised learning is BadEncoder [10], which targets a pre-trained clean model. It fine-tunes the model with a small dataset and trigger, causing samples with added triggers close to the target class features in the feature space, thereby enabling a backdoor attack. Subsequently, Saha et al. introduced SSL_Backdoor [7], which implements backdoor attacks via data poisoning. In this method, triggers in the form of patches are added to the samples of the target class, causing the model to recognize the trigger as an important feature of that class. However, the attack success rate of SSL_Backdoor is relatively low, mainly due to the random data augmentation typically performed during the self-supervised learning phase, which may result in the patch triggers being cropped. Additionally, the inconsistency between the trigger forms during the training and inference phases further weakens the attack's effectiveness. To address this issue, CTRL [8] proposed a data poisoning method based on global optical triggers. Unlike SSL_Backdoor, the global optical triggers used by CTRL are more robust under data augmentation, allowing for higher attack success rates. However, the limitation of CTRL is that its triggers are single-target, making it suitable only for single-target attacks and not scalable to multi-target attack scenarios. Furthermore, ESTAS [9] introduced a strategy of adding triggers after data augmentation. This approach ensures that the triggers are not cropped during the augmentation process and maintains consistency between the training and inference phases. However, the drawback of ESTAS is that it requires the attacker to have control over the training process. Our goal in designing a backdoor defense method is to remove the backdoor from multiple types of attack methods.

2.3 Limitations of Related Backdoor Defense

Existing backdoor defense methods face numerous challenges when detecting backdoors in self-supervised learning (SSL) encoders. Researchers have proposed DECREE [22], a method specifically designed for backdoor detection in SSL, which performs well in detecting backdoor attacks using patch-based triggers. However, its detection accuracy significantly decreases when dealing with globally invisible triggers. PatchSearch [11] and ASSET [12] are primarily used to remove backdoor samples from the training dataset during the training phase. However, in real-world applications, we often cannot access the full dataset, and we lack the computational resources to retrain an entirely new model. More importantly, the challenge we face is how to remove the backdoor from the model itself. Existing backdoor removal methods mainly include knowledge distillation [7], pruning methods [15, 16], and trigger reconstruction-based approaches [13, 14]. Knowledge distillation significantly reduces model performance, while pruning methods are prone to targeted attacks due to the specific pruning strategies employed. Trigger reconstruction-based methods are more generalizable for backdoor removal, but the currently available SSL-specific trigger reconstruction method, SSL_Cleanse [13], has certain drawbacks. First, it assumes that the defender has access to a dataset containing the target class of the backdoor attack. However, since SSL relies on large amounts of unlabeled data, it is difficult to ensure that the small subset of data available to the defender contains all possible classes. Secondly, SSL_Cleanse focuses on forgetting the backdoor trigger during the fine-tuning process but does not consider maintaining the discriminability between different features, which results in a significant loss of model performance.

3   Methodology

3.1   Threat Model

I. Attacker's Goal: The goal of the attacker is to train a self-supervised learning (SSL) image encoder with a backdoor such that any downstream classifier trained using this encoder inherits the backdoor functionality. Specifically, the downstream classifier should maintain normal classification performance on clean inputs, but when the input contains a trigger, the model should consistently classify it into a predefined target class.

II. Attacker's Capabilities and Knowledge: The attacker, as the trainer of the SSL encoder, has full control over the entire training process. This includes selecting the self-supervised learning method, choosing the encoder architecture, having access to the complete training dataset, and the freedom to design and choose triggers, as well as control over the training process.

Defender's Goals and Capabilities:The defender's goal is to use the trained SSL image encoder as safely as possible, without knowing whether it contains a backdoor. The defender *possesses a small amount of unlabeled clean data and has the ability to fine-tune the SSL image encoder. The aim is to* effectively reduce the success rate of backdoor attacks while minimizing any significant degradation in model performance.

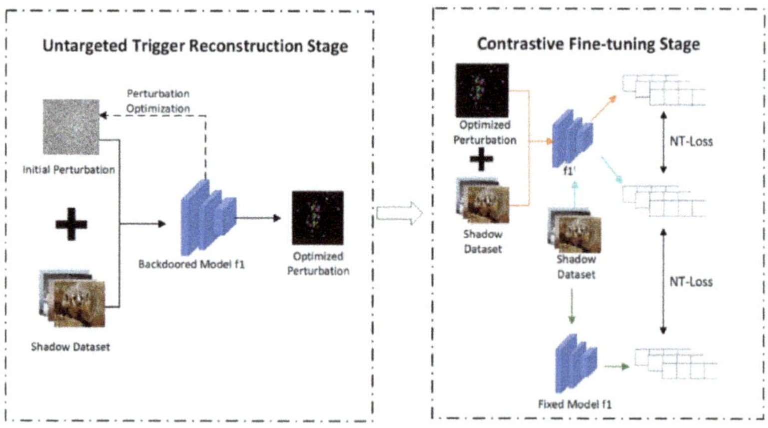

Fig. 1. Overview of UTRBD.

3.2 UTRBD: Untargeted Trigger Reconstruction for Backdoor Defense

In backdoor attacks on self-supervised learning, the goal of the attacker is to bring the feature vectors of backdoor samples closer to the feature vectors of a specific target class. Based on this, existing trigger reconstruction methods usually employ optimization strategies to generate a perturbation that, when added to any sample, causes its feature to cluster toward the target class. However, this approach has a significant limitation: as defenders, we cannot foresee the target class chosen by the attacker. The traditional solution is to reconstruct triggers for all classes, that is, to optimize triggers for different classes separately. However, this method relies on a premise: we must have samples of the target class to guide the perturbation optimization using its feature. In practice, however, data from the target class may not be available, which presents a challenge for trigger reconstruction.

The research by DECREE demonstrates that, compared to normal samples, backdoor samples exhibit highly clustered characteristics in the feature space. Based on this conclusion, we propose a target-free trigger reconstruction strategy: optimizing the model's trigger by using the clustering degree of the perturbed samples in the feature space as the objective. This design allows us to effectively reconstruct the trigger for the target class even when the target class samples are unavailable, making it possible to reconstruct the trigger effectively without the target class samples.

Regarding the fine-tuning of the model to forget the reconstructed triggers, current algorithms design the forgetting objective to keep the feature distance between clean samples and perturbed samples (with added triggers) sufficiently close. The goal of this design is to ensure that adding triggers does not cause a drastic shift in the features of the samples, thus rendering the backdoor attack ineffective. However, this design only considers the model's desensitization effect and does not deliberately maintain the feature distinguishability during the forgetting fine-tuning process, while distinguishability is a key factor that determines model performance. Therefore, it can lead to some loss of model accuracy. To address this, we propose using a contrastive loss during the fine-tuning process of forgetting triggers, which helps maintain feature distinguishability, thereby maximizing the model's performance while removing the backdoor.

The overall framework of the backdoor removal method proposed in this paper is shown in Fig. 1, consisting of two stages. The first stage reconstructs the trigger, and the second stage fine-tunes the model to forget the reconstructed trigger, thereby achieving the removal of the backdoor.

Untargeted Trigger Reconstruction

In this papper, we reconstruct the model trigger by optimizing the clustering degree of perturbed samples in the feature space. In the absence of target class samples, we optimize the trigger by minimizing the clustering property of backdoor samples while imposing a regularization constraint on the magnitude of the perturbation. Let the original clean sample be x, and the perturbation be composed of a mask m and a trigger t. The perturbed backdoor sample x_{bd} can be expressed as:

$$x_{bd} = (1 - m) \odot x + m \odot t$$

$\odot$ denotes element-wise multiplication. The feature representation of the backdoor sample after passing through the model is given by $z_{bd} = f(x_{bd})$. To quantify the clustering degree of backdoor samples in the feature space, we minimize the mean cosine similarity between sample pairs, formulated as:

$$T_1 = -\frac{1}{N^{i,j}}\cos\left(z_{bd}^{(i)}, z_{bd}^{(j)}\right)$$

$cos(a, b)$ denotes the cosine similarity between feature vectors. Additionally, to control the magnitude of the perturbation, we incorporate an l_1 norm regularization term as the optimization objective T_2:

The final objective function is defined as:

$$T = T_1 + \lambda T_2$$

λ is a regularization weight that balances the minimization of feature clustering and the sparsity constraint of the perturbation. This optimization strategy ensures that even in the absence of target class samples, we can effectively reconstruct the trigger, leading to high clustering of backdoor samples in the feature space.

Contrastive Fine-Tuning

In this papper, we propose utilizing contrastive fine-tuning to unlearn the reconstructed trigger. Specifically, given a backdoored model f, we first duplicate its parameters to obtain a new model f'. During training, the parameters of f remain unchanged, while only f' is optimized.

For a batch of samples $\{x_i\}_{i=1}^n$, we construct their perturbed versions as x_i^+. Then, we feed both x_i and x_i^+ into the two models and obtain their feature representations: $z_i = f(x_i)$, $z_i' = f'(x_i)$, $z_i^+ = f'(x_i^+)$. To optimize f', we define the following objective function:

$$L = max \sum_{i=1}^n \left(\mathrm{sim}(z_i, z_{i'}) + \mathrm{sim}\left(z_{i'}, z_i^+\right)\right)$$

$\mathrm{sim}(\cdot, \cdot)$ is measured using NT-loss1, a commonly used similarity metric in contrastive learning. This objective ensures that f' maintains feature consistency with f on clean samples while also encouraging stability in feature representations even when the input is perturbed. As a result, the impact of the trigger is mitigated. Through this fine-tuning process, we can effectively remove the backdoor while preserving the model's original performance as much as possible.

4 Experiments

III. Experimental Setup

Attack Methods: To comprehensively evaluate the effectiveness of our proposed method, we compare it with four representative backdoor attack methods: SSL_Backdoor, CTRL, ESTAS, and BadEncoder. Among them, SSL_Backdoor and CTRL are data poisoning-based backdoor attack methods, while ESTAS and BadEncoder involve manipulating the training process to implant backdoors.

Experimental Environment: All experiments are conducted on an Ubuntu 20.04.5 operating system. The computational resources include a Tesla V100 GPU to accelerate the training and inference of deep learning models.

Datasets: We evaluate our method on the CIFAR-10 [17] and ImageNet-100 datasets [18]. CIFAR-10 is a widely used dataset for image classification tasks, containing 10 categories of 32 × 32 color images. It consists of 60,000 images in total, with 50,000 images for training and 10,000 images for testing. ImageNet-100 is a subset of the original ImageNet dataset, randomly selecting 100 categories. Each category contains a large number of high-resolution color images. Compared to CIFAR-10, ImageNet-100 has greater visual complexity and a broader category distribution. The dataset contains approximately 110,000 images in total, with around 100,000 images for training and 10,000 images for testing.

Model Selection: In our experiments, we adopt two mainstream self-supervised learning models, SimCLR and BYOL, for comparative analysis. SimCLR trains models using contrastive learning, whereas BYOL learns representations without explicit negative pairs. Both methods are representative approaches in the field of self-supervised learning.

Encoder: We select different encoders for feature extraction on different datasets. For CIFAR-10, we use ResNet-18 [19] as the encoder.For ImageNet-100, we employ EfficientNet V2 Small [20] as the encoder.

Evaluation Metrics: We use Attack Success Rate (ASR) and Benign Accuracy (BA) as the evaluation metrics.

Defense Methods: We compare our method with several existing backdoor defense techniques, including CLP, I-BAU, and SSL Cleanse.

A. Experimental Results

The experimental results are presented in Tables 1 and 2. Table 1 presents the experimental results on the CIFAR-10 dataset, while Table 2. Shows the results on the ImageNet-100 dataset. The results demonstrate that our defense method UTRBD exhibits significant advantages in mitigating backdoor attacks. Specifically, our method achieves a lower performance drop while effectively reducing the attack success rate. On CIFAR-10, the average BA decreases by only 1% compared to the original model, while the average ASR is reduced to below 10%, which is significantly better than other backdoor defense methods.

Table 1. Defense Performance of UTRBD and Other Methods on CIFAR10

SSL Methods	Attack Methods	None Defense		CLP		I-BAU		SSL_Cleanse		UTRBD	
		BA	ASR	BA	ASR	BA	ASR	BA	ASR	BA	ASR
SimCLR	SSL_BackDoor	77.19	16	76.12	9.13	72.83	2.16	71.62	2.58	77..6	3.66
	CTRL	74.94	89.4	73.94	6.57	65.94	0.33	66.92	44.6	75.31	6.11
	BadEncoder	77.67	59.98	76.67	40.09	69.37	1.17	70.54	21.69	77.09	2.29

(continued)

Table 1. (*continued*)

SSL Methods	Attack Methods	None Defense		CLP		I-BAU		SSL_Cleanse		UTRBD	
		BA	ASR	BA	ASR	BA	ASR	BA	ASR	BA	ASR
	ESTAS	76.22	99.19	76.86	5.01	70.07	1.97	71.55	49.53	75.8	5.08
BYOL	SSL_BackDoor	72.42	18.03	70.11	8.23	45.90	1.01	59.41	11.92	71.75	4.05
	CTRL	73.12	89.79	74.05	4.9	51.30	0.44	54.387	43.5	72.26	4.6
	BadEncoder	72.95	87.83	71.65	21.76	52.29	0.47	54.33	62.94	70.58	17.93
	ESTAS	75.10	99.98	74.56	56.52	54.34	1.07	54.8	29.68	74.26	5.2
AVE		74.95	70.03	74.25	19.03	60.26	1.08	62.94	33.31	73.86	6.12

Table 2. Defense Performance of UTRBD and Other Methods on Imagenet100

SSL Methods	Attack Methods	None Defense		CLP		I-BAU		SSL_Cleanse		UTRBD	
		BA	ASR	BA	ASR	BA	ASR	BA	ASR	BA	ASR
SimCLR	SSL_BackDoor	59.39	38.22	54.82	32.83	37.19	2.4	51	26.66	56.71	6.72
	CTRL	58.22	38.98	53.23	19.46	34.1	0.61	50.83	17.93	55.91	4.96
	BadEncoder	60.03	61.11	48.99	35.19	40.87	2.23	51.16	14.02	58.64	15.77
	ESTAS	62.88	98.17	43.29	37.44	43.58	9.27	54.18	10.51	59.73	11.68
BYOL	SSL_BackDoor	61.25	37.88	50.07	26.33	46.69	8.5	52.47	16.57	60.34	10.54
	CTRL	59.45	42.33	57.98	20.09	43.80	5.71	53.84	9.59	56.73	8.4
	BadEncoder	58.33	64.51	41.7	20.31	40.9	9.88	54.26	12.6	58.65	11.97
	ESTAS	61.57	99.33	47.05	59	47.36	7.28	54.77	11.22	60.18	13.52
AVE		60.14	60.07	49.64	31.33	41.81	5.74	52.81	14.89	58.36	10.45

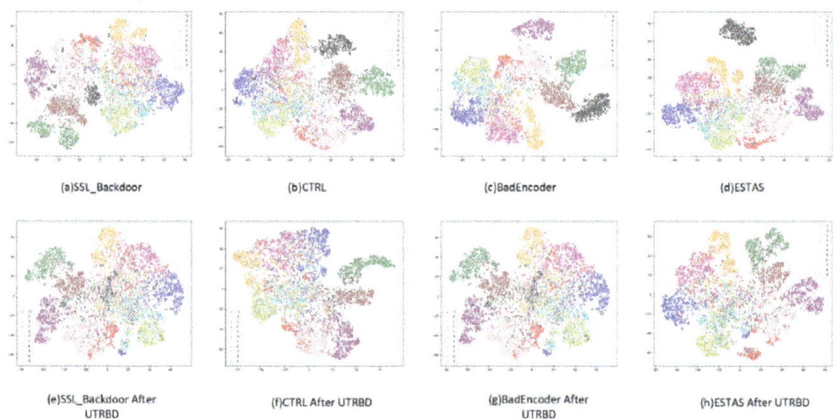

Fig. 2. t-SNE Visualization of Feature Distributions Before and After UTRBD

The CLP method performs well against various attacks on the CIFAR-10 dataset, but its effectiveness drops significantly on the ImageNet-100 dataset. Additionally, CLP struggles to defend against BadEncoder, indicating a lack of generalization. I-BAU and SSL_Cleanse demonstrate strong performance in reducing the attack success rate (ASR), with I-BAU achieving the best ASR reduction. However, this comes at the cost of greater accuracy (BA) degradation, as I-BAU sacrifices more model precision and relies on labeled data. SSL_Cleanse also suffers from significant performance loss. It requires iterative reconstruction of the trigger for each cluster, leading to higher computational costs. Furthermore, it demands access to samples from the target class, which limits its practicality in real-world scenarios.

To further analyze the effectiveness of our UTRBD defense method, we visualize the feature distributions before and after applying UTRBD using t-SNE [21]. The visualization is conducted on the CIFAR-10 dataset with SimCLR as the SSL model. Figure 2(a)–(d) illustrate the feature distributions of SSL_Backdoor, CTRL, BadEncoder, and ESTAS attacks before defense, while Fig. 2(e)–(h) depict the corresponding distributions after applying UTRBD.

In these visualizations, black dots represent backdoor samples. Before applying UTRBD, as shown in Fig. 2(a)–(d), the backdoor samples are tightly clustered in the feature space, which enables the backdoor attack to be successful. However, after applying UTRBD (Fig. 2(e)–(h)), the black dots become more widely dispersed rather than concentrated in a specific region. This indicates that our defense method effectively modifies the feature space, disrupting the structured clustering of backdoor features and making the model resistant to backdoor attacks.

5 Conclusion

In this paper, we propose UTRBD, a novel backdoor defense method based on untargeted trigger reconstruction and a contrastive fine-tuning strategy, which effectively mitigates backdoor attacks in self-supervised learning (SSL) models. Unlike existing defense methods that rely on labeled data or incur high computational costs, UTRBD achieves strong defense performance using only a small amount of unlabeled data and minimal computational resources. Extensive experiments on CIFAR-10 and ImageNet-100 demonstrate that UTRBD effectively defends against multiple backdoor attack methods across different SSL models and encoder architectures. Our results show that UTRBD not only significantly reduces the backdoor attack success rate but also maintains high model utility, achieving a better balance between defense effectiveness and model performance. These findings highlight the robustness and practicality of UTRBD as a promising solution for backdoor defense in self-supervised learning.

References

1. Chen, T., Kornblith, S., Norouzi M., et al. A simple framework for contrastive learning of visual representations. International Conference on Machine Learning. PMLR, pp. 1597–1607 (2020)

2. Grill, J.B., Strub, F., Altché, F., et al.: Bootstrap your own latent-a new approach to self-supervised learning. Adv. Neural. Inf. Process. Syst. **33**, 21271–21284 (2020)

3. Chen, X., Fan, H., Girshick, R., et al.: Improved baselines with momentum contrastive learning (2020). arXiv preprint arXiv:2003.04297

4. Chen, X., He, K.: Exploring simple siamese representation learning. Proceedings of the IEEE/CVF Conference on Computer Vision and Pattern Recognition. pp. 15750–15758 (2021)

5. Jing, L., Tian, Y.: Self-supervised visual feature learning with deep neural networks: a survey. IEEE Trans. Pattern Anal. Mach. Intell. **43**(11), 4037–4058 (2020)

6. Elnaggar, A., Heinzinger, M., Dallago, C., et al.: Prottrans: toward understanding the language of life through self-supervised learning. IEEE Trans. Pattern Anal. Mach. Intell. **44**(10), 7112–7127 (2022)

7. Saha, A., Tejankar, A., Koohpayegani, S.A., et al.: Backdoor attacks on self-supervised learning. Proceedings of the IEEE/CVF Conference on Computer Vision and Pattern Recognition. pp. 13337–13346 (2022)

8. Li, C., Pang, R., Xi, Z., et al.: An embarrassingly simple backdoor attack on self-supervised learning. Proceedings of the IEEE/CVF International Conference on Computer Vision. pp. 4367–4378 (2023)

9. Xue, J., Lou, Q.: Estas: effective and stable trojan attacks in self-supervised encoders with one target unlabelled sample (2022). arXiv preprint arXiv:2211.10908

10. Jia, J., Liu, Y., Gong, N.Z.: Badencoder: backdoor attacks to pre-trained encoders in self-supervised learning. 2022 IEEE Symposium on Security and Privacy (SP). IEEE, pp. 2043–2059 (2022)

11. Tejankar, A., Sanjabi, M., Wang, Q., et al.: Defending against patch-based backdoor attacks on self-supervised learning. Proceedings of the IEEE/CVF Conference on Computer Vision and Pattern Recognition. pp. 12239–12249 (2023)

12. Pan, M., Zeng, Y., Lyu, L., et al.: {ASSET}: Robust backdoor data detection across a multiplicity of deep learning paradigms. 32nd USENIX Security Symposium (USENIX Security 23). pp. 2725–2742 (2023)

13. Zheng, M., Xue, J., Wang, Z., et al.: Ssl-cleanse: Trojan detection and mitigation in self-supervised learning. European Conference on Computer Vision. Cham: Springer Nature Switzerland, pp. 405–421 (2024)

14. Zeng, Y., Chen, S., Park, W., et al.: Adversarial unlearning of backdoors via implicit hypergradient (2021). arXiv preprint arXiv:2110.03735

15. Zheng, R., Tang, R., Li, J., et al.: Data-free backdoor removal based on channel lipschitzness. European Conference on Computer Vision. Cham: Springer Nature Switzerland, pp. 175–191 (2022)

16. Liu, K., Dolan-Gavitt, B., Garg, S.: Fine-pruning: defending against backdooring attacks on deep neural networks. International Symposium on Research in Attacks, Intrusions, and Defenses. Cham: Springer International Publishing, pp. 273-294 (2018)

17. Ho-Phuoc, T.: CIFAR10 to compare visual recognition performance between deep neural networks and humans (2018). arXiv preprint arXiv:1811.07270

18. Deng, J., Dong, W., Socher, R., et al.: Imagenet: a large-scale hierarchical image database. 2009 IEEE conference on computer vision and pattern recognition. IEEE pp. 248–255 (2009)

19. Chen, Z., Jiang, Y., Zhang, X., et al.: ResNet18DNN: prediction approach of drug-induced liver injury by deep neural network with ResNet18. Briefings Bioinformatics **23**(1), bbab503 (2022)

20. Tan, M., Le, Q.: Efficientnetv2: smaller models and faster training[C]//International conference on machine learning. PMLR **2021**, 10096–10106 (2021)
21. Van der Maaten, L., Hinton, G.: Visualizing data using t-SNE. J. Mach. Learn. Res. **9**(11) (2008)
22. Feng, S., Tao, G., Cheng, S., et al.: Detecting backdoors in pre-trained encoders. Proceedings of the IEEE/CVF Conference on Computer Vision and Pattern Recognition. pp. 16352–16362 2023

The Impact of Generative AI on Learning: Discussion and Reflection Based on the Multi-Store Model of Memory

Yong Liu[1,2]([envelope]) and Xiaoling Li[3]

[1] School of Computer and Intelligence Education, Lingnan Normal University, Zhanjiang, People's Republic of China
liuyong@lingnan.edu.cn

[2] School of Software Engineering, South China University of Technology, Guangzhou, People's Republic of China

[3] Office of International Exchange and Cooperation, Lingnan Normal University, Zhanjiang, People's Republic of China
lixiaoling@lingnan.edu.cn

Abstract. With the rapid development of generative AI technology, its application in the field of education has gradually attracted attention. Generative AI is profoundly changing the traditional teaching model and learning process through functions such as natural language processing, image generation, and creative content creation. Based on the perspective of the multi-storage model of memory, this paper explores the impact of generative AI on the learning process and analyzes its positive and negative effects in learning. The study found that generative AI technology can not only improve learning efficiency, promote personalized learning and the cultivation of high-order thinking ability, but also may lead to over-dependence, false confidence and negative learning attitude. With the widespread application of generative AI technology in education, teachers should balance the use of technological tools with the cultivation of students' autonomous learning ability when designing teaching activities to avoid the abuse and misuse of AI tools. This paper also discusses the ethical and legal issues of AI in education and emphasizes the responsibilities and challenges of educators in the face of AI technology. In short, generative AI technology brings great potential to education, but it is also accompanied by a series of challenges. Educators need to respond cautiously to ensure that AI technology promotes the all-round development of students while avoiding its possible negative effects.

Keywords: Generative AI · The Multi-store Model of Memory · Educational Technology · Instructional Design

1 Introduction

As human civilization enters the digital age, the accumulation and dissemination of knowledge has shown exponential growth, and scientific and technological progress is driving social development at an unprecedented speed. Renowned scholar Ray Kurzweil

© The Author(s) 2026
P. Siarry et al. (Eds.): WCNA 2024, LNEE 1550, pp. 276–285, 2026.
https://doi.org/10.1007/978-981-95-6946-5_28

pointed out in his research that when the speed of knowledge innovation approaches the critical point of the exponential curve, Artificial Intelligence (AI) is expected to have formal creation and innovation capabilities, marking that human civilization will cross thousands of development trajectory and move towards a new knowledge paradigm. [1] This breakthrough change is called the "technological singularity." He predicted that the advancement of artificial intelligence will become the core driving force for the development of human society.

Against this backdrop, in the fall of 2022, OpenAI launched ChatGPT, a generative artificial intelligence based on natural language processing and deep learning technology. In just a few months, ChatGPT quickly attracted global attention with its powerful language understanding ability and almost omniscient response level. This technology can not only automatically generate various types of copywriting such as news, novels, and scripts, but also provide creative solutions in fields such as science, mathematics, and programming. Amazingly, ChatGPT can also work with other third-party applications such as DALL-E and Midjourney to generate creative works such as images, characters, briefings, picture books, animations, and even films through natural language dialogues. The development of artificial intelligence has not only promoted breakthroughs in fields such as data analysis, speech recognition, and image recognition, but also expanded the boundaries of creative fields such as literature and art, and inspired extensive discussions in the education community on the application of generative AI in classroom teaching.

However, the rapid development of artificial intelligence in the field of education has also triggered deep thinking about its potential impact on human learning and cognition. Should we be blindly optimistic about this change and think that it will only bring positive effects? More and more scholars have begun to question the impact of the application of emerging technologies on cognitive memory and creative thinking. Studies have shown that over-reliance on online search tools may cause students to mistakenly believe that obtaining information is learning, ignoring the role of in-depth thinking and memory, forming the "Google Effect". [2] When information acquisition becomes too easy, students' sense of self-efficacy may be over-amplified, forming a superficial understanding of knowledge rather than a deep grasp. Furthermore, the phenomenon of Digital Amnesia has also emerged in this process, especially when learners are accustomed to relying on digital tools and ignore the internalization process of information. [3] At the same time, the rise of generative AI such as ChatGPT has further changed the traditional way of learning: unlike traditional search engines, generative AI can directly provide learners with answers to questions in the form of natural language, and in some cases, it even saves learners the steps of information screening and judgment.

Therefore, can the application of generative AI in education effectively improve learning efficiency and creativity, or will it lead to the inertia of thinking and the weakening of cognitive depth? In the context of the increasing popularity of this technology, will generative AI inspire deeper exploratory learning, or will it cause scholars to have negative views on the "redundancy" of memory training and practice? This article attempts to explore the possible impact of generative AI in the learning process from the perspective of a multi-storage model, focusing on how it affects students' cognitive processing, knowledge internalization, and the generation of creative thinking. From this perspective, this article aims to provide theoretical support and practical guidance for the application of generative AI in educational practice, helping us better understand how this technology can promote learning and creativity while avoiding its potential negative effects.

2 The Impact of Technological Development on Cognitive Learning and Information Processing Models

2.1 The Multi-Store Model of Memory

Atkinson provides an important theoretical framework for understanding the human information processing process. The model regards memory as a system composed of multiple stages. [4] According to the length of time that information is stored in the system, it can be divided into three different subsystems: sensory memory (or instantaneous memory), short-term memory, and long-term memory. The learner's sensory memory first captures external information and transfers part of the information into short-term memory through the screening effect of attention. In this process, learners may strengthen the retention of information through continuous review, association, and reasoning. As knowledge becomes more and more proficient, information can eventually be transferred to long-term memory and become a relatively stable and easy-to-retrieve knowledge reserve, so that when encountering similar situations in the future, relevant information and experience can be effectively retrieved from long-term memory for application. This multi-store model provides a theoretical basis for subsequent cognitive learning and teaching design, helping us understand the entire cognitive process of individuals from sensory input to long-term knowledge storage.

In Atkinson's model, the key factors that affect learning and teaching outcomes include: (1) whether teaching can effectively attract learners' attention and help them filter out useful content from a large amount of sensory information; (2) whether teaching activities provide sufficient opportunities for learners to interact and practice with information repeatedly so as to transfer the knowledge learned from short-term memory to long-term memory; (3) whether the concepts formed by learners in the learning process are accurate and whether there are any misunderstandings or vague knowledge points; (4) whether the newly learned knowledge can be effectively connected with the existing knowledge system, so as to form a knowledge network that is easy to retrieve in future learning and problem solving; (5) whether learners can extract relevant knowledge and experience from long-term memory when needed to assist in problem solving; and (6) whether short-term memory can effectively filter out useful clues in the face of a large amount of information and complete learning tasks through teacher guidance or peer interaction.

Although Atkinson's multi-store model provides a detailed description of the individual cognitive process, it fails to explore the interaction and co-construction of knowledge between people and between people and the environment. Therefore, with the gradual development of social group activities, the individual's cognitive learning process gradually shifted to consider the influence of the social environment, giving rise to new learning paradigms such as social learning theory and social constructivism. [5] These theories emphasize the role of interactive forms such as observational learning, demonstration learning, and cooperative learning, enriching teaching strategies and methods, and providing new directions for the education system. However, the focus of this article is not to review these classic teaching design methods, but to focus on the impact of technological progress on learning and teaching paradigms, especially how generative AI changes human learning patterns.

2.2 The Impact of Search Engines and Transactive Memory on Human Learning and Memory

With the advancement of science and technology, the accumulation and dissemination of information has become increasingly rapid, and the updating of knowledge has become more frequent. This change has brought increasing challenges to the way of imparting knowledge in the traditional education system, especially when the way to acquire new knowledge has become more convenient, learners often rely on external tools to supplement their personal memory. The emergence and popularization of search engines is a typical manifestation of this trend. Nowadays, when faced with any problem, people often use search engines such as Google and Facebook to quickly find answers, and this way of quickly acquiring information seems to have gradually relegated traditional memory and knowledge storage functions to the second line. Technological tools have replaced some memory functions, especially in knowledge retrieval and problem solving, and have become a new memory-aiding mechanism.

The phenomenon of "Ask Google everything" reflects the Internet generation's reliance on information acquisition and questioning of traditional memory methods. Modern learners are gradually accustomed to obtaining instant answers through the Internet and social media, rather than relying on knowledge accumulated over a long period of time through recitation and memorization in school education. Traditional textbooks, due to their relatively simple and boring presentation methods, are often considered to lack interactivity and interest, and are difficult to compare with information sources in social media and self-media. In this context, learners have questioned the efficiency of school education: in today's rapidly updated knowledge, is it more important to quickly obtain the required information through platforms such as social media rather than memorizing it by rote?

A related concept is "Transactive Memory", which means that team members transfer memory and knowledge to each other through cooperation and sharing to solve problems together. [6] In complex social life, individuals often rely on external resources and the expertise of others to make up for their own memory deficiencies. For example, drivers use navigation apps to obtain real-time traffic information while driving, or team members share office materials through cloud storage. This memory model that relies on external information and collaborative sharing has become an indispensable part of modern society.

Through the transactive memory mechanism, individuals no longer need to store all information in their personal long-term memory, but share the memory burden through collaboration and sharing, thereby improving overall efficiency. This mechanism is essentially similar to the individual multi-storage model, and still includes the three stages of information encoding, storage, and retrieval. [7] However, unlike the traditional individual memory model, transactive memory places more emphasis on cooperation and division of labor among group members. Each member undertakes the task of information storage and retrieval according to his or her own professional field, and other members only need to know when, where, and how to obtain the required information.

With the rise of generative AI (such as ChatGPT), this interactive memory model based on collaboration and sharing has been further deepened. Under the framework of generative AI, artificial intelligence becomes an "omniscient being", replacing the role of humans in information storage and retrieval. Learners no longer need to store a large amount of knowledge through traditional recitation and memorization, but through interaction with AI, input specific tasks, situations and problems, and quickly obtain close to correct answers. This process seems to have gradually faded the traditional memory function, and the focus of learning has shifted to how to efficiently use these technological tools to solve problems.

However, can the intervention of generative AI completely replace human cognitive learning? The application of knowledge in the real world is far more than simple memory retrieval. High-level application transformation, hypothesis verification, reflective adaptation and complex problem solving still require careful consideration and critical thinking. Although generative AI can provide instant answers, whether it can replace human judgment, value trade-offs and decision-making processes in the face of situations is still a question worth exploring.

In short, the application of generative AI is gradually changing the traditional cognitive learning model, pushing the focus of learning from memory and information storage to thinking and application. In the future, learners may no longer need to rely on traditional methods to perform a large number of memory exercises, but instead focus on how to use technology to assist memory and cultivate higher-order thinking skills. However, the far-reaching impact of generative AI on learning models still needs further research, especially on how to balance the relationship between knowledge acquisition and critical thinking and creative application.

3   The Impact of Generative AI on the Learning Process

The concept of "Digital Natives" [8] proposed by Prensky describes the younger generation who grew up in a digital technology environment, in contrast to "digital immigrants" who only come into contact with digital technology in adulthood. With the birth of large language models, generative AI has gradually been integrated into the field of education, especially in basic education, where teachers apply it to drawing, language learning, composition and other aspects. With this trend, a new learning group - "AI natives" has gradually formed. These students have grown up under the influence of generative AI technology since childhood, and their cognition and learning methods are naturally deeply influenced by artificial intelligence tools.

AI natives not only rely on traditional teaching, but also learn to interact with generative AI, using it to improve learning efficiency, stimulate creativity, and solve problems that are difficult to deal with traditional methods. Teachers are also constantly adjusting their teaching methods to adapt to the characteristics of this group, using generative AI to achieve personalized learning, automatic homework correction and other functions, thereby changing the way students learn. The rise of AI natives marks a profound change in education and learning methods. How to effectively integrate generative AI technology has become a key challenge for the future development of education.

3.1 The Positive Impact of Generative AI on the Learning Process

Dual-track learning model of long-term and short-term memory and transactive memory.

Although the Brain-Computer Interface has not yet made a breakthrough to fully integrate external memory with brain nerves, human learning and memory still mainly rely on the brain's long-term and short-term memory system and technology-assisted transactive memory dual-track parallel processing. However, with the widespread application of generative AI technology, students are increasingly accustomed to using external tools as an extension of memory and integrating them into their personal learning model. This transactive memory can not only assist in the storage and retrieval of information, but also help students effectively process a large amount of knowledge, reduce the burden of memory, and improve learning efficiency. As the application of technology gradually becomes the basic literacy of learning, future education will no longer focus only on traditional memory ability, but will emphasize more on higher-order thinking abilities, such as scientific process skills such as scientific inquiry, computational thinking and problem solving. This shift means that traditional teaching models and assessment methods, especially memory-based evaluations guided by standard answers, will face more severe challenges. Technology-assisted learning methods provide students with a more flexible and personalized learning experience, enabling them to explore and think more effectively in a dynamic knowledge network.

Learning Assistants with Natural Language Interaction

Throughout its history, school education has focused more on the popularization and standardization of education, aiming to quickly train the workforce, which often ignores the individual needs and interests of students in the education process. [9] Due to the heavy workload of teachers and the high student-teacher ratio, individualized teaching and immediate personalized feedback are often difficult to achieve. However, the natural language processing capabilities of generative AI provide a possible solution to this problem. With simple prompts, teachers can transform large language models into personalized learning assistants in the classroom, answering students' questions in real time and providing differentiated feedback. AI can not only answer students' immediate doubts, but also help students verify the accuracy of knowledge and guide them to think critically about their answers. In this way, generative AI can help students learn knowledge more autonomously, promote their deep thinking and verification capabilities, and promote a more personalized and adaptable learning experience. [10].

Learning Partners with AI Mechanisms

Technology integrated with AI mechanisms in education is far more than providing tools or auxiliary functions. It can also appear as a learning guide, counselor, learning partner, and other multiple roles. [11] From the perspective of social learning, learners will not only learn from the teacher's demonstration in the learning process, but also get support through the interaction and imitation of their peers. This is the concept of "Peer Modeling". With the development of generative AI technology, the Intelligent Tutoring

System (ITS) has gradually matured in the past few decades. Generative AI can not only provide personalized guidance as a virtual tutor, but also accompany students as a learning partner. By simulating the interaction between peers, AI not only provides learners with instant feedback, but also helps them apply and integrate knowledge in the learning process. The addition of AI peers provides learners with a more flexible and interactive learning environment, which encourages students to actively participate in the learning process and improves their learning motivation and effectiveness.

Richness and Accessibility of Cross-Language and Cross-Domain Resources
Language is the cornerstone of cognitive development and a bridge between thought expression and cross-cultural learning. [12] However, language and cultural barriers have long restricted the dissemination of knowledge and cross-cultural communication. Generative AI, through its powerful natural language processing capabilities and the use of a large training corpus, can provide rich cross-language and cross-domain learning resources. Whether it is the translation of academic documents or the interpretation of cultural background, AI can provide support quickly and accurately, breaking the language barrier. In addition, AI can effectively integrate knowledge from different languages and fields, provide cross-disciplinary dialogue and application, and enable learners to explore in a broad field of knowledge, thus breaking through the limitations of traditional education. In this process, students no longer need to focus only on the surface meaning of language, but can focus more on understanding and thinking about the content, improving their critical and cross-cultural thinking skills. Furthermore, the image, video and creative generation capabilities of generative AI enable learners to be inspired by content in different forms, expand their learning horizons, and truly achieve cross-disciplinary and cross-domain learning and innovation.

In summary, the positive impact of generative AI in the learning process is multi-faceted, especially in promoting personalized learning, improving learning efficiency, breaking through the boundaries of language and culture, and stimulating students' innovation and critical thinking. With the continuous development of generative AI technology, the education model and learning process will undergo a profound transformation, and students' learning experience will become richer, more flexible and more efficient.

3.2 Negative Impact of Generative AI on the Learning Process

Over-Reliance on Negative Learning Attitudes
The "Google Effect" refers to the fact that when learners rely on technology-assisted tools (such as search engines) to obtain information, although they can remember the storage location and retrieval method of the information, their memory of the specific content becomes vague. This phenomenon shows that over-reliance on AI tools may cause learners to have too high confidence in their cognitive abilities, but may not actually improve their learning achievements. Especially in the process of AI technology being widely used in education, if the teaching design overemphasizes the convenience of quickly obtaining answers through technological tools, students may mistakenly believe that the learning process is too easy, thereby reducing their enthusiasm for learning. If teaching only focuses on generating perfect answers or images through precise prompts,

then when the entertainment and novelty of AI fade, whether students can maintain in-depth learning of knowledge content becomes a problem. Therefore, the teaching design using generative AI should recognize the multi-storage model of learning, emphasizing not only the extraction of information and the generation of results, but also how to use technological tools to transform and verify knowledge, and help students build or reconstruct a knowledge system for solving problems. Teachers need to clarify the objectives of teaching tasks and think about how to balance the auxiliary role of AI with the cultivation of students' independent learning ability, so as to avoid over-dependence, false self-confidence and negative learning attitude caused by AI.

Over-Reliance on Negative Learning Attitudes

Generative AI provides a large number of instant answers through natural language dialogue. Although this convenience improves learning efficiency, it may weaken students' ability to solve complex problems and conduct high-level thinking. According to the learning theory of the multi-store model, the cultivation of problem solving and high-level thinking skills often depends on repeated trial, error and practical operation. Over-reliance on AI tools to provide quick answers may hinder students from developing these abilities through autonomous exploration and deep thinking. Therefore, the teaching of generative AI should not focus too much on the optimization of questioning skills or prompts, but should focus on how to use technological tools to cultivate students' scientific and technological thinking and help them use AI tools to solve complex problems in real life. With the advancement of generative AI technology, more personalized dialogue intention optimization algorithms may appear in the future, and the existing need for prompt teaching may gradually disappear. But in any case, in the multi-store model where dual-track memory works in parallel, how to promote students' critical thinking, communication skills, teamwork, creativity and problem-solving ability will become the core goal of AI generation education design.

Over-Reliance on Negative Learning Attitudes

The operation of generative AI is based on machine learning and probabilistic algorithms, and it generates answers through training on large amounts of data. However, this generation mode has certain limitations, especially when generating answers with clear knowledge, AI may make mistakes of "putting the wrong name on the wrong head". For example, when generating literature or reference materials, AI may generate seemingly reasonable but actually fictitious literature based on the algorithm, causing learners to mistakenly believe in inaccurate information. In addition, due to the bias of training data or the limitations of the field, generative AI may also have biases in gender, race, etc. Therefore, in the teaching design of generative AI, teachers need to be aware of the limitations of AI and cultivate students' attitude of maintaining critical thinking about AI-generated information. Teaching activities should encourage students to verify and validate the content provided by AI to avoid relying on inaccurate or false information. The guiding role of teachers is crucial, and students need to be helped to form effective methods of information identification and verification.

Over-Reliance on Negative Learning Attitudes

Generative AI has great potential in providing personalized learning support, especially in cross-domain and cross-language knowledge retrieval. However, in actual teaching, the design of many generative AI applications is more inclined to personalized learning functions, and less emphasis is placed on promoting opportunities for teamwork and collective creation. In some AI-assisted learning environments, students tend to interact with AI rather than discuss and cooperate with teachers or peers, which may lead to reduced social interaction and weakened teamwork ability. Although AI tools can effectively reduce students' anxiety about asking questions to teachers and increase learning participation, over-reliance on AI may lead to a reduction in real interaction between students and teachers or peers, which in turn affects the experience of collective cooperation and the opportunity for social learning. Therefore, when designing classroom tasks, teachers should balance the ratio of personalized learning and teamwork, so as to not only use AI technology to promote personalized learning, but also stimulate students' collective wisdom through teamwork and promote the unity of internal cognition and external interactive memory. Through this balance, students can not only improve their personal learning ability, but also cultivate communication, collaboration and problem-solving skills in cooperation.

4   Conclusion

This article explores the profound impact of generative AI on the learning process from the perspective of a macroscopic The Multi-store Model of Memory. Research shows that generative AI technology has not only changed learners' memory methods and information processing processes, but also triggered a transformation in learning methods and teaching design. This change requires educators to reflect on and adjust teaching content, methods, and evaluation methods to meet the needs of the new generation of AI students. In the future, as generative AI technology continues to advance and mature, the education community must actively explore how to effectively integrate artificial intelligence and traditional teaching resources, both to leverage the advantages of AI and improve learning efficiency, and to avoid its potential negative effects, to ensure that students gain a more comprehensive and profound learning experience with the assistance of AI.

In short, generative AI technology is bringing unprecedented opportunities and challenges to education. In this wave of technology, educators should maintain keen insight, constantly adjust teaching strategies and methods, ensure that AI applications can truly serve students' cognitive development and the cultivation of creative thinking, and promote the overall progress of education. At the same time, we also need to pay more attention to the ethical and legal issues caused by the development of AI technology to ensure that the application of AI in education can be reasonably regulated and guided.

Acknowledgment. Supported by the projects grant from Guangdong Philosophy and Social Sciences Planning Project (Grant No. GD22CXW01, GD24CXW01), Research Platforms and Projects of Colleges and Universities in Guangdong (Grant No. 2022KTSCX071).

References

1. Kurzweil, R.: Superintelligence and singularity. Machine Learning and the City: Applications in Architecture and Urban Design pp. 579–601 (2022)
2. Sparrow, B., Jenny, L., Daniel, M.W.: Google effects on memory: Cognitive consequences of having information at our fingertips. Science **333**(6043), 776–778 (2011)
3. Ferial, C., Mounir D.: The impact of artificial intelligence dependency in research: analyzing the advantages, drawbacks and ethical concerns. Diss. Ibn Khaldun University-Tiaret (2024)
4. Atkinson, R.: Human memory: a proposed system and its control processes. The Psychology of Learning and Motivation (1968)
5. Bada, Steve Olusegun, and Steve Olusegun. Constructivism learning theory: a paradigm for teaching and learning. J. Res. Method Educ. **5**(6), 66–70 (2015)
6. Wegner, D.M.: Transactive memory: a contemporary analysis of the group mind. Theories of Group Behavior/Springer-Verlag (1987)
7. Wegner, D.M., Erber, R., Raymond, P.: Transactive memory in close relationships. J. Pers. Soc. Psychol. **61**(6), 923 (1991)
8. Prensky, M.: Digital natives, digital immigrants part 2: do they really think differently? On the Horizon **9**(6), 1–6 (2001)
9. Baker, D.P.: The educational transformation of work: towards a new synthesis. J. Educ. Work. **22**(3), 163–191 (2009)
10. Xu, W., Ouyang, F.: A systematic review of AI role in the educational system based on a proposed conceptual framework. Educ. Inf. Technol. **27**(3), 4195–4223 (2022)
11. Fenuku, S.D.: Language, culture, and mentality: the three-dimensional axis of language studies and effective communication. Int. J. Lang. Transl. Stud. **4**(1), 76–103 (2024)
12. Ward, A.F.: People mistake the internet's knowledge for their own. Proceedings of the National Academy of Sciences **118**(43), e2105061118 (2021)

A Spacecraft Thermal Balance Test Method and Simulation Analysis for Space Debris Damage

Yumei Zhang[1]([✉]), Chaofan Cai[1], Runze Wang[2], and Ji Li[1]

[1] Aerospace Dongfanghong Satellite Co. Ltd., Beijing, China
`8917829@qq.com`
[2] Beijing Satellite Institute of Environmental Engineering, Beijing, China

Abstract. A new thermal equilibrium test method is proposed for spacecraft with space debris damage. It consists of two thermal equilibrium tests before and after space debris damage. Ground tests were carried out using this test method. The simulation and analysis method of the satellite temperature field after the test is given and simulated. The test and simulation results compare well, and an error analysis is carried out.

Keywords: Thermal equilibrium test · space debris damage · ground-based simulation · simulation analysis

1 Introduction

With the increase of space debris, thermal design and verification for space debris damage have become a new need. Therefore, it is necessary to carry out research on spacecraft thermal balance test methods and simulation analysis for space debris damage. Domestic and foreign research on thermal balance test and simulation analysis has a wide range of research results. The details are as follows.

Liu F et al. [1] started from the heat conduction equation, solved the equilibrium temperature approximation solution, and predicted the equilibrium temperature in the thermal equilibrium test. Overseas generally on the spacecraft thermal analysis model and test results correlation validation work put forward requirements, the U.S. military standard MIL-STD-1540D requirements of the thermal analysis model and test results are generally within ± 3 °C [3], the European Space Agency standard ECSS-E-10-03A is also the same attention to the correlation of the thermal analysis model [4]. Literature [5, 6] clarifies the concept of validity of spacecraft AIT models and tests, and literature [6, 7] proposes a method for evaluating the validity of models and tests.

Meng Fankong [8] proposed a method for determining the temperature stability criterion for spacecraft heat balance test. Zhu Xi et al. [9] proposed a particle swarm algorithm for spacecraft thermal equilibrium temperature prediction. Liang Li [10] designed a thermal control system for NIT-1 CubeSat and conducted thermal tests and analysis. Fang

© The Author(s) 2026
P. Siarry et al. (Eds.): WCNA 2024, LNEE 1550, pp. 286–295, 2026.
https://doi.org/10.1007/978-981-95-6946-5_29

Hongjun et al. [11] investigated the optimization method of thermofluid meter placement for the heat balance test of a micro-nano-satellite. [12] proposed a windy thermal balance test environment simulation technique for Mars rover.

The above research results are for spacecraft in normal operation in orbit, and this paper proposes a heat balance test method and a simulation and analysis method for space debris damage. The differences with the traditional methods are.

(1) Two heat balance tests are required; and
(2) The infrared cage needs to be switched on and off during the test.
(3) The satellite needs to be rotated during the test.
(4) In the simulation analysis, the thermal control parameters of the star meter need to be adjusted according to the test results.

2 Thermal Equilibrium Test Method for Space Debris Damage

2.1 Purpose of the Test

1) The purpose of the thermal equilibrium experiment before space debris damage is to obtain the performance parameters and temperature field distribution of the satellite and to verify the thermal design.
2) Thermal equilibrium experiments after space debris damage are aimed at verifying the changes in the temperature field distribution inside the satellite after the satellite surface has been damaged by space debris, as well as the electronic effects on the single machine on board, and verifying the thermal design.

2.2 The Test States

1) The temperature field in the star is mainly measured by thermocouples, and a small number of thermistors are used for temperature control and comparison.
2) Satellite surface, deploy thermocouples to measure temperature changes.
3) Infrared cage baffle removal; removal of the zonal heat flow meter corresponding to the target bulkhead for debris testing.

3 Test System

The tests consisted of a thermal equilibrium test, a space debris simulation test and a second thermal equilibrium test, as shown in table 1.

The participating systems included a satellite, an environmental simulation system, a temperature measurement system and a test system. The environmental simulation system includes, in addition to the test equipment for the conventional thermal equilibrium test, an infrared cage rotation mechanism and a turntable. In order to facilitate the debris test, it is necessary to design the infrared cage rotation mechanism on one side of the infrared cage frame, and the infrared cage on that side can be opened. The infrared cage rotating mechanism selects servo motor as the driving mechanism, drives the infrared cage rotation through the turbine worm gear mechanism, designs a number of bearings in the infrared cage frame to ensure the coaxiality of the main shaft, and does the active temperature control method such as pasting film heater on the motor to control

Table 1. The experimental process

Serial number	Key link	Main content
1	Close the tank door, and test conditions will be established	Satellite shutdown, establishment of a vacuum system
2	Heat balance and Testing	(1) Establishment of low and high temperature conditions for the heat balance of the satellite and completion of the satellite state test using the ground integrated test system (2) Discussion of the results of the heat balance test (3) Debris test technology status confirmation
3	Space Debris Simulation Experiment (SDSE)	Rotate the satellite to the side of the satellite where the space debris test is to be performed, open the infrared cage and perform the test
4	Continuing to thermally equilibrate high-temperature operating conditions	(1) Conducting post-space-debris-test heat-balance high-temperature operational tests (2) Analysis of the results of the heat balance test after the space debris test
5	Thermally balanced low-temperature service	(1) post-debris-test heat-balance low-temperature operating test (2) The rotary table rotates the low-temperature cable test (3) Analysis and discussion of test results
6	The end of the experiment	(1) A brief summary of the overall status of the experiment
7	Satellite out of the vacuum tank	(1) Appearance inspection and photographing of satellites before they leave the tank (2) Out of the tank, open the infrared cage, satellite appearance inspection, photographs (3) Replacement cable, low-temperature rotating cable condition check (4) satellite testing

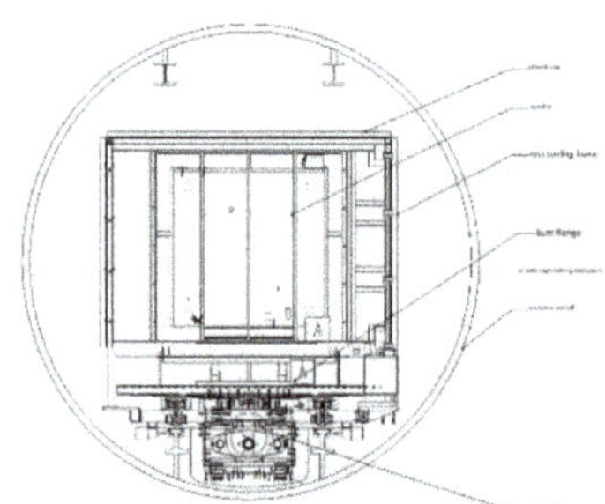
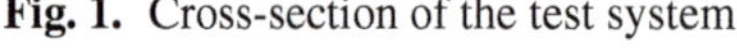

Fig. 1. Cross-section of the test system

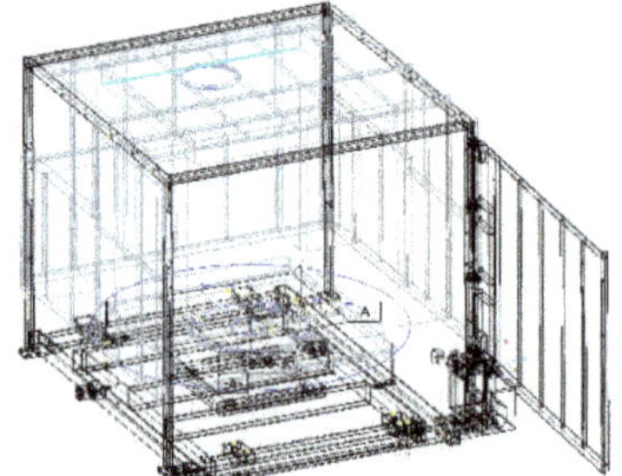

Fig. 2. Schematic diagram of the infrared cage

the temperature range of the motor. For the rotating mechanism, special tests under high and low temperatures were carried out, and the test results proved that it could rotate stably. The test system section and infrared cage are shown in Figs. 1 and 2.

All sides of the satellite face the problem of space debris test, so in the ground simulation, the vacuum tank needs to be installed in the turntable, so that the satellite can be rotated in the vacuum tank, in order to ensure that the four sides of the circumferential direction of the test can be carried out. The satellite rotary table is shown in Fig. 3.

According to the above test program, the test verification was carried out, and during the whole test process, each working condition satisfied the stability criterion, and the test data were real and effective. The heat sink temperature during the test is shown in Fig. 4.

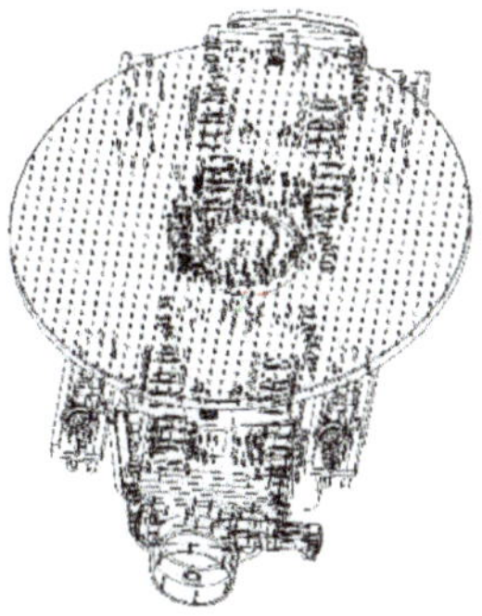

Fig. 3. Schematic diagram of the satellite transponders

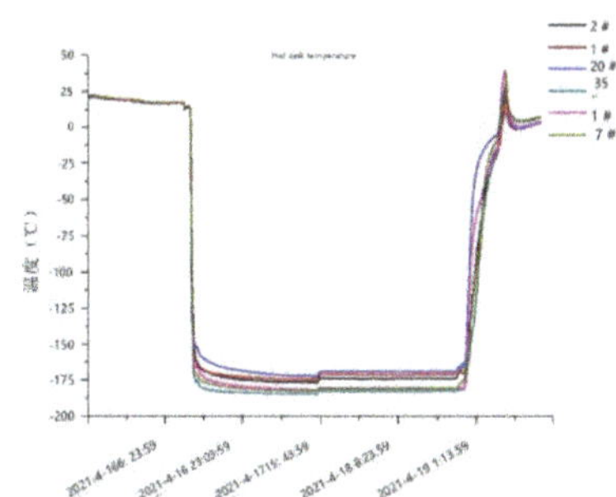

Fig. 4. Heat sink temperature during the test

4 Two Simulations of Thermal Equilibrium Satellite Temperature Field Variations

The simulation method for the temperature field of the thermal equilibrium test before the space debris test is the conventional satellite temperature field analysis method. The thermal equilibrium temperature field simulation method after the space debris test is based on the thermal control performance evaluation model, injecting thermal control parameters on the surface of the satellite after the space debris test, combining with models of external heat flow changes, configuration and internal heat source distribution, obtaining the temperature field of the entire satellite through simulation, and then evaluating the temperature change of the stand-alone unit and its influence, as shown in Fig. 5 below.

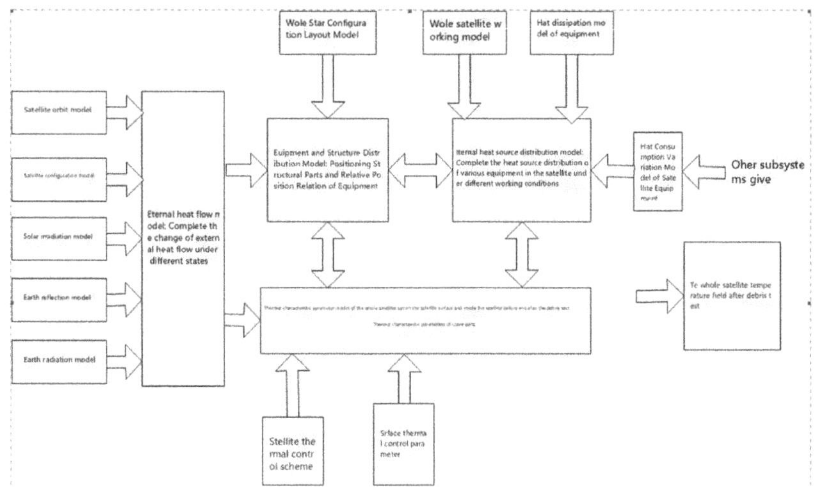

Fig. 5. Methods for analyzing the temperature field of a satellite after a space debris test

The self-developed thermal control simulation software was used to mathematically model the thermal control of the whole star and simulate the temperature change of the whole star after the space debris test. Table 2 shows the comparison of the difference between the software simulation and the real thermal equilibrium test. From the comparison results, the maximum deviation of the minimum value of the simulation is the camera electronic control box, which is lower than the real test by -3.0 °C, and the maximum deviation of the simulation is the momentum wheel S, which is higher than the real test by 4.1 °C, and the deviation is within 5 °C, so that it is considered that the simulation accuracy meets the requirements.

Table 2. Comparison of simulation temperatures with high temperature test temperatures after debris test

Temperature measurement points	Difference	
	Min	Max
Measurement point 1	0.7	0.4
Measurement point 2	0.3	1.7
Measurement point 3	-2.8	-0.9
Measurement point 4	-1.9	-0.5
Measurement point 5	-2.9	-1.9
Measurement point 6	-0.3	1.6
Measurement point 7	-0.4	3.0
Measurement point 8	1.0	1.7
Measurement point 9	-0.3	4.1

(continued)

Table 2. (*continued*)

Temperature measurement points	Difference	
	Min	Max
Measurement point 10	0.3	2.3
Measurement point 11	0.7	1.3
Measurement point 12	1.5	2.3
Measurement point 13	-0.3	3.9
Measurement point 14	1.3	-0.2
Measurement point 15	1.0	0.7
Measurement point 16	-3.0	-2.3

Table 3 shows the comparison of the low-temperature thermal balance difference between the software simulation and the real thermal balance test. From the comparison results, the maximum deviation of the minimum value of the simulation is the stack assemblage 1, which is higher than that of the real test by 2.1 °C, and the maximum deviation of the simulation is the momentum wheel S, which is higher than that of the real test by 3.9 °C, with a deviation of 5 °C or less, and it is considered that the accuracy of the simulation meets the requirements.

Table 3. Comparison of simulation temperatures after debris tests with low-temperature test temperatures

Temperature measurement points	Difference	
	Min	Max
Measurement point 1	-0.3	-0.7
Measurement point 2	1.4	2.5
Measurement point 3	-0.3	0.7
Measurement point 4	0.3	0.7
Measurement point 5	0.1	0.9
Measurement point 6	1.9	2.9
Measurement point 7	1.2	2.6
Measurement point 8	1	1.9
Measurement point 9	1.4	3.9
Measurement point 10	0.6	1.8
Measurement point 11	0.7	1.3
Measurement point 12	2.1	3.1
Measurement point 13	-0.5	0.7
Measurement point 14	-0.28	-0.38

(*continued*)

Table 3. (*continued*)

Temperature measurement points	Difference	
	Min	Max
Measurement point 15	0.2	-0.2
Measurement point 16	0.5	1.4

5 Error Analysis

The error between simulation and experimental results is analyzed as follows.

5.1 External Heat Flow Simulation Errors

1) Electrical power measurement error.
   The measurement error is an accidental error, and the heating power error of the analog device is negligible.
2) Simulation error of heat flow outside the infrared cage.

Infrared cage added external heat flow by the heat flow meter to measure, so the error that is the heat flow meter (5% of the measurement error.

According to the aforementioned, the -Z1 fixed solar wing partition cannot be reduced to the target temperature by the influence of the background heat flow, and after analyzing, the influence of the difference of the external heat flow on the instrumentation near the -Z side panel is not more than 1°C, and the influence on the temperature of the rest of the whole star equipment is not more than 0.5 °C.

Since the whole star multi-layer uses infrared cage to simulate its external heat flow, for the external heat flow on the multi-layer, there is no systematic infrared measurement error, except for the influence of the inhomogeneity of the infrared cage. However, the infrared cage is difficult to meet the control requirements for lower target heat flow meter temperatures, and the effect is more obvious when there are partitions with higher heating requirements near the infrared cage.

5.2 Temperature Measurement Error

Usual temperature measurement system deviation (T1 = (0.5 °C (except for errors in installation) [QJ1846-90].

Thermistor temperature measurement than thermocouple temperature measurement error is slightly larger, usually within (2 °C. Unstable error (T2 by (2 °C.

5.3 Heat Leakage and Boundary Simulation Errors in Wire and Bracket

Heat leakage from conductors.

Since there is no temperature measurement point on the cable, it can only be determined according to the experience of previous models. According to other models of the relevant thermal balance test data and previous test experience can be seen, in the lead out of the star outside the thermocouple, heat flow meter and heating the return line of the 2-point temperature difference of 200mm apart from the maximum of not more than 15 °C, the cross-sectional area of the copper conductor core of 13.32mm ^{2} ((0.2mm, including 79 thermocouples, heat flow meter 28, so the cable a total of 107. The cross-sectional area is $F = 107 \times 3.14 * (0.2/2)^2 = 3.3598 mm^2$)

The amount of heat leakage is

$$\frac{\lambda F \Delta T}{L} = \left(400 \times \frac{3.3598 \times 10^{-6} \times 15}{0.2} \right) = 0.100785 W$$

About 280 cables lead out from the star, the area of a single one is not more than 0.5mm ^{2}, according to the data measured in the heat balance test of the CE-1 satellite, the temperature difference of the low-frequency cables within 100mm is not more than 15°C.

Heat leakage. $\frac{\lambda F \Delta T}{L} = \left(400 \times \frac{280 \times 0.5 \times 10^{-6} \times 15}{0.1} \right) = 8.4W$.

Therefore, the total heat leakage of thermal conductivity is 8.5W.

2) heat leakage from the bracket.

Since the bracket heating tracks the satellite flange temperature, and the temperature difference between the bracket and the satellite flange is not more than 1 °C, this heat leakage $\Delta Q_2 = h_c A \Delta T = 10.0 \times 1 \times \frac{1}{4} \times \pi \times \left(1.2^2 - 1.16^2 \right) = 0.8W$.

5.4 Heat Leakage Errors

The heat leakage error is mainly composed of cable heat leakage and butt ring heat leakage, and it can be seen from the above that the total heat leakage is less than 10W, and the temperature level error of the whole star electronic equipment due to the heat leakage of the conductor and the bracket heat leakage error is not more than 1.5 °C. The temperature level error of the whole star electronic equipment is not more than 1.5 °C, and it can be seen from the above that the total heat leakage is less than 10W.

5.5 Summary of Test Errors

Therefore, in addition to the heat leakage error mentioned above, the error of the test is mainly reflected in the simulation error of heat flow outside the heat sink of the whole star. According to the previous analysis, it can be seen that the effect of this error on the temperature of the whole star instrumentation is negligible.

6 Conclusion

In this paper, the thermal balance test method of spacecraft for space debris damage is proposed, simulation analysis is carried out and experimental verification is performed. The results show that the thermal balance test method is feasible and the simulation analysis method meets the accuracy requirements. For the thermal control design of spacecraft with the risk of space debris damage, the test results can be used to improve the design, such as arranging multiple layers in the module to avoid surface thermal control damage and strengthening the isothermalization design to avoid the overall impact caused by local damage. It is proposed to carry out further research as follows.

1) Conducting more heat balance tests for different space debris damage requirements, accumulating data and revising simulation parameters.
2) Further improve the thermal balance test method, the current test method, the satellite can only rotate, can not roll.

References

1. Liu, F.: Prediction of spacecraft temperature field and its ultimate equilibrium state using transition theory.AIAA 16th Thermophysics Conference,California,USA,23—25 June
2. Welch, W.J.: Assessment of thermal balance test criteria requirements on test objectives and thermal design. 46th International Conference on Environmental Systems. Vienna (2016)
3. Test requirements for lunch,upper-stage,and space vehicles :MIL-STD-1540D[S]. Los Angeles:Space and Missile System Center (1999)
4. Space engineering-testing: ECSS-E-10–03A[S]. Noordwijk: European Cooperation ECSS for Space Standardization (2002)
5. Messidoro, P., Roumeas, R.: MATED: improving ESA's model and test philosophies. ESA Bull. **109**, 122–127 (2002)
6. ECSS-E-10-02A Space engineering: Verification (1998)
7. Meng, F., Zhao, L., Qing, H., et al.: A method for determining the temperature stability criterion of spacecraft heat balance test. Astronautical J. **37**(10) (2016)
8. Zhu, X., Guo, G., Liu, S., et al.: Particle swarm algorithm for spacecraft thermal equilibrium temperature prediction. Astronautical J. 37(11) (2016)
9. Li, L.: Passive thermal control design and its experimental research of NanTech One Cube Star. Nanjing University of Science and Technology (2016)
10. Hongjun, F., Zhiming, X., Dongpo, N., et al.: Optimization method of heat flow meter placement for heat balance test of micronano-satellite. Spacecraft Eng. **30**(2) (2021)
11. Gao, Q.H., Li, P., Liu, J.B., et al.: Mars rover airborne heat balance test environment simulation technology. Spacecraft Environ. Eng. **36**(6) (2019)
12. Xianwen, N., Fan, J., Yang, C., et al.: Chang'e 5 detector thermal balance test program design and realization. China Space Sci. Technol. **41**(6) (2021)
13. Jinlong, L., Tenfu, Q., Yifei, P., et al.: Impact analysis of additional external heat flow for heat balance test using solar simulator. Spacecraft Environ. Eng. **2** (2015)

Research on the Effectiveness of Generative AI-Driven Virtual Patients in Nursing Clinical Thinking Training

Xiaoxi Liu, Wei Meng[(✉)], and Chunhui Li

School of Nursing and Welfare, Changchun Humanities and Sciences College, Changchun, China
402276235@qq.com

Abstract. Objective: This study aims to develop and evaluate an AI chatbot virtual patient system dedicated to nursing, addressing the shortcomings of traditional nursing clinical thinking training, such as high costs and limited case diversity. Methods: Based on the Transformer architecture, the chatbot was fine-tuned using a nursing-specific dataset (NANDA-I nursing diagnoses, clinical guidelines) to simulate three typical case scenarios (chronic diseases, acute and critical illnesses, elderly care). Sixty nursing students were randomly assigned to the experimental group (n = 30) or the control group (n = 30). The Lasater Clinical Judgment Rating Scale (LCJR) was used to assess clinical judgment ability, and the SEGUE framework was utilized to evaluate communication skills. Mixed methods analysis included statistical testing using SPSS and thematic analysis of interviews. Results: The total LCJR score for clinical judgment in the experimental group (28.72 ± 3.21) was significantly higher than that in the control group (23.52 ± 2.82) (t = 6.32, p < 0.01), and the SEGUE communication skills score in the experimental group (19.86 ± 3.05) was significantly better than that in the control group (11.16 ± 2.14) (t = 14.787, p < 0.05). Qualitative interviews revealed that 82% of the experimental group students believed that the system effectively improved their consultation skills. Student satisfaction survey results indicated that 93.17% of students felt that the "case design was close to clinical practice," 87.26% expressed a "willingness to continue using it," and 76.35% believed that it "can assist traditional teaching." Conclusion: The AI virtual patient system can effectively enhance the clinical thinking and communication skills of nursing students, promoting their clinical work abilities. Additionally, artificial intelligence technology provides scalable new teaching ideas and technologies for nursing education reform with limited resources.

Keywords: Artificial intelligence in nursing education · Virtual patient simulation · Clinical judgment training · Natural language processing · Chatbot-based learning

© The Author(s) 2026

P. Siarry et al. (Eds.): WCNA 2024, LNEE 1550, pp. 296–304, 2026.

https://doi.org/10.1007/978-981-95-6946-5_30

1 Introduction

With the rapid development of nursing discipline towards precision and specialization, clinical thinking ability has become the core competency of nursing talents. The World Health Organization (WHO) points out that over 60% of medical errors stem from clinical decision-making errors, and nursing staff play a key role in patient monitoring, risk assessment, and emergency response. However, current clinical nursing education is facing significant challenges: on the one hand, traditional clinical training is limited by hospital resource allocation, patient privacy protection, and the impact of sudden public health emergencies (such as the COVID-19 pandemic), resulting in a sharp decline in students' practical opportunities; On the other hand, standardized patient (SP) training has bottlenecks such as high cost, single case types, and poor reproducibility, while existing virtual patient systems rely heavily on pre-set scripts, making it difficult to simulate the dynamics and complexity of real clinical situations [1]. How to break through the limitations of time and space and build high fidelity clinical thinking training scenarios has become a key proposition for nursing education innovation.

In recent years, breakthroughs in Generative AI technology have provided a new path for medical education. Technologies represented by Large Language Models (LLMs) and Generative Adversarial Networks (GANs) can dynamically generate realistic case descriptions, simulate natural dialogue interactions, and support multi-path inference of disease evolution. Research has shown that AI based virtual training systems have demonstrated potential in medical diagnosis teaching, but their application in the nursing field is still in the exploratory stage [2]. Nursing clinical thinking emphasizes the continuous evaluation of the overall state of patients, the combination of humanistic care and evidence-based decision-making, which puts higher demands on the situational construction, interactive logic, and feedback mechanism of virtual patients. Therefore, exploring the effectiveness of generative AI driven virtual patients in nursing clinical thinking training is not only an important attempt at educational technology innovation, but also an urgent need to improve the quality of nursing talent cultivation. The latest research indicates that virtual patient systems based on artificial intelligence have the potential to overcome these limitations and provide new solutions for nursing education [3]. In the field of medical education, AI chatbots have shown significant potential, as chatbots can effectively support diagnostic reasoning training for medical students [4]. However, these studies mostly focus on medical diagnosis and overlook the unique clinical thinking patterns of nursing education. Nursing clinical thinking not only requires accurate diagnosis of the condition, but also emphasizes overall nursing evaluation, therapeutic communication, and humanistic care. This specificity makes it difficult for general medical chatbots to meet the needs of nursing education.

Upon reviewing existing literature, it was found that a few researchers have begun exploring AI teaching tools specifically for nursing. The virtual patient system developed by some researchers has confirmed the feasibility of technology enhanced learning in nursing education, but its interaction mode is still limited to preset paths and lacks true intelligent dialogue ability [5, 6]. At the same time, the breakthrough progress of large-scale language models, such as the GPT series, provides new opportunities for developing more intelligent nursing teaching tools. However, how to adapt these advanced technologies to the specific scenarios of nursing education is still an urgent problem to be solved.

The aim of this study is to develop and validate a nursing specific AI virtual patient system, with a focus on addressing the following key issues: (1) how to construct a dialogue model that meets the training needs of nursing clinical thinking; (2) The impact of this system on the clinical judgment ability of nursing students; (3) Advantages and limitations compared to traditional teaching methods. By integrating natural language processing technology and nursing education theory, this study aims to provide empirical evidence for the development of intelligent nursing education.

2 Object and Method

2.1 Research Object

This study adopted a randomized controlled trial design and selected 60 undergraduate nursing students from a certain university as the research subjects. Inclusion criteria: (1) Completed basic nursing courses; (2) I have not received virtual patient training. The subjects were divided into an experimental group (n = 30) and a control group (n = 30) using a random number table method. There was no statistically significant difference between the two groups of students in terms of age, gender, and previous grades (p > 0.05).

2.2 Research Method

1) system development
a) Core engine architecture

This system adopts Meta's open-source LLaMA-2-7B model as the infrastructure, and achieves nursing scene adaptation through a three-stage fine-tuning strategy Pre training of general medical knowledge: Conduct domain adaptation training based on nursing related literature open in PubMed Central to learn medical terminology expression patterns. Nursing specific fine-tuning:Use a self built Chinese nursing dialogue dataset (simulating real nurse patient dialogue records) for supervised fine tuning, with a focus on enhancing the following abilities:Nursing Assessment Framework (Gordon Functional Health Model);Therapeutic communication skills (such as open-ended questioning and empathetic expression);Clinical Decision Prioritization (Maslow Hierarchy of Needs Theory).Security layer reinforcement:Using rule-based content filters, combined with Constitutional AI principles, automatically intercept the following risky content:A dialogue path that does not conform to nursing ethics;Recommendations for non-standard procedures outside of clinical guidelines;Possible expressions that may trigger medical disputes.

b) *Knowledge base construction process*

To overcome the limitations of the general medical model, a multidimensional nursing knowledge system has been constructed.Structured knowledge source:Diagnostic

Criteria Library: Integrating NANDA-I Nursing Diagnosis 2021–2023 Edition Diagnostic Criteria;Intervention measures library: extracting standardized intervention items from the Nursing Measures Classification (NIC).Unstructured knowledge processing:Use Bi LSTM + CRF model to extract entities from "Internal Medicine Nursing" (6th edition) and construct a nursing knowledge graph.Dynamic update mechanism:Establish a closed-loop optimization system based on Active Learning, which automatically triggers a manual review process when a user's question exceeds the scope of the knowledge base, ensuring the continuous evolution of the knowledge base.Interaction Design:Support multiple rounds of natural language conversations;Including 5 typical clinical scenarios (diabetes management, acute heart failure attack, etc.).

2) *Intervention plan*

The experimental group used an AI virtual patient system for training, twice a week for 45 min each time, lasting for 8 weeks. The control group used traditional case-based teaching method with the same class hours. The training content includes: medical history collection, nursing evaluation, development of intervention measures, and health education.

3) *Evaluation*
a) *Lasater Clinical Judgment Rating Scale (LCJR)*

This scale consists of 11 items, divided into 4 dimensions: observation (3 items), explanation (2 items), response (4 items), and reflection (2 items). The coefficients of each dimension are Cronbach's alpha ranging from 0.615 to 0.722, and the Cronbach's alpha of the entire scale is 0.896. The scoring is based on the Likert 4-point scoring method, with scores ranging from 1 to 4 in order of unfamiliarity, proficiency, proficiency, and proficiency. The higher the score, the better the clinical judgment ability[7].

b) *SEGUE Communication Skills Rating*

The SEGUE scale consists of 5 main dimensions and a total of 25 evaluation items. Including preparation stage, information collection, information provision, understanding patients, and concluding consultations. Each entry adopts a binary evaluation of "yes/no" or "unable to answer", and some entries need to be deducted points based on specific behaviors [8].

c) *Learner satisfaction questionnaire*

Design a self-designed survey questionnaire to investigate students' satisfaction with virtual patient applications.

3) *Statistical method*

Statistical analysis was conducted using SPSS 26.0. The measurement data is expressed as `x ± S and t-test is used for inter group comparison; Count data is expressed as a percentage (%). $P < 0.05$ indicates a statistically significant difference.

3 Result

A. *Comparison of clinical judgment ability between two groups of students*

After intervention, the total score of LCJR in the experimental group (28.72 ± 3.21) was significantly higher than that in the control group (23.52 ± 2.82) ($t = 6.32$, $p < 0.01$). The comparison of scores for each dimension is shown in Table 1.

Table 1. Comparison of LCJR scores between two groups after intervention ($x \pm S$, points)

Dimension	experimental group(n = 30)	control group(n = 30)	t	p
follow with interest	7.21 ± 0.82	6.12 ± 0.71	4.56	< 0.01
explain	7.52 ± 0.91	6.34 ± 0.80	5.12	< 0.01
respond	7.83 ± 1.11	6.51 ± 0.92	4.89	< 0.01
reflect	6.21 ± 0.73	5.62 ± 0.61	3.45	< 0.05
Total score	28.72 ± 3.21	23.52 ± 2.82	6.32	< 0.01

B. Comparison of communication skills between two groups of students

The SEGUE score of the experimental group (19.86 ± 3.05) was significantly better than that of the control group (11.16 ± 2.14) ($t = 14.787$, $p < 0.05$). Qualitative interviews showed that 82% of the experimental group students believed that the system had effectively improved their consultation skills.

C. *Student satisfaction status*

93.17% of students believe that "case design is close to clinical practice", 87.26% of students express "willingness to continue using", and 76.35% of students believe that "it can assist traditional teaching".

4 Discuss

A. *Virtual patient system can effectively enhance students' clinical judgment ability*

This study integrated nursing program theory and cognitive apprenticeship theory to construct an educational intervention framework for AI virtual patients. The research confirmed that the AI virtual patient system can effectively improve the clinical judgment ability of nursing students, especially in the dimensions of "explanation" and "response". This is basically consistent with the research results of other scholars [9]. But this study further validated its specific value in nursing education. In traditional nursing teaching, it is difficult for teachers to find targeted cases, but virtual patients can effectively solve this problem. By training students to collect medical history, conduct nursing evaluations, develop intervention measures, and provide health education, their clinical thinking can be enhanced, thereby effectively improving their clinical judgment ability. In addition, virtual patients have repeatability, allowing students to repeatedly train, which helps deepen their mastery of professional knowledge and enhance their ability to apply knowledge.

B. Interdisciplinary Implications of Technological Innovation
1) Breakthrough in Domain Adaptive Technology

This study proposes an innovative "three-stage progressive domain adaptation" approach to address the unique nature of nursing education scenarios.The technical framework effectively solves the terminology comprehension bias and nursing issues faced by general medical.AI models in the context of nursing core issues such as inadequate adaptation to operational scenarios. Specifically, the first stage (domain knowledge pre training): Based on nursing textbooks, clinical nursing records, and nursing operation guidelines, a specialized corpus is constructed to reinforce the model's embedding representation of professional knowledge such as nursing evaluation indicators (such as Braden pressure ulcer score, pain grading scale) and nursing measure terminology (such as closed suction and PICC maintenance) through comparative learning strategies; Phase 2 (Situational Fine tuning): Introduce nursing simulation teaching video transcription data and real nurse patient dialogue records, adopt Curriculum Learning strategy, gradually improve the modeling ability of the model for nursing workflow and decision logic from basic nursing scenarios (such as vital sign monitoring) to complex scenarios (such as comprehensive nursing of patients with multiple organ failure); The third stage (task driven optimization): Combining reinforcement learning framework, using nursing clinical thinking training objectives (such as risk assessment accuracy and intervention rationality) as reward functions, dynamically optimize the response quality of the model in virtual patient interaction. Experiments have shown that the three-stage adaptive model achieves an accuracy rate of 93.7% in nursing terminology recognition, which is significantly improved compared to the general medical model (78.2%), especially in key tasks such as translating patient complaints (such as accurately mapping "palpitations" to "palpitations") and generating nursing diagnoses (such as distinguishing between "ineffective airway cleaning" and "impaired gas exchange").

2) *Revolutionary improvement in knowledge representation*

The constructed nursing knowledge graph adopts a triple structure of "entity relationship evidence level", which enables the system to generate suggestions based on evidence-based nursing principles, avoiding the risk of "black box decision-making" of traditional AI tools.

C. Deep reflection on the limitations of this study
1) Mechanical nature of emotional interaction

The mechanical defects of emotional interaction are mainly reflected in the system's too single recognition dimension of emotional clues. Although current artificial intelligence systems can accurately identify explicit emotional words (such as "unbearable pain" and "tears of joy" with clear emotional orientations) through keyword matching technology, there are still significant limitations when dealing with more complex non-verbal emotional expressions. This limitation is manifested in three aspects: firstly, the system is difficult to effectively capture the silent gaps in dialogue - intentionally prolonged pauses in human communication may imply hesitation, sadness, or thinking, but the system usually treats them simply as dialogue interruptions; Secondly, the lack of analytical ability for subtle changes in tone and intonation (such as pitch tremors, sudden changes in speech speed, and other paralingual features) results in the inability to recognize emotional fluctuations that may be hidden behind seemingly calm statements like "I'm fine"; More importantly, system responses often follow fixed emotional response templates, making it difficult to flexibly adjust comfort strategies based on the context [10].

2) Lack of cultural sensitivity

The current version does not take into account regional cultural differences (such as the health beliefs of ethnic minority patients), which may affect the universality of consultation training.

3) Limitations of evaluation tools

Relying on the LCJR scale may lead to a "test oriented" clinical judgment, and in the future, it is necessary to combine behavioral observation with real clinical environments.

D. *The educational value reflected in this study*
1) *Dimension of Educational Technology Innovation*

The current nursing clinical education is facing three major pain points: the shortage of clinical practice resources leads to insufficient training opportunities (especially during public health emergencies); Traditional standardized patient training is expensive and the types of cases are limited; The existing virtual patient system has defects such as rigid interaction and solidified case library. This study introduces generative AI technology to construct a dynamic and intelligent virtual patient system, breaking through the limitations of time and space to generate diverse clinical scenarios, achieving human-machine natural language interaction, and providing innovative solutions for nursing clinical thinking training. This has breakthrough value in educational technology, which can promote the transformation of nursing education models from "experience imparting" to "intelligent training".

2) *Dimensions of clinical competency development*

Clinical thinking, as a core nursing ability, urgently requires repeated practice in real-life situations for its cultivation. This study uses an immersive training environment constructed by AI virtual patients to enable learners to: (1) engage in dialogue exercises for medical history collection and condition assessment in a zero risk scenario; (2) Handling multimodal information of complex disease progression; (3) Provide immediate feedback and correction for nursing decisions. This dynamic case system based on generative AI can better cultivate the situational perception, clinical reasoning, and emergency response abilities of nursing staff compared to traditional static case teaching, and has a direct promoting effect on improving the quality of nursing talent training.

3) *Dimensions of healthcare education reform*

The research results will have a dual demonstration effect: at the technical application level, verify the feasibility and effectiveness of generative AI in medical education, and provide empirical evidence for the development of intelligent education products; At the level of educational equity, building a replicable virtual training platform can help alleviate the problem of uneven distribution of nursing education resources between regions, especially for talent cultivation in grassroots medical institutions, which has universal value. The research will also establish a multidimensional evaluation system for the effectiveness of virtual patient training, providing methodological support for the intelligent transformation of nursing education.

The innovation of this study lies in the deep integration of generative AI dialogue capabilities with virtual patient systems for the first time, breaking through the limitations of traditional systems' script based interaction; Constructing a dynamic disease evolution model to achieve non-linear training in the nursing process; Develop intelligent evaluation algorithms to achieve quantitative analysis of clinical thinking abilities. The research results will provide theoretical basis and practical examples for empowering nursing education with artificial intelligence, and have significant academic value and social benefits.

5   Conclusion

The AI virtual patient system developed in this study can significantly improve the clinical judgment ability and communication skills of nursing students, with good acceptance and satisfaction. At the same time, this study achieved a paradigm upgrade of nursing education AI tools from "knowledge transmission" to "thinking shaping" through technological innovation and theoretical reconstruction. The main contributions are reflected in three dimensions: (1) Methodology level: Created the first multimodal AI training framework based on nursing procedure theory, and its dynamic scaffolding mechanism can be extended to other clinical skills training fields. (2)On the technical level, the development of domain adaptive fine-tuning solutions significantly improves the professionalism of the model, and the evidence-based structured design of knowledge graphs provides new ideas for the interpretability research of medical AI. (3)At the societal level, it has been confirmed that low-cost AI solutions are feasible for narrowing the regional education gap, providing a feasible technological path for the grassroots nursing talent training goal of "Healthy China 2030". Future research should focus on establishing a cross institutional nursing AI education resource sharing platform and exploring the ethical boundaries and regulatory framework of generative AI in nursing education, so as to further empower and promote the development of higher education teaching with artificial intelligence.

Acknowledgment. Supported by a Research Project on Teaching Reform of Welfare Characteristics Special Project at Changchun University of Humanities" Research and Practice on the Digital Teaching Reform of Health Assessment Course from the Perspective of Welfare" (2024FZ 019).

References

1. Luping, X., Miaofen, Z., Tianjun, Y., et al.: Research and analysis on the construction and implementation of nursing clinical thinking training mode in nursing student training and teaching. China High. Med. Educ. China **9**, 141–143 (2024)
2. Weifan, H., Wenting, H., Ying, Z., et al.: Design and application of a virtual training system for bone marrow puncture based on VR technology. China Med. Educ. Technol. China **5**, 588–590 (2021)
3. Wenjun, Y., Zijuan, J., Xin, X., et al.: Application research of preoperative visit system based on virtual simulation in improving the psychological nursing effect of Parkinson's surgery patients. General Nursing China **22**, 3422–3425 (2024)
4. Roubai, P., Rong, T.: A brief discussion on the impact of chatbots based on large language models on medical education. Internal Med. Theor. Pract. China **18**, 439–446 (2023)
5. Wenxiao, Z., Yanli, L.: The application effect of virtual patient joint simulation practice teaching in the cultivation of clinical reasoning ability for nursing undergraduate students. Nursing Res. China **19**, 3511–3513 (2021)
6. Dan, X., Liqing, H.: Analysis of the effectiveness of the combined teaching mode of micro video, virtual patient, and scenario simulation in clinical reasoning teaching. J. Inner Mongolia Med. Univ. China **43**, 100–103 (2021)
7. Min, W., Jie, C., Chen, C.: Revision and evaluation of lasater clinical judgment evaluation scale. Nursing Pract. Res. China **17**, 8–10 (2020)
8. Lu, Z., Ziying Z., Junwen B., et al.: Application of SEGUE scale in evaluating medical students' communication skills between doctors and patients. J. Inner Mongolia Med. Univ. China **46**, 24–27 (2024)
9. Weiwei, G., Tao, G., Jinwei, Q.: Application of virtual standardized patients in medical students' emergency internship. China Continuing Med. Educ. China **116**, 104–108 (2024)
10. Mandong, Z.: Integration of intelligence and emotion: a three-dimensional approach to emotion calculation in ideological and political education in universities based on artificial intelligence. J. Qiqihar Univ. (Philosophy Soc. Sci. Ed.), China **3**, 165–168 (2025)

Obstacle Avoidance Path Planning Technology for Fully Autonomous Tunnel Inspection Drone Based on SLAM and IMU Fusion

Tian Guo[1]([✉]), Yang Zhao[1], Shuo Wang[1], Han Luo[1], and Yiming Ding[2]

[1] State Grid Beijing Electric Power Company, Beijing 100031, China
gghh210024@163.com
[2] Beijing Excellent Power Construction Co., Ltd., Beijing 101399, China

Abstract. In fully autonomous tunnel inspection, only obstacle feature points can be used for path planning, resulting in low inspection efficiency. One of the key technologies of fully autonomous tunnel inspection UAV is obstacle avoidance path planning technology. Therefore, a fully autonomous tunnel inspection UAV obstacle avoidance path planning technology based on SLAM and IMU fusion is proposed. The node image coordinates and depth information are obtained through visual SLAM, and the coordinates are unified to the same coordinate system through coordinate transformation to construct a fully autonomous tunnel inspection node graph. Based on IMU fusion, the feature point search range is limited to the reasonable area of the current motion trend of the UAV, and a comprehensive evaluation matrix is constructed to extract obstacle feature points in the tunnel. The initially extracted obstacle feature points are preprocessed, and the key and reliable feature points for obstacle avoidance path planning are screened out by setting thresholds and Euclidean distance constraints, and the feature point cost is calculated by comprehensively considering the cumulative cost and heuristic cost. By constructing a dynamic model of the tunnel environment, using a path search algorithm fused with IMU, designing a comprehensive cost function to guide path generation, and real-time monitoring of environmental changes and triggering a path replanning mechanism, dynamic obstacle avoidance path planning of UAVs in complex tunnel environments is realized. Experimental results show that the proposed technology plans routes that are more direct and shorter, can flexibly respond to sudden obstacles, and the time cost of the proposed technology path is always lower than other methods, which fully proves that the technology can improve inspection efficiency.

Keywords: SLAM technology · IMU fusion · Tunnel obstacles · Inspection drones · Obstacle avoidance path planning

1 Introduction

In the field of infrastructure maintenance and safety monitoring, tunnel inspection is a vital and challenging task. Traditional manual inspection methods are not only inefficient and costly, but also difficult to effectively guarantee the safety of inspectors in complex

P. Siarry et al. (Eds.): WCNA 2024, LNEE 1550, pp. 305–317, 2026.
https://doi.org/10.1007/978-981-95-6946-5_31

and dangerous environments. One of the key technologies of fully autonomous tunnel inspection drones is drone obstacle avoidance path planning technology. However, the tunnel environment has its own particularities, such as narrow space, insufficient light, and a large number of obstacles, which brings great difficulties to the autonomous flight and obstacle avoidance of drones. During the tunnel inspection process, drones need to perceive the surrounding environment in real time, accurately identify obstacles, and quickly plan a safe and efficient obstacle avoidance path to ensure the smooth completion of the inspection task.

In related research, Jones M et al. [1] conducted a comprehensive review of the path planning of UAVs in complex environments, pointed out the impact of environmental complexity on path planning, and proposed the advantages and disadvantages of various path planning methods. They emphasized that in complex environments, UAVs need to be able to perceive the surrounding environment in real time and quickly plan the optimal path. However, this path planning method is often difficult to adapt to the dynamic changes of the environment, resulting in low inspection efficiency. Ait Saadi A et al. [2] conducted an in-depth study on the optimization method in UAV path planning. They pointed out that the quality of path planning can be significantly improved through optimization algorithms. However, in complex tunnel environments, a single optimization method is often difficult to simultaneously meet the requirements of inspection efficiency. Puente-Castro A et al. [3] reviewed the application of artificial intelligence in the path planning of UAV swarms. They pointed out that artificial intelligence algorithms have significant advantages in dealing with complex environments and dynamic changes. However, the application of artificial intelligence algorithms to obstacle avoidance path planning of fully autonomous tunnel inspection UAVs still faces many challenges, such as low real-time and stability of inspection efficiency. Li J et al. [4] proposed a path planning method for the target coverage task of UAVs in dynamic environments. They achieved efficient inspection of UAVs in dynamic environments through optimization algorithms. However, the inspection efficiency of this method in complex environments such as tunnels is low and still needs further verification.

In view of the low inspection efficiency in complex tunnel environments in the above literature, a tunnel fully autonomous inspection drone obstacle avoidance path planning technology based on SLAM and IMU fusion is proposed. SLAM technology can build environmental maps in real time and realize the precise positioning of drones, while IMU can provide high-precision attitude and speed information. This technology realizes real-time and accurate perception and positioning of the tunnel environment by integrating the environmental perception capability of SLAM technology and the high-precision attitude information of IMU. At the same time, combined with optimization algorithms and artificial intelligence algorithms, it can quickly plan the optimal obstacle avoidance path and improve the inspection efficiency of drones.

2 Design of Obstacle Avoidance Path Planning Technology for Fully Autonomous Tunnel Inspection Drone

2.1 Building a Fully Autonomous Tunnel Inspection Node Map Based on SLAM

This paper constructs a SLAM-based environmental perception model, which aims to provide real-time environmental information for tunnel autonomous inspection drones to support path planning and obstacle avoidance tasks. The model mainly relies on the SLAM sensor node network [5] to collect and filter data related to the tunnel environment. The SLAM sensor node network is a distributed system composed of multiple sensor nodes with SLAM functions. Through communication and collaboration between nodes, the collection and processing of tunnel environmental information is realized. In terms of constructing the tunnel autonomous inspection node map, the environmental information obtained by the SLAM-based environmental perception model is used to construct a three-dimensional simulation map of the tunnel autonomous inspection nodes. The node map can accurately reflect the spatial structure and obstacle distribution in the tunnel, providing a basis for the design of path planning technology.

In the process of constructing the tunnel autonomous inspection node map, we first use visual SLAM to obtain the image coordinates and depth information of the autonomous inspection nodes [6]. Assume that the image coordinates are (u, v), the corresponding depth information is z, the camera's intrinsic parameter matrix is K, the three-dimensional coordinates (X, Y, Z) of the fully autonomous inspection node in the camera coordinate system can be calculated by the following formula (1):

Furthermore, in order to unify the coordinates of the fully autonomous inspection nodes obtained at different times into the same coordinate system, it is necessary to use the camera's external parameters (rotation matrix R and translation vectors t) to transform the coordinates. Assume that the coordinates of the fully autonomous inspection node in the camera coordinate system at a certain moment are (X_1, Y_1, Z_1), the coordinates in the camera coordinate system at another moment are (X_2, Y_2, Z_2), and the external parameter relationship between the cameras at two moments is known, then the following formula can be used to convert (X_2, Y_2, Z_2) transform to (X_1, Y_1, Z_1) in the same coordinate system:

$$\begin{bmatrix} X \\ Y \\ Z \end{bmatrix} = zK^{-1} \begin{bmatrix} u \\ v \\ 1 \end{bmatrix} \tag{1}$$

$$\begin{bmatrix} X_2' \\ Y_2' \\ Z_2' \end{bmatrix} = R \begin{bmatrix} X_2 \\ Y_2 \\ Z_2 \end{bmatrix} + t \tag{2}$$

Through the above method, an accurate tunnel autonomous inspection node map can be constructed based on SLAM, providing strong support for the path planning and obstacle avoidance of drones.

2.2 Extraction of Obstacle Feature Points in Tunnels Based on IMU Fusion

In the process of constructing a fully autonomous tunnel inspection node map based on SLAM technology, due to the particularity of the tunnel environment, such as long distance, weak light, sparse feature points, and possible dynamic interference, it is often difficult to ensure the accuracy and robustness of the map by relying solely on visual SLAM. In particular, in areas where there are insufficient feature points or large changes in lighting conditions, the SLAM system may have positioning drift or incomplete map construction. Therefore, in order to improve the accuracy and reliability of tunnel inspection, it is necessary to introduce IMU (inertial measurement unit) data for fusion processing. Through IMU fusion, obstacle feature points in the tunnel can be effectively extracted. The acceleration and angular velocity information provided by the IMU can be used to assist visual SLAM in more stable positioning and map construction in the absence of features or poor lighting, thereby enhancing the environmental adaptability and autonomy of the inspection system. Therefore, in the process of extracting obstacle feature points, the high-precision attitude and velocity information provided by the IMU is used to limit the search range of feature points to a reasonable area determined by the current motion trend of the drone [7], so as to improve the efficiency and accuracy of feature point extraction.

Figure 1 shows a schematic diagram of the feasible area for extracting obstacle feature points based on IMU fusion.

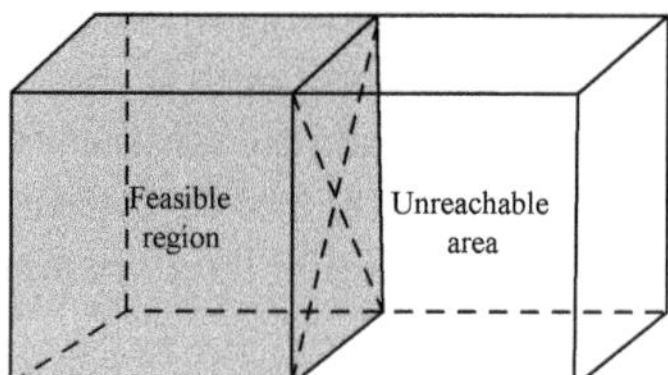

Fig. 1. Schematic diagram of the feasible area for extracting obstacle feature points based on IMU fusion

In Fig. 1, the current movement direction of the UAV is determined by the attitude and speed information measured by the IMU. Assuming that the current movement direction is to the upper right, the search range of obstacle feature points is limited to a rectangular area on the right side of the current movement direction, and points outside the rectangular area are considered to be currently unreachable, so as to reduce unnecessary attitude adjustments of the UAV in subsequent path planning and improve inspection efficiency.

The feature extraction part of the algorithm in this paper is based on the data processing method of IMU fusion. By fusing the acceleration, angular velocity and other information measured by IMU and the existing environmental information in the tunnel fully autonomous inspection node map, a comprehensive evaluation matrix is constructed to evaluate the possibility of each pixel area becoming an obstacle feature point. Pixel areas with larger evaluation values are considered to be obstacle feature points with significant motion characteristics or environmental characteristics.

Assume the acceleration vector measured by IMU is a, the angular velocity vector is ω, the environmental feature vector of the corresponding pixel area in the tunnel fully autonomous inspection node map is e, then the comprehensive evaluation matrix M can be expressed as:

$$M = w_1 a^T a + w_2 \omega^T \omega + w_3 e^T e \tag{3}$$

where: T is the threshold value, w_1, w_2, w_3 are the weight coefficient, which is used to adjust the contribution of different information in the comprehensive evaluation.

Furthermore, in order to screen out significant obstacle feature points from the comprehensive evaluation matrix, when the evaluation value M_q corresponding to a certain pixel area is greater than threshold T, the pixel area is determined to be an obstacle feature point. At the same time, in order to more accurately describe the properties of obstacle feature points, the feature point extraction formula is:

$$S = \frac{M_q - T}{\max(M) - T} \tag{4}$$

where: $\max(M)$ is the maximum value in the comprehensive evaluation matrix.

2.3 Screening of Characteristic Points of Obstacle Avoidance Paths for Inspection Drones

Due to the possible noise and cumulative errors in IMU data, as well as the complexity and diversity of the tunnel environment, the extracted obstacle feature points may contain a large amount of redundant or inaccurate information, which will increase the computational complexity of the path planning algorithm and may even plan a non-optimal or unfeasible obstacle avoidance path. Therefore, in order to optimize the obstacle avoidance path planning of the inspection drone, the initially extracted obstacle feature points in the tunnel are preprocessed to remove noise interference, and on this basis, the threshold is set to screen out the more critical and reliable feature points for obstacle avoidance path planning, thereby improving the efficiency and accuracy of the algorithm.

Considering the high requirements of the stability and reliability of feature points in the complex environment of tunnels, this paper adds constraints based on Euclidean distance to the feature points. In a tunnel environment, if the distance between two feature points is too small, they may be mismatched due to factors such as reflection, occlusion or sensor noise from the tunnel wall, thus affecting the accuracy of path planning. Therefore, these feature points are screened by setting a distance threshold to ensure that only points with more dispersed spatial distribution and higher stability are selected as obstacle avoidance path feature points [8]. Assume that the set of obstacle feature points in a certain frame of the inspection drone obstacle avoidance path image is $S_h = [s_1, s_2, \cdots, s_n]$, where the coordinates of each feature point s_i can be expressed as (x_i, y_i, z_i) (Considering three-dimensional space), the Euclidean distance between feature points can be expressed as:

$$d(s_i, s_j) = \sqrt{(x_i - x_j)^2 + (y_i - y_j)^2 + (z_i - z_j)^2} \tag{5}$$

If the distance threshold is set to d_{th}, then the feature point pairs that meet the distance constraint can be defined as satisfying the point $d(s_i, s_j) > d_{th}$.

In screening the feature points of the obstacle avoidance path of the inspection drone, the cost value F of the feature points that meet the distance constraint is calculated based on the cumulative cost G and the heuristic cost H, as shown in the following formula:

$$F = w_G \times G + w_H \times H \tag{6}$$

Among them, the cumulative cost G represents the flight time or energy consumption from the starting point to the current feature point, and the heuristic cost H represents the estimated time or distance cost from the current feature point to the target point. w_G and w_H are the weight coefficient, which is used to balance the contribution of the two in the cost calculation.

Considering the motion characteristics of the drone, when it reaches the current feature point, the flight time is affected by the horizontal angle $\Delta\psi$, vertical corner $\Delta\theta$ and height change value Δh. Therefore, the calculation method of the cumulative cost G and the heuristic cost H is:

$$\begin{cases} G = (\Delta\psi + \Delta\theta)\Delta h \\ H = \dfrac{L_h}{v_{\max}} \end{cases} \tag{7}$$

where: L_h is the estimated distance from the current feature point to the target point, $v_{\max}$ is the maximum flight speed of the drone.

Taking into account the relative position relationship between the current feature point and the target node in three-dimensional space, the obstacle avoidance path feature screening formula of the inspection UAV is:

$$S_b = \sqrt{F(G - H)^2 L_h} \tag{8}$$

Through the above formula, the most critical feature points for obstacle avoidance path planning can be effectively screened out from the obstacle feature points in the tunnel, providing strong guarantees for the safe and efficient inspection of drones.

2.4 Planning the Dynamic Obstacle Avoidance Path of the Fully Autonomous Tunnel Inspection Drone

After the SLAM and IMU fusion technology is used to process and select the characteristic points of the inspection drone's obstacle avoidance path, the specific process of completing the dynamic obstacle avoidance path planning of the tunnel's fully autonomous inspection drone is as follows:

1) A dynamic model of the tunnel environment is constructed using 3D point cloud data combined with SLAM technology. This model not only contains static obstacle information[9], but also can update the drone attitude and position data provided by the IMU in real time, thereby accurately determining the starting and ending points of the inspection path. Suppose the tunnel environment model is the set $P =$

$\left[p_1, p_2, \cdots, p_m\right]$ of feature points of the obstacle avoidance path of the inspection drone selected, where each point p_m contains its spatial coordinates (x_m, y_m, z_m) and obstacle attribute information.

2) Starting from the starting point, the fused IMU is used to search for the path. The algorithm uses the real-time posture provided by the IMU to adjust the search direction and combines the environment map built by SLAM to expand the nodes to ensure the real-time and accuracy of the path. Assume that the current node is D_c, its adjacent node set is $N(D_c)$, then the node expansion process can be expressed as finding adjacent nodes that meet certain conditions (such as no obstacles, low cost).

3) Design a comprehensive cost function to guide the optimal path generation. This cost function combines the Euclidean distance (considering the straight-line distance in space), the dynamic obstacle avoidance factor [10] (based on SLAM real-time detection), and the path smoothness function (reducing the sudden change of the drone's posture). Let the cost function be $C(m)$, for any node m, its cost can be expressed as:

$$C(m) = w_1 d(m, O) + w_2 f(m) + w_3 r(m) \tag{9}$$

where: $d(m, O)$ is the Euclidean distance from node m to the target point, $f(m)$ is the dynamic obstacle avoidance factor, $r(m)$ is the evaluate of the path smoothness.

4) The drone starts inspection according to the planned path and monitors environmental changes in real time through SLAM and IMU fusion technology. If the drone reaches the target point, the path planning ends; otherwise, proceed to the next step.

5) When SLAM technology detects a new obstacle or the IMU senses that the drone's posture is abnormal, the path replanning mechanism is immediately triggered and returns to step 2 to replan the path using the updated environmental model and drone status information.

Through the above process, combined with SLAM and IMU fusion technology, the UAV can achieve efficient and safe dynamic obstacle avoidance path planning in complex and changeable tunnel environments.

3 Experiments

3.1 Experimental Platform

The simulation experiment environment is set as follows: the operating system uses Windows 11, the processor is upgraded to Intel Core i5-9400F, the main frequency reaches 2.90 GHz, equipped with 16 GB memory, and the compilation tool uses Matlab R2020a. To ensure the efficiency and stability of command transmission, the experiment uses a high-performance wireless network module, combined with multiple high-precision sensor nodes, and transmits control commands to the central processing unit of the drone in real time through an optimized network protocol. The operator sends inspection tasks and path planning instructions through the background management system. Based on the complex environment characteristics of the tunnel, a three-dimensional high-precision map model is constructed, as shown in Fig. 2, to simulate the real tunnel scene and verify the effectiveness of the obstacle avoidance path planning algorithm.

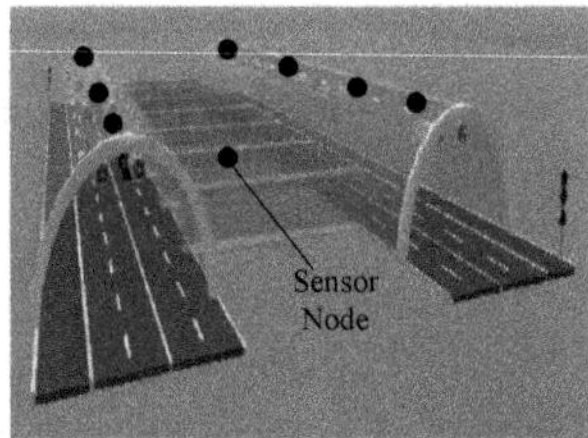

Fig. 2. Tunnel 3D high-precision map model

The experimental platform of the obstacle avoidance path planning technology for fully autonomous tunnel inspection UAV based on SLAM and IMU fusion is mainly composed of UAV body, lidar device, IMU inertial measurement unit and onboard computer, and the specific parameters are shown in Table 1.

Table 1. Specific parameters of the experimental platform

No	Component	Parameter Details
1	UAV Body	Design Type: Quadcopter; Motor Type: High-torque Brushless Motor; Number of Motors: 4; Electronic Speed Controller: Independent Electronic Speed Controller; Main Controller: STM32H7 Microcontroller; Communication Interface: High-speed Serial Communication Interface
2	LiDAR Device	Installation Position: Underneath the UAV Body; Beam Design: 16-beam; Scanning Frequency: 10 Hz Detection Range: Up to 50 m; Resolution: $0.1°$
3	IMU Inertial Measurement Unit	Model: ADIS16470; Built-in Sensors: Triaxial Accelerometer, Triaxial Gyroscope; Dynamic Accuracy: $0.1°$; Static Accuracy: $0.002°$; Bandwidth Range: 5–200 Hz
4	Onboard Computer	Configuration: Intel NUC 11 Pro; Processor: Intel i7 Memory: 16G; Solid State Drive: 512G; Operating System: Ubuntu 20.04; Software Development Environment: ROS Noetic + PCL Library

In the experimental platform, the drone body adopts a quad-rotor design, equipped with a high-torque brushless motor and an independent electronic speed regulator to achieve flexible flight control. The laser radar device scans the environment in real time and builds a three-dimensional map; the IMU provides stable attitude data; the onboard computer is equipped with a high-performance processor and runs an advanced operating system, providing strong support for obstacle avoidance path planning.

3.2 Experimental Preparation

In order to verify the effectiveness of the obstacle avoidance path planning algorithm based on SLAM and IMU fusion in the tunnel environment, this section will use a tunnel engineering measured data set for quantitative analysis. This data set is collected synchronously using a high-precision 3D laser scanner and an IMU sensor in a real tunnel scene, and is equipped with true value trajectories provided by professional measurement equipment for accuracy verification. The experiment selects a dynamic tunnel sequence containing complex obstacles as test data, in which moving engineering vehicles and floating dust exist as dynamic interference factors. An example of a tunnel dynamic scene data set is shown in Fig. 3, which fully simulates the complex environment that may be encountered in actual inspections.

Fig. 3. Tunnel dynamic scene dataset

This section uses 8 sets of data from the tunnel scene dataset for experiments. The slow-related dataset contains slow-moving obstacles, such as slowly moving inspection vehicles, which are low-dynamic interference scenarios; the fast-related dataset contains fast-moving objects, such as shuttle drones or high-speed trains, which constitute high-dynamic interference scenarios. The dataset suffix fixed indicates that the obstacle avoidance perception platform is relatively fixed, the environment structure is relatively simple, and the positioning challenge is relatively small; the suffix complex indicates that the obstacle avoidance perception platform moves along a complex curved path and is accompanied by multi-angle rotation, which significantly increases the difficulty of positioning.

The obstacle avoidance perception platform is composed of multispectral sensors. The selection of sensors provides solid and reliable hardware support for obstacle identification in complex tunnel environments, which directly affects whether the UAV can accurately and efficiently plan obstacle avoidance paths. The platform mainly uses the UAV-MS-06 model multispectral sensor, and the specific parameters are shown in Table 2.

The experimental scene layout of the fully autonomous inspection drone in the tunnel is to paste reflective signs in a grid-like manner on the inner wall of the tunnel at equal intervals according to the preset coordinates. The spacing between adjacent signs is set to 3 m, and the starting sign and the drone inspection path are clearly defined. In order

Table 2. Specific parameters of UAV-MS-06 multispectral sensor

No.	Parameter Name	Parameter Value
1	Spectral Range	400–1000 nm
2	Resolution	640 × 512 pixels
3	Frame Rate	30 fps
4	Dynamic Range	> 120 dB
5	Accuracy	± 2% reflectance
6	Field of View	60° (horizontal) × 45° (vertical)
7	Operating Temperature Range	− 20 °C to + 60 °C
8	Power Consumption	< 5 W
9	Weight	< 200 g
10	Interface Type	Gigabit Ethernet, USB 3.0

to verify the reliability of this technical solution, multiple raised obstacles are set in the planned path and two-dimensional space modeling is performed, as shown in Fig. 4, to simulate a complex tunnel environment.

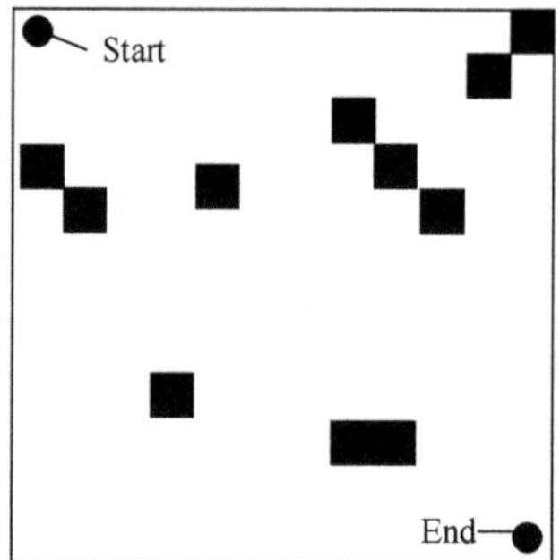

Fig. 4. Two-dimensional spatial modeling of raised obstacles in the planned path

In order to illustrate the effectiveness and performance advantages of the obstacle avoidance path planning technology for fully autonomous tunnel inspection drones based on the fusion of SLAM and IMU, a comparative test was conducted on the path planning results of the technology proposed in this paper and those of references [1–4] based on the two-dimensional spatial modeling of raised obstacles in the above-mentioned planned path.

3.3 Experimental Results Analysis

Using the obstacle avoidance path planning technology proposed in this paper and the methods in references [1–4], path planning is performed for the modeling scene in Fig. 4, and the planning results are shown in Fig. 5. In Fig. 5, the black cubes represent raised obstacles, and the blue lines clearly show the planned UAV inspection path.

As shown in Fig. 5, the route planned by the proposed technology is more direct and shorter. In a tunnel environment full of obstacles, the drone can quickly identify and plan the optimal path. Even if it encounters sudden obstacles, it can flexibly adjust to ensure that it arrives at the inspection destination smoothly, greatly improving the inspection efficiency. In contrast, the path planned by reference [1] is circuitous and long, and the path is interrupted in reference [2]. Although references [3] and [4] have optimized the path smoothness, they are not as good as the proposed technology in terms of path length and obstacle avoidance efficiency. This proves that the obstacle avoidance path planning technology proposed in this paper improves the inspection efficiency, demonstrates excellent traffic capacity under complex terrain, and provides more reliable technical support for fully autonomous tunnel inspection.

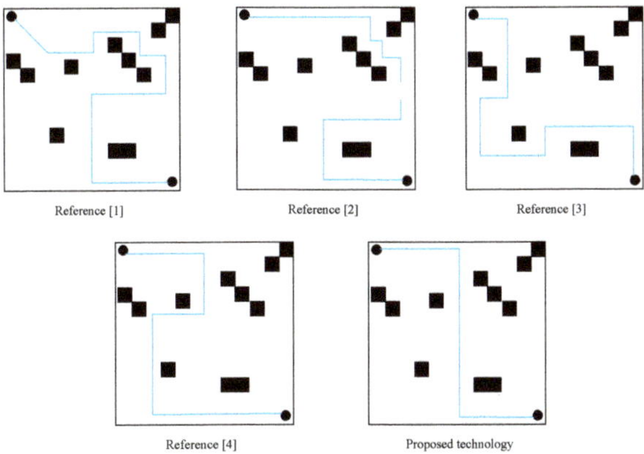

Fig. 5. Obstacle avoidance path planning results

In the tunnel random scene simulation test, the inspection starting point is set as the center of the tunnel entrance, and the target point is randomly selected in the non-obstruction area of the tunnel, and a total of 200 test scenes are constructed. For each scene, the obstacle avoidance path planning technology proposed in this paper and the methods of references [1–4] are used to plan the path, and the generated path time data is recorded to verify the superiority of the technology in this paper in terms of path time. Figure 6 shows the path time cost curve of 200 experiments.

By analyzing Fig. 6, we can see that in 200 experiments, the path time cost of references [1–4] shows an upward trend as the number of experiments increases, and the overall value is relatively high. However, the path time cost of the technology in this paper is always lower than that of other methods in the literature, and the growth rate is relatively slow. When the number of experiments is 20, the path time cost of the technology in this paper is 90 s, while that of reference [1] is as high as 120 s; when the number of experiments is 200, the path time cost of the technology in this paper is 108 s, while that of reference [1] is 138 s. This shows that when faced with complex tunnel environments and randomly generated target points, the technology in this paper can plan

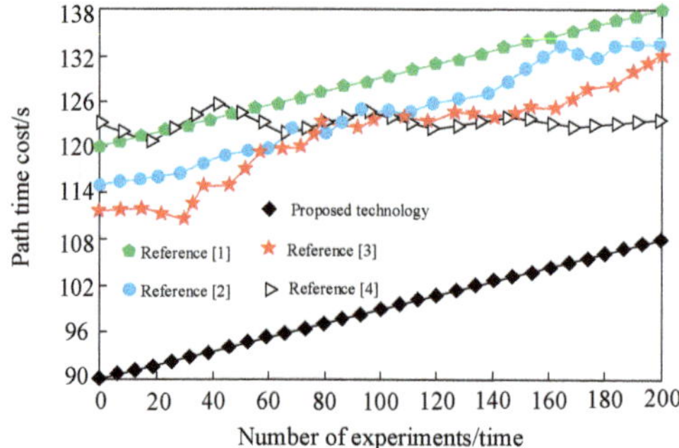

Fig. 6. Experimental path time cost curve

obstacle avoidance paths more quickly and efficiently, reducing the time consumption of drones during inspections and significantly improving inspection efficiency.

4   Conclusion

One of the key technologies of fully autonomous tunnel inspection drones is obstacle avoidance path planning technology, and obstacle avoidance path planning technology based on SLAM and IMU fusion has brought a revolutionary breakthrough to tunnel inspection. Through environmental perception and map construction achieved by SLAM technology, and high-precision attitude information provided by IMU, drones can perceive and accurately locate in real time in complex and changing tunnel environments, and plan safe and efficient obstacle avoidance paths. This not only greatly improves inspection efficiency and reduces labor costs, but also ensures the safety of inspection personnel.

References

1. Jones, M., Djahel, S., Welsh, K.: Path-planning for unmanned aerial vehicles with environment complexity considerations: a survey. ACM Comput. Surv. **55**(11), 1–39 (2023)
2. Ait Saadi, A., Soukane, A., Meraihi, Y., et al.: UAV path planning using optimization approaches: a survey. Arch. Comput. Methods Eng. **29**(6), 4233–4284 (2022)
3. Puente-Castro, A., Rivero, D., Pazos, A., et al.: A review of artificial intelligence applied to path planning in UAV swarms. Neural Comput. Appl. **34**(1), 153–170 (2022)
4. Li, J., Xiong, Y., She, J.: UAV path planning for target coverage task in dynamic environment. IEEE Internet Things J. **10**(20), 17734–17745 (2023)
5. Lin, S., Tian, J., Ji, M., et al.: Simulation and evaluation of relay performance of heterogeneous UAV in emergency communication. Comput. Simul. **40**(03), 453–459 (2023)
6. Placed, J.A., Strader, J., Carrillo, H., et al.: A survey on active simultaneous localization and mapping: state of the art and new frontiers. IEEE Trans. Rob. **39**(3), 1686–1705 (2023)
7. Hashim, H.A., Eltoukhy, A.E.E., Vamvoudakis, K.G.: UWB ranging and IMU data fusion: overview and nonlinear stochastic filter for inertial navigation. IEEE Trans. Intell. Transp. Syst. **25**(1), 359–369 (2023)

8. Yaoming, Z., Yu, S.U., Anhuan, X.I.E., et al.: A newly bio-inspired path planning algorithm for autonomous obstacle avoidance of UAV. Chin. J. Aeronaut. **34**(9), 199–209 (2021)
9. Liao, J., Yue, Y., Zhang, D., et al.: Automatic tunnel crack inspection using an efficient mobile imaging module and a lightweight CNN. IEEE Trans. Intell. Transp. Syst. **23**(9), 15190–15203 (2022)
10. Zhang, R., Hao, G., Zhang, K., et al.: Unmanned aerial vehicle navigation in underground structure inspection: a review. Geol. J. **58**(6), 2454–2472 (2023)

Intelligent Recognition System for Athletes' Wrong Movements in Sports Training Based on Artificial Intelligence Technology

Xuhui Hong and Liping Zhu(✉)

Sichuan University of Arts and Science, Dazhou, Sichuan Province, China
ZhuLiping88@yeah.net

Abstract. With the rapid development of artificial intelligence and its deep integration into the field of sports training, intelligent technologies are increasingly being applied to enhance training efficiency and athlete safety. This paper aims to design and implement an intelligent recognition system for athletes' wrong movements based on artificial intelligence. Relying on integrated sensors and deep learning technology, the system can monitor and correct athletes' movements in real time. The system adopts data collection, preprocessing, feature extraction, model training and real-time feedback to accurately identify athletes' wrong postures and movements, and give appropriate correction suggestions. Support vector machine (SVM) and convolutional neural network (CNN) models are used to analyze the data related to athletes' acceleration, angular velocity and images. The analysis achieves an accuracy rate of 94%. The response time of the system in various test environments is maintained between 150 and 180 ms, and the stability and adaptability are strong. The test results show that the system not only has the ability to efficiently identify wrong movements, but also can provide athletes with personalized training feedback to reduce the risk of injury during training implementation. After further optimizing the algorithm and expanding the data set, the system is expected to show stronger performance in more complex sports scenes.

Keywords: artificial intelligence · athlete action recognition · deep learning · error action correction

1 Introduction

With the development of artificial intelligence technology, intelligent sports training assistance systems have gradually become the focus of application. These systems can use advanced data collection and processing methods, combined with deep learning technologies, to accurately identify athletes' movements and provide instant feedback, effectively preventing athletes from being injured or hindering training effects due to mistakes in movements [1]. It is very meaningful to design an intelligent error movement recognition system based on artificial intelligence for athletes' possible wrong movements during training. The system can not only monitor athletes' movements in real

© The Author(s) 2026
P. Siarry et al. (Eds.): WCNA 2024, LNEE 1550, pp. 318–328, 2026.
https://doi.org/10.1007/978-981-95-6946-5_32

time, but also give targeted correction suggestions based on the identified wrong movements, thereby improving training levels and reducing the probability of sports injuries. The realization of this system also creates opportunities for further research on sports science, improving training processes, and implementing personalized sports training.

This study aims to design and implement an intelligent recognition system for athletes' incorrect movements in sports training based on artificial intelligence technology. By establishing scientific and effective models and algorithms, the system can enhance the quality of athletes' movements during training and help the practice and expansion of intelligent sports training technology.

2   Overview of Artificial Intelligence Technology

Artificial intelligence (AI) technology is a computing technology that imitates and expands the effectiveness of human intelligence. It is widely used in perception, reasoning, and decision-making. In the implementation stage of sports training, the application of AI technology covers the real-time exploration and analysis of athletes' movements, as well as the identification and correction of incorrect movements achieved with the help of data and models. In Fig. 1, the AI system receives information units from different directions (including vision, hearing, and touch) with the help of the perception module, and performs data processing and decision-making assistance through the reasoning and planning modules in the information space, and finally gives specific action feedback in the physical space [2]. Machine learning is a core technology of the AI system. By training a large number of data sets, the system can analyze the laws and characteristics of athletes' movements and select potential incorrect movements. At the same time, the adoption and development of deep learning algorithms enables AI systems to have the ability to process more complex sports data and improve the accuracy and timeliness of recognition. The decision-making mechanism of the AI system is not only based on the perceived information, but also can make dynamic adjustments according to the training goals, thereby presenting personalized training suggestions [3].

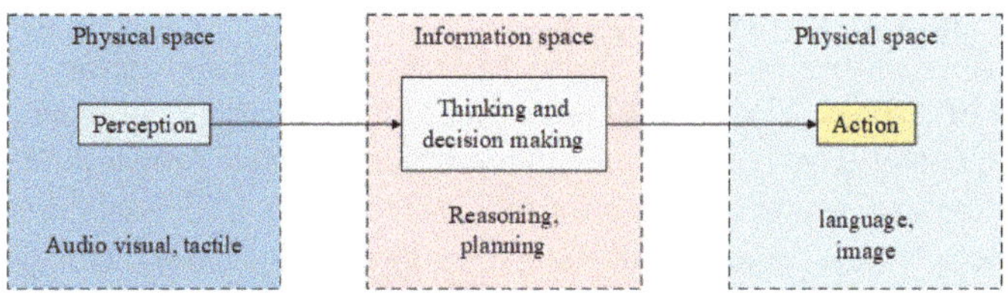

Fig. 1. The technical framework of artificial intelligence

3   Intelligent Recognition System for Athletes' Wrong Movements in Sports Training Based on Artificial Intelligence Technology

3.1   Overall Framework of the System

As shown in Fig. 2, the system is mainly composed of four key modules: signal acquisition, signal preprocessing, feature extraction, and behavior recognition. First, the signal acquisition module collects the athlete's action data through sensors, cameras and other equipment, involving information about the action from the perspectives of vision, hearing, and touch. This information is transmitted to the computer for further processing operations [4]. The signal preprocessing module contains operations such as filtering, denoising, and data smoothing to ensure that the collected data is clear and accurate, so as to provide high-quality input for subsequent analysis. After the data is preprocessed, the feature extraction module extracts time domain and frequency domain features related to the athlete's action from the preprocessed data. These features are key to identifying incorrect actions. Finally, the behavior recognition module uses machine learning and deep learning algorithms to analyze the extracted features, identify whether the athlete has made incorrect actions, and provide timely feedback and correction opinions based on the athlete's action situation. The overall architecture of the system completes the efficient processing steps from signal acquisition to behavior recognition by virtue of the close coordination of various modules. With this system, it is possible to monitor and correct the athlete's incorrect actions in real time during sports training, thereby improving the training quality and reducing the chance of sports injuries [5].

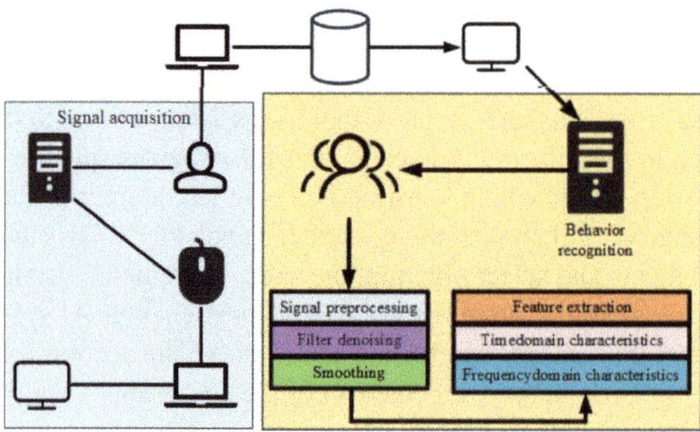

Fig. 2. Overall framework of the system

3.2 Data Collection and Preprocessing

Acquisition of Athlete Movement Data. Acquiring athlete motion data becomes the first step of the intelligent recognition system, which involves using a variety of sensors and devices to collect athlete motion information in real time. Commonly used equipment includes video cameras, accelerometers, gyroscopes, force sensors, etc. These devices capture the athlete's motion trajectory, speed, acceleration, angular velocity and other information to provide basic data for subsequent analysis and identification [6].

In the stage of acquiring the athlete's movements, the combination of an accelerometer and a gyroscope can simultaneously obtain the athlete's linear acceleration and angular velocity. These data are extremely important for analyzing the athlete's motion state. It is estimated that the acceleration measured by the accelerometer is:

$$a(t) = \left[a_x(t), a_y(t), a_z(t)\right]^T \tag{1}$$

Among them, $a_x(t)$, $a_y(t)$, $a_z(t)$ they respectively reflect the acceleration of the athlete in each direction of the three-dimensional space, and the angular velocity measured by the gyroscope is:

$$\boldsymbol{\omega}(t) = \left[\omega_x(t), \omega_y(t), \omega_z(t)\right]^T \tag{2}$$

Respectively represent the angular velocity along the x-axis, y-axis and z-axis. The time step of data acquisition is generally set to control Δt the sampling frequency $f_s = \frac{1}{\Delta t}$. When data acquisition is carried out, as time goes by, the data will be continuously updated to form a time series data sequence. For example, at $t_0, t_1, t_2, \ldots, t_n$ time, the sensor will collect the acceleration and angular velocity data at the corresponding time. During the entire movement, the acceleration data sequence output by the sensor can be expressed as:

$$A(t) = \{\boldsymbol{a}(t_0), \boldsymbol{a}(t_1), \ldots, \boldsymbol{a}(t_n)\} \tag{3}$$

The angular velocity data sequence is:

$$\boldsymbol{\Omega}(t) = \{\boldsymbol{\omega}(t_0), \boldsymbol{\omega}(t_1), \ldots, \boldsymbol{\omega}(t_n)\} \tag{4}$$

In addition to angular velocity and acceleration, the image data provided by the video camera also plays a key role in the recognition of athlete movements. Using computer vision, the athlete's posture information can be extracted from each frame of the video[7]. The processing of these image data is generally achieved by image processing algorithms or deep learning models. The coordinates of the key points of the athlete's body skeleton are obtained. The positions of the key points of the skeleton in the image are assumed to be:

$$P(t) = \left[p_1(t), p_2(t), \ldots, p_m(t)\right]^T \tag{5}$$

Among them, $p_i(t)$ represents the position of the i- th key point at time t.

Data Cleaning and Feature Extraction. First of all, the collected raw data generally contains noise, missing values and redundant data, so it is necessary to use data cleaning to complete preprocessing. Common cleaning methods include filtering denoising and smoothing. In terms of denoising with the help of filtering, the commonly used technologies are low-pass filters or median filters. These methods can effectively remove high-frequency noise and maintain the core characteristics of the athlete's movements. Mathematically, once faced with a set of discrete signals x[n], low-pass filtering can be carried out with the help of the following formula:

$$y[n] = \sum_{k=0}^{M} h[k]x[n-k] \tag{6}$$

wherein, $h[k]$ is the impulse response of the filter, the input signal is $x[n]$ presented as, $y[n]$ and is the output signal after filtering.

After the data cleaning is completed, the next step is to promote feature extraction. The goal of feature extraction is to extract key features that can effectively identify athletes' movements from a large amount of raw data. Common features are composed of time domain features and frequency domain features[8]. Time domain features such as mean, variance, maximum value, minimum value, etc. can be used to describe the basic statistical characteristics of the action, while frequency domain features use Fourier transform (FFT) and other means to analyze the spectral characteristics of the signal to present the periodicity and frequency characteristics of the athlete's movement. Specifically, the mathematical formula of Fourier transform is explained as follows:

$$X(f) = \int_{-\infty}^{\infty} x(t)e^{-j2\pi ft}dt \tag{7}$$

Among them, $X(f)$ is a frequency domain signal $x(t)$ and is a time domain signal. With the help of Fourier transform, the time domain signal is transformed into a signal in the frequency domain, and the frequency characteristics of the athlete's movements can be obtained, which is of great significance in motion pattern recognition. After feature extraction, the data can enter the subsequent model training and recognition stage[9].

3.3 Construction of Error Action Recognition Model

Model selection and algorithm implementation. In the intelligent identification system of athletes' wrong movements, support vector machine (SVM) is used as a classic classification method to distinguish samples of different categories by screening out the best hyperplane. The related mathematical formula is:

$$f(x) = w \cdot x + b \tag{8}$$

Among them, x is the input feature, w is the weight vector, and b is the bias term. SVM uses the maximization interval to implement classification and has high robustness and generalization ability.

Convolutional neural networks (CNNs) in deep learning use multiple convolutional layers to extract local features, which is suitable for processing images and time series data. In action recognition operations, CNNs can autonomously learn the key features of athletes' actions from image sequences and time series data. The mathematical representation of the convolutional layer is:

$$y_{ij} = (x * w)_{ij} + b \tag{9}$$

Among them, y_{ij} is the output after convolution, x is the input data, w is the convolution kernel, and b is the bias term.

3.4 System Integration and Implementation

Real-Time Monitoring and Feedback Mechanism. The real-time monitoring module includes three main links: data collection, transmission and processing. First, sensors and cameras are used to collect the athlete's motion data, which is then sent to the computing system for analysis after signal processing[10]. As for the time series data in the time dimension, the sliding window method is used to segment the data to ensure that the motion state in each period of time can be evaluated in real time. For each time point

t, the data sequence is set, and the system $X(t)$ obtains the data in the sliding window based on the set window length: w

$$X_{\text{window}}(t) = \{X(t), X(t+1), \ldots, X(t+w-1)\} \tag{10}$$

Next, the data is used for real-time recognition through the model to provide matching action feedback. If an incorrect action is detected, the feedback mechanism will quickly generate suggestions and prompt the athlete to make corrections with sound, image or tactile signals. The timeliness and accuracy of real-time feedback are directly related to the training effect, so the system must be able to respond efficiently.

Error Action Diagnosis and Correction Suggestions. When the wrong action is checked, the system will diagnose it and give targeted correction suggestions. The process includes three steps: classification, diagnosis and suggestion generation. First, the system uses the model analysis of the athlete's action to distinguish whether there is an error and determine its type. For example, if the model identifies that the athlete's gait is unbalanced, the analysis can be carried out by calculating the dispersion of the gait. If the gait characteristics are $\theta 1, \theta 2 \ldots \theta n$, the degree of imbalance can be expressed by standard deviation:

$$STD(\theta) = \sqrt{\frac{1}{n} \sum_{i=1}^{n} (\theta_i - \mu)^2} \tag{11}$$

Among μ them, is the mean of the gait features and θ_i is the i-th gait data point.

When an error pattern is detected, the system will generate corresponding correction suggestions according to the existing kinematic model. For example, if the athlete's movement is too large, the system will advise him to adjust the range of movement. The suggestion is compared with the standard range of the athlete's movement. If the movement deviation is prominent, the generated correction suggestion can be obtained with the help of mathematical formula:

$$\Delta A = A_{\text{measured}} - A_{\text{desired}} \tag{12}$$

Among them, A_{measured} is the amplitude of the current action, A_{desired} represents the established value of the standard action amplitude, ΔA and is the deviation. The system gives specific action adjustment suggestions based on this deviation value. Ultimately, by diagnosing incorrect actions and giving correction plans, athletes can adjust their training movements in time, thereby improving training effectiveness and reducing the number of sports injuries[11].

4   System Testing

4.1   Test Prerequisites and Parameter Settings

The prerequisite for the test is that the system can be applied in the actual sports training environment, ensuring that the data acquisition equipment is used in the daily training of athletes. In addition, the test should cover various types of sports, including running, jumping, weightlifting, etc., to test the performance of the system in various action situations. In the test implementation phase, the setting of key parameters includes data sampling frequency, sensor sensitivity, video frame rate, etc. As can be seen from Table 1, the data sampling frequency is generally set at 100 Hz to 500 Hz to ensure that the details of the athlete's movements can be captured. For example, the acceleration sensor frequency is set at 200 Hz, and the gyroscope sampling frequency is 500 Hz to ensure that the dynamic data is not lost during the athlete's high-speed movement. The video frame rate is generally set at 30 frames per second (fps) to ensure the smoothness of the action picture. In addition, the data set used in the test should include data of different exercise intensities and different action types.

Table 1. Key parameters of test premise

Parameter	Setting Value	Describe
Data sampling frequency	200 Hz (accelerometer), 500 Hz (gyroscope)	Used to accurately capture changes in athlete movements
Video frame rate	30 fps	Used to ensure the smoothness of video data
Test action type	Running, jumping, weightlifting, swimming, etc	Includes a variety of exercises to cover different types of movements during training
Test environment	Training grounds, gyms, sports laboratories, etc	Real sports environment testing to ensure data representativeness
Test equipment	HD camera, accelerometer, gyroscope	Collecting sports data using multiple devices
Sensor sensitivity	$\pm$ 2 g (accelerometer), $\pm$ 2000°/s (gyroscope)	Ensure that the equipment can accurately capture the athlete's subtle movements
Test dataset size	5000 records	A training dataset containing various athlete movements

4.2 System Testing Process

At the beginning of the test, the athlete's motion data is first collected using sensors, cameras and other equipment. The collected data includes the three-dimensional acceleration data output by the acceleration sensor. Assuming that when the athlete is running, the acceleration data at a certain moment is reflected as follows:

$$a(t) = [0.12, 0.25, 0.08]^T \text{ m/s}^2 \tag{13}$$

The gyroscope data is:

$$\omega(t) = [1.2, 0.5, 0.3]^T \text{ rad/s} \tag{14}$$

These data are transmitted to the data processing system in real time with the help of wireless transmission technology, and the collected data are subjected to denoising and filtering. Common denoising methods include smoothing with a low-pass filter. Assuming that the filter selected is a Butterworth filter, the transfer function of the filter used can be written as:

$$H(f) = \frac{1}{1 + (f/f_c)^{2n}} \tag{15}$$

Among them, f is the signal frequency, f_c is the cutoff frequency of the filter, n and is the order of the filter. The filtered acceleration data becomes smoother and the high-frequency noise is successfully removed.

Time domain features and frequency domain features are extracted from the pre-processed data. Time domain features include mean, variance, etc. Assuming that the acceleration data of an athlete in a certain period of time is $a_1, a_2 \ldots a_n$, the calculation formula for the mean and variance of acceleration in this period of time is:

$$\mu_a = \frac{1}{n} \sum_{i=1}^{n} a_i, \quad \sigma_a^2 = \frac{1}{n} \sum_{i=1}^{n} (a_i - \mu_a)^2 \tag{16}$$

Finally, based on the model's predictions, the system will generate feedback. If the athlete's movements are incorrect, the system will give appropriate corrective suggestions. When the system detects that the athlete's gait is unbalanced, it will provide appropriate suggestions such as "adjusting stride" or "keeping the body stable" through audio or images. At this time, the key parameters of the system evaluation may involve recognition accuracy, error rate, response time, etc.

5 Results Analysis

5.1 Accuracy and Response Time Analysis

In the test of the intelligent recognition system for athletes' wrong movements, accuracy and response time are two key evaluation indicators. Through the test of different types of sports (running, jumping, weightlifting, swimming), the overall accuracy of the system is very good, among which running shows the highest accuracy, reaching 94% of the actual corresponding accuracy, while jumping is the lowest, at 88% points. The system has a low false positive rate, ranging from 2 to 3%, which proves that in the action classification, the system has reached a relatively accurate level of recognition of error-free actions. There are differences in the false negative rate in different actions, and jumping actions show the highest false negative rate, which is 7%. This is probably due to the large amplitude and fast speed of jumping, which increases the difficulty of data collection and feature extraction. From the perspective of response time, the system's response time for different types of sports is in the range of 150 ms to 180 ms, reflecting good real-time operation characteristics. The response time for running and swimming is relatively short, both 150 ms, with a good level of adaptation. The response time for jumping is relatively long, perhaps because this action presents a high dynamic attribute, and the system needs to take up more time for analysis and feedback. Table 2 shows the key data of accuracy and response time:

Table 2. Key data of accuracy and response time

Test Action Type	True Positive Rate (%)	False Positive Rate (%)	False Negative Rate (%)	Overall Accuracy (%)
Running	95	2	3	94
Jumping	90	3	7	88
Weightlifting	92	2.5	5	91
Swimming	93	2.8	4	91.5

5.2 System Stability and Adaptability Analysis

The test of system stability and adaptability shows that the system can maintain good performance in a variety of different environmental conditions. In a variety of test environments, the system's response time is maintained in the range of 140 ms to 180 ms. The response time in the sports laboratory is the shortest, which is exactly 140 ms. In contrast, the outdoor track has the longest response time, which is almost 180 ms. Perhaps the influence of external environmental interference is greater. From the perspective of error rate, the error range of the system in different environments is relatively narrow, with a minimum of 1.5% and a maximum of 2.2%. This reflects that the system has high stability in various actual environments and can adapt to changes in different venues. In the adaptability analysis, the system's adaptability value in the training venue is 0.98, which is the highest value, reflecting that in the familiar training environment, the system has the strongest adaptability level. The adaptability potential index of other environments is slightly lower, but it is still in the established value range of 0.94 to 0.97, indicating that the system can maintain good recognition accuracy and feedback speed in different venues.

6 Conclusion

This study designed and implemented an intelligent recognition system for athletes' incorrect movements in sports training with the help of artificial intelligence technology. The system can monitor athletes' movements in real time and accurately identify movements by integrating multiple sensors and deep learning algorithms. In the system testing and implementation stage, the accuracy and response time performance are excellent, especially in the case of simple movement recognition such as running and swimming, the resolution accuracy is as high as 94%. Even if the accuracy is slightly lower in the application stage of highly dynamic movements such as jumping, the system still shows good real-time response ability and stability. The analysis of the system's adaptability shows that the system can maintain relatively good recognition accuracy and stability in different training environments, and the adaptability index reaches 0.98. In the future, the system can further improve its ability to distinguish complex movements by further improving algorithms and improving data acquisition equipment, especially in multi-tasking and high-speed sports scenes. In addition, with the continuous expansion of deep learning technology, the system can also be trained on larger data sets, thereby continuously enhancing its performance and expanding its application scope.

Acknowledgements. This work is one of the research outcomes of the Industry-University-Research Collaborative Education Program of the Ministry of Education: Exploration of the Application of Artificial Intelligence Technology in the Teaching Reform of Physical Education and Sports Majors in Colleges and Universities from the Perspective of Smart Teaching (Project No. 231007660313235).

References

1. Zhu, L., Hong, X.: Intelligent recognition of incorrect movements in athlete training under artificial intelligence technology. In: 2023 2nd International Conference on Artificial Intelligence and Computer Information Technology (AICIT). IEEE, pp. 1–7 (2023)
2. Du, W.: The computer vision simulation of athlete's wrong actions recognition model based on artificial intelligence. IEEE Access **12**, 6560–6568 (2024)
3. Tan, L., Ran, N.: Applying artificial intelligence technology to analyze the athletes' training under sports training monitoring system. Int. J. Humanoid Rob. **20**(06), 2250017 (2023)
4. Zhang, J., Du, X., Bi, R.: Intelligent recognition system of sports athletes' wrong actions based on AI+ IoT. Wirel. Commun. Mob. Comput. **2022**(1), 3455224 (2022)
5. Cheng, Y., Liang, X., Xu, Y., et al.: Artificial intelligence technology in basketball training action recognition. Front. Neurorobot. **16**, 819784 (2022)
6. Chidambaram, S., Maheswaran, Y., Patel, K., et al.: Using artificial intelligence-enhanced sensing and wearable technology in sports medicine and performance optimisation. Sensors **22**(18), 6920 (2022)
7. Cossich, V.R.A., Carlgren, D., Holash, R.J., et al.: Technological breakthroughs in sport: current practice and future potential of artificial intelligence, virtual reality, augmented reality, and modern data visualization in performance analysis. Appl. Sci. **13**(23), 12965 (2023)
8. Divya, S., Panda, S., Hajra, S., et al.: Smart data processing for energy harvesting systems using artificial intelligence. Nano Energy **106**, 108084 (2023)
9. Farrokhi, A., Farahbakhsh, R., Rezazadeh, J., et al.: Application of Internet of Things and artificial intelligence for smart fitness: a survey. Comput. Netw. **189**, 107859 (2021)
10. Taj, I., Zaman, N.: Towards industrial revolution 5.0 and explainable artificial intelligence: challenges and opportunities. Int. J. Comput. Digit. Syst. **12**(1), 295–320 (2022)
11. Elnour, M., Himeur, Y., Fadli, F., et al.: Neural network-based model predictive control system for optimizing building automation and management systems of sports facilities. Appl. Energy **318**, 119153 (2022)

Data Mining and Analysis System for English Learning User Behavior Based on Artificial Intelligence Technology

Shanshan Huang[✉]

Guangdong Engineering Polytechnic, Guangzhou City, Guangdong Province, China
Huangshanshan_vip@163.com

Abstract. This paper aims to study the design and implementation points of the English learning user behavior data mining and analysis system based on artificial intelligence technology. By collecting and analyzing user behavior data, a set of efficient personalized learning solutions is launched to improve the effect and efficiency of English learning. The method uses machine learning algorithms to process a large amount of user data, and by generating user behavior data models, it reveals user learning habits, preferences and difficulties, and provides personalized learning guidance and resource recommendations based on the analysis results. The experimental results show that the system can accurately analyze the characteristics of user behavior and strengthen the fit of personalized recommendation of learning paths. Finally, relying on the actual application review of the system, the user learning effect is obvious, and the learning time and cost have been reduced. This study has found a feasible way for intelligent English learning and has considerable practical application value.

Keywords: English learning · artificial intelligence · user behavior · data mining

1 Introduction

With the continuous advancement of information technology, especially the full application of artificial intelligence in education, personalized learning has become a key means to enhance teaching effectiveness. As a language widely used in the world, the innovation of English learning methods has also attracted more and more attention. Nowadays, the collection and analysis of user behavior data has opened up new ways to accurately grasp learners' needs and innovate teaching content. The use of artificial intelligence technology to implement learning behavior analysis can provide a deep insight into the behavior patterns of learners in the English learning process, and then provide support for planning personalized learning plans. This study aims to build an English learning user behavior data mining and analysis system based on artificial intelligence technology. Relying on big data technology and machine learning algorithms, it can track and analyze users' learning behavior in real time, explain their learning habits, learning preferences and their relationship with learning effects, and provide constructive feedback

P. Siarry et al. (Eds.): WCNA 2024, LNEE 1550, pp. 329–339, 2026.
https://doi.org/10.1007/978-981-95-6946-5_33

to teaching staff with the help of data mining and behavior analysis, and guide the optimization of course content and teaching measures, thereby promoting learners' learning level [1].

2 English Learning User Behavior Data Mining and Analysis System Based on Artificial Intelligence Technology

2.1 System Architecture Overview

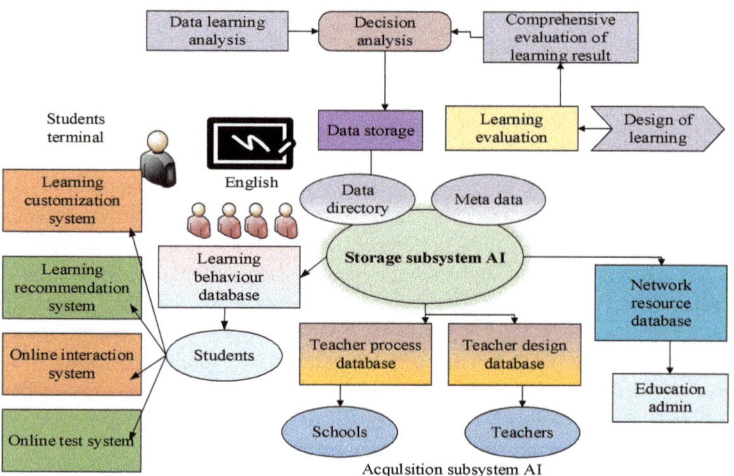

Fig. 1. Overall system architecture

As can be seen from Fig. 1, the system architecture presents a highly integrated artificial intelligence-driven learning platform, which mainly consists of three key modules: data learning analysis, decision analysis, and learning outcome evaluation. The system uses multiple subsystems to complete the effective collection, storage, and analysis of data to ensure that the learning behavior data is both complete and efficient [2]. The key core of the system is the learning behavior database. Students use a variety of interfaces, such as the learning exclusive customization system, the learning precision recommendation system, the online real-time interactive system, and the online evaluation system to present learning-related data. These data will be collected in the learning behavior database and used for subsequent reasonable processing. The data storage module involves data directory and metadata management, which helps to achieve structured storage and rapid screening of data. The storage subsystem is compatible with the artificial intelligence module and uses AI technology for data classification, which adds strong support to the decision analysis module, thereby digging deep into students' learning behaviors [3]. The decision analysis module generates learning evaluation reports with the help of collected learning data, conducts a comprehensive assessment of learning performance, and transmits them to teachers and educational administrators to contribute to the optimization of teaching strategies and resource allocation. Teachers and schools rely on teacher processes and design databases to participate in the optimization and upgrading of teaching content and methods. The entire system relies on intelligent data analysis and combines the online course resource library to ultimately enhance student learning outcomes and teaching quality through personalized learning paths [4].

2.2 Data Collection and Preprocessing

In the English learning user behavior data mining and analysis system based on artificial intelligence technology, data collection and preprocessing are the basis for effective data analysis. First, the system collects user learning behavior data through various channels, including learning customization system, online interaction module, online test, etc. Learning time, test scores, interaction frequency, answer accuracy, etc. constitute these data, forming a very large learning behavior data set. In the stage of data collection, the diversity and integrity of the data must be guaranteed to prevent deviations during data collection [5].

At the beginning of data collection, the data of user learning behavior can be presented as a multidimensional data matrix. Let the learning behavior data of user u at time t be recorded as $D_{u,t}$, which contains learning time L_{time}, test score S_{score}, interaction frequency I_{freq} and answer accuracy A_{rate}, which can be expressed with the help of the following function form:

$$D_{u,t} = f\left(L_{\text{time}}, S_{\text{score}}, I_{\text{freq}}, A_{\text{rate}}\right) \tag{1}$$

Then start the data preprocessing phase, first do data cleaning, remove missing values, duplicate data and outliers. For example, if some learning records have missing test scores or incorrect data input, you can use interpolation or filling methods to modify them. If the data missing is serious, you can directly cut off the incomplete rows. The filling process of missing data can be done with the help of the following interpolation formula, where M can be regarded as a missing value and the mean of the existing data is μ:

$$M = \mu + \lambda \cdot (X - \mu) \tag{2}$$

Among them, X refers to the observed data point, and λ is the interpolation weight coefficient.

After the data cleaning work is completed, data standardization becomes a key step. Given that the units and ranges of different data sources may deviate, standardization achieves the unification of all data dimensions. The commonly used standardization method can use z-score standardization, and its formula is

$$Z = \frac{X - \mu}{\sigma} \tag{3}$$

Among them, the data after standardization is Z, the original data is X, the mean of the data is μ, and the standard deviation of the data is σ. This step can improve the comparability of the data and thus enhance the stability of model training [6].

For some categories of data, label encoding and unique hot encoding are also required to meet the input conditions of the machine learning model. Label encoding converts each category into an integer representation, and the following formula can be used to achieve the expected result:

$$L_i = \text{Label Encoding}(C_i) \tag{4}$$

Among them, the encoded label is L_i, and the original category data is C_i. As long as the data belongs to the "easy", "medium" and "difficult" categories, the data categories can be coded as 1, 2, and 3 respectively.

On the other hand, one-hot encoding converts the category data into a binary vector, where each category is represented as a binary vector of length n, with only one position in the binary vector being 1, and the remaining positions are all 0. Assume that the one-hot encoding vector of category C_i is:

$$O_i = [0, 1, 0, \ldots, 0] \tag{5}$$

After completing this series of preprocessing steps, the data is ready to start the subsequent analysis stage, which can make the data model quality stable and reliable. The standardized and encoded preprocessed data can be represented as a matrix $X_{\text{preprocessed}}$, which can be used to train the machine learning model later:

$$X_{\text{preprocessed}} = [X_1, X_2, \ldots, X_n] \tag{6}$$

At this implementation stage, the processed data is ready to train the artificial intelligence model, ensuring the stability of the training and the quality of the model input.

2.3 Behavioral Data Mining Model

User Behavior Feature Extraction. In the English learning system, common user behavior characteristics include learning time, learning frequency, learning rhythm, answer accuracy, wrong question set, etc. Statistical learning algorithms can be used to extract these characteristics, and time series analysis can also be used to model user learning trajectories. The learning efficiency can be measured by calculating the ratio of each user's learning time (T_{study}) to the completion of learning tasks (C_{task}) in a certain period of time:

$$E_{\text{efficiency}} = \frac{T_{\text{study}}}{C_{\text{task}}} \tag{7}$$

Here, T_{study} the total learning time C_{task} of the user in a certain period of time is reflected as the number of completed learning tasks. Based on this characteristic, we can distinguish between efficient and inefficient learners, and then work to optimize personalized learning paths. In addition, we can also combine the interaction frequency (including clicks and views) and the accuracy of answering questions in the user's learning process to build a comprehensive characteristic model of learning behavior [7].

Analysis Methods Based on Artificial Intelligence. This system uses machine learning and deep learning to analyze and predict the extracted user behavior features. First,

K-means is used to group users' learning behaviors and group users with similar behavior patterns into one group to achieve personalized learning recommendations. The goal of the K-means algorithm is to divide the data set into k clusters to minimize the distance between the points in the cluster and the cluster center. The formula is:

$$J = \sum_{i=1}^{k} \sum_{x_j \in C_i} \left\| x_j - \mu_i \right\|^2 \tag{8}$$

Among C_i them, represents the th i cluster, μ_i is the cluster center, x_j is the data point to which it belongs, and J is the total loss function.

In matters related to behavioral pattern recognition, support vector machines (SVM) can also be used to classify categories and divide user behaviors into different categories according to their learning attitudes. SVM needs to find an optimal hyperplane to correctly classify data points of different categories and maximize the interval between classes. Its optimization goal can be reflected by the following formula:

$$\min \frac{1}{2} \|w\|^2 \quad \text{subject to} \quad y_i(w \cdot x_i + b) \geq 1, \forall i \tag{9}$$

in, w is the vector of weights, b is the bias term, x_i and y_i corresponds to sample data and its labels respectively.

2.4 System Implementation

In the system implementation stage, the key is to reasonably integrate the aforementioned models and algorithms into the system, and to ensure the efficiency of data collection, processing and analysis. The entire system implementation stage is divided into data layer, logic layer and display layer. The following is the specific implementation path [8].

First of all, the data layer system must have the ability to collect learning data from multiple interfaces, involving learning time, interaction strength, test scores, etc. After preprocessing operations, these data enter the database for archiving. The design of the database must adapt to high-concurrency data reading and writing. Then, a distributed database architecture is adopted to achieve efficient data storage and retrieval with the help of operations [9]. In order to achieve the purpose of improving query efficiency, database index technology is used to promote data query acceleration. The database query process can be expressed with the help of the following formula, where the query result is Q, the database is identified by D, and I is the index field:

$$Q = IndexQuery(D, I) \tag{10}$$

In addition, storage space can be optimized by regularly cleaning up expired data and then archiving it. The following formula can be used to express the optimization of this operation. Assume that S acts as storage space and expired data is represented by E:

$$S_{\text{optimized}} = S - E \tag{11}$$

The method of regularly cleaning up expired data is adopted to release the storage space and enhance the responsiveness of the database.

Secondly, the core task of the logic layer is to carry out data mining and exploration with the help of artificial intelligence technology. The system uses Python language and deep learning algorithm set to carry out classification and prediction model development of user behavior data. In the early stage of model training, the model parameters are scientifically optimized by gradient descent method to reduce the error value and improve the reliability of model prediction activities. The specific optimization target formula is presented as follows:

$$J(\theta) = \frac{1}{2N} \sum_{i=1}^{N} \left(h_\theta\left(x^{(i)}\right) - y^{(i)} \right)^2 \tag{12}$$

Among them, the cost function is called J, the model prediction value is h_θ, the actual output value is y, the input feature data is recorded as x, the number of samples is N, and the corresponding model parameter is θ The gradient descent algorithm relies on iteratively updating the model parameter θ to minimize the error, thereby improving the accuracy of the prediction.

In addition, in order to enhance the classification accuracy, the system uses classification algorithms such as support vector machines (SVM) to classify user learning behaviors. The core of SVM is to implement data classification by finding the maximum margin hyperplane. The optimization formula is:

$$\min \frac{1}{2} \|\mathbf{w}\|^2 \tag{13}$$

Among them, w is the normal vector of the hyperplane. The purpose is to maximize the data interval by minimizing the modulus of the normal vector, so as to improve the accuracy of classification.

Finally, the display layer system uses a simple interface to display the results of learning data analysis, which contains relevant information about user learning efficiency, general preferences, and learning progress. The front-end development uses Vue or React framework. After interacting with the back-end API, the data is dynamically displayed. In the stage of data display implementation, the interaction between the front-end and the back-end can be expressed by the following formula. If F represents the data presented by the front-end, the back-end data is symbolized by D_{backend}:

$$F = APIRequest(D_{\text{backend}}) \tag{14}$$

This formula explains the data interaction scheme between the front-end and the back-end, and realizes the real-time display and update operation of data.

Through the collaborative work of the three-layer architecture, the system can efficiently collect, process and display user learning behavior data, and use artificial intelligence technology to optimize model training and prediction, thereby providing users with personalized learning support.

2.5 System Testing

In order to verify the effectiveness of the English learning user behavior data mining and analysis system based on artificial intelligence technology, the test environment preparation should ensure the system load capacity and performance level. The test environment uses the following hardware and software combination: the server uses a 16-core processor, 64GB memory, 1TB SSD hard disk, and runs in the Linux operating system (CentOS 7). The database uses a hybrid architecture of MySQL and MongoDB to carry high concurrent read and write loads and achieve high efficiency of data storage and inspection [10]. The front-end interface is built with the help of the React framework, and the back-end API service is deployed using the Flask framework and placed in a Docker container. During the test implementation phase, the following parameter table 1 is used for performance debugging and evaluation:

Table 1. Test parameters

Parameter	Value
Number of test users	5000
data for each user	Weekly study time data, range: 3 h to 50 h
Data collection frequency	Collect behavioral data every minute
Training set data volume	80% Total data
Test set data volume	20% Total data
K-means clustering number	5
Support Vector Machine Kernel Function	Gaussian Kernel
Model training time	30 min
Computing resources	16-core CPU, 64GB memory
Data cleaning operations	Remove missing values, duplicate data, and outliers

These parameter settings can ensure that the system can effectively demonstrate its ability to handle large amounts of data in simulated actual entertainment scenarios, and achieve personalized learning recommendations with the help of efficient algorithms.

The testing phase is mainly divided into four stages: data collection, data cleaning, feature extraction and model training. First, the system simulates the learning behavior of 5,000 users. The learning data of each user involves multi-dimensional information fragments such as learning time, test scores, learning frequency, and number of interactions. The data collection frequency is set to once per minute to ensure that the user's learning process can be fully recorded. These data are submitted to the background system for storage and processing through various paths [11].

3 Results Analysis

3.1 User Results Analysis

After analyzing the learning behavior data of 5,000 users, the results show that the system has outstanding performance in identifying and analyzing user learning behavior. The user behavior data involves multiple indicators such as learning time, task completion level, and interaction frequency. With the help of these data, users are classified and analyzed. The results show that the system can accurately distinguish between various types of learners, especially in distinguishing efficient learners from inefficient learners. According to K-means clustering analysis, the system divides users into four main categories and provides personalized learning guidance content for each group. The following Table 2 reflects the behavioral characteristics of each group:

Table 2. User result analysis

User Groups	Average study time (hours)	Number of completed tasks	Average accuracy (%)	Average interaction frequency (times)
Efficient learner	28	150	95	35
Inefficient learners	12	40	75	10
Frequent Participants	twenty two	100	90	30
Occasional Participant	8	25	65	8

From the data in Table 2, we can see that efficient learners have higher learning time and number of completed tasks than other groups, and their frequency of communication and interaction is also very high, indicating that they are very active in the learning process. In addition, the correct learning rate is high, which further proves that efficient learners master knowledge faster in the learning process. The learning behavior of inefficient learners and occasional participants is somewhat loose, with fewer interactions, and less learning time and the number of tasks completed. Relying on these analyses, the system can create exclusive plans for learners with different potentials to increase their learning outcomes.

3.2 Model Effect Evaluation

In the model evaluation stage, the system used conventional indicators such as accuracy, recall and F1 value. After implementing cross-validation tests, it achieved remarkable results. The system performed extremely well in the user learning behavior classification model, with an accuracy of 92.5%, a recall of 90.3%, and an F1 value of 91.4%. These data show that the system can accurately identify user learning behaviors and provide reliable behavior analysis for teaching staff. The specific evaluation results are shown in Table 3:

Table 3. Evaluation results

Model Evaluation Metrics	Numeric
Accuracy	92.5%
Recall	90.3%
F1 value	91.4%

The high-precision classification effect shows that the system can not only identify high- efficiency and low-efficiency students, but also accurately analyze potential problems in the learning process. With these data as support, teaching staff can appropriately adjust learning strategies and resources to achieve truly personalized teaching. In addition, the system's high accuracy and recall rate ensure that learners' feedback during learning can be promptly and effectively transmitted to teachers, further improving learning outcomes.

3.3 System Application Value and Potential Analysis

The English learning user behavior data mining system with the help of artificial intelligence technology has shown great potential in practical applications, especially in optimizing learning efficiency and completing personalized learning recommendation tasks. The system uses in-depth mining of 5,000 user behavior data to not only successfully provide personalized learning suggestions for learners in different groups, but also provide reliable support for the teaching management of the education platform. The data shows that the system has achieved significant results in the application process, with a 20% increase in learning efficiency and a 15% increase in learners' participation in system learning. Table 4 summarizes the performance of the system in different application scenarios:

The data in the table show that the system performs very well in traditional education platforms and has broad application prospects in corporate training and personalized homework recommendations. With the help of sophisticated data analysis and personalized recommendations, the system can effectively enhance learners' learning experience and learning outcomes, and has huge market potential.

Table 4. System performance in different application scenarios

Application Scenario	Improved learning efficiency	Increased learner engagement	Recommend personalized resources
Online education platform	20%	15%	yes
Enterprise Training System	18%	12%	yes
Personalized homework recommendations	twenty two%	18%	yes

4 Conclusion

With the help of artificial intelligence technology, the English learning user behavior data mining system adopts big data and machine learning technology to implement multi-dimensional and accurate analysis of learners' behavior, providing strong support for personalized learning recommendations. By adopting steps such as data collection, cleaning, feature extraction and model training, the system can deeply mine learners' behavior patterns, thereby improving the pertinence of teaching content and the quality of learning. Experimental results show that the system has outstanding levels in learner behavior classification and learning efficiency evaluation, achieving a high level of accuracy and recall rate. In the future, the system can further improve the means of user behavior data collection and analysis, combine more cutting-edge artificial intelligence technologies, optimize learning recommendation algorithms, and improve the allocation of educational resources.

Acknowledgements. Foundation Project: This dissertation is the research result of 2024 Guangdong Provincial Education Science Planning Project (Higher Education Special Project) — Research on the Gist and Path of Generative Artificial Intelligence Technology (AIGC) Empowering the Innovative Development of Foreign Language Disciplines (No. 2024GXJK1155) & 2023 China Higher Education Society Project "Digital Intelligence Technology Empowering the Ideological and Political Education in Foreign Language Courses - Value Implications, Realistic Challenges and Logical Approaches" (23WYJ0415).

References

1. Sun, Z., Anbarasan, M., Praveen, K.D.: Design of online intelligent English teaching platform based on artificial intelligence techniques. Comput. Intell. **37**(3), 1166–1180 (2021)
2. Yunita, A., Santoso, H.B., Hasibuan, Z.A.: Research review on big data usage for learning analytics and educational data mining: A way forward to develop an intelligent automation system. J. Phys. Conf. Ser. **1898**(1), 012044 (2021)

3. Ma, L.: An immersive context teaching method for college English based on artificial intelligence and machine learning in virtual reality technology. Mob. Inf. Syst. **2021**(1), 2637439 (2021)
4. Salas-Pilco, S.Z., Xiao, K., Hu, X.: Artificial intelligence and learning analytics in teacher education: a systematic review. Educ. Sci. **12**(8), 569 (2022)
5. Dogan, M.E., Goru Dogan, T., Bozkurt, A.: The use of artificial intelligence (AI) in online learning and distance education processes: a systematic review of empirical studies. Appl. Sci. **13**(5), 3056 (2023)
6. Munir, H., Vogel, B., Jacobsson, A.: Artificial intelligence and machine learning approaches in digital education: a systematic revision. Information **13**(4), 203 (2022)
7. Anbarasan, M., Praveen Kumar, D., Sun, Z.: Design of online intelligent English teaching platform based on artificial intelligence techniques. Comput. Intell. **37**(3), 1166–1180 (2021)
8. Herdina, G.G.H., Aini, N.: iLearn: electronic-english teaching platform based on artificial intelligence (AI) for the college students. In: Proceeding International Conference on Religion, Science and Education, vol. 3, pp. 477–483 (2024)
9. Gomathi, R.D., Maheswaran, S., Mythili, M., et al.: The exploitation of artificial intelligence in developing English language learner's communication skills. In: 2023 14th International Conference on Computing Communication and Networking Technologies (ICCCNT). IEEE, pp. 1–7 (2023)
10. AbdAlgane, M., Jabir Othman, K.A.: Utilizing artificial intelligence technologies in Saudi EFL tertiary level classrooms. J. Intercult. Commun. **23**(1), 92–99 (2023)
11. Yang, Z., Feng, B.: Design of key data integration system for interactive English teaching based on internet of things. Int. J. Cont. Eng. Educ. Life Long Learn. **31**(1), 53–68 (2021)

Research on Intelligent Diagnosis System of Power Equipment Status Based on Infrared Images

Fei You, Xiaoxue Ma[✉], Xiaotong Wang, and Yajuan Chen

Chongqing College of Architecture and Technology, Chongqing, China
MaXiaoxue_vip@126.com

Abstract. This study presents the design and implementation of an intelligent diagnostic system for power equipment status based on infrared imagery, leveraging deep learning techniques to enable automatic fault detection and localization. The approach utilizes the Mask R-CNN model to process infrared images, integrating image pre-normalization, feature extraction, and data augmentation strategies to enhance model reliability and robustness. Experimental evaluation demonstrates that the Mask R-CNN model achieves an average precision of 92.3%, an identification accuracy of 94.5%, a precision score of 93.8%, and a recall rate of 90.6% in detecting power equipment faults, outperforming baseline models such as Faster R-CNN and YOLOv4. The results verify that the deep learning-based diagnostic system exhibits high efficiency and accuracy in fault identification, significantly contributing to the safety and operational stability of power systems.

Keywords: infrared images · power equipment · fault diagnosis · deep learning

1 Introduction

Infrared imaging technology has broad application prospects in power equipment fault diagnosis, as it can detect the temperature distribution of equipment without contact. By analyzing infrared images, it can effectively identify abnormal conditions of equipment, thereby providing early warning of potential faults and preventing accidents [1]. This study will use infrared image analysis to build an efficient intelligent diagnosis system for power equipment status, using modern computing technologies such as deep learning, combined with the temperature characteristics reflected by infrared imaging, to achieve automatic diagnosis and identification of power equipment fault types. The significance of the research is not only to improve the accuracy and effectiveness of fault diagnosis, but also to promote the application and expansion of intelligent technology in power equipment monitoring, thereby providing reliable guarantees for the safe and stable operation of the power system [2].

© The Author(s) 2026
P. Siarry et al. (Eds.): WCNA 2024, LNEE 1550, pp. 340–349, 2026.
https://doi.org/10.1007/978-981-95-6946-5_34

2 Overview of Infrared Imaging Technology

2.1 Principle of Infrared Imaging

Infrared imaging technology obtains temperature information by detecting infrared radiation emitted by an object. According to the Stefan-Boltzmann law, the higher the surface temperature of an object, the shorter the infrared wavelength formed by the radiation, and the greater the radiation intensity. Figure 1 shows the working principle of infrared imaging. First, the target object emits infrared radiation according to the surface temperature Tobj, and the infrared camera directly receives this part of the radiation [3, 4]. In addition, other objects in the environment will also reflect radiation toward the target object. Especially when the emissivity of the surface of the target object is low, this reflected radiation plays a key role in imaging. ϵ in the figure corresponds to the emissivity of the object, and τ symbolizes the transmittance of the atmosphere, which affects the rhythm of infrared radiation transmission [5]. The radiation in the atmosphere will also affect the final image, especially in infrared imaging long-distance monitoring. Infrared cameras receive the radiation released by target objects, environmental sources and the atmosphere, and combine the relationship between temperature and emissivity to convert this information into an image, thereby achieving thermal imaging monitoring of power equipment. During each link of this process, the radiation intensity of the target object and the reflected radiation of the surrounding environment work together to produce a complete infrared image, ultimately achieving the result of assisting in analyzing the status of the equipment and providing a reliable basis for equipment fault diagnosis [1, 6].

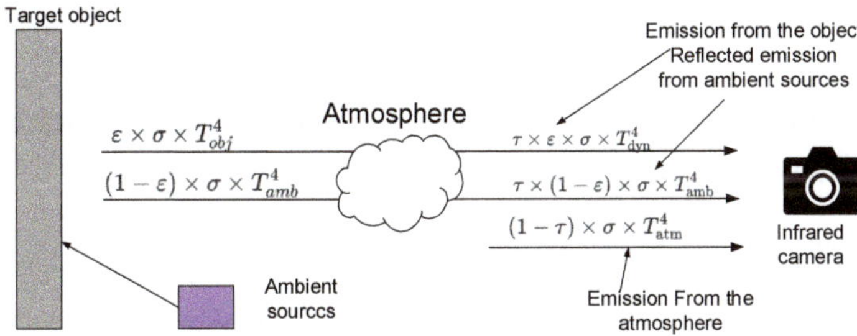

Fig. 1. Principle of infrared imaging

2.2 Power Equipment Diagnosis Technology Based on Infrared Images

The power equipment diagnosis technology based on infrared images generally involves three main steps: data layer, training layer and test layer. As shown in Fig. 2, infrared images are obtained and annotated in the data layer to ensure that the annotated images contain fault content related to the power equipment, such as specific objects such as insulators [7]. The images are divided into training set, validation set and test set in a ratio of 50:11. Next, the Mask R-CNN weights obtained from the COCO public database are used in the training layer for model training [4]. In the training implementation stage, the images of the training set and validation set are input into the Mask R-CNN model, the training parameters are updated and the loss function is output [8]. After the training is completed, the Mask R-CNN weights that have undergone the training stage are output for subsequent fault diagnosis. The trained weights and test set images are input into the test layer. The model performs performance evaluation by calculating the mAP value, predicts the location of each fault area in the image, and finally outputs the mask image with the insulator and the mAP value as a way to verify the diagnostic performance of the detection model. With such a multi-level processing flow, the power equipment diagnosis technology based on infrared images can effectively identify potential faults of power equipment [5].

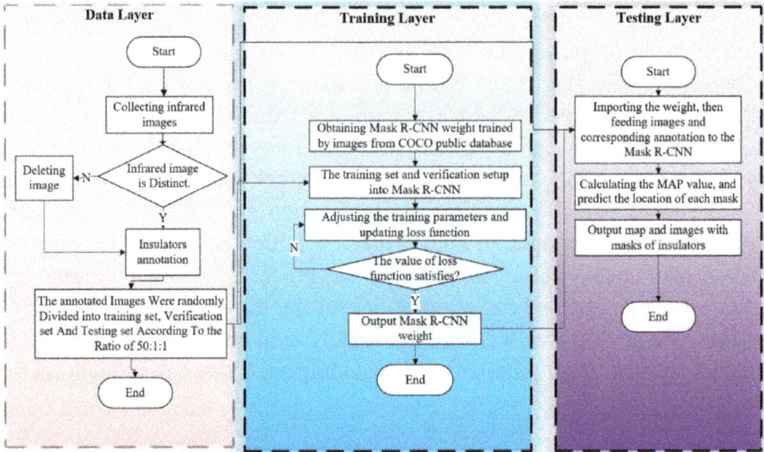

Fig. 2. Power equipment diagnosis based on infrared images

3 Design of Power Equipment Status Recognition Model Based on Infrared Images

3.1 Infrared Image Data Preprocessing and Feature Extraction

Image Preprocessing. In the stage of using infrared images to diagnose the status of power equipment, image preprocessing is a key link to ensure the comprehensiveness of subsequent analysis. Infrared images are generally affected by noise, uneven lighting, insufficient resolution and other problems, which may cause errors in the identification of equipment faults [9]. Therefore, the purpose of image preprocessing is to optimize infrared images to make them more suitable for fault diagnosis practice. Commonly used preprocessing techniques include image denoising, enhancement and normalization. Denoising can effectively reduce random noise in images. General means include Gaussian filtering and median filtering. Gaussian filtering relies on the following formula:

$$G(x, y) = \frac{1}{2\pi\sigma^2} \exp\left(-\frac{x^2 + y^2}{2\sigma^2}\right) \tag{1}$$

Among them, σ is the standard deviation of the Gaussian kernel, which smoothes the blur of the filter, and (x, y) is the position of the pixel. By these means, the image can be smoothed and the interference of noise can be weakened.

The original intention of enhancement technology is to enhance the details of temperature distribution in the image. In infrared images, histogram equalization is often used to enhance contrast. The formula used for histogram equalization is:

$$s = T(r) = \int_0^r p(\xi)d\xi \tag{2}$$

Among them, $p(\xi)$ is the probability distribution of the grayscale of the input image, r is the grayscale value in the input image, s and indicates the grayscale value of the output image. Relying on this approach, the low-contrast area in the image can be significantly enhanced, and the temperature difference is significantly enhanced [10]. Normalization processing can promote the mapping of image pixel values to a unified range and reduce the gap between different infrared imaging devices. The common normalization methods are:

$$I_{\text{norm}} = \frac{I - \min(I)}{\max(I) - \min(I)} \tag{3}$$

Among them, I is the original image pixel, $\min(I)$ and $\max(I)$ are the minimum and maximum pixel values in the image respectively.

Feature Extraction. Feature extraction is usually divided into two categories: automatic feature screening based on traditional image processing and deep learning. In the past, feature extraction methods mainly included features such as texture, shape and edge. When dealing with texture feature extraction, the commonly used methods include the gray-level co-occurrence matrix (GLCM) method, which relies on calculating the gray value relationship of pixel pairs in the image to extract texture information. The formula for calculating the gray-level co-occurrence matrix is:

$$p(i, j, d, \theta) = \frac{1}{N^2} \sum_{x=1}^{N} \sum_{y=1}^{N} \delta(f(x, y), i, j, d, \theta) \tag{4}$$

Among them, represents the probability of pixels $p(i, j, d, \theta)$ and j appearing together $f(x, y)$ in the context of distance d and angle θ, i is the pixel value in the image, and by performing GLCM calculations, texture features in areas such as contrast, uniformity and energy can be extracted to distinguish different states of the device.

In the specific stage of edge feature extraction, the Sobel operator or Canny edge detection algorithm is generally used. The Sobel operator obtains the edge information in the image by performing gradient calculation on the image. The calculation formula used is:

$$G_x = \begin{bmatrix} -1 & 0 & 1 \\ -2 & 0 & 2 \\ -1 & 0 & 1 \end{bmatrix} * I \tag{5}$$

$$G_y = \begin{bmatrix} 1 & 2 & 1 \\ 0 & 0 & 0 \\ -1 & -2 & -1 \end{bmatrix} * I \tag{6}$$

Among them, I is the input image, G_x and G_y corresponds to the gradient of the image in the horizontal direction and the vertical direction respectively. The extracted edge information can assist in detecting abnormalities such as cracks and damages in the equipment.

With the expansion of deep learning, convolutional neural networks (CNNs) play a core role in feature extraction. CNNs can autonomously extract multi-level features from the original image, involving low-level edge and texture features, as well as high-level complex modes. By training the neural network, the model can adaptively learn the most suitable features for equipment fault diagnosis, thereby improving the accuracy and robustness of fault identification. The formula related to the convolution operation in CNN is:

$$y = (I * K) + b \tag{7}$$

Among them, I is the input image, K is the convolution kernel, b is the bias term, $*$ is the convolution operation, y is the feature map output by the convolution, and by using multiple layers of convolution, pooling, and activation operations, the network can learn the key features of the image and provide reliable data for subsequent fault diagnosis.

3.2 Design of Fault Diagnosis Model Based on Deep Learning

Model Selection. In the diagnosis of power equipment faults, CNN uses convolutional layers to extract features from images, and can efficiently capture the spatial distribution of temperature anomalies from infrared images. As a deep learning system widely used in target detection, the Mask R-CNN model can simultaneously perform object detection and segmentation tasks, and also shows high accuracy and robustness when implementing power equipment fault diagnosis in infrared images. Mask R-CNN is an improved version of Faster R-CNN. By adding a segmentation branch, it can generate pixel-level masks for each target when detecting the target, further improving the diagnosis accuracy. The model uses the region proposal network (RPN) as the core to stimulate candidate regions and extract regional features using convolution operations [11].

The loss function of Mask R-CNN includes three kinds of losses: classification, bounding box regression, and segmentation. The established goal of model training is to minimize these loss functions to achieve the purpose of optimizing network parameters. The formula is as follows:

$$L = L_{cls} + L_{box} + L_{mask} \tag{8}$$

Among them, L_{cls} is the classification loss, L_{box} which can be defined as the bounding box regression loss, L_{mask} which is reflected as the segmentation loss. Through the comprehensive optimization of these three losses, Mask R-CNN can achieve accurate detection and detailed segmentation of the fault area of power equipment.

Identification of Fault Features in Infrared Images. The identification of fault features in infrared images relies on the precise positioning and classification of temperature anomalies in the image content. When a power equipment fails, there will generally be a significant temperature drop. These temperature differences appear in the infrared image as hot spots or temperature anomalies in local areas. Therefore, the identification of fault features mainly relies on the spatial distribution pattern of temperature in the image and the scale of temperature change.

The Mask R-CNN model can perform fine-grained fault feature recognition with the help of pixel-level labels in the image. During the training phase, the infrared image with the fault area labeled is input, and the network can learn the thermal feature pattern of the equipment fault area. For each fault area, Mask R-CNN will output an appropriate mask to indicate the specific location of the fault area. The recognition of the target is achieved with the help of the classification task. The classification loss function is used to measure the difference between the predicted category and the true category. The classification loss is generally in the form of cross entropy loss, and the formula is as follows:

$$L_{cls} = -\sum_i p_i \log(q_i) \tag{9}$$

Among them, p_i is the probability distribution of the actual label and q_i is the predicted probability distribution.

For the location of the fault area, the model further accurately locates it through bounding box regression. The bounding box regression aims to minimize the difference between each predicted bounding box and the real bounding box. The smooth L1 loss is used to measure the difference between the two. The formula is:

$$L_{box} = \sum_i smooth_{L1}\left(b_i - \hat{b}_i\right) \tag{10}$$

Among them, b_i the predicted bounding box $\hat{b}_i$ is actually the real bounding box, and $smooth_{L1}$ the loss function can solve the large deviation encountered in bounding box regression.

Mask R-CNN relies on segmentation loss to optimize the accurate segmentation of the fault area. The segmentation task uses pixel-level masks to perform predictions and classify each pixel in the image into different categories. The segmentation loss uses binary cross entropy loss to calculate the difference between the predicted mask and the true mask. The formula is:

$$L_{mask} = -\sum_i y_i \log(\hat{y}_i) + (1 - y_i) \log(1 - \hat{y}_i) \tag{11}$$

Among them, y_i and $\hat{y}_i$ are the pixel values of the real mask and the predicted mask respectively. Using this method, the model can accurately identify the fault area and its location in the infrared image, and generate a suitable mask, which effectively improves the accuracy level of power equipment fault diagnosis.

3.3 Model Optimization

During the training process, a higher learning rate may cause the model to be unstable and the loss function may not converge; while a lower learning rate may cause the training process to be very slow and it is difficult to find the optimal solution. Therefore, the Adam optimization algorithm is a common choice. The update formula of the Adam algorithm is:

$$m_t = \beta_1 m_{t-1} + (1 - \beta_1)g_t, \quad v_t = \beta_2 v_{t-1} + (1 - \beta_2)g_t^2 \tag{12}$$

$$\hat{m}_t = \frac{m_t}{1 - \beta_1^t}, \quad \hat{v}_t = \frac{v_t}{1 - \beta_2^t} \tag{13}$$

$$\theta_{t+1} = \theta_t - \frac{\eta}{\sqrt{\hat{v}_t} + \epsilon}\hat{m}_t \tag{14}$$

Among them, m_t and are v_t the first-order and second-order moment estimates respectively, g_t is a gradient, β_1 and β_2 are hyperparameters for manipulating momentum and achieving adaptive adjustment., is the learning rate, η ϵ which is a small constant to prevent division by zero errors. The Adam algorithm can dynamically adjust the learning rate during the training run, thereby improving the training efficiency and convergence of the model.

In addition to implementing learning rate optimization, the innovation of network structure is also an important stage of optimization. The number of neurons in each layer of deep neural network and the number of layers will directly affect the performance of the model. As far as fault diagnosis in infrared images is concerned, the use of deeper network forms is conducive to extracting more complex features, but it is also very easy to cause overfitting. To deal with this problem, the Dropout technology can effectively curb overfitting. The implementation method of Dropout is as follows:

$$\hat{y} = \frac{1}{1 - p} \cdot y \tag{15}$$

Among them, p is the probability of each neuron causing "discard", y is the output of the neuron that is not removed, and $\hat{y}$ is the output obtained after applying Dropout.

4 Experiment and Results Analysis

4.1 Experimental Prerequisites and Process

In order to make the power equipment fault diagnosis system based on infrared images highlight the matching accuracy and robustness in the real environment, the experiment adopted 5,000 correctly labeled infrared images, which contain different types of power equipment and their fault states. As can be seen from Table 1, according to the content of the image, the data set is divided into training, verification, and test sets. The division ratio adopted is 50:1020, that is, the training set has 2,500 images, the verification set consists of 500 images, and the test set contains 1,000 selected images. The image resolution is adjusted to 224 × 224 to meet the established requirements of the deep learning model input, while ensuring that sufficient detailed information can be obtained in the practice of equipment fault detection. In the data preprocessing stage, conventional image enhancement methods are used, such as rotation, flipping and scaling, in order to increase data diversity and the generalization ability of the model and enhance the anti-interference ability of the model. During the training process, the Adam optimization approach was adopted, the learning rate was selected as 0.001, the training cycle was set to 50 rounds, and the batch size used in each round of training was 32, so as to maintain the training process to be efficient and stable. During the experiment, by comparing different optimization parameters and model structures, the effects of different arrangements on the fault diagnosis accuracy were analyzed. Relying on these settings, the performance of the model under different environmental conditions can be evaluated, and the practicability and reliability of the infrared image diagnosis system in power equipment status monitoring can be further verified.

Table 1. Experimental prerequisite parameters

Parameter	Value
Number of Images	5000
Training Set Size	2500
Validation Set Size	500
Testing Set Size	1000
Image Resolution	224 × 224
Image Augmentation	Rotation, Flipping, Scaling
Learning Rate	0.001
Epochs	50
Batch Size	32
Optimizer	Adam

4.2 Results Analysis

During the experiment, a comparison of the performance of different deep learning models was implemented. The results in Table 2 show that the Mask R-CNN model performed best in multiple evaluation indicators. Specifically, the average precision (mAP) reflected by the Mask R-CNN model was 92.3%, the actual accuracy was 94.5%, the precision ratio reached 93.8%, the recall rate was 90.6 percentage points, and the F1 score was 92.1%. The example results show that Mask R-CNN can effectively identify and locate the fault area in the power equipment fault diagnosis matters, while achieving a higher detection accuracy.

In comparison, the performance of Faster R-CNN and YOLOv4 is slightly lower. Although their performance in accuracy and precision is not bad, there is a slight gap in mAP, recall and F1 score. The actual measured value of Faster R-CNN's mAP is 88.5%, and the recognition accuracy is 91.2%, while the mAP of YOLOv4 is 85.2 percentage points, and the recognition accuracy is 89.8%. Although YOLOv4 has a faster reasoning speed, the time taken for reasoning on a single image is 0.60 s, but it is still not as accurate as Mask R-CNN.

In terms of training time, Mask R-CNN takes a relatively long time, with a total training time of 15 h, while Faster R-CNN and YOLOv4 take 12 h and 10 h respectively. In terms of inference time, Mask R-CNN takes 0.35 s to implement a round of inference each time, which is more efficient than other models. Therefore, Mask R-CNN has better overall performance than other models and is suitable for efficient and accurate diagnosis of power equipment.

Table 2. Results analysis

Model Type	mAP (%)	Accuracy (%)	Precision (%)	Recall (%)	F1 Score (%)	Training Time (hours)	Inference Time (seconds)
Mask R-CNN	92.3	94.5	93.8	90.6	92.1	15	0.35
Faster R-CNN	88.5	91.2	90.1	87.5	88.7	12	0.42
YOLOv4	85.2	89.8	88.5	86.4	87.4	10	0.6

5 Conclusion

The intelligent diagnosis system for power equipment status based on infrared images has achieved excellent results in equipment fault detection and diagnosis with the help of deep learning technology. The exploration results show that the Mask R-CNN model can accurately identify and lock the fault area of power equipment, and is better than other common models in multiple performance indicators, such as Faster R-CNN and YOLOv4, and has high diagnostic accuracy and reasoning efficiency. The system is implemented using technical means such as image preprocessing, feature extraction and data enhancement, which effectively improves the robustness and reliability of the diagnosis process. In the future, with the further optimization and upgrading of infrared imaging technology and deep learning algorithms, the fault diagnosis system based on infrared images will have the ability to be applied in more complex power equipment and environments. Subsequent research can focus on optimizing the model structure, improving the reasoning rate, and relying on adversarial training and other technologies to enhance the system adaptability to meet the requirements of the ever-increasing monitoring of power systems.

Acknowledgements. This work was supported by the Scientific and Technological Research Projects of Chongqing Municipal Education Commission (KJQN202205203) and (KJZD-K202205201).

References

1. Xia, C., Ren, M., Wang, B., et al.: Infrared thermography-based diagnostics on power equipment: state-of-the-art. High Volt. **6**(3), 387–407 (2021)

2. Ou, J., Wang, J., Xue, J., et al.: Infrared image target detection of substation electrical equipment using an improved faster R-CNN. IEEE Trans. Power Deliv. **38**(1), 387–396 (2022)
3. Singh, L., Alam, A., Kumar, K.V., et al.: Design of thermal imaging-based health condition monitoring and early fault detection technique for porcelain insulators using machine learning. Environ. Technol. Innov. **24**, 102000 (2021)
4. Glovacz, A.: Fault diagnosis of electric impact drills using thermal imaging. Measurement **171**, 108815 (2021)
5. Choudhary, A., Mian, T., Fatima, S.: Convolutional neural network based bearing fault diagnosis of rotating machine using thermal images. Measurement **176**, 109196 (2021)
6. Kudelina, K., Vaimann, T., Asad, B., et al.: Trends and challenges in intelligent condition monitoring of electrical machines using machine learning. Appl. Sci. **11**(6), 2761 (2021)
7. Meng, F., Yang, S., Wang, J., et al.: Creating knowledge graph of electric power equipment faults based on BERT–BiLSTM–CRF model. J. Electr. Eng. Technol. **17**(4), 2507–2516 (2022)
8. Kurukuru, V.S.B., Haque, A., Khan, M.A., et al.: A review on artificial intelligence applications for grid-connected solar photovoltaic systems. Energies **14**(15), 4690 (2021)
9. Shakiba, F.M., Azizi, S.M., Zhou, M., et al.: Application of machine learning methods in fault detection and classification of power transmission lines: a survey. Artif. Intell. Rev. **56**(7), 5799–5836 (2023)
10. Mohammadi, E., Alizadeh, M., Asgarimoghaddam, M., et al.: A review on application of artificial intelligence techniques in microgrids. IEEE J. Emerg. Sel. Top. Ind. Electron. **3**(4), 878–890 (2022)
11. Elsisi, M., Tran, M.Q., Mahmoud, K., et al.: Effective IoT-based deep learning platform for online fault diagnosis of power transformers against cyberattacks and data uncertainties. Measurement **190**, 110686 (2022)

MEMSEL: Memory Monitoring Strategy Selection in Userspace

Lei Song, Ruibo Wang[✉], Zhenwei Wu, and Wenzhe Zhang

College of Computer Science and Technology, National University of Defense Technology,
Changsha, People's Republic of China
`{song.423,ruibo,zhenweiwu,wenzhezhang}@nudt.edu.cn`

Abstract. Memory access monitoring aims to obtain the memory access behavior of programs and has extremely wide applications in the fields of system software and architecture research. This paper deeply explores various memory access monitoring methods including–signal handling, Userfaultfd, and MPK (Memory Protection Keys), comprehensively tests the performance and scalability of each method, and deeply analyzes the advantages and disadvantages of them. Through the comparative analysis of the evaluation results, the optimal memory access load scenarios for each of the three different memory access monitoring methods are identified. Subsequently, the most suitable memory access monitoring method is selected according to different scenarios, and an innovative memory access monitoring method that integrates UFFD and signal handling is proposed to enhance the overall performance of memory access monitoring. Experimental results show that compared with single methods, the performance of the integrated UFFD and Signal-handing method can be improved by up to 72.37%.

Keywords: memory access monitoring · Intel MPK · page protection · userfaultfd

1 Introduction

With the continuous development of modern computer systems, the increasing number of research has confirmed memory is one of the core resources [18, 19], and its memory access behavior directly affects the performance and stability of programs. The main purpose of memory access monitoring technology is to reveal the running characteristics of programs, identify potential problems, and provide a basis for performance analysis and optimization, program debugging, and security protection by observing and analyzing the memory access behavior of programs during their execution. Common methods include memory access monitoring methods based on the page protection mechanism, memory access monitoring methods based on the UFFD method, and memory access monitoring methods based on the hardware characteristics of MPK.

As one of the core technologies of memory management in modern computer systems, the memory paging mechanism plays a crucial role in efficient memory utilization and system performance guarantee. Its operating principle is to divide both physical

P. Siarry et al. (Eds.): WCNA 2024, LNEE 1550, pp. 350–360, 2026.
https://doi.org/10.1007/978-981-95-6946-5_35

memory and virtual memory into pages of the same size to achieve memory management. Each virtual page is mapped to the corresponding physical page through a page table, which is stored in memory and jointly managed by hardware and the operating system. Under this architecture, programs access memory through virtual memory addresses [5–8].

The page protection mechanism is an indispensable part of the operating system's security strategy, which is generally implemented by setting protection bits in the page table entries. These protection bits define in detail the access rights of the corresponding pages. When a program attempts to access memory beyond the established rights, the processor will immediately generate an exception, and at this time, the operating system can take corresponding measures according to preset rules [9, 10]. This mechanism effectively ensures the security of memory access, strongly prevents illegal memory access behaviors among programs, and maintains the stable operation of the system.

The memory access monitoring technology based on page protection skillfully utilizes the hardware paging characteristics. By means of the page protection mechanism of the operating system, it sets the access rights of pages, triggers page faults, and realizes the memory access monitoring of programs through the handling of page faults.

This method can accurately obtain the memory access information of the monitored program without making any modifications to the monitored program or obtaining its source code. However, during the implementation of memory access monitoring based on page protection, certain performance overhead will be incurred. This part of the overhead mainly stems from the privilege-level switching caused when modifying the page access right flag bits of the page table entry (PTE) and the refresh operation of the translation look-aside buffer (TLB) [14].

The userfaultfd feature in the Linux kernel has been supported since the Linux 4.3 version. It was initially designed to support real-time VM migration [1]. The Userfaultfd technology gives user-space the ability to directly handle various memory page faults, a capability that was previously only available to kernel code. Specifically, the user-space can create and initialize userfaultfd, and register one or more virtual memory regions. Once a page fault occurs within these registered regions, relevant messages will be quickly transmitted to userfaultfd, and then the user-space will be notified. When a thread encounters a fault in this region, the kernel sends the fault information to the file descriptor and blocks the thread. The handler thread monitors the file descriptor and handles the received page faults, and wakes up the faulty thread through ioctl.

The Userfaultfd technology provides a flexible and effective solution for memory access monitoring. However, it also has inherent performance issues. First, the communication between the faulty thread and the handler thread essentially forms an IPC mechanism, which incurs additional overhead. Second, Userfaultfd becomes a bottleneck for serialization, limiting the scalability of the system [2–4].

Memory Protection Keys, as an extension of the existing page-based memory permissions, hold a unique position and value in the field of memory management [11]. The original design intention of the MPK hardware extension is to enhance the memory access security of programs. Traditional page-table-based memory permission management often incurs high system call costs and causes TLB invalidation when changing permissions. In contrast, pkeys provide a mechanism that does not require modifying the page table every time the permission is changed, greatly improving the efficiency and flexibility of permission management. Therefore, the pkeys method can be used to implement memory access monitoring [13, 15]. Once this is done, the application can easily remove the write access permission, or even all access permissions, to the marked pages just by changing the content of the register. pkeys work in conjunction with the existing PROT_READ, PROT_WRITE, and PROT_EXEC permissions passed through system calls such as mprotect and mmap, and always serve to further restrict these traditional permission mechanisms. When a process performs an access operation that violates the pkey restrictions, it will receive a SIGSEGV signal.

However, to use the pkeys feature, not only does the processor need to support this characteristic, but the kernel must also include support for it on a specific processor. Although pkeys have the potential to add an extra layer of security and reliability to applications, it is not primarily designed as a security feature. For example, WRPKRU is a completely unprivileged instruction. Therefore, in cases where an attacker can control the PKRU register or execute arbitrary instructions, pkeys will be rendered ineffective.

Overall, each of the three methods has its own advantages and disadvantages. The userfaultfd-based memory access monitoring method does not involve heavy-weight structures such as vmas during operation, thus avoiding the performance bottleneck caused by handling a large number of vmas. However, the existence of IPC leads to unsatisfactory performance and scalability. The page-protection-based memory access monitoring method does not require any modification to the monitored program or obtaining its source code. Nevertheless, this method may cause performance degradation due to the privilege-level switching triggered when modifying the page access right flag bits of the page table entry (PTE) and the refresh operation of the translation look-aside buffer (TLB). MPK (Memory Protection Keys) provides user-mode access permission configuration through hardware, reducing system calls and thus optimizing the signal processing overhead. However, this method is dependent on hardware features.

The objectives of this paper are as follows: thoroughly test the performance and scalability of the three memory access monitoring methods, determine the optimal scenarios for each method based on the test results, integrate multiple methods to monitor the memory access of the same program, so as to take advantage of each method and maximize the performance of memory access monitoring.

2 Background

2.1 Memory Access Monitoring Method Based on Page Protection Mechanism

During the operation of a computer system, page faults are a common phenomenon. When the physical page corresponding to the virtual address accessed by a program is not in memory, or the access rights do not meet the requirements, a page fault will be triggered. The page fault handling mechanism is designed to deal with this situation. The memory access monitoring technology based on page protection skillfully utilizes the hardware paging characteristics, and by means of the page protection mechanism of the operating system, sets the access rights of pages, triggers page faults, and realizes the memory access monitoring of programs through the handling of page faults [17].

Specifically, as shown in Fig. 1 below, the access rights of the target virtual memory area are pre-set to non-accessible or read-only. This makes it impossible for the read-write access or write access of the monitored program to the target virtual memory area to pass the memory access right check of the MMU, resulting in a hardware interrupt. At the same time, the kernel sends a segmentation fault signal SIGSEGV to the process that triggered the exception. Taking memory write access monitoring as an example, its basic principle is to set write protection for the target memory area, that is, configure the access rights of this area as read-only. In this way, when the monitored program issues a write request, an access rights violation will be triggered. Subsequently, in the custom-defined SIGSEGV signal handling function, the memory address that the monitored program attempts to write to can be obtained. After successfully recording the address of the accessed memory page, the SIGSEGV signal handling function can lift the write protection of this memory page by calling the mprotect system call, ensuring that after the signal handling function returns, the interrupted write access operation can resume execution. Similarly, if the access rights of the target memory area are set to non-accessible, any read-write request initiated by the monitored program to this area will trigger an access rights violation exception, thereby achieving comprehensive memory read-write access monitoring.

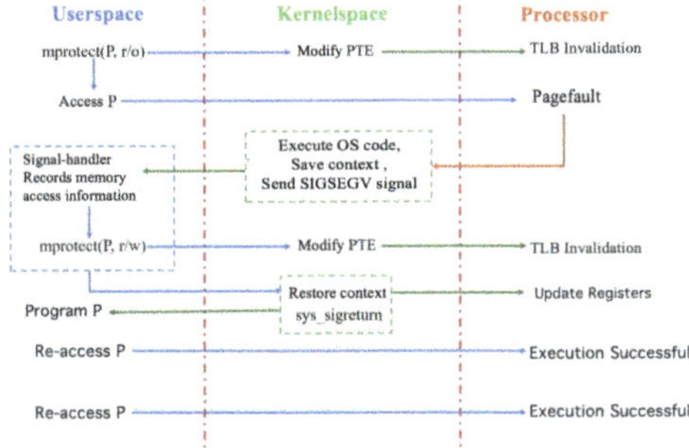

Fig. 1. Flowchart of Memory Access Monitoring Based on Signal-handling Mechanism

2.2 Memory Access Monitoring Method Based on UFFD Technology

Userfaultfd was initially designed to provide technical support for the real-time migration of VMs (Virtual Machines). During the migration of virtual machines, UFFD plays a crucial role. It can effectively handle memory-related transactions, strongly guarantee the consistency and integrity of memory data during the migration process, and thus achieve smooth virtual machine migration. An important feature of UFFD is to give the user-space the ability to control various memory page faults, which lays a solid foundation for memory access monitoring [12]. The specific process is as follows: after creating and initializing UFFD in the user-space, one or more virtual memory regions are registered. When a page fault occurs within the registered regions, the corresponding message will be immediately transmitted to UFFD, and then the user-space will be notified.

Specifically, as shown in Fig. 2, the application first selects the memory area to be monitored and submits core information such as its starting address and range size to the kernel. After successfully completing the registration process, it obtains the file descriptor userfaultfd for subsequent communication. While using ioctl to monitor a specific memory area, a polling thread, namely the uffd monitor thread, needs to be specially started. This thread continuously polls using the poll() function until a page-fault exception occurs. When a thread triggers a page-fault exception within this memory area, the faulting thread enters the kernel to handle the exception, and the kernel then calls the handle_userfault() function to hand it over to userfaultfd for processing. Thereafter, the faulting thread enters a blocked state and sends a uffd_msg to the monitor thread, waiting for the processing to be completed. The monitor thread calls ioctl to handle the page-fault exception, for example, performing the UFFDIO_COPY operation to copy user-defined data to the faulting page. After the processing is completed, the monitor thread sends a signal to wake up the faulting thread to continue running, thus completing a full cycle of uffd memory access monitoring.

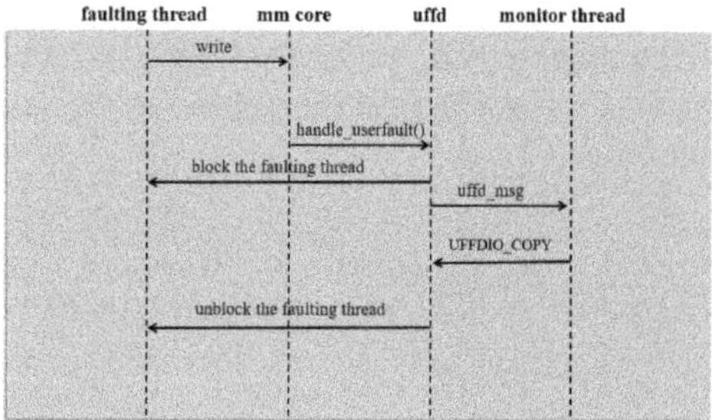

Fig. 2. Flowchart of Memory Access Monitoring Based on UFFD Technology

2.3 Memory Access Monitoring Method Based on MPK Hardware Characteristics

Intel has innovatively introduced a hardware feature named MPK (Memory Protection Keys) [11] on the 64-bit CPUs of the new-generation x86–64 architecture. The initial design goal of this feature focuses on achieving lightweight memory isolation in user mode. MPK adopts a unique group-level permission control strategy, which is specifically accomplished by marking memory pages with protection keys (pkeys). Its implementation principle is to use 4 bits that were originally unused in the page table entry (PTE) to store memory protection keys. Based on this, up to 16 different page groups can be constructed. This mechanism is in sharp contrast to the traditional page protection mechanism. In the traditional mechanism, during the process of permission changes, high-performance-consuming system calls are often required, and problems such as TLB invalidation will be triggered. However, MPK has the advantage of lightweight permission switching, which can significantly improve operation efficiency.

In the CPU architecture that supports MPK, each core is equipped with a 32-bit user-space register, namely PKRU, whose structure is shown in Fig. 3. These 32 bits are divided into 16 pairs in sequence, and each pair of bits represents a unique key value. Each key contains two key permission control bits: the WD bit (Write Disable) and the AD bit (Access Disable). When the WD bit is set to 1, the corresponding memory area will be prohibited from write operations; when the AD bit is set to 1, this memory area cannot be accessed. It is worth noting that PKRU is a user-mode register. When a user program reads and writes it, it only needs to execute the non-privileged RDPKRU/WRPKRU instructions, without initiating a system call. This feature effectively avoids the overhead caused by context switching. It is estimated that its operation overhead is only about 20 cycles. In addition, the protection keys in the page table have a one-to-one correspondence with the 16 keys in the PKRU register, ensuring precise control of memory access permissions.

When a page protected by MPK is accessed, the system will automatically trigger the Page Fault mechanism. In the operating environment of the Intel x86–64 architecture, once the exception is triggered, the CPU will quickly push the error code related to the page exception onto the stack. The structure of the error code is shown in Fig. 4. The error code is 32 bits in total, and the fifth bit is the PK bit closely related to MPK. When the page exception is caused by the access or modification of an MPK-protected page, the fifth PK bit of the error code will be automatically set to 1. The system-built-in page exception handling function can accurately determine whether the exception is related to MPK according to the status of the PK bit in the error code, and take corresponding handling measures accordingly.

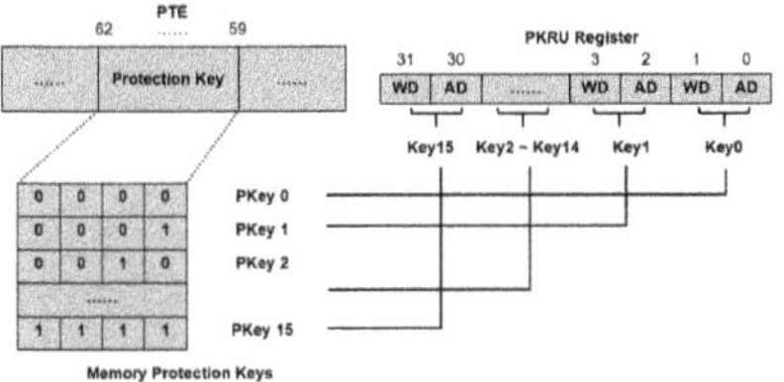

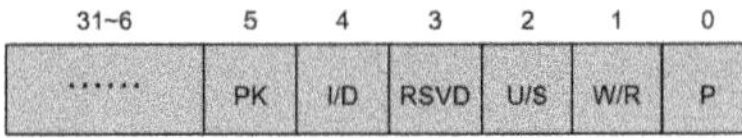

Fig. 4. Schematic Diagram of Error Codes

Fig. 3. Schematic Diagram of MPK

The specific operation process is as shown in Fig. 5. First, a memory protection key value pkey needs to be applied for from the operating system kernel. Then, with the pkey_mprotect system call, the pkey is marked on the page table entry of the memory area planned to be monitored. Virtual memory pages with the same pkey will naturally form a memory page group, simply referred to as the pkey group. After the binding operation between the pkey and the virtual page is completed, by executing the WRPKRU instruction, the access-permission-description bits corresponding to the pkey in the PKRU register can be modified, thus achieving restrictions on the monitored program's access to the pkey group, that is, imposing page protection measures. When the monitored program attempts to access the pkey group, a page fault will be triggered, causing the program execution to be interrupted. In response to this situation, the operating system kernel will immediately send a SIGSEGV signal to the monitored program and store the pkey value corresponding to the accessed pkey group in the signal stack. In the SIGSEGV signal handling function designed in this paper, the pkey corresponding to the pkey group that the monitored program is trying to access can be extracted from the signal stack. By adjusting the access-permission control bits corresponding to the pkey in the PKRU register saved in the interrupted scene, the access restrictions on the pkey group are lifted, enabling the monitored program to resume accessing the pkey group. When the program accesses the Pkey group again, since the access permissions have been adjusted, no page fault will be triggered, thus successfully implementing the memory access monitoring function for the program.

3 MEMSEL: Design and Experimental Evaluation

3.1 Experimental Environment

In this paper, MEMSEL is implemented on the Linux kernel version 6.8.0-48. We conduct our experiments on an Intel i7-13700KF [16] machine. This machine is equipped with 8 performance cores and 8 energy-efficient cores, has a base frequency of 3.4GHz, a total of 24 threads, a maximum turbo frequency of 5.4GHz, 64GB of DDR5 RAM, and a 30MB L3 cache. Our evaluation focuses on answering the following two research questions:

- Q1: Test the performance of three different memory access monitoring methods and their scalability under multi-threading.
- Q2: Compare the memory access monitoring methods of Signal-handling and UFFD, and test how the performance of the integrated memory access monitoring method of the two is specifically improved.

3.2 Experimental Design

In this paper, five different memory access monitoring methods are first implemented, namely: 1) Memory access monitoring method based on the page protection mechanism. 2) Memory access monitoring method based on UFFD technology. 3) Memory access monitoring method based on the hardware characteristics of MPK. 4) Memory access monitoring method based on Signal-handing + UFFD.

In the experiments of this paper, the number of user-defined program pages is first used to simulate the memory access frequency of the program. By applying for read-only pages and performing write operations on these pages, the memory access behavior of the program to memory objects can be monitored. The number of pages ranges from 2, 4, 8, 16, and 2097152, to observe the impact of the program's memory access load on the memory access monitoring methods. Then, 2, 4, 8, 16, and 32 threads are used to analyze the scalability of the first three memory access monitoring methods. Finally, the performance comparison between the last two methods and the single-method memory access monitoring is tested in scenarios where the number of pages is 2, 64, 2048, 65536, and 2097152. In addition, all the data in this experiment are the average values of 5 executions.

3.3 Performance Evaluation

1) Performance Evaluation of Three Memory Access Monitoring Methods

As can be seen from Fig. 6, the MPK method performs the most stably and optimally overall, with the latency always remaining below 15 µs. This is due to the fact that the hardware provides user-mode access permission configuration. To impose/remove page protection (modify the access permissions of page groups), it only requires configuring the corresponding permission bits in the user-mode PKRU register. Configuring the PKRU register is achieved through the processor's WRP-KRU user-mode instruction, resulting in low overhead. In contrast, under the traditional MMU-provided hardware page protection mechanism, modifying page access permissions requires switching to the privileged mode to modify the page table (i.e., execute the mprotect system call), incurring a significant overhead.

Secondly, the overall performance latency of the Signal-handing method remains stable at 25–30 μs, which is weaker than that of the MPK-based method. This is because in this memory access monitoring method, the mprotect operation is required to modify the page access permissions. This leads to frequent context switches, incurring significant overhead and thus causing a decline in performance.

The overall average page memory access latency of the UFFD method shows a downward trend as the number of program memory accesses increases. Compared with the conventional mremap/mprotect virtual memory management methods, the UFFD memory access monitoring method does not involve heavy-weight structures such as vmas during operation. This avoids the performance bottleneck caused by handling a large number of vmas, making it suitable for monitoring memory accesses in the GB and TB levels. However, for a small number of memory accesses, the associated initialization and IPC lead to poor performance.

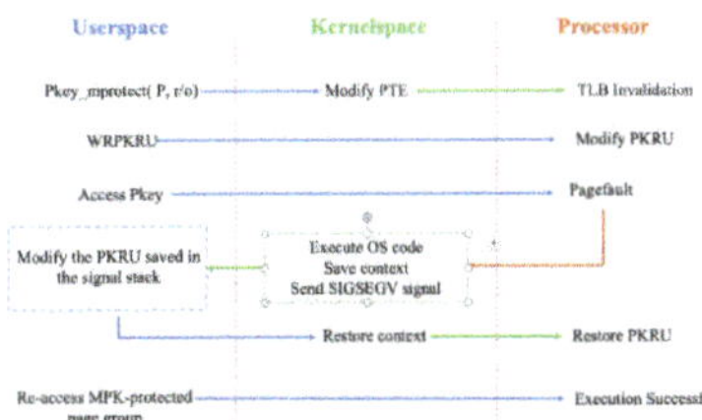

Fig. 5. Flowchart of Memory Access Monitoring Based on MPK Hardware Characteristics

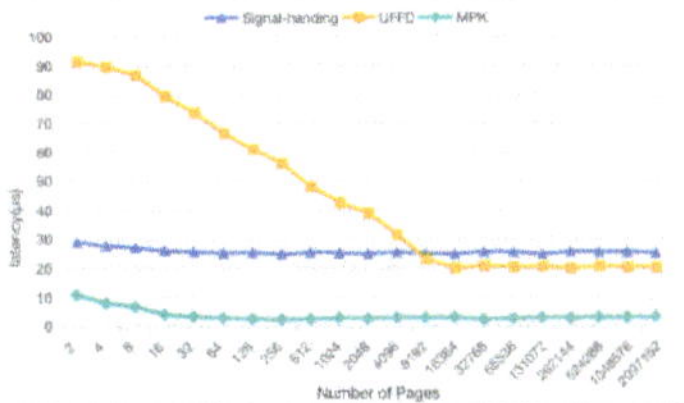

Fig. 6. Average PF handing latency Query ID="Q3" Text="Caption of Figs. 6 and 8 seems to be identical. Please check and correct if necessary."

2) Scalability Analysis of the Three Methods

Figure 7 shows the performance of three different memory access monitoring methods, namely Signal-handing, UFFD, and MPK, under different numbers of threads. Specifically, we analyze the scalability of these three memory access monitoring methods by using 2, 4, 8, 16, and 32 threads for execution. We run a fixed number of operations, set the number of pages to 4096, measure the latency of each memory access achieved by all threads, and calculate the average page-fault latency.

Among them, due to the round-trip IPC in the fault-handling path, the UFFD memory access monitoring method has poor performance scalability. In contrast, the other two memory access monitoring methods, Signal-handing and MPK, show better scalability. The latter is an optimization of the former's signal processing, so the overall performance of the latter is better. The former is based on modifying the read-write permissions of pages, and its scalability will be affected by context switching. The MPK method relies on a specific register, PKRU, to control page access permissions. When the number of threads increases, multiple threads may compete for access to these registers, resulting in a decrease in scalability.

3.4 Design and Performance Evaluation of Integrated Memory Access Monitoring Methods

1) Design of Integrated Memory Access Monitoring Methods

As can be seen from Fig. 8, when the number of pages reaches 8192, the performance of the UFFD and Signal-handling memory access monitoring methods is nearly the same. When the number of pages is less than 8192, the Signal-handling method outperforms UFFD; when the number of pages is greater than 8192, the UFFD method is superior to Signal-handling.

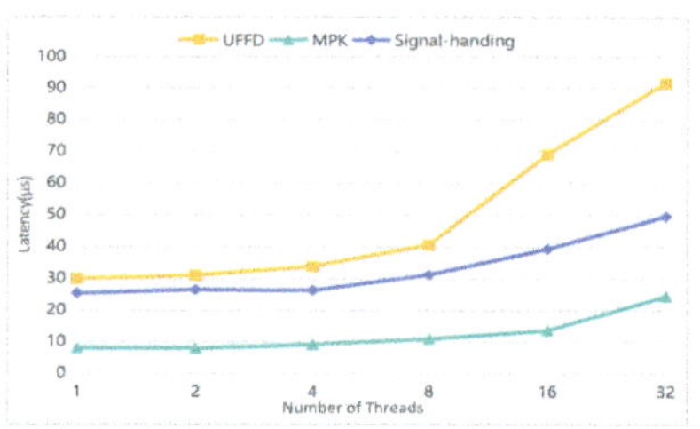

Fig. 7. Average PF handing latency under different degrees of concurrency.

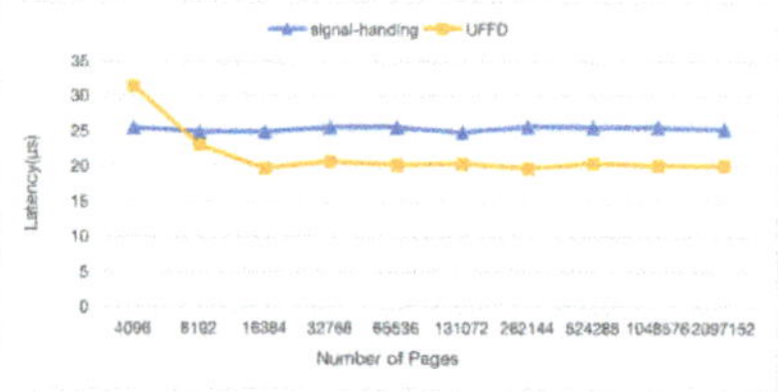

Fig. 8. Average PF handing latency

Therefore, based on the above experimental analysis, we select 8192 as the dividing point for the number of pages and integrate the two memory access monitoring methods of UFFD and Signal-handling. When the number of memory-accessed pages in the program is greater than this point, the UFFD memory access method is used; when it is less than this point, the Signal-handling memory access monitoring method is adopted.

Figure 9 shows the performance graphs of the UFFD memory access monitoring method, the Signal-handling memory access monitoring method, and the UFFD + Signal-handling memory access monitoring method.

2) Performance Analysis of Integrated Memory Access Monitoring Methods

Figure 10 presents the performance comparison between the UFFD method, the Signal-handling method, and the UFFD + Signal-handling integrated method under scenarios where the number of pages is 2, 64, 2048, 65536, and 2097152. By measuring the average time consumption for memory access of each page, it is calculated that the UFFD + Signal-handling integrated method improves performance by 72.37%, 62.04%, and 34.66% compared to the single UFFD method in scenarios with 2, 64, and 2048 pages. In scenarios with 65536 and 2097152 pages, the performance is improved by 17.81% and 20.23% respectively compared to the single Signal-handling method.

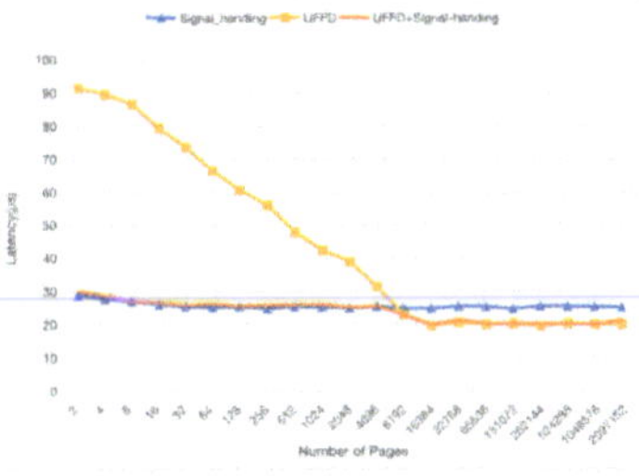

Fig. 9. Average PF handing latency

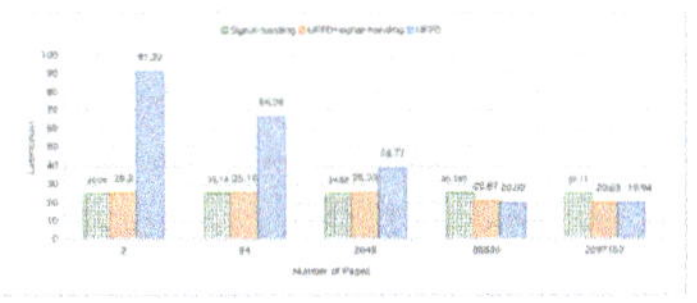

Fig. 10. Average PF handing latency

4 Conclusion

This paper conducts an in-depth study on memory access monitoring methods. It explores various memory access monitoring methods based on the Signal-handing, UFFD, and MPK mechanisms, comprehensively analyzing their performance, scalability, advantages, and disadvantages. The performance of the three methods under different memory access loads and multi-threading scenarios is tested to determine the optimal memory access load scenarios for each method. On this basis, this paper proposes integrated memory access monitoring methods of UFFD + Signal-handing. Experimental data shows that compared with single methods, the UFFD + Signal-handing integrated method can achieve a maximum performance improvement of 72.37%.

References

1. "Userfaultfd Kernel Documentation"[Online]. Linux Kernel Documentation. [Accessed 14 Jan. 2025]. https://www.kernel.org/doc/html/latest/admin-guide/mm/userfaultfd.html
2. Caldwell, B., Goodarzy, S., Ha, S., Han, R., Keller, E., Rozner, E., Im, Y.: Fluidmem: full, flexible, and fast memory disaggregation for the cloud. In: 2020 IEEE 40th International Conference on Distributed Computing Systems (ICDCS), pp. 665–677 (2020)
3. Raybuck, A., Stamler, T., Zhang, W., Erez, M., Peter, S.: Hemem: Scalable tiered memory management for big data applications and real nvm. In: Proceedings of the ACM SIGOPS 28th Symposium on Operating Systems Principles, pp. 392–407 (2021)
4. Wang, C., Qiao, Y., Ma, H., Liu, S., Chen, W., Netravali, R., Kim, M., Xu, G.H.: Canvas: isolated and adaptive swapping for Multi-Applications on remote memory. In: 20th USENIX Symposium on Networked Systems Design and Implementation (NSDI 23), pp. 161–179 (2023)
5. Daley, R.C., Dennis, J.B.: Virtual memory, processes, and sharing in Multics. Assoc. Comput. Mach. **11**(5), 306–312 (1968)
6. Rashid, R., Tevanian, A., Younget, M., Golub, D., Baron, R., Black, D.: Machine-independent virtual memory management for paged uniprocessor and multiprocessor architectures. IEEE Trans. Comput. **37**(8), 896–908 (1988)
7. Tomasulo, R.M.: An efficient algorithm for exploiting multiple arithmetic units. IBM J. Res. Dev. **11**(1), 25–33 (1967)
8. Villavieja, C., Karakostas, V., Vilanova, L.: DiDi: Mitigating the performance impact of TLB shootdowns using a shared TLB directory. In: 2011 International Conference on Parallel Architectures and Compilation Techniques, pp. 340–349 (2011)

9. Saltzer, J.H.: Protection and control of information sharing in multics. In: Proceedings of the fourth ACM symposium on Operating system principles (SOSP '73), p. 119 (1973)
10. Witchel, E., Cates, J., Asanović, K.: Mondrian memory protection. ACM SIGPLAN Notices **37**(10), 304–316 (2002)
11. Park, S., Lee, S., Xu, W., Moon, H., Kim, T.: libmpk: Software Abstraction for Intel Memory Protection Keys. Preprint, 2018
12. Sepehr, J., Shaurya, P., Milad Rezaei, H., Margo, S., Alexandra, F.: ExtMem: enabling application-aware virtual memory management for data-intensive applications. In: Proceedings of the 2024 USENIX Annual Technical Conference, ATC 2024, pp. 397–408 (2024)
13. Wang, R.B., Wu, Z.W., Zhang, W.Z., Wu, H.J., Zhang, Y.S.Q., Lu, K.: Fine-grained memory access monitoring based on memory protection keys. Comput. Eng. Sc. **46**(1), 21–27 (2024)
14. Wu, Z.W., Lu, K., Nisbet, A., Zhang, W.Z., Lujan, M.: PMThreads: persistent memory threads harnessing versioned shadow copies. In: ACM-SIGPLAN Symposium on Programming Language Design and Implementation (2020)
15. Zhang, Y.S.Q., Lu, K., Wu, Z.W., Zhang, W.Z.: PMemTrace: lightweight and efficient memory access monitoring for persistent memory. In: Algorithms and Architectures for Parallel Processing (2022)
16. "Intel® Core™ i7–13700KF Processor-30M Cache, Up to 5.40 GHz-Specifications" [Online]. Intel China Official Website, [Accessed 14 Jan. 2025]. https://www.intel.cn/content/www/cn/zh/products/sku/230489/intel-core-i713700kf-processor-30m-cache-up-to-5-40-ghz/specifications.html
17. Lu, K., Zhang, W.Z., Wang, X.P., Lujan, M., Nisbet, A.: Flexible page-level memory access monitoring based on virtualization hardware. ACM SIGPLAN NOTICES **52**(7), 201–213 (2017)
18. Gao, W.R., Fang, J.B., Huang, C., Xu, C.F., Wang, Z.: WrBench: comparing cache architectures and coherency protocols on ARMv8 many-core systems. J. Comput. Sci. Technol. **38**(6), 1323–1338 (2023)
19. Fang, J.B., Liao, X.K., Huang, C., Dong, D.Z.: Performance evaluation of memory-centric ARMv8 many-core architectures: a case study with phytium 2000+. J. Comput. Sci. Technol. **36**(1), 33–43 (2021)

Research on Medical Image Analysis Based on 5G and Deep Learning

Hong Liang, Jipeng Sun, Yong Fang, and Zhengbo Zhang[(✉)]

Department of Medical Innovation and Research, Chinese PLA General Hospital, Beijing, China
zhengbozhang@126.com

Abstract. With the continuous improvement of medical digitalization, new technologies such as 5G and artificial intelligence are more and more widely used in the medical industry, especially deep learning technology has shown great application potential in medical image analysis. This paper proposes a medical image analysis scheme based on 5G and artificial intelligence technology, which can realize the collection, accurate analysis, rapid transmission, and processing of complex medical images through the four-layer architecture. The architecture include data acquisition and edge computing layer, cloud computing and data management layer, intelligent analysis and deep learning layer, application and decision support layer, so as to improve the efficiency and accuracy of medical services and promote the remote and personalized medical services.

Keywords: 5G · artificial intelligence · deep learning · medical imaging

1 Introduction

At present, China's medical resource supply is still facing challenges, which can not fully meeting the current demand, and still falling short of the average of most developed countries. Faced with this situation, the medical industry is actively exploring innovative solutions to improve the quality of services.

In this context, medical industry is working on the application of new technologies, such as mobile health, to transform from a traditional treatment model to a more comprehensive health management model. It is expected that in the future, smart healthcare and smart hospitals will benefit from the development of 5G technology [1–3]. With its high speed, low latency, strong mobility, and big data analysis capabilities, 5G technology will enable everyone to enjoy convenient and timely smart medical services and meet people's new expectations for the future of healthcare, including telemedicine, remote first aid, remote clinics, smart operating rooms, smart wards and smart guidance [4–7].

Medical image analysis plays a vital role in modern medicine, which is an important basis for disease diagnosis and treatment decisions. With the proliferation of medical imaging data, traditional methods of transmission and processing are challenging. As a new generation of communication standards, 5G technology has brought revolutionary changes to the medical field [8]. With its high-speed bandwidth, low latency, and wide range of connectivity capabilities, 5G technology provides unprecedented speed and

P. Siarry et al. (Eds.): WCNA 2024, LNEE 1550, pp. 361–370, 2026.
https://doi.org/10.1007/978-981-95-6946-5_36

efficiency for the transmission of medical image data. 5G enables medical images to be transmitted from medical devices to remote servers or doctors in near realtime, providing critical support for emergency and telemedicine. At the same time, low latency ensures that physicians can quickly acquire and analyze patients' medical images, speeding up the diagnosis and treatment process. The application of 5G technology indicates a qualitative leap forward in the field of medical imaging, providing more efficient and innovative solutions for medical services.

With the advent of the 5G era, medical imaging data is no longer limited by geographical location, and telemedicine has been greatly expanded. Through 5G networks, doctors will be able to access patients' medical images in real time, making remote diagnosis a reality, which is of great significance for patients in remote areas, emergency medical care, and outbreak response.

This paper focuses on the synergistic effect of 5G technology and artificial intelligence (AI) in medical image analysis, aiming to improve the transmission speed, accuracy and intelligence of medical services [9–11]. AI technology can quickly analyze a large amount of medical imaging data, such as X-rays, CT and MRI scans, through deep learning algorithms [12–15], to assist doctors in disease diagnosis. AI technology significantly improves the accuracy and efficiency of diagnosis, especially in tumor detection, where AI can help doctors identify and distinguish between malignant and benign tumors to guide further treatment plans [16–18]. Combined with the high-speed bandwidth and low latency characteristics of 5G technology, the application of AI in medical imaging will become more intelligent, achieving more real-time and efficient diagnosis and treatment support [19–21].

2   System Architecture

This paper proposes a four-layer architecture which integrates 5G, cloud computing, edge computing, and artificial intelligence technologies to form a complete process from data acquisition to intelligent analysis, to application and decision support, aiming to improve the efficiency and accuracy of medical image analysis, while using 5G technology to achieve fast data transmission and telemedicine services, as shown in Fig. 1.

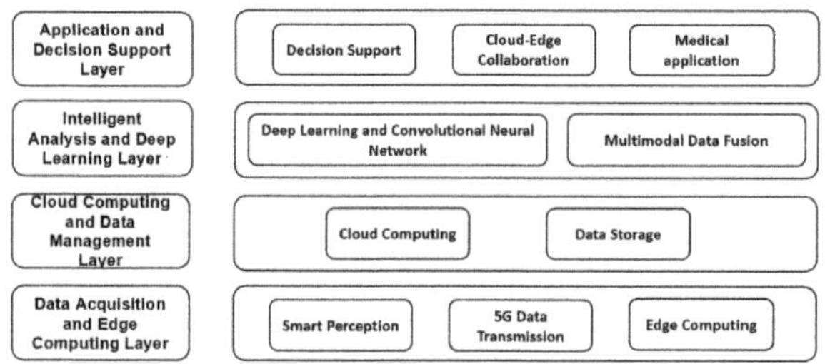

Fig. 1. System Architecture

2.1 Data Acquisition and Edge Computing Layer

Data is sensed and collected by medical devices, and data transmission is completed through 5G technology. At the same time, it can also use edge computing capabilities to provide value-added services for data. It mainly includes the following functions:

Smart sensing: It is composed of medical perception equipment and data acquisition equipment, which is responsible for the perception and collection of medical data.

Data transmission: Through 5G technology, the data collected by the intelligent perception layer is transmitted at high speed.

Edge computing: Edge computing technology provides value-added capabilities on the CT side and IT side, including AI, video encoding and decoding, IoTPaaS, etc., and provides characteristic application capabilities according to the needs of different medical scenarios.

2.2 Cloud Computing and Data Management Layer

This layer mainly implements data storage and processing, as well as device-cloud collaboration with medical terminal equipment, and its main functions are as follows:

Cloud computing: The function of processing the integrated data transmitted by the data transmission into the information required by doctors through intensive computing. For example, neuronal algorithms, artificial intelligence, machine learning, big data processing, etc., are used to process user data.

Data storage: Inn the cloud computing environment, build a virtual database to store all kinds of medical information.

2.3 Intelligent Analysis and Deep Learning Layer

This layer mainly uses various deep learning technologies to process medical images, and at the same time fuses multi-modal medical data for further fusion analysis, and the main functions are as follows:

Deep learning and convolutional neural networks (CNN): Deep learning technologies, including LSTM(Long Short-Term Memory) [22], ResNet (Residual Network) [23], CNN [24, 25],etc., are used to extract features from medical images for classification, detection, segmentation and other tasks.

Multimodal data fusion: Multiple types of medical data fusion (such as images, sounds, and physiological signals) into a whole to improve diagnostic accuracy and reduce the rate of misdiagnosis.

2.4 Application and Decision Support Layer

Based on the information provided by the previous three layers, this layer can provide decision support for various medical applications, and also provide interfaces for external services in order to achieve more medical service collaboration.

Medical application: Combined with specific needs of the medical industry, to achieve intelligence process, mainly used to solve problems such as information processing and human-computer interaction.

Decision support system: Based on the results of AI analysis, it provides diagnostic suggestions and treatment solutions to assist doctors in decision-making.

Cloud-edge Collaboration: It supports collaboration interfaces with enterprise private clouds, industry clouds, and Internet public clouds, and cooperates with integrated services to integrate computing, network, storage, platform capabilities, and ecological applications.

3 Key Technology Application and Analysis

3.1 Cloud Computing

To improve the efficiency of data processing and reduce the burden on terminal equipment, this paper uses cloud computing technology to realize remote collaborative analysis, as follows:

- Cloud data storage: medical image data is stored in a cloud data center to ensure data security and reliability. To prevent data loss and ensure the long-term availability of medical data through distributed storage and redundant backup strategies. Advanced data encryption technology is used to ensure that medical image data is secure during storage and transmission in the cloud. At the same time, privacy protection measures have been implemented and relevant regulations and norms are followed to ensure that patient privacy is protected.
- Cloud data processing: efficient processing of medical image data is achieved by using the large-scale data processing capability of the cloud computing platform. Parallel computing and distributed computing technologies are used to accelerate the analysis of large-scale data. By using cloud GPU resources for image processing tasks, the processing speed of medical images is improved, and faster diagnosis is realized.
- Cloud collaborative analysis: Support multiple users to access cloud data at the same time to achieve collaborative work between doctors. This helps remote doctors to participate in the analysis and diagnosis of medical images, and improves the professionalism and accuracy of medical decision-making. By introducing real-time collaborative editing technology, doctors are able to share, annotate, and modify medical images in real time, enhancing communication between physicians and facilitating remote teamwork.

With construction of cloud computing technology, the smart medical care realizes the unified storage and processing of medical data and promotes the cross-domain sharing of medical data. By deploying medical service applications in the cloud, it can achieve dynamic and elastic expansion according to the number of visits, avoiding the problem of medical services being unable to run due to excessive system load. In addition, the application of cloud services can also effectively save a lot of hardware resource procurement and operation and maintenance costs, and reduce the physical redundancy of hardware. In the event of natural disasters, hacker intrusion and other unexpected situations, the powerful disaster recovery capabilities of cloud computing technology can effectively protect medical data stored in the cloud.

3.2 Edge Computing

Edge computing is a distributed information technology architecture in which client data is processed at the perimeter of the network, as close to the origin of the data as possible. The development of this architecture has been driven by mobile computing, with the reduction in the cost of computer components and the increase in the number of networked devices in the Internet of Things.Depending on how the edge computing architecture is implemented, time-sensitive data can be processed by a smart device at the origin or sent to an intermediate server which is geographically closest to the client for processing. For less time-sensitive data, it can be sent to the cloud for historical analysis, big data analysis, and long-term storage.

The combination of 5G and edge computing provides richer capacity, flexibility, and functionality for information service offerings. The ultra-low latency and ultra-high reliability of 5G communication make edge computing an inevitable choice. The combination of the two technologies is mutually beneficial, with the adoption of 5G and edge computing, meeting higher user expectations in terms of low latency and always-on connectivity, while edge computing provides low-latency support for 5G, which in turn provides faster speeds for edge computing.

In the healthcare field, the rapid response of data is critical to the quality of service. Edge computing enables sensors in healthcare to respond quickly, contributing to more convenient diagnosis in telemedicine. At the same time, collaboration at the edge is also very important in healthcare. For example, when the flu breaks out, the hospital acts as a fringe node to share data with multiple nodes such as pharmacies, pharmaceutical companies, governments, and the insurance industry, and shares information such as the number of infected people with the current flu, the symptoms of the flu, and the cost of treating the flu. Such collaboration can lead to a more effective response to the health crisis. This integrated healthcare information network architecture not only improves data processing speed and response time, but also strengthens collaboration between different medical institutions and related fields forming an efficient and collaborative healthcare ecosystem.

3.3 AI Technology

In the field of medical image analysis, encoder-decoder technology is employed, which is a deep learning method that combines CNN and LSTM. This unique combination is designed to realize a more comprehensive analysis and understanding of medical image data by taking into account both spatial and temporal series information when processing medical images.

Convolutional neural network excel in image processing, and are able to effectively capture spatial features of images. Through the operation of convolutional layers, CNN can extract edges, textures and other local features in the image, so as to accurately capture spatial information. In medical image analysis, CNN can effectively identify the characteristics of the lesion area and provide a basis for the diagnosis of the disease [26, 27]. However, in medical image analysis, it is necessary to pay attention not only to the spatial structure of still images, but also to take into account the dynamic changes in time.

In order to cope with the time series information in medical images, the LSTM is introduced [28–30], which is a variant of the Recurrent Neural Network (RNN) suitable for processing sequence data, which is designed to better capture and memorize long-term dependencies in sequences. This network structure allows the model to better understand the temporal evolution process in medical image data, providing more accurate modeling and prediction of the dynamic changes of diseases. In medical image analysis, LSTM can be used to analyze dynamic medical images, such as MRI sequences, as well as to process image-related text data, such as radiology reports.

The adoption of CNN and LSTM technology makes medical image analysis more comprehensive and in-depth. By fusing sensitivity to spatial and temporal information, this deep learning method provides a more refined and detailed image interpretation for the medical field, helping to improve the accuracy and timeliness of disease diagnosis. The potential of this technology lies in providing medical professionals with more comprehensive image information to support personalized treatment and health management of patients.

Residual Network, is a deep convolutional neural network structure proposed by Kaiming He in 2015. The design concept of ResNet is to solve the common problems of gradient vanishing and gradient explosion in deep neural networks, thus making it easier and more stable to train deeper neural networks. In medical image analysis, ResNet has been widely used in the clinical auxiliary diagnosis of major diseases such as lung tumors, breast cancer, skin diseases, and cardiovascular and cerebrovascular diseases, and has achieved good results.

The advantages of ResNet are mainly reflected in the following aspects: first, it has a strong horizontal connection, which promotes the transfer of information and the backpropagation of gradients through skip connections; Secondly, ResNet can train very deep neural networks, which effectively solves the problems of gradient vanishing and gradient explosion. Furthermore, ResNet has achieved significant performance improvements in multiple tasks such as image classification and object detection. In this paper, ResNet is applied to process medical images to extract the feature information of the images. These features are then fed into the LSTM network for further diagnostic analysis. By introducing ResNet, the design scheme can better capture the complex features in medical images, provide richer and more accurate information for the model, and thus enhance the performance and reliability of medical image analysis.

3.4 System Flow

The collaborative scheme of 5G technology and deep learning in medical image analysis proposed is shown in Fig. 2. Firstly, data is collected through medical devices, and then data preprocessing, including denoising and enhancement. Secondly, the 5G network is used to realize the high-speed transmission of image data to facilitate the initial processing on edge devices. The processed data is stored in the cloud to support further analysis. In the intelligent analysis stage, artificial intelligence technology is used, the ResNet model is used for feature extraction to identify key information in the image, and then the LSTM network is applied to classify, detect and diagnose the image. As a result, the analysis results are used to assist medical decision-making and improve service efficiency, and are transmitted to other medical applications through interfaces.

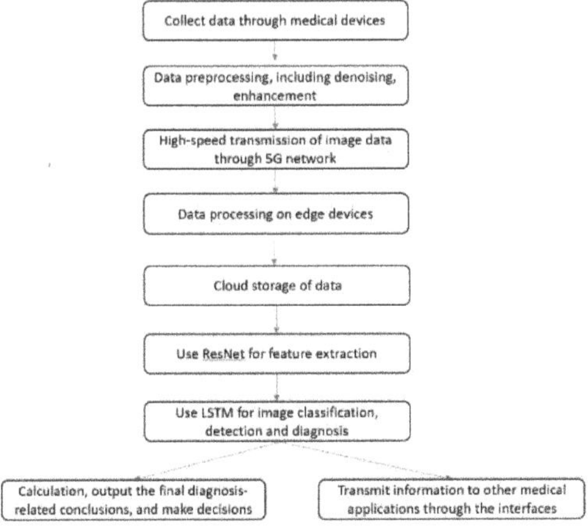

Fig. 2. The system flow of medical image process

ResNet is used to process the medical images, extract the feature information, and then pass the extracted feature information to the LSTM network as input. The LSTM network will analyze the dynamic changes of these features in the time dimension based on its advantages in sequence data processing, and finally assist in the diagnosis and analysis of diseases.

The formula used in ResNet is as follows:

$$X_{out}(i,j,k) = \sum_{m=0}^{k_h-1} \sum_{n=0}^{k_w-1} \sum_{l=0}^{c_{in}-1} X_{in}(i+m, j+n, l) \times K(m, n, l, k) \tag{1}$$

where X_{out} is the output feature image, X_{in} is the input image feature image, (i,j) represents a certain position, k represents the index of the output channel. k_h stands for the height of the convolution kernel, k_w stands for the width of the convolution kernel), c_{in} represents the number of channels. m is the height dimension of the convolution kernel K, n is the width dimension of the convolution kernel K, l is index variable for both the input feature image X_{in} and the convolution kernel K in the channel dimension.

In dynamic medical imaging, the feature images corresponding to different time points are arranged in chronological order, to form a feature sequence as the input of the LSTM network. The core unit in the LSTM network consists of input gates, forgotten gates, and output gates. The core update formula is as follows:

Forgotten gates use the follow formula:

$$f_t = \delta(W_f \bullet [h_{t-1}, x_t] + b_f) \tag{2}$$

where, δ is an activation function (such as the sigmoid function), W_f is the weight matrix corresponding to the forget gate, h_{t-1} is the hidden state at the previous time step, x_t is the input at the current time step, and b_f is the bias term.

Input gates use the formula bellow:

$$i_t = \sigma(W_i \bullet [h_{t-1}, x_t] + b_i) \tag{3}$$

$$\tilde{C_t}= \tanh(W_c \bullet \left[h_{t-1}, x_t\right] + b_c) \tag{4}$$

where $\tilde{C_t}$ is the candidate cell state at the current time step, W_i and W_c are the corresponding weight matrices, b_i and b_c are the bias terms.

State Update uses formula (5):

$$C_t = f_t \times C_{t-1} + i_t \times \tilde{C_t} \tag{5}$$

where C_t is current state, C_{t-1} is the previous state.

Output Gate uses formula (6) and (7):

$$o_t = \sigma(W_o \bullet \left[h_{t-1}, x_t\right] + b_o) \tag{6}$$

$$h_t = o_t \times tanh(C_t) \tag{7}$$

where, h_t is the hidden state output at the current time step, which will participate in subsequent calculations and output the final analysis results. W_o and b_o are the weight matrix and bias term corresponding to the output gate.

By continuously updating the state and hidden state on the time series, the LSTM can capture the long-term dependence of the feature sequence in the time dimension, and when processing the sequence data, the update of key components including the forgetting gate, input gate, state and output gate, that is, to understand the evolution of medical image features over time, such as the development trend of the lesion area, to help the model capture the change rule of image features in the time dimension, and finally outputs the result based on the analysis of the entire feature sequence. This result contains information about the temporal dynamics, which can be used for the subsequent more accurate diagnosis of the disease.

4 Conclusion

By combining 5G communication technology with advanced medical image analysis methods, this paper aims to improve the processing efficiency, data transmission speed, and accuracy of clinical decision-making of medical images. By using CNN and LSTM technologies in deep learning, this paper successfully captures the spatiotemporal information in medical images, and provides doctors with more accurate diagnosis and treatment plan support. ResNet is introduced as the basic structure of deep neural network, and its powerful network depth and lateral connection mechanism make the feature extraction of medical images more robust and reliable. In terms of data transmission, this paper designs optimized cloud computing and edge computing technologies to realize the rapid transmission and collaborative analysis of medical image data.

In future, further optimization and improvement of the technology, including the improvement of the technical level, the expansion of application fields, and the opportunities for interdisciplinary research. The results of this research are expected to promote innovation in the medical field, provide useful references for future medical research and practice, provide safer and more efficient medical services for patients, and more comprehensive patient information to support more accurate clinical decision-making for doctors.

Acknowledgment. Research Project: 145BHQ090003000X15.

References

1. Rao, S., Kumari, N.: 5G based Edge Computing for Internet of Medical Things in Fast Healthcare Data Processing and Management. In: 2022 Second International Conference on Advanced Technologies in Intelligent Control, Environment, Computing & Communication Engineering (ICATIECE), Bangalore, India, pp. 1–6 (2022)
2. Chang, X., Han, Y., Jia, W., Xiao, N.: Application research of telemedicine combined with 5G in the era of big data. In: 2021 3rd International Conference on Machine Learning, Big Data and Business Intelligence (MLBDBI), Taiyuan, China, pp. 435–438 (2021)
3. Weisong, S., Xingzhou, Z., Yifan, W., et al.: Edge computing: current situation and prospects. J. Comput. Res. Dev. **56**(1), 69–89 (2019)
4. Abir, S.M.A.A., Abuibaid, M., Huang, J.S., Hong, Y.: Harnessing 5G networks for health care: challenges and potential applications. In: 2023 International Conference on Smart Applications, Communications and Networking (SmartNets), Istanbul, Turkiye, pp. 1–6 (2023)
5. Zhang, H., Zhang, Q., Diao, Z., Sun, F.: The sensorless nursing system based on 5G Internet of Things. In: 2021 3rd International Symposium on Smart and Healthy Cities (ISHC), Toronto, ON, Canada, pp. 1–4 (2021)
6. Shabbir, M., et al.: Enhancing security of health information using modular encryption standard in mobile cloud computing. IEEE Access **9**, 8820–8834 (2021)
7. Sun, Jiang, X., Ren, H., Guo, Y.: Edge-cloud computing and artificial intelligence in internet of medical things: architecture, technology and application. IEEE Access **8**, 101079–101092 (2020)
8. Li, C., Liu, X., Zhang, H.: 5G network slicing for medical Internet of Things: a survey. IEEE Internet Things J. **9**(24), 22474–22491 (2022)
9. Wang, Y., Zhang, Y.: 5G and AI enabled smart healthcare: a survey. IEEE Trans. Veh. Technol. **72**(12), 15879–15894 (2023)
10. Dou, Q., Chen, H., Heng, P.A.: A survey on deep learning in medical image analysis. Med. Image Anal. **42**, 60–88 (2019)
11. Shen, D., Wu, G.: Artificial intelligence in medical imaging: a review. IEEE Trans. Med. Imaging **38**(8), 1934–1947 (2019)
12. Rajpurkar, P., Irvin, J., Zhu, K., et al.: CheXNet: Radiologist - Level Pneumonia Detection on Chest X - Rays with Deep Learning (2017). arXiv preprint arXiv:1711.05225
13. Litjens, G., et al.: A survey on deep learning in medical image analysis. Med. Image Anal. **42**, 60–88 (2017)
14. Shin, H.C., et al.: Deep convolutional neural networks for computer - aided detection: CNN architectures, dataset characteristics and transfer learning. IEEE Trans. Med. Imaging **35**(5), 1285–1298 (2016)
15. Esteva, A., et al.: Dermatologist - level classification of skin cancer with deep neural networks. Nature **542**(7639), 115–118 (2017)
16. Giger, M.L.: Artificial intelligence in medical imaging. Phys. Med. Biol. **62**(11), 31–68 (2017)

17. He, K., Zhang, X., Ren, S., Sun, J.: Deep residual learning for image recognition. In: Proceedings of the IEEE Conference on Computer Vision and Pattern Recognition, pp. 770–778 (2016)
18. Szegedy, C., Liu, W., Jia, Y., et al.: Going deeper with convolutions. In: Proceedings of the IEEE Conference on Computer Vision and Pattern Recognition, pp. 1–9 (2015)
19. Ronneberger, O., Fischer, P., Brox, T.: U - Net: convolutional networks for biomedical image segmentation. In: Proceedings of the International Conference on Medical Image Computing and Computer - Assisted Intervention, pp. 234–241 (2015)
20. Tian, J., Liu, G., Gu, S., et al.: Research and challenges of deep learning methods for medical image analysis. Acta Automatica Sinica **44**(3), 401–424 (2018)
21. Lan, X., Wei, R., Cai, H., et al.: Application of machine learning algorithm in medical field. Chin. Med. Equip. J. **40**(3) (2019)
22. Yu, Y., Si, X., Hu, C., et al.: A review of recurrent neural networks: LSTM cells and network architectures. Neural Comput. **31**(7), 1235–1270 (2019)
23. Alaeddine, H., Jihene, M.: Deep residual network in network. Comput. Intell. Neurosci. **2021**, 1–9 (2021)
24. LeCun, Y., Bottou, L., Bengio, Y., Haffner, P.: Gradient - based learning applied to document recognition. Proc. IEEE **86**(11), 2278–2324 (1998)
25. Krizhevsky, A., Sutskever, I., & Hinton, G.E..: ImageNet classification with deep convolutional neural networks. Adv. Neural Inf. Process. Syst., 25 (2012)
26. Turaga, S.C., et al.: Convolutional networks can learn to generate affinity graphs for image segmentation. Neural Comput. **22**(2), 511–538 (2010)
27. Hochreiter, S., Schmidhuber, J.: Long short-term memory. Neural Comput. **9**(8), 1735–1780 (1997)
28. Gers, F.A., Schmidhuber, J., Cummins, F.: Learning to forget: continual prediction with LSTM. Neural Comput. **12**(10), 2451–2471 (2000)
29. Graves, A., Schmidhuber, J.: Framewise phoneme classification with bidirectional LSTM and other neural network architectures. Neural Netw. **18**(5–6), 602–610 (2005)
30. Sundermeyer, M., Schluter, R., Ney, H.: LSTM neural networks for language modeling. Interspeech, 2546–2549 (2012)

Research on Automation of Rural Landscape Design Based on Convolutional Neural Network and Generative Adversarial Network

Ling Wang[✉]

School of Art and Design, Nanning University, Nanning City 530000,
Guangxi Zhuang Autonomous Region, China
18776923587@163.com

Abstract. Driven by the rural revitalization strategy, rural landscape design is entering a new stage. It needs to face the dual challenges of innovation and sustainability. This study explores the application of deep learning in rural landscape design and analyzes its potential in constructing design patterns and optimizing design concepts. The natural environment, cultural heritage and historical evolution data of the countryside were collected. The data were preprocessed and the CNN and GAN technologies were used to construct an automatic generation model for landscape design. The results of the experiment showed that the quality of GAN image generation exceeded that of the traditional CNN model. The output images were clearer, more detailed, and more similar in structure, which more accurately reflected the individual characteristics of the rural scenery.

Keywords: deep learning · rural landscape design · convolutional neural network · generative adversarial network

1 Introduction

Deep learning has achieved major breakthroughs in all walks of life. Its application in image recognition and automated design has opened a new chapter for landscape design. In particular, in rural landscape layout, deep learning relies on the analysis and processing of massive data to help designers efficiently produce creative and personalized design drawings, and coordinate multiple dimensions such as ecology, history and culture [1]. This study focuses on the implementation skills of deep learning technology in rural landscape design, implements efficient design processes, improves design quality and efficiency, implements systematic design processes and experimental analysis, and strives to open up new technical paths for rural landscape design and jointly promote rural sustainable development and cultural heritage protection.

© The Author(s) 2026
P. Siarry et al. (Eds.): WCNA 2024, LNEE 1550, pp. 371–379, 2026.
https://doi.org/10.1007/978-981-95-6946-5_37

2 Overview of Deep Learning

2.1 Convolutional Neural Network (CNN)

Figure 1 shows a typical convolutional neural network layout. The input layer uses a 28 × 28 pixel grayscale image, which represents a handwritten digit. The image undergoes two convolutions and one pooling step at the initial stage. The convolution layer extracts local feature fragments of the image. The width and height of the convolution kernel are 5 grids each. Appropriate padding is used. The edges are not padded when the image is input to reduce the dimension of the feature map [2]. After each convolution operation, each is followed by a pooling layer, which uses a 2 × 2 maximum pooling method. The purpose is to reduce the space occupation of the image, maintain key attributes, and simultaneously reduce the amount of computation and the risk of overfitting. The convolution layer can extract feature sets from multiple channels. As the network is gradually extended, the size of the feature map gradually decreases, but the number is expanded, which helps to identify more complex feature association clusters. After the convolution layer and pooling layer processing, the network flattens the extracted features and uses the fully connected layer as a platform for classification. The network outputs a layer consisting of 10 units, a set of ten handwritten digit categories (0 to 9).

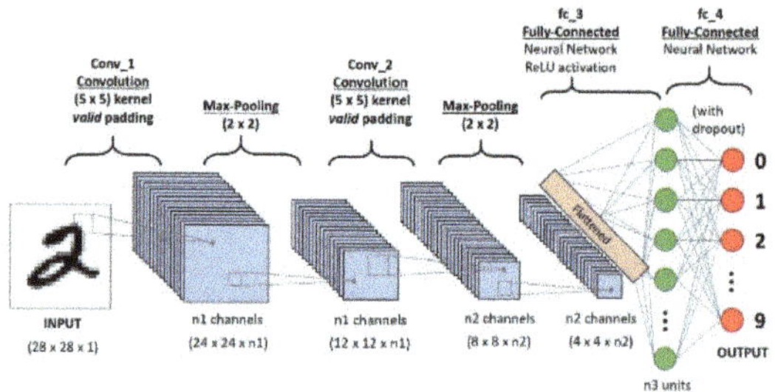

Fig. 1. The basic architecture of a convolutional neural network

2.2 Recurrent Neural Networks (RNNs)

Recurrent neural networks are widely used in time series and sequence prediction. The input layer nodes are complex and each corresponds to each dimension of the input data. The characteristics of recurrent neural networks can be summarized as follows: each hidden layer node receives the input data at the current moment, and the hidden state data at the previous moment has been obtained. This structure gives RNN the ability to store previous input information, so it can handle time series dependent data, such as text records, voice records, financial data, etc. In terms of network structure, the initial input data first enters the hidden layer, and the nonlinear transformation results are passed to the next layer. The recursive connection mechanism allows the network to grasp the contextual information of the sequence through feedback at each time stage, track data evolution, and the final output (y) of the network output is used for classification prediction [3].

2.3 Generative Adversarial Networks (GANs)

As shown in Fig. 3, this network consists of two major components: the construction network (G) and the discrimination network (D). The network generator extracts sample data sets from the low-dimensional latent space to generate a multi-dimensional data set. The discriminant network is responsible for comparing the generated image with the real image and verifying its authenticity [4]. In pursuit of generating the most realistic image effect, the discriminator strives to accurately distinguish the generated image from the actual image. The structure adopts an adversarial training method to enable the two networks to conduct adversarial learning during the training phase and continuously enhance their respective performance levels. In the network mentioned above, the G network extracts samples from the low-dimensional surface of the latent space. Such samples generally belong to the category of random noise vectors, and the generator network produces fake photos. The discriminant network (D) proofreads the generated fake image with the original image and generates a binary classification result: the generator network attempts to deceive the discriminant network. The generated fake image strives to be highly similar to the real image so that the discriminator cannot recognize the difference between them. The training progresses gradually, the generator and the discriminator are continuously optimized and upgraded, and the image generation effect becomes more and more realistic [5] (Fig. 2).

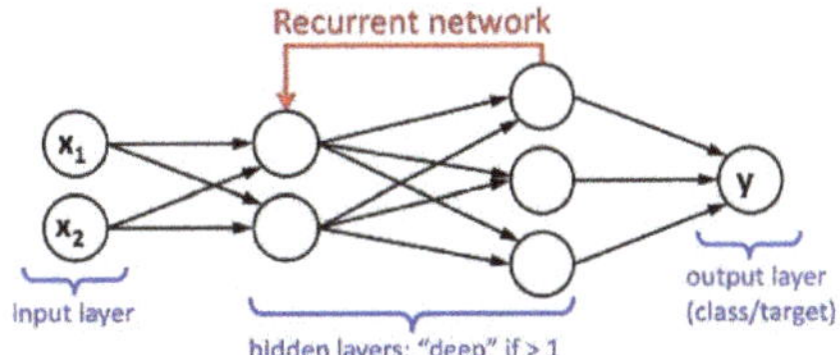

Fig. 2. Composition of a recurrent neural network

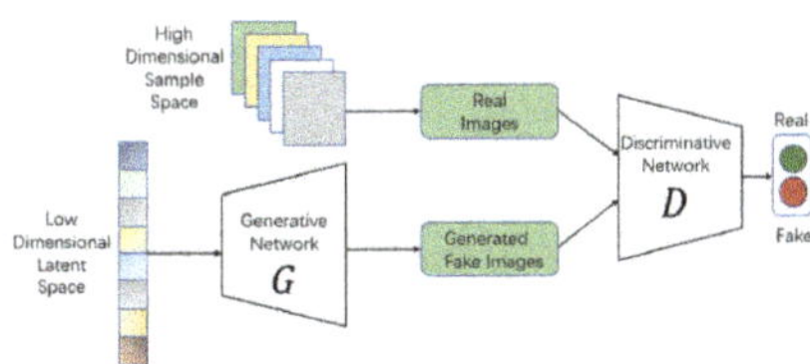

Fig. 3. The composition of the generative adversarial network

3 Rural Landscape Design Based on Deep Learning

3.1 Framework and Steps of the Design Process

3.1.1 Design Requirements Analysis

The demand analysis of rural landscape design is based on the systematic data collection based on the unique environment, culture and historical background of the village to ensure that the plan is closely integrated with the regional characteristics. Initially, a quantitative study of the rural environment should be carried out, covering terrain, climate, land use and other aspects. GIS technology can be used to analyze the geographical spatial attributes of various rural areas, collect the altitude, slope, land use and other attributes of the area, use interpolation to fill in the missing data, and implement the Kriging interpolation algorithm to supplement the null values in the elevation data [6]. The formula is as follows:

$$Z(x) = Z_0 + \sum_{i=1}^{n} \lambda_i(x_i - x) \tag{1}$$

Among them, $Z(x)$ is the point x interpolation result, Z_0 that is, the numerical value of the known data point, x_i is the given spatial position coordinate value, λ_i that is, the weight factor, and n is the quantity range of the known points. Using this formula, we can accurately fill the geographic information gaps in a specific area and assist in the subsequent landscape design decision-making implementation [7].

3.1.2   Data Preprocessing

In the rural landscape planning stage, accurate identification and classification of landscape elements, data preprocessing is the basis, which is directly related to the subsequent deep learning model results. Remote sensing technology and drone photography technology are used to collect image data of rural areas, and laser radar scanning is used to collect three-dimensional point clouds and image data of buildings, vegetation, water bodies and other elements in rural landscapes. Such data generally need to complete a multi-step processing procedure: information purification, noise removal, labeling and standardization, identification of architectural styles, and analysis of laser radar point cloud data. The K-means clustering algorithm is used to perform spatial segmentation operations on buildings. The K-means algorithm attempts to cluster point cloud data into k categories:

$$J = \sum_{i=1}^{n} \sum_{k=1}^{k} w_{ik} \|x_i - \mu_k\|^2 k \tag{2}$$

Among them, the objective function is J named, which w_{ik} reveals x_i the membership ratio of the sample point μ_k in the cluster. k It is the center coordinate of cluster k, x_i μ_k and the absolute value of the difference represents the distance from the sample point to the cluster center. It can accurately identify the shape boundary of rural buildings, vegetation and water body indexing technology. The convolutional neural network (CNN) technology of remote sensing images can be used for automatic classification. If the vegetation attributes extracted from the remote sensing image are recorded as x = (x1, the support vector machine model completes the classification task according to this formula:

$$f(x) = sign\left(\sum_{i=1}^{n} \alpha_i y_i K(x, x_i) + b\right) \tag{3}$$

Among them, x is the input variable, α_i is the weight coefficient of the support vector, y_i is used as the label of the support vector, x_i belongs to the category of kernel function, b is a constant term, can accurately classify various types of vegetation, water bodies and other objects on the surface, and reserve reliable data resources for subsequent design, aiming to maintain data uniformity and efficient model operation, and implement data standardization. The formula is used:

$$x' = \frac{x - \mu}{\sigma} \tag{4}$$

Among them, x' after standardization, μ the mean of the sample is defined, σ which represents the mean square error of the sample. After normalization, the data level is guaranteed to be uniform, which is conducive to improving the stability and accuracy of deep learning model training [8].

3.2 Construction and Training of Deep Learning Models

3.2.1 Generating Landscape Design Images Using Convolutional Neural Networks (CNNs)

First, formulate the requirements, then input the rural geographical environment map, use the 5×5 convolution kernel, perform feature map analysis, and the feature map output by each convolution layer must undergo the dimension reduction operation of the maximum pooling layer, so as to retain its core elements and use multi-level convolution and pooling technology to mine deep spatial attributes. The image resolution is 28×28 pixels, the image size output by the first layer of convolution operation is 24×24 pixels, and the output image size after the second layer of convolution processing is 12×12, which finally realizes the one-dimensional processing of the data. The image data is sent to the fully connected layer for generation. The CNN model uses the back-propagation method to optimize the generation effect and achieve the designed target value. The mean square error loss function is the basis for evaluating the difference between the generated image and the target image:

$$L = \frac{1}{N} \sum_{i=1}^{N} (y_i - \hat{y}_i)^2 \tag{5}$$

Among them, y_i is the pixel value of the target image, $\hat{y}_i$ that is, the pixel value of the image generated, N that is, the total number of image pixels. With the goal of minimizing the loss function, the image generation effect can be improved [9].

3.2.2 Optimization During Training

The network parameters are adjusted by minimizing the loss function, using the Adam optimizer as a reference. The update criteria are as follows:

$$\begin{aligned}
m_t &= \beta_1 m_{t-1} + (1 - \beta_1)\nabla_\theta J(\theta) \\
v_t &= \beta_2 v_{t-1} + (1 - \beta_2)(\nabla_\theta J(\theta))^2 \\
\hat{m}_t &= \frac{m_t}{1-\beta_1^t}, \quad \hat{v}_t = \frac{v_t}{1-\beta_2^t} \\
\theta &= \theta - \eta \frac{\hat{m}_t}{\sqrt{\hat{v}_t}+\epsilon}
\end{aligned} \tag{6}$$

Among them, the gradient of the parameter θ to the loss function J can be recorded as $\nabla_\theta J$, m_t and v_t are the estimates of the first-order and second-order moments, β_1 and β_2 are the key hyperparameters for adjusting the decay rate, η that is, the learning rate value, $\in$ which is the minimum parameter set to avoid zero division errors. Repeatedly perform this optimization process, the CNN model continuously adjusts the weights in each iteration, gradually reduces the image generation error value, and then produces a landscape map that is more in line with the design intent. In order to reduce overfitting, regularization methods are required during training. Among the regularization techniques, the dropout method and L2 normalization are the main methods. The Dropout mechanism is used to randomly discard a certain proportion of neurons to reduce the dependence on specific features, thereby enhancing the universality of the model [10].

3.3 Design Generation and Optimization

In the implementation design stage, the output landscape images should first be post-processed, using image enhancement technology, including contrast improvement and edge enhancement, to enhance the resolution and detail level of the image. The generated image design needs to be integrated with the actual landscape, and the architectural style and vegetation layout should be sorted out in detail. Genetic algorithms can be used to implement multiple rounds of optimization for the design. The optimization process of genetic algorithms is as follows:

$$f(x) = \sum_{i=1}^{n} w_i x_i \tag{7}$$

Among them, $f(x)$ the mathematical notation definition of the objective function x_i involves the various variables in the design plan, w_i that is, the weight of the variable. The selection, hybridization and mutation technology of the genetic algorithm can screen out the most suitable plan for design needs in each generation, and gradually enhance the advantages and disadvantages of the design plan. The optimized design drawings perfectly meet the aesthetic standards, taking into account the unique ecological environment and cultural characteristics of the countryside, and scientifically guaranteeing the implementation of rural landscape planning.

4 Testing and Result Analysis

4.1 Experimental Setup

The core parameters of the training process are set in the experimental stage. These parameters have a great impact on the model training effect and the generated results. The size of the experimental image input is 256 pixels × 256 pixels, which can coordinate the computing pressure and image fineness. The number of batches is set to 32, striving to use the GPU efficiently during the training stage to prevent the consequences of memory overflow caused by too large batches. The number of training rounds is arranged to be 50 rounds, striving to fully train the model and avoid the model from over-learning the training set. The learning rate is selected as 0.0001, which has a significant impact on the optimizer update rate and training stability. The Adam optimization tool is used. This adaptive optimization algorithm is frequently used and only requires a small amount of hyperparameter adjustment to achieve ideal results. This model uses the mean square error as the optimization method of the loss function. This loss function is suitable for the regression task solution stage and can accurately quantify the difference between the generated image and the target image. The experimental parameter list is shown in Table 1:

Table 1. Experimental parameters

Parameter	Value
Image Size	256×256
Batch Size	32
Epochs	50
Learning Rate	0.0001
Optimizer	Adam
Loss Function	MSE

4.2 Experimental Results Analysis

Experimental data show that the CNN basic model, enhanced CNN model and GAN have different performances in terms of image generation quality, training time, peak signal-to-noise ratio and structural similarity. As can be seen from the table, the image generation effect based on GAN significantly exceeds the traditional CNN model, with the lowest MSE result (0.021), PSNR (34.2 dB) and SSIM (0.91) both significantly ahead of other models. The images generated by GAN have enhanced clarity and more delicate details. Although GAN training takes a long time, up to 18 h, its excellent image quality offsets this shortcoming. Compared with the combination of CNN and data enhancement, GAN shows positive growth in all indicators, especially in terms of image quality. Table 2 and Fig. 4 detail the in-depth data mining of the experimental data:

Table 2. Experimental results analysis

Model	Generated Image Quality (MSE)	Training Time (hours)	PSNR (dB)	SSIM (Score)
CNN (Baseline)	0.035	15	30.5	0.85
CNN + Data Augmentation	0.028	14	32.1	0.88
GAN	0.021	18	34.2	0.91

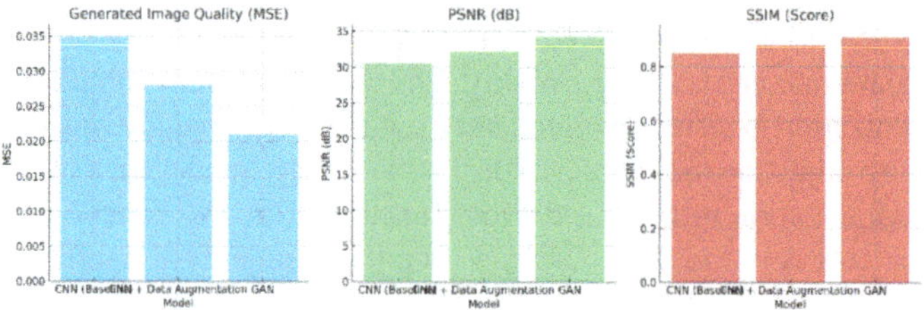

Fig. 4. Visualization of experimental results

4.3 Results and Discussion

The results show that the effect of using GAN in the field of rural landscape design exceeds that of traditional CNN and CNN combined with data enhancement. In terms of image generation quality, GAN has the lowest mean square error of 0.021, and the difference between its generated image and the target image is minimized. GAN technology can generate more detailed rural landscape design drawings. Compared with other models, GAN's PSNR and SSIM indicators are significantly better, achieving values of 34.2 decibels and 0.91 respectively. Research data show that the images generated by GAN are closer to real images in quality, and their structural textures are more delicate and colorful. Although GAN training takes a long time, up to 18 h, compared with 15 h for the CNN basic model and only 14 h for the CNN enhanced data model, GAN image quality is significantly improved, and extending the training time is more meaningful. CNN and its data enhancement version have poor results in PSNR and SSIM indicators, probably because they cannot effectively enhance the realism and detail of the image with adversarial training like GAN.

The GAN model is good at producing images that are both diverse and high-quality. When implementing creative and personalized rural landscape design, GAN allows designers to easily produce realistic and detailed images. The training cost is not low. The application prospects of GAN in the field of rural landscape design are bright, which will help rural landscape design innovation and move towards sustainable development goals.

5 Conclusion

This study uses deep learning technology to analyze the implementation path of rural landscape design. The results show that the GAN model performs well in terms of generated image quality, detail richness, and matching degree with real landscape elements, which is significantly better than traditional CNN and data-enhanced CNN models. Combining rural environment, culture, historical background and data, deep learning technology improves landscape design, individual creativity is taken to a higher level, and rural-urban integration of rural landscape design innovation and sustainable development is implemented. The application potential of deep learning in rural landscape design has not yet been fully explored, especially in terms of multimodal data integration, comprehensive consideration of ecological and cultural factors, and optimization of generation models, focusing on improving design precision and its practical value.

References

1. Stupariu, M.S., Cushman, S.A., Pleşoianu, A.I., et al.: Machine learning in landscape ecological analysis: a review of recent approaches. Landscape Ecol. **37**(5), 1227–1250 (2022)
2. Wang, L., Han, X., He, J., et al.: Measuring residents' perceptions of city streets to inform better street planning through deep learning and space syntax. ISPRS J. Photogramm. Remote. Sens. **190**, 215–230 (2022)
3. Ning, H., Yu, Z., Zhang, Q., et al.: An in-memory computing architecture based on a duplex two-dimensional material structure for in situ machine learning. Nat. Nanotechnol. **18**(5), 493–500 (2023)
4. Bot, K., Borges, J.G.: A systematic review of applications of machine learning techniques for wildfire management decision support. Inventions **7**(1), 15 (2022)
5. Ramírez, T., Hurtubia, R., Lobel, H., et al.: Measuring heterogeneous perception of urban space with massive data and machine learning: an application to safety. Landsc. Urban Plan. **208**, 104002 (2021)
6. Chen, R., Zhao, J., Yao, X., et al.: Generative design of outdoor green spaces based on generative adversarial networks. Buildings **13**(4), 1083 (2023)
7. Lu-Yao, W., Yi-Ping, H.: Environmental landscape art design based on visual neural network model in rural construction. Ecol. Chem. Eng. **30**(2), 267–274 (2023)
8. Lu, F., Chenwen, Y.U., Yuting, S.U.N., et al.: Generative design of plantscape based on generative adversarial network: a case study of the generation of flower border plan. Lands. Arch. **31**(9), 59–68 (2024)
9. Lu, Y., Chen, D., Olaniyi, E., et al.: Generative adversarial networks (GANs) for image augmentation in agriculture: a systematic review. Comput. Electron. Agric. **200**, 107208 (2022)
10. Adamiak, M., Będkowski, K., Bielecki, A.: Generative adversarial approach to urban areas NDVI estimation: a case study of Łódź, Poland. Quaestiones Geographicae **42**(1), 87–105 (2023)

Cigarette Packet Appearance Defect Detection Algorithm Based on Improved YOLOv8

Chunhui Huang[1], Sixiao Chen[2], Haihua Lu[2(✉)], Ying Bao[2], Xudong Wang[1], and Liang Chen[2]

[1] Cigarette Manufacturing Department, Xiamen Tobacco Industrial Co., Ltd., Xiamen, China
[2] Ningbo Cigarette Factory, China Tobacco ZheJiang Industrial Co., Ltd., Ningbo, China
{Chensx,Baoy,Chenlinag}@zjtobacco.com, 25747316@qq.com

Abstract. Aiming at the demand of cigarette enterprises for the inspection of the appearance quality of cigarette packets in the production process, a method for detecting surface defects of cigarette packs based on machine vision was designed. The EML-YOLO used in this paper is an industrial surface defect detection algorithm based on improved YOLOv8. It improves the feature extraction capability of the model by designing an efficient large convolution module (ELK) and provides multi-scale feature representation while preserving spatial information. The modules of image acquisition, image preprocessing and defect classification are built; the self-fitting brightness adjustment algorithm is proposed to complete the pixel value statistics and obtain a clear defect feature image. The experimental results show that the detection accuracy of the EML-YOLO algorithm is 98.84%, which has good application potential in industrial defect detection scenarios, and can accurately and efficiently realize the classification and recognition of defect information on the surface of cigarette packets.

Keywords: machine vision · cigarette packet defect · online detection · EML-YOLO

1 Introduction

Cigarette packets inevitably produce various defects during the manufacturing process, such as packaging breakage, extrusion deformation, lid cracks, scratches, surface contamination, etc. Algorithms with high versatility and execution efficiency are needed to cope with various forms of defects on the product surface. Previous defect detection algorithms mainly relied on image processing-based methods, such as threshold segmentation, edge detection, and region growing, to distinguish and segment the defective part from the non-defective part of the image through image processing techniques, and then process and analyze the defective part [1]. Although these traditional methods have achieved some application results in industrial inspection, they are usually limited by the selection of features, light and image noise, and their detection accuracy is not high. it is also difficult to balance high precision and lightweight, which is difficult to apply to ultra-high-speed cigarette production lines [2].

© The Author(s) 2026
P. Siarry et al. (Eds.): WCNA 2024, LNEE 1550, pp. 380–387, 2026.
https://doi.org/10.1007/978-981-95-6946-5_38

In this paper, the proposed cigarette packets surface defect detector EML-YOLO is a novel model based on YOLOv8n, which improves the receptive field and enhances the feature extraction and fusion capability through a parallel large convolutional kernel structure ELK [3]. Secondly, a multi-branch parallel multi-scale context module MCM is proposed, which is capable of target sensing at different scales and effectively utilizes contextual information to enhance detection accuracy and robustness. Finally, the original Neck is optimized to reduce the computational complexity and number of parameters, maintain the performance, and find a balance between lightweight and high detection accuracy [4].

2 Network Structure Improvement

In order to improve the defect detection accuracy of cigarette packets in high-speed production environment, the improved EML-YOLO algorithm proposed in this paper adds ELK module in the backbone network to reduce redundant information and enhance the judgment ability of subtle defects [5]. The MCM module with multi-branch parallel structure is designed to obtain a more comprehensive and rich feature representation by target information on different scales. The VoVGSCSP module is used in Neck to reduce the computational complexity and parameter amount. While ensuring detection accuracy, it has a smaller number of model parameters and faster detection speed [6].

2.1 Backbone Network Improvements

In the field of target detection, increasing the effective receptive field of a model is an important issue. Although the expansion rate of the convolution kernel is adjusted to increase the receptive field, a certain degree of local feature loss will be introduced [7]. This loss may lead to grid-like missing in the prediction map, which reduces the accuracy and detail representation of the prediction results. This paper designs a parallel large convolution kernel structure. This structure can help the model capture target features in a wider spatial range, reduce the redundant information in the feature map and enhance the fine granularity judgment ability of the model (Fig. 1).

which effectively solves the problem of background interference in the detection process. The specific structure is shown in Fig. 2.

In convolutional neural networks, there is usually a large amount of redundant information in the output feature maps of the convolutional layer, and some of the feature maps may be highly similar. In order to reduce the feature redundancy, the input feature map is decomposed into two sub-feature maps, which are realized by processing each channel separately to reduce the redundant information between the feature maps [8]. Feature extraction is performed on one of the feature maps using the LK block. The LK block uses depth-separable convolution using a convolutional kernel of 27×27 size, which can capture target features with a wide spatial range and more effectively acquire contextual information around the target, which can more fully perceive the environmental context in which the target is located, eliminate the confusion problem between the target and the background, and improve the accuracy of detection. In another feature map, the EC (Efficient Conv) block is used for processing, applying the regular Conv for

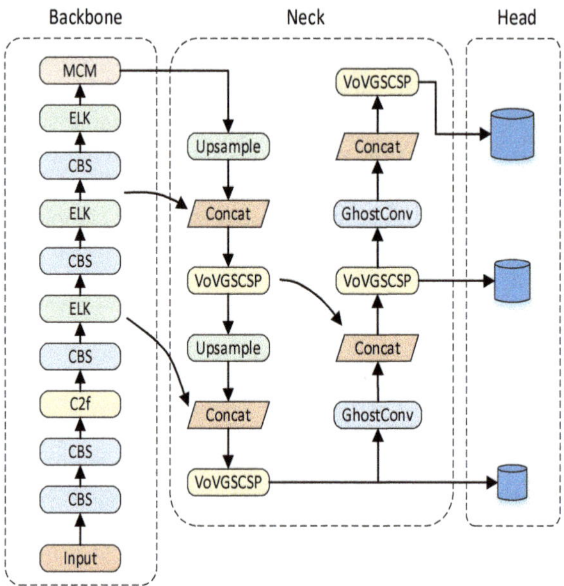

Fig. 1. EML-YOLO network structure diagram.

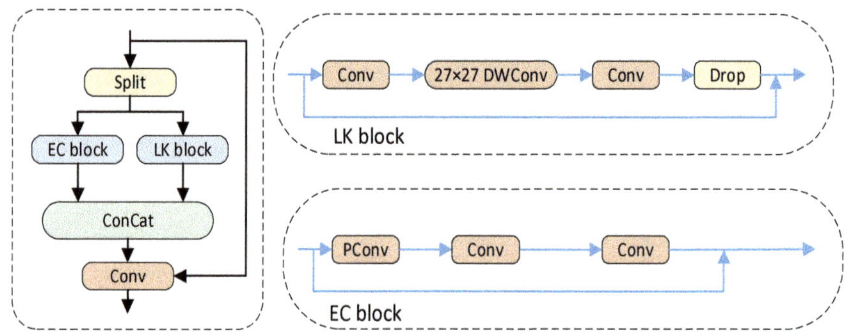

Fig. 2. ELK structure diagram.

spatial feature extraction to some input channels, while keeping the rest of the channels unchanged. By focusing on feature extraction in local areas, detailed information such as the boundary and texture of the target object can be better processed, and the model's fine granularity judgment ability of the target can be enhanced [9]. The combination of LK block and EC block can be used to extract the features at different levels: the LK block focuses on the target features and context information in the wide spatial range, while the EC block focuses on localized EC block focuses on local details. This combination can capture the various features of the target more comprehensively and improve the detection accuracy of the model. The calculation formula is:

$$H \times \omega \times K^2 \times c_{in}^2 \tag{1}$$

where C_{in} is the number of input channels, C_{out} is the number of output channels, K is the size of the convolution kernel, H and ω are the height and width of the feature map.

2.2 Multi-scale Context Module

In order to reduce the computational and parametric quantities of the MCM module, model compression techniques such as parameter sharing, channel compression, depth compression, and knowledge distillation are applied to improve the efficiency and portability of the model while maintaining high performance. The MCM module has a structure of multiple branches in parallel, each with different parameter settings, and is able to perform feature extraction on the input data at different scales, thus providing a more comprehensive and richer feature representation than the traditional single view [10]. In order to comprehensively utilize the information of the multi-scale module in the spatial and channel dimensions, the coordinate attention (CA) block is introduced and paralleled into the multiscale module to introduce contextual information into the feature representation by taking into account the spatial location information of the input feature maps, thus better capturing the global contextual information and improving the performance of the model [11]. For inputs first each channel is encoded along the horizontal and vertical coordinate directions using pooling kernels of dimensions (H,1) and (1,W).

The output Z of the i-th channel with height H is expressed by (2):

$$Z_c^h(h) = \frac{1}{W} \sum_{0 \leq 1 \leq W} A_c(h, i) \tag{2}$$

The output Z of the c-th channel with width ϖ is expressed by (3):

$$Z_c^\varpi(\varpi) = \frac{1}{H} \sum_{0 \leq 1 \leq W} A_c(j, \varpi) \tag{3}$$

By encoding the feature maps in both horizontal and vertical directions, the spatial information of the feature maps for each channel can be integrated, so that the feature map of each direction can capture the contextual information in the input feature map. Also, this encoding operation preserves the positional information. The feature maps in both directions are fused through (4) to highlight the representation of interest and enhance the task-relevant information in the input feature map.

$$Z_c = \frac{1}{H \times W} \sum_{i=1}^{H} \sum_{j=1}^{W} x_c(i, j) \tag{4}$$

where H and W represent the feature maps in the horizontal and vertical directions.

The feature map F obtained after the fusion operation is transformed using 1×1 convolution to generate the intermediate feature map f, which is then sliced into two separate tensors f^h and f^ϖ as shown in (5)–(7):

$$f = \sigma\left(F_1\left(\left[Z^h, Z^\varpi\right]\right)\right) \tag{5}$$

$$g^h = \sigma\left(F_h\left(f^h\right)\right) \tag{6}$$

$$g^{\varpi} = \sigma\left(F_{\varpi}\left(f^{\varpi}\right)\right) \tag{7}$$

After expanding g^h and g^{ϖ} as attention weights, the final output of the CA module is:

$$y_c(i,j) = x_c(i,j) \times g_c^h(i) \times g_c^{\varpi}(j) \tag{8}$$

The MCM structure fuses features at different scales and uses coordinate attention to provide contextual information to provide a richer feature representation to improve the performance of target detection. The final output is:

$$out = concat\begin{pmatrix} DWconv3 \times 3, DWconv5 \times 5 \\ DWconv7 \times 7 \\ x_c(i,j)g_c^h(i) \times g_c^{\varpi}(j) \end{pmatrix} \tag{9}$$

2.3 Lightweight Neck Structure

In order to reduce the computational complexity of the model this paper uses the LW-Neck structure [12], replacing the 3x3 ordinary convolution in the Neck with GhostConv. To reduce the computational complexity of the model and improve the performance and training, the VoVGSCSP module is used, which consists of a 1×1 convolution and GSConv, and includes a residual connection using the 1×1 convolution. Which helps to enhance information transfer and mitigate the gradient vanishing problem (Fig. 3).

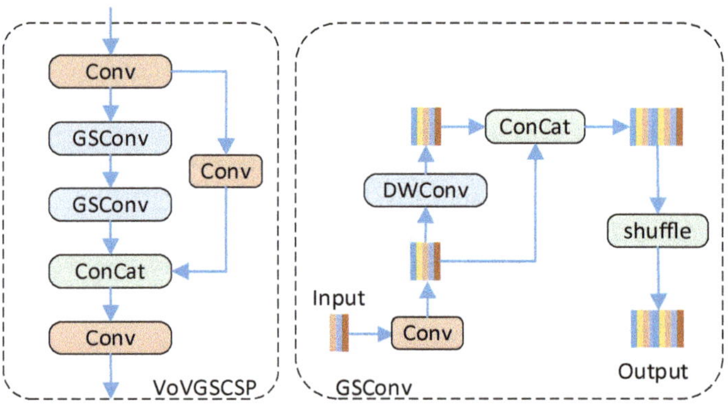

Fig. 3. VoVGSCSP structure diagram.

GSConv combines the features of Group Convolution and Shuffle operation to interact and integrate information in the feature map. Group Convolution divides the input feature map into multiple groups, performs convolution operations on the channels in each group, and then splices the output of each group, thus reducing the amount of calculation and improving the efficiency of the model. Shuffle operation, on the other hand, groups the channels and shuffles the channels of different groups, thus increasing the interaction between the channels.

3 Experiment Analysis

3.1 Data Set

In this experiment, 6 types of defects collected from cigarette production line were organized and summarized as a dataset, with image size 2048 × 1280. The defect types include missing box lid, folded corner reversal, creases, breakage, stains, and deformations, which are totaled to 2,360.They were divided into training sets and test sets according to the ratio of 8:2.

3.2 Evaluation Indicators

The experiments were evaluated by using metrics such as Precision, Recall, Average of Defects (AP), and Average of Defects across all categories (mAP).

$$Precision = \frac{TP}{TP + FP} \tag{10}$$

$$Recall = \frac{TP}{TP + FN} \tag{11}$$

$$AP = \int_0^1 P(r)dr \tag{12}$$

$$mAP = \frac{1}{N} \sum_{i=1}^{N} AP_i \tag{13}$$

3.3 Backbone Network Ablation Experiment

The results of the ablation experiments show that the use of the ELK module in the improved backbone network proposed in this paper has significant advantages over other backbone network models, exhibiting higher detection accuracy. The experimental results show that the network using the ELK module outperforms the YOLOv8n base backbone network in terms of detection accuracy, and the number of parameters and computation of the ELK module outperforms FasterNet and ReplkNet. The mAP metrics are improved by 1.9% and 1.1%, which achieves a significant result in terms of improving the performance of target detection (Table 1).

Table 1. Backbone network ablation experiment

Models structure	Params(M)	FLOPs(G)	mAP
YOLOv8n-EfficientNet	3.2	8.3	70.5
YOLOv8n-EfficientNet	2.2	7.1	69.2
YOLOv8n-FasterNet	2.9	8.3	72.7
YOLOv8n-ReplkNet	3.0	8.9	73.5
YOLOv8n-ELK module	2.7	7.7	74.6

3.4 Comparative Experiment

The EML-YOLO algorithm was tested on the GC10-DET dataset for a detailed comparison with other algorithms.

From Table 2, it is clear that EML-YOLO improves the mAP to 98.84% while the number of parameters is only 2.7.

Table 2. Backbone network ablation experiment

Model	Params(M)	FLOPs(G)	mAP	worn	deletion	dogear	deform	deletion
Faster-RCNN	41.3	251.4	97.50	97	97.8	98.7	97.5	97.70
YOLOv5s	7.2	16.8	97.68	96.9	97.8	97.5	98.3	97.9
YOLOX	9.5	26.1	97.16	95.2	96.3	97.9	98.1	98.3
YOLOv7-tiny	6.2	13.1	93.72	96.2	87.8	93.9	96.2	94.5
YOLOv8n	11.4	8.3	96.80	97.5	98.2	95.4	96.0	96.9
EML-YOLO	2.7	7.0	98.84	99.2	98.8	98.3	99.0	98.9

Compared with Faster-RCNN, EML-YOLO's mAP increased by 1.34%, and the computational complexity and parameter amount decreased by 35.9 times and 15.3 times respectively. Compared with YOLOv8n, the mAP is improved by 2.04% and the number of parameters and computational complexity are much lower than that of YOLOv8n.

In addition, the mAPs of EML-YOLO are all better than those of YOLOv5s, YOLOX, YOLOv7-tiny, and YOLOv8n, which proves that EML-YOLO not only performs well in maintaining high-precision detection, but also possesses smaller storage requirements and occupies fewer computational resources. EML-YOLO can complete detection tasks quickly and efficiently.

4 Conclusion

In order to achieve efficient and accurate industrial surface defect detection, this paper proposes an improved YOLOv8 industrial surface defect algorithm, which enhances the feature extraction capability by introducing an efficient convolution module, ELK, and designs an effective feature fusion module to obtain rich feature information and global context information. The ELK module improves the efficiency and accuracy of feature extraction by optimizing the convolution operation. It not only can perform defect detection efficiently, but also has a lower amount of parameters and calculations, making it easier to deploy and apply. Future research directions will focus on further improving the detection accuracy of the surface defect detection model and reducing the influence of external environmental factors on the algorithm to enhance its application value.

References

1. Wu, L., Hao, H., Song, Y.: Review of industrial metal surface defect detection based on computer vision. J. Autom. 1–24 (2025)

2. Lu, G.: Research on Image Scene Classification and Weak and Small Target Segmentation Methods Based on Deep Learning. Dalian Maritime University (2023)
3. Su, J., Jia, Z., Qin, Y., Zhang, J.: Improved YOLOv8 algorithm for industrial surface defect detection. Comput. Eng. Appl. **60**, 1–15 (2024)
4. Li, J.: Research on Sports Vehicle Detection Algorithm in Real Scenes. Shenyang University of Technology (2023)
5. Kang, T.: Research on High-Speed Railway Suspension Insulator Fault Detection Method Based on Optimized Deep Learning Algorithm. Shijiazhuang Tiedao University (2024)
6. Ouyang, B.: Research on Multi-vehicle Detection and Tracking Algorithm Based on UAV Platform. Zhejiang University of Science and Technology (2024)
7. Wang, Z.: Research on Fault Identification and Surface Reconstruction Based on Deep Learning. Hebei University of Geosciences (2024)
8. Gao, L., Zhao, B., Jian, W.: Research on vehicle traffic sign detection algorithm based on Faster-YOLOv8 network model. J. Chongqing Jiaotong Univ. (Nat. Sci. Ed.) 1–9 (2025)
9. Zhang, N.: Research on Small Target Detection Algorithm Based on Feature Fusion and Attention Residual Network. Northeast Forestry University (2023)
10. Zhang, T.: Research on Ship Detection and Identification Technology in SAR Images Based on Deep Learning. University of Electronic Science and Technology of China (2022)
11. Tian, B.: Research on Fabric Defect Detection Algorithm Based on Context-Coordinated Visual Saliency. Zhongyuan Institute of Technology (2023)
12. Feng, F.: Research on Audio-Visual Event Localization and Recognition Based on Cross-Modal Learning. Beijing University of Posts and Telecommunications (2023)

Early Request System for Auxiliary Materials in Linked Feeding for Tobacco Packaging Units

Qi Xu, Jie Li, Meng Shu, Haibin Yao, Chengting Zhang, Guorui Li, Long Chen, Haihua Lu, and Weilin Cao[✉]

Ningbo Cigarette Factory, China Tobacco Zhejiang Industrial Co., Ltd., Ningbo, Zhejiang, China
wlcao@qq.com

Abstract. There used to be a problem with cigarette packaging in the tobacco industry, and an important factor is auxiliary materials. Such request signal-based methods entail a lot of time and manual effort resulting in disruptions and hence affecting the continuous production. This paper outlines enhancing an Auxiliary Material Early Request System with a real-time monitoring and early warning system in addition to an Automated Guided Vehicle System for efficient and automatic replenishing of the auxiliary materials. The system comprises four core modules: Real-Time Monitoring and Early warning, Automatic call System, Scheduling of AGV, and Real-Time Database. Due to such technological advancements as the Isolation Forest for anomaly detection and LSTM for consumption forecast, this system greatly increases efficiency, decreases labor input, and increases the productivity of cigarette packaging lines.

Keywords: Auxiliary materials · Tobacco industry · Automated guided vehicles · Real-time monitoring · Safety inventory · Early warning system · Isolation Forest · LSTM forecasting

1 Introduction

In the tobacco industry where changes are occasioned frequently, it is essential to have constant production lines to meet production goals. A key issue relates to the optimal handling of support items used for the packaging of cigarettes. The use of conventional signals which involve manual material request signals may lead to the formation of a bottleneck which in turn slows the system. This paper presents an efficient new Auxiliary Material Early Request System which will help in rectifying these problems via the use of monitoring, early warning signs, and AGV-based logistics. Using such techniques as Isolation Forest and LSTM, this system provides a rather efficient solution for proper material restocking and continuous production.

2 System Design

The proposed system is structured around four core modules: The proposed system is structured around four core modules:

© The Author(s) 2026
P. Siarry et al. (Eds.): WCNA 2024, LNEE 1550, pp. 388–395, 2026.
https://doi.org/10.1007/978-981-95-6946-5_39

Real-Time Monitoring: Real-time tracking of inventory and material usage with the help of sensors and IoT gadgetry.

Early Warning and Automatic Call: Ensuring that there is an early warning system that when certain conditions are reached, automatically generates a material requisition.

AGV Scheduling: The efficient way and timing of how the AGVs are to transport the raw materials without any hindrance to normal operations.

Real-Time Database: Storing as well as processing data in real-time for decision making as well as to modify the system.

The algorithms that are relevant to this system are the Isolation Forest for anomaly identification as well as the LSTM networks for the consumption of materials forecast. The Isolation Forest algorithm identifies outliers, which may point to instances of shortages or overages, while the LSTM network predicts future requirements of the materials based on past use.

3 Methodology

The material call method described in this invention can be described as an assertive material call technique that seeks to make certain that cigarette auxiliary material is restocked as soon as it is realized that the current stock is almost depleted. The following part provides an outline of the procedures and sub-procedures of the method.

Material Quantity Monitoring

The system starts with constantly checking the current stock of the auxiliary materials on hand at each of the workstations. This is done through a material monitoring module having weight sensors for the purpose. These sensors offer information on the amount of material still in use and this is very important in estimating and planning for replenishment.

Caculation of Average Time Per Station

To forecast the consumption of material more effectively, the operation time of the material in each station is computed. This average time is based on the records of the time it has taken for the various materials to be consumed at a particular station. The calculation incorporates such aspects as productivity rate, frequency of material consumption, and differences in processing time.

Estimation of Depletion Time

From the known total amount of material and the average time spent at a station, the system determines the time that is estimated to be taken to deplete the material at each station. This estimation is important since it indicates when the stock of the materials will be depleted so that ways of restocking it can be sought. The estimated time at which the resource will be depleted fully is calculated as follows:

$$Et = Cq/Ar$$

where Et (Estimated Depletion Time) refers to the estimated time when the material will be depleted, indicating the point in time when the material is expected to run out.

Cq(Current Material Quantity) refers to the current amount of material, representing the current quantity of cigarette accessories at the workstation.

Ar(Average Usage Rate) refers to the average time it takes for each use of the material, indicating the time required for each use of the material.

This formula is used to calculate the estimated depletion time for cigarette accessories.

Calculation of Advance Calling Time

The system also determines an advance calling time, that is, the time at which a material call should be made so that materials are ordered before they run out. This advance calling time depends on the Advance Calling Time for every workstation as defined in the system and takes into account lead-time for receiving materials and includes other buffer times. The advance calling time is given by the formula:

$$At = Ct + Pt$$

where At(Advance Calling Time) refers to a scheduled time in the future when a specific action or process should be initiated.

Ct(Current Time)refers to the present time or the time at which the calculation is being made.

Pt(Predefined Advance Time)refers to a specified amount of time that is added to the current time to determine when to initiate a particular action or process.

This formula is used to calculate a future time when an action should be initiated, based on the current time and a predetermined interval.

Comparison and Decision-Making

The last process is a comparison of the estimated depletion time with advance calling time. If the advance calling time is greater than or equal to an estimated depletion time, the system makes a material call. This comparison ensures that raw materials are replenished before they are depleted in the production line hence avoiding a halt.

The system also records the Isolation Forest algorithm used in the detection of the anomalous data which is a good method of outlier detection. This serves to reduce the impact of such factors as equipment failure that might cause wrong entry of data into the system, offering more reliable forecasts.

The Isolation Forest's only use case is Anomaly Detection, and it works by building a random binary search tree. Mathematically, it is possible to express the principle of the algorithm with the help of the following formula:

$$IF = \frac{E[h(x)] - c(n)}{b(n)}$$

where IF is the anomaly score of the Isolation Forest, h(x) is the average path length of the data point, x in the random tree, c(n) is the expected path length in a tree with n nodes, and b(n) is a normalization constant used to adjust the range of the anomaly score.

Additionally, the Long Short-Term Memory (LSTM) network is used for forecasting the consumption of materials over time. By analyzing historical data, LSTM can capture consumption patterns and trends, allowing it to predict future demand.

The update equations for an LSTM unit can be expressed as:

$$i_t = \sigma(W_{xi}x_t + W_{hi}h_{t-1} + b_i)$$

$$f_t = \sigma(W_{xf}x_t + W_{hf}h_{t-1} + b_f)$$

$$o_t = \sigma(W_{xo}x_t + W_{ho}h_{t-1} + b_o)$$

$$\tilde{c}_t = \tanh(W_{xc}x_t + W_{hc}h_{t-1} + b_c)$$

$$c_t = f_t \cdot c_{t-1} + i_t \cdot \tilde{c}_t$$

$$h_t = o_t \cdot \tanh(c_t)$$

Here, i_t, f_t and o_t are the activation values of the input gate, forget gate, and output gate respectively, while $\tilde{c}_t$ is the candidate cell state. c_t is the cell state, and h_t ht is the hidden state. σ is the sigmoid function, and is the sigmoid function, and tanh is the hyperbolic tangent function. W and b represent weights and biases.

In operation, the system continuously monitors material consumption at each workstation. Through IoT sensors deployed on the production line, real-time data on inventory and consumption is collected and transmitted to a centralized real-time database for processing and storage. Once the estimated depletion time (Et) is calculated, it is compared with the advance calling time (At). If At is greater than or equal to Et, indicating there is sufficient buffer time to initiate a material request before depletion, the system will automatically trigger a replenishment request.

Simultaneously, upon receiving the replenishment request, the AGV scheduling system calculates the optimal path and scheduling plan. This ensures that the AGVs can complete material delivery in the shortest possible time, taking into account the current position of the AGVs, traffic conditions within the production facility, and the priority levels of various tasks. This enhances material flow and at the same time reduces interference with the normal production processes.

In a nutshell, the system encompasses real-time monitoring, early warning, and AGV-based logistics to ensure timely provisioning and hence uninterrupted production. They also use Isolation Forest and LSTM to improve the speed, minimize effort, and increase the performance of lines for cigarette packing.

4 Implementation

System Architecture

The proposed system is composed of four interrelated modules:

Real-Time Monitoring Module: This module is in charge of the constant collection of the usage and stock status of supporting materials at several feeding points in the packaging subsectors. In this regard, IoT sensors are installed in appropriate places to track the status of the stock in real time. It has various sensors that send data to the Real-Time Database that is conversely responsible for the processing and storage of the data.

Early Warning and Automatic Call Module: This module uses sophisticated algorithms to compare real-time data with certain safety limits. It employs historical data and Machine learning algorithms to forecast usage patterns and precede alarm and consequent automatic re-ordering. It can adapt the safety limits promptly depending on production plans and expected rates of use, thus being preventive in terms of the material supply.

AGV Scheduling System Module: This system decides the path and schedule for AGVs after having received the replenishment requests from the Early Warning module. Such factors as the current position of AGVs, traffic conditions in the production facility, and priorities of different tasks are taken into account for effective organization of AGV dispatching and routing. The module also guarantees that the flow of materials is smooth while trying not to interfere with the other production processes.

Real-Time Database System: This module serves as the archival database of all the material-related data ranging from the current positions, the usage rate, and history. It enables data analysis and the machine learning algorithms used by the functionality Early Warning module. The database guarantees that first, second, third, and fourth Party data is comprehensive, and up-to-date to support decision-making.

Data Processing and Analysis

The data processing workflow is designed to ensure the accuracy and efficiency of the system's operations: The data processing workflow is designed to ensure the accuracy and efficiency of the system's operations:

Data Preprocessing: The IoT sensor data is preprocessed to remove noise and outliers of the raw data collected. For this purpose, the Isolation Forest algorithm is used that can help to filter out such records that may hurt the analysis of data.

```python
import pandas as pd
from sklearn.ensemble import IsolationForest
# Sample code snippet for anomaly detection using Isolation Forest.
df = pd.DataFrame({'consume': [d1, d2, d3...]})
model = IsolationForest(n_estimators=100, contamination=0.1)
model.fit(df[['consume']])
df['scores'] = model.decision_function(df[['consume']])
df['anomaly'] = model.predict(df[['consume']])
cleaned_data = df[df['anomaly'] == 1]  # Filter out anomalies
```

Intelligent Early Warning: After cleaning the data, Long Short-Term Memory (LSTM) networks are applied to forecast future consumption rates. LSTM is chosen for its ability to handle time-series data and capture long-term dependencies. The forecasts help in dynamically adjusting the safety inventory levels and trigger early warnings if the predicted usage is likely to exceed the threshold (Fig. 1).

```python
import numpy as np
from keras.models import Sequential
from keras.layers import LSTM, Dense
# Sample code for consumption forecasting using LSTM
data = cleaned_data['consume'].values.reshape(-1, 1)
train_size = int(len(data) * 0.8)
test_size = len(data) - train_size
train, test = data[0:train_size, :], data[train_size:len(data), :]
def create_dataset(dataset, look_back=1):
    X, Y = [], []
    for i in range(len(dataset) - look_back - 1):
        a = dataset[i:(i + look_back), 0]
        X.append(a)
        Y.append(dataset[i + look_back, 0])
    return np.array(X), np.array(Y)
look_back = 3
trainX, trainY = create_dataset(train, look_back)
testX, testY = create_dataset(test, look_back)
model = Sequential()
model.add(LSTM(4, input_shape=(look_back, 1)))
model.add(Dense(1))
model.compile(loss='mean_squared_error', optimizer='adam')
model.fit(trainX, trainY, epochs=20, batch_size=1, verbose=2)
```

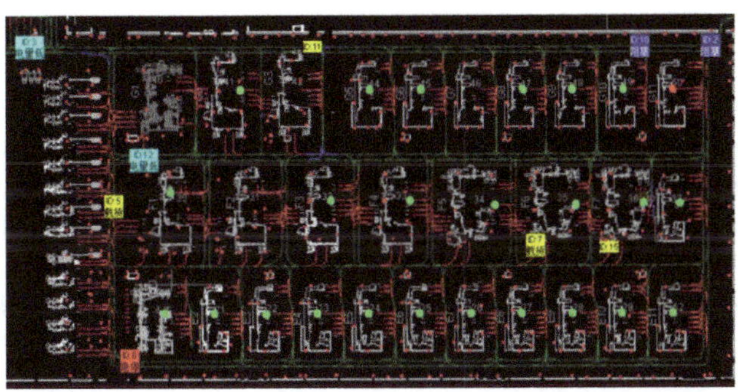

Fig. 1. Live Demonstration of Autonomous Vehicle Operation in Tobacco Industry.

5 Conclusion

The current research offers a new perspective on the handling of additional items in the course of packing cigarettes in the tobacco sector. Introducing the real-time monitoring, early warning, and AGV-based logistics of the proposed Auxiliary Material Early Request System. The advantages of the system that has been adopted include; Increased production rates, a decrease in the intensity of work and labor, and better decision-making through the provision of data analysis tools.

Our results highlight the importance of the application of sophisticated machine learning strategies and digital tools to increase the efficiency of the usage of resources.

Future developments will be concentrated on the improvement of these methods and the application of them in other industries. That is why the principles and methodologies outlined in this study provide a solid basis for addressing similar problems in other industries with high production demand.

To evaluate the effectiveness of the system several KPIs were selected and compared before and after the implementation of the system. The specific comparative data are as follows (Fig. 2):

Content	Before Implementation	After Implementation
Downtime	4hours/month	2 hours/month (50% reduction)
Production Cycle	3 hours/batch	2.5 hours/batch (10% reduction)
Manual Interventions	100 times/week	70 times/week (30% reduction)
Forecast Accuracy	85%	95% (10% increase)
False Alarm Rate	10%	2.5% (75% reduction)
Replenishment Cycle	7 days	5 days (29% reduction)

Fig. 2. Table of Before and After Effect Implementation

The comparative data above illustrate the fact that the introduction of the system led to the improvement of production efficiency, better forecasting, and, therefore, a decrease in material losses, and the shortening of the replenishment lead time. These improvements not only increased the overall efficiency of the production line but also significantly improved resource utilization.

Through these quantitative data, it is clear that the system has brought significant benefits in multiple aspects, demonstrating its important value and effectiveness in auxiliary material management.

References

1. Yang, T., Su, C.T., Hsu, Y.R.: Systematic layout planning: a study on semiconductor wafer fabrication facilities. Int. J. Oper. Prod. Manag. **20**(11), 1359–1371 (2000)
2. Wiyaratn, W., Watanapa, A., Kajondecha, P.: Improvement plant layout based on systematic layout planning. Int. J. Eng. Technol. 76–79 (2013). (0975–4024)
3. Tang, X.Z.: Plane layout planning of the distribution center of passenger car general assembly. Logist. Sci.-Tech. (2013)
4. Zhang, H., Wu, D., Yao, T.: Research on AGV trajectory tracking control based on double closed-loop and PID control. In: Journal of Physics: Conference Series, vol. 1074, no. 1, p. 012136. IOP Publishing (2018)

5. Li, X., Liao, K., Chen, W.: Two-stage Multi-AGV path planning based on speed pre-allocation. In: 2019 IEEE 3rd Advanced Information Management, Communicates, Electronic and Automation Control Conference (IMCEC), pp. 657–663. IEEE (2019)
6. Tavares, R.S.: Simulated annealing with adaptive neighborhood: a case study in off-line robot path planning. Expert Syst. Appl. **38**, 2951–2965 (2011)
7. Srinivas, N., Deb, K.: Muiltiobjective optimization using nondominated sorting in genetic algorithms. Evol. Comput. **2**(3), 221–248 (1994)
8. Nishi, T., Tanaka, Y.: Petri net decomposition approach for dispatching and conflict-free routing of bidirectional automated guided vehicle systems. IEEE Trans. Syst. Man Cybern.-Part A: Syst. Hum. **42**(5), 1230–1243 (2012)
9. Xu, W., Wang, Q., Yu, M., et al.: Path planning for multi-AGV systems based on two-stage scheduling. Int. J. Performabil. Eng. **13**(8), 1347 (2017)
10. Ropke, S., Pisinger, D.: An adaptive large neighborhood search heuristic for the pickup and delivery problem with time windows. Transp. Sci. **40**(4), 455–472 (2006)
11. Atashpaz-Gargari, E., Lucas, C.: Imperialist competitive algorithm: an algorithm for optimization inspired by imperialistic competition. In: 2007 IEEE Congress on Evolutionary Computation, pp. 4661–4667. IEEE (2007)
12. Barer, M., Sharon, G., Stern, R., et al.: Suboptimal variants of the conflict-based search algorithm for the multi-agent pathfinding problem. In: Proceedings of the International Symposium on Combinatorial Search, vol. 5, no. 1, pp. 19–27 (2014)

Advancing National Park Security Deployment of IoT and Deep Learning in Electromechanical Systems

Huang Qin[✉]

Chengdu Vocational College of Agricultural Science and Technology, Sichuan 611130, China
hezhilinec@163.com

Abstract. This research aims to elevate the operational efficiency of electromechanical systems within national nature reserves, focusing on enhancing visitor safety and ecological preservation through cutting-edge computer technology. Central to our systematic methodology—spanning design, implementation, and evaluation—is the optimization of monitoring mechanisms and sensor technology, integrated with advanced data processing capabilities. We introduced an innovative deployment of high-resolution imaging and infrared sensors, orchestrated through an automated system empowered by the Internet of Things (IoT). This setup ensures seamless data transmission and sophisticated processing, with a significant emphasis on computer science principles. Additionally, we leveraged deep learning algorithms to refine image analysis and manage data flows more effectively, which are crucial for the real-time processing needs of the system. These computational enhancements have dramatically improved the system's responsiveness, reducing event processing time from 15 to 5 min and achieving a 95% problem resolution rate. The frequency of system failures has also notably decreased, leading to a marked increase in tourist satisfaction. This study demonstrates that leveraging computer technology, particularly through the application of big data analytics and automated data processing systems, can significantly improve the efficiency and stability of safety systems in national parks, providing a stronger technical foundation for both environmental conservation and the safety of park visitors.

Keywords: component · park safety · electromechanical technology system · data processing technology · system efficiency improvement

1 Introduction

Against the backdrop of the continuous expansion of tourism activities, the security and protection system of nature reserves around the world has become particularly critical. With the increase in population mobility and the higher requirements for environmental protection, the existing security system has encountered multiple challenges and urgently needs to be innovated and strengthened with the help of advanced scientific and technological forces. As an efficient tool, the electromechanical technology system

© The Author(s) 2026
P. Siarry et al. (Eds.): WCNA 2024, LNEE 1550, pp. 396–403, 2026.
https://doi.org/10.1007/978-981-95-6946-5_40

that integrates electronic technology and mechanical technology has opened up a new way to improve the efficiency of the national park's security and prevention system [1]. This study is committed to in-depth analysis of the current security and protection status and existing problems of national parks, and then by optimizing the architectural design of the electromechanical technology system, improve its comprehensive efficiency and rapid response capabilities in the actual operation link. This study focuses on optimizing the park's security and prevention system through a systematic methodology, and emphasizes the suitability of technology applications to the environment and its economic returns, aiming to ensure the balanced coexistence of tourists' personal safety and natural ecology [2].

2 Analysis of the Current Situation and Problems of National Park Security

2.1 Overview of National Park Security System

In order to improve the security and prevention capabilities of national parks, their systems must be carefully designed and optimized. This task requires comprehensive consideration of many global influencing factors, including the decline of biodiversity, the continuous change of climate, and the transformation of social structure, cultural model, economic layout and information technology involved in the process of globalization. As shown in Fig. 1, global driving forces have an impact on the human and natural resources within the national park area, which in turn affects the operation and effectiveness of the security and protection system. In the field of human composition, many roles are involved, such as government agencies, private enterprises, non-governmental organizations, and community participants on site. These participants include long-term residents, new immigrants, temporary residents, and tourists from home and abroad. In the public space of the park, the interaction between various stakeholders has constructed its unique social structure and operation mechanism [3]. In order to ensure the harmonious coexistence of artificial structures and natural environments in national parks, as well as the long-term sustainability of resources, it must be comprehensively considered in its security and protection measures. In the process of improving the electromechanical technology system of national parks, it is crucial to build a flexible and comprehensive security solution that can adapt to multiple complex elements to ensure that the natural resources of the park are properly protected while meeting the safety needs of tourists and local residents [4].

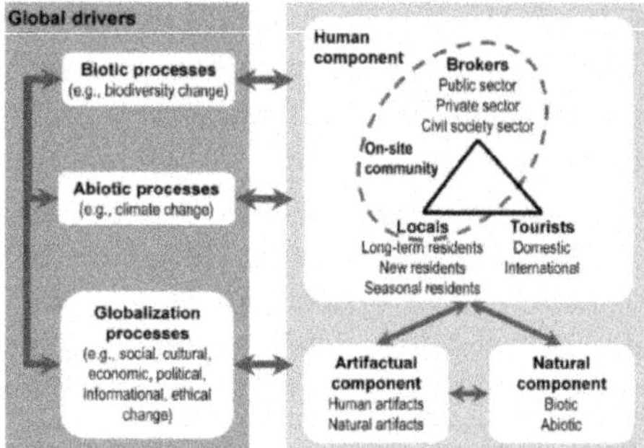

Fig. 1. Composition of the National Park System

2.2 Problem Analysis

In the actual operation process, although many global factors have been incorporated into the design of the security system of national parks, many challenges still stand out. First, the global evolution of society and the frequent changes in culture have greatly increased the challenges at the management level. In this context, traditional electromechanical safety systems are often unable to adapt quickly to current changes, which in turn has a negative impact on immediate safety protection and overall effectiveness [5]. Climate change and the frequent natural disasters it triggers have exposed the lack of flexibility and responsiveness of existing safety systems in the face of sudden natural events. Due to the limitations of technology and financial funds, some national parks have omissions in the layout of the security system, resulting in blind spots in monitoring. In the humanities field, the lack of stakeholders to coordinate the interests of all parties often leads to an imbalance in resource sharing and inefficient implementation of emergency strategies. These problems have led to a decrease in security efficiency and increased the risk level at the park management level, posing a threat to the safety of tourists and biodiversity. In order to improve the security system of national parks, it is necessary to conduct a detailed analysis of the existing problems and formulate practical solutions [6].

3 Optimal Design of Electromechanical Technology System

3.1 System Design Principles

When designing a national park security system, a series of established principles must be strictly followed to ensure its effective operation and long-term reliability. First, when designing the system, the principle of comprehensiveness must be followed. This principle requires comprehensive consideration of global drivers, including various elements of both human and natural aspects, while taking into account the specific needs of different groups involved. An efficient system must be built on a highly flexible and adaptable basis that can quickly adapt to changes in environmental and management needs [7]. In terms of system design, it is key to focus on scalability and maintainability in order to adapt to the development of technology and the evolution of park needs, and ensure that the system can be upgraded and maintained as necessary. In system design, maximizing monitoring and response speed, as well as minimizing operating and maintenance costs, are key considerations for efficiency issues. When designing the final security system, another key factor must be considered - environmental friendliness, ensuring that the technical solutions used have the least impact on the natural ecological environment of the park. The use of a series of carefully formulated design guidelines will greatly promote the improvement of a park's safety and efficiency while ensuring its environmental sustainability [8].

3.2 Key Technology Selection

For the security system of national parks, in the process of selecting key technologies, the research focuses on the integration and optimization of monitoring and control systems. In terms of technology selection, the construction of automatic video surveillance systems is mainly considered, including high-resolution and night vision cameras, as well as supporting advanced image data analysis algorithms. For large amounts of image data, deep learning convolutional neural networks (CNNs) are used to achieve accurate processing and analysis [9]. The basic operations of convolutional neural networks (CNNs) can be expressed by the following convolution operation formula:

$$a_{ij}^{(l+1)} = f\left(\sum_m \sum_n w_{mn}^{(l)} \cdot a_{i+m,j+n}^{(l)} + b^{(l)}\right)$$

where is $a_{ij}^{(l+1)}$ the activation value of $w_{mn}^{(l)}$ the first layer, $l + 1$ is the weight of the convolution kernel, $b^{(l)}$ is the bias term, f is the activation function. In deep learning, the choice of activation function is crucial, among which ReLU and its variants and the sigmoid function are common Two options, using this technology, can significantly improve the accuracy of identifying monitoring targets, thereby speeding up the response speed and accuracy of the monitoring system. In the field of control system optimization, automated control technology and remote monitoring systems are adopted to To improve the performance of control systems, IoT technology is used to integrate sensors and cameras for real-time capture and in-depth processing of data. The following formula can be used to optimize communication between IoT devices:

$$R = B \log_2\left(1 + \frac{S}{N}\right)$$

In communication systems, R is the data transmission rate, B is the bandwidth, S is the signal power, and N is the noise power, which are key parameters that affect the signal transmission quality. These formulas are beneficial to the design and optimization of wireless communication networks, so that Ensuring the stability and immediacy of information transmission and the comprehensive use of these advanced technologies can significantly enhance the national park's security system, allowing it to respond more quickly to potential threats. At the same time, it also improves the stability and economic benefits of the system [10].

3.3 System Integration and Optimization

In the security system of national parks, the efficient operation and flexible response of the system based on electromechanical technology depends on the overall construction and continuous optimization of the system. In this study, the cutting-edge integration technology and its optimization strategy are emphasized to ensure the seamless collaboration of various monitoring, perception and control systems. First, by configuring a centralized management platform, the data fusion of multiple types of monitoring instruments and perception elements is realized. In an efficient central processing module, this system implements a multi-layer architecture strategy and uses real-time data stream processing technology to ensure efficient data processing and transmission. In the process of processing data streams, data fusion is a crucial link, which covers the comprehensive processing of information from multiple sources. When the mathematical model aims to achieve data fusion, it can be characterized by the weighted average method. The weight coefficient of each data source in this method reflects its reliability and accuracy. The corresponding mathematical formula is as follows:

$$x_{\text{est}} = \frac{\sum_{i=1}^{n} w_i x_i}{\sum_{i=1}^{n} w_i}$$

where x_i is the observed value of the i th data source, w_i is the corresponding weight, and x_{est} is the estimated value after fusion. Furthermore, in order to optimize the performance and resource allocation of the system, an optimization algorithm based on linear programming is introduced for decision support and resource allocation. For example, the optimization algorithm can be expressed as:

$$\text{Minimize } Z = c^T x$$
$$\text{Subject to } Ax \le b$$
$$x > 0$$

to optimize the decision variable vector c under the premise of minimizing the total cost, where A represents the cost vector, b represents the technical coefficient matrix, b is the resource constraint vector, and Z is our objective function. In order to adapt to environmental changes and operational requirements, the system is regularly evaluated for performance, and heuristic algorithms such as simulated annealing are used to adjust resource allocation and configuration, and then optimize the process. With the help of the simulation of the physical annealing process, a series of algorithms gradually explore and lock the global optimal solution. The process can be described by the formula:

$$P(e, e', T) = \exp\left(-\frac{e' - e}{kT}\right)$$

where e and e' T are the energies of the current solution and the new solution respectively, k is the Boltzmann constant, and P is the probability of accepting the new solution.

4   System Implementation and Effect Evaluation

4.1   Implementation Plan

In this academic study, a mountain forest park known for its rich biodiversity and large number of tourists was selected as a case to verify the implementation effectiveness of the national park safety prevention system. The average annual number of tourists exceeds 500,000. The first phase of the plan involves installing a cutting-edge surveillance system at strategic locations in the park, consisting of high-definition cameras and Composed of infrared activity sensors. Each camera is equipped with ultra-high-definition 4K resolution configuration parameters with a monitoring range of 300 square meters, and has the night vision ability to clearly capture images in no light environment. In some key locations, such as the entrance to the park, the main Sensor equipment is installed on paths and dangerous areas where accidents occur frequently. Its function is to collect real-time data on the movement of wild animals and the flow of tourists. In the process of system integration, Internet of Things technology is used to realize the interconnection of various devices. Interoperable and unified access to the central data processing facility at the core. In order to cope with the real-time processing requirements of a large number of data streams, an optical fiber network with a network bandwidth of 1 Gbps is used to realize real-time transmission of data. In the data center, in order to cope with the processing and storage of large-scale data, a dual ten-core CPU is deployed, high-performance server with 64 GB RAM and 10 TB SSD hard drive, real-time image and sensor data analysis is achieved by fusing machine learning algorithms with automated analysis software, which processes the data. By analyzing past information patterns, the algorithm can independently identify anomalies such as unauthorized entry, lost visitors, or proximity to risk areas. Once identified, staff responsible for park management and security will be notified immediately. This process is carried out through a mobile app. Application implementation. One component of this solution is a comprehensive emergency response mechanism. This mechanism can automatically call the nearest security forces or implement corresponding pre-plan measures for various security incidents. Once a fire situation is detected, the automated system The nearest fire-fighting equipment will be activated, and the park's public broadcasting network will be used to issue warnings to all visitors. It is foreseen that through the gradual implementation of the plan, the park's safety maintenance level will be greatly enhanced, safety accidents caused by human factors and natural phenomena will be reduced, and visitors will be ensured safety and protection of park assets.

4.2   Effect Evaluation

After the park implemented a comprehensive system of safety precautions, a detailed evaluation of its effectiveness was conducted in order to estimate the real effect of the system improvements on the park's safety and operational efficiency. In the process of comprehensive evaluation of system performance, key performance indicators play a vital role. These indicators include not only event response time and event handling efficiency, but also cover aspects such as system stability and visitor satisfaction. The central data processing center provides a full year of data records and analysis results on safety incidents, which is necessary for effectiveness evaluation.

Table 1. Comparison before and after implementation

index	Before implementation	After implementation
Average incident response time (minutes)	15	5
Incident resolution rate (%)	75	95
System failure rate (occurrences per month)	4	1
Tourist satisfaction rating (out of 5 points)	3.5	4.5

After in-depth analysis of the information in Table 1, it revealed a significant optimization process, which reduced the original average incident response time from 15 min to 5 min, reflecting that the system response speed has been significantly enhanced through efficient data processing and alarm delivery., the new system has significantly increased the incident resolution efficiency from 75% to 95%, which shows that the system's performance in identifying and responding to potential risks has been significantly enhanced. In addition, the reliability of the system has been significantly improved, and maintenance work Improvements have also been seen. The number of monthly failures has been reduced from four to one, and tourists' satisfaction scores have increased from 3.5 to 4.5. This change shows that their perception and satisfaction with the safety system have significantly improved. This data set not only It verified the effective deployment of the security protection system at the technical level, and revealed its significant role in enhancing management efficiency, reducing security risks, and optimizing tourists' experience.

5 Summary and Outlook

This academic research conducted an in-depth analysis and enhancement of the electromechanical safety protection system within the national park, significantly boosting both the system's overall performance and its response agility. Central to our approach was the practical application of advanced computer technologies and the strategic foresight in selecting implementation plans for these technical solutions. System integration was meticulously managed to ensure both functionality and future scalability. Evaluation results have demonstrated that the integration of cutting-edge computer technology and automated processes substantially improved the efficiency and accuracy of incident response, while also fortifying the system's stability. Consequently, there was a marked increase in tourist satisfaction.

Looking forward, as technology continues to advance and environmental adaptability improves, we plan to explore a range of innovative applications. For instance, leveraging the robust capabilities of artificial intelligence, particularly through deep learning, will enhance data processing efficiency. This will enable us to address increasingly complex security threats more effectively. Continuous improvements and upgrades to the national park's security system are planned, focusing on sophisticated computer algorithms and automated electromechanical designs to provide a robust technical foundation for the protection of natural resources and the safety of visitors.

References

1. Adegbite, A.O., Nwasike, C.N., Nwaobia, N.K., et al.: Innovative power management in electro-mechanical systems: exploring the new paradigms of energy efficiency and system longevity. Eng. Sci. Technol. J. **4**(6), 401–417 (2023)
2. Matveev, S.A., Korotkov, E.B., Zhukov, Y.A., et al.: Diagnostic and monitoring system for technical condition of electromechanical section of thermal control systems in spacecraft. Int. J. Math. Eng. Manag. Sci. **5**(1), 181 (2020)
3. Zheng, S., Fu, Y., Wang, D., et al.: Investigations on system integration method and dynamic performance of electromechanical actuator. Sci. Prog. **103**(3), 0036850420940923 (2020)
4. Kim, S., Park, K.J., Lu, C.: A survey on network security for cyber–physical systems: from threats to resilient design. IEEE Commun. Surv. Tutor. **24**(3), 1534–1573 (2022)
5. Li, C., Zhuo, G., Tang, C., et al.: A review of electro-mechanical brake (EMB) system: structure, control and application. Sustainability **15**(5), 4514 (2023)
6. Ali, G., Robert, W., Mijwil, M.M., et al.: Securing the Internet of Wetland Things (IoWT) using machine and deep learning methods: a survey. Mesopotamian J. Comput. Sci. **2025**, 17–63 (2025)
7. Ohalete, N., Aderibigbe, A., Ani, E., et al.: Challenges and innovations in electro-mechanical system integration: a review. Acta Electronica Malaysia (AEM) (2024)
8. Delwar, T.S., Aras, U., Mukhopadhyay, S., et al.: The intersection of machine learning and wireless sensor network security for cyber-attack detection: a detailed analysis. Sensors **24**(19), 6377 (2024)
9. Hashim, M., Tabassum, B., Khan, T., et al.: Advancing natural resources management through the environmental IoT-based model economy. In: IoT-Based Models for Sustainable Environmental Management: Sustainable Environmental Management, pp. 165–195. Springer, Cham (2024)
10. Dimililer, K., Dindar, H., Al-Turjman, F.: Deep learning, machine learning and internet of things in geophysical engineering applications: an overview. Microprocess. Microsyst. **80**, 103613 (2021)

Construction of a Library Intelligent Book Selection Decision Support System: Based on Data Mining of the RFID System

Yanxiang Lu[✉]

Library of Jianghan University, Wuhan, China
327579500@qq.com

Abstract. In the digital environment of 2025, intelligent libraries have achieved efficient management and service through RFID technology. However, the value of RFID data has not yet been fully realized. This paper takes the library of Jianghan University as an example to construct an intelligent book selection decision support system based on RFID circulation data. The system analyzes book borrowing data, generates keywords, and performs clustering to ultimately achieve personalized recommendations. The system can effectively enhance the scientific nature of book selection and the utilization rate of resources, providing intelligent support for library acquisition decisions. In the future, with the application of technologies such as smart bookshelves, the system will be further optimized to better meet the needs of readers.

Keywords: Smart library · RFID technology · Data mining · Intelligent book selection · smart bookshelf

1 Introduction

In the digital landscape of 2025, the smart library centered on RFID (Radio Frequency Identification) technology has become the central node of the urban knowledge ecosystem. Through the deep integration of miniature RFID tags with books, each book has become a dynamic data carrier. Integrated with numerous IoT sensor nodes, the RFID smart library management system can achieve functions such as book positioning, self-service borrowing and returning, book navigation, and self-service inventory management.Artificial intelligence algorithms, based on readers' borrowing trajectories and the dynamic big data of the library's collection, have realized personalized recommendations and intelligent suggestions.

The earliest known application of RFID technology in libraries was at the National Library of Singapore, which began using high-frequency RFID tags to replace barcodes for document management in 1998 [1]. In China, libraries started to introduce RFID technology in 2006, gradually achieving functions such as self-service borrowing and returning of books, self-service inventory, document navigation, intelligent recommendations, and security against theft. As RFID technology has matured and the cost of

© The Author(s) 2026

P. Siarry et al. (Eds.): WCNA 2024, LNEE 1550, pp. 404–413, 2026.
https://doi.org/10.1007/978-981-95-6946-5_41

equipment and tags has decreased, it has become widely applied in the library field, effectively improving the work efficiency and service quality of libraries. Increasingly, libraries are leveraging innovative applications of RFID, such as smart bookshelves, inventory robots, intelligent compact storage systems, and AGV-based sorting, to achieve scientific and intelligent management and services [2]. Despite the vigorous application of RFID technology in libraries, its data value has not yet been fully explored and utilized. In the context of mobile internet, as electronic resources gradually become the primary resources utilized by readers, the borrowing volume of printed documents in libraries continues to decline year by year, and the budget for purchasing printed documents keeps shrinking. As a result, enhancing the scientific nature and intelligence of purchasing printed resources has become one of the key tasks for libraries (Table 1).

Table 1. Key Equipment in the Implementation Plan of RFID Technology for University Libraries.

Libraries	RFID Tags	RFID Shelf Labels	Self-Service Borrowing and Returning Equipment	Security Gates	Librarian Workstation	Mobile/Portable Inventory Devices	Smart Bookshelves	Reservation Book Cabinets	Book Sorting System
Sun Yat-sen University Library	√	√	√	√	√	√	√	√	×
Wuhan University Library	√	√	√	√	√	√	√	√	√
Shanghai Jiao Tong University Library	√	√	√	√	√	√	×	×	√
China Agricultural University Library	√	√	√	√	√	√	×	√	√
Shenzhen Technology University Library	√	√	√	√	√	√	×	√	√
The Chinese University of Hong Kong, Shenzhen Library	√	√	√	√	√	√	×	×	√
Jianghan University Library	√	√	√	√	√	√	×	×	×

Due to the lack of comprehensive and intuitive data on the utilization of documents, selectors often have to rely on their own knowledge structure, work experience, and established procurement standards to make subjective selection decisions. Consequently, the purchased books fail to accurately meet the readers' needs and thus remain largely untouched.

Therefore, deeply integrating RFID circulation data into the selection process and thoroughly mining readers' borrowing behavior data—such as categories of borrowed books, borrowing frequency, and number of renewals—can be highly beneficial. By leveraging machine learning techniques for analysis and modeling, and constructing an intelligent selection decision support system, it is possible to enhance the scientific nature and efficiency of selection decisions. This approach can better meet user needs and improve the utilization rate of resources.

The application of cutting-edge technologies such as big data, artificial intelligence, and machine learning is constantly driving innovation in book acquisition and literature resource construction. By analyzing vast amounts of data, these technologies can accurately predict readers' needs and optimize book purchasing strategies, thereby improving the efficiency of book selection. At the same time, they also assist librarians in making more scientific decisions, ensuring that the selected books not only meet the interests of current readers but also anticipate future trends. This approach enhances the overall quality of literature resource construction. The development of library resources is thus moving towards a high-quality direction characterized by data-driven, intelligent, and precise management [3].

2 Research Status of Intelligent Acquisition System

The intelligent acquisition and selection system has been widely studied and applied in the industry, involving procurement personnel, acquisition and selection algorithms, and acquisition and selection experiences. Muslikhin et al. proposed an automatic picking system based on artificial intelligence (AI) while studying the online store picking system and developed the YOLO algorithm, in which robots replace manual tasks for picking [4]. Many scholars have explored the application of big data and artificial intelligence in book acquisition and selection to promote the innovation of book acquisition and selection modes.

R. H. Wang, Y. Tang, G. Q. Liu, and others proposed a book purchase method based on a genetic neural network. The model adjusts the threshold according to different weights, thereby predicting whether to purchase a book. Simulation experiments have also shown that the book purchase model has good predictability and generalization ability. Compiling lists of books to be purchased through data processing can ensure the efficiency of new book purchases and the quality of the books. Zhu Huashun believes that using digital information technology and text mining algorithms to build a user-based book acquisition system can realize functions such as book resource provision, collection and analysis, and interactive processing [5].

Tu Jiaqi, Li Weichao, Fan Di, Wang Hong, and others have respectively studied the system architecture, recommendation algorithm, mode, and flow of the intelligent acquisition and selection system [6]. Based on understanding readers' needs, they strengthened the automation of acquisition and selection, improved the efficiency of librarians' acquisition and selection, and optimized the library's collection structure. These studies theoretically verify that big data and artificial intelligence technologies, as well as the extended intelligent interview system, can deeply analyze readers' needs, infer the quality of books, and effectively achieve the optimal allocation of resources. However, the research is still in the theoretical or design stage and has not been applied in practice.

In practice, some libraries have already implemented book acquisition and selection support systems. For example, the decision support system for book acquisition and selection developed by Fudan University Library constructs a dimension model of the book lending data warehouse and clickstream data warehouse based on the library's existing data. It enables multi-dimensional statistical analysis of various types of data, including collection data, circulation data, OPAC search logs, e-book usage, and comparisons between Chinese and English printed books and e-books. Meanwhile, Nanchang University Library has added a data collection system to its Huiwen system to facilitate the decision support system for Chinese book acquisition and selection. The data provided by this system include the author's resume and background, the number of papers published by the author in the field and their citations (based on the CNKI platform), as well as the frequency and number of previous works borrowed by readers. After the new book list is imported into the system, it offers decision support for Chinese book acquisition librarians through author association. Statistics show that the zero borrowing rate of books selected via the decision support system has decreased by 30%. These auxiliary acquisition systems provide librarians with references for acquisition decision-making through data statistical analysis [7].

3 Construction of an Intelligent Book Selection Decision Support System Based on RFID Circulation Data

3.1 Data Sources and Analysis Procedure

The data for this study are derived from the RFID system of Jianghan University Library. By collecting, analyzing, and mining the borrowing data within the RFID system, a book selection decision support system has been established. It is based on the analysis of book titles, authors, publishers, publication years, and borrowing frequencies. The system automatically screens for high-quality books that align with readers' needs and are closely related to disciplinary construction (Fig. 1). This approach provides intelligent support for acquisition librarians to make scientific and efficient acquisition decisions.

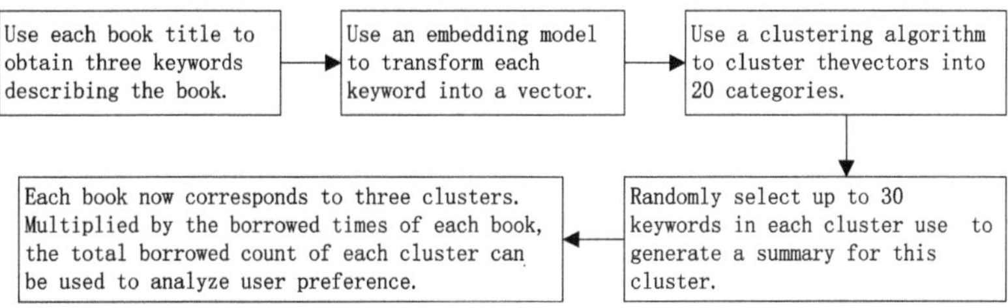

Fig. 1. Flowchart for the Construction of an Intelligent Book Selection Decision Support System.

3.2 Data Processing and Analysis

Select the book borrowing rankings for the years 2021, 2022, and 2023. Compile a list of the top 1000 books based on borrowing frequency and cluster them by title, author, publisher, publication year, and borrowing frequency (Table 2).

Table 2. Top 1000 Books Borrowed in 2023

Title	Author	Publisher	Publication Year	Borrowing Frequency
Plastic Mold Design and Manufacturing Training Manual	Zhang Jingli	Chemical Industry Press	2009	9
C Primer Plus	Stephen Prata	People's Posts and Telecommunications Press	2016	9
Data Structures: An Object-Oriented Approach Using C + +	Yin Renkun	Tsinghua University Press	2021	9
…	…	…	…	…

By employing the TF-IDF algorithm, three keywords that can accurately encapsulate the core content of the book were extracted based on its title. These keywords are: Keyword1, Keyword2, Keyword3 (Table 3).

Table 3. Keyword1 Keyword2 Keyword3

Keyword 1	Keyword 2	Keyword 3
Plastic Molds	Design and Manufacturing	Training Manual
C Language	Programming Basics	Programming Fundamentals
Data Structures	Object-Oriented	C + + Language
…	…	…

An embedding model is used to convert the Chinese keywords into corresponding numerical vectors for statistical analysis. This process utilizes Conan-embedding-v1 model (Table 4).

Table 4. Embedding_Keyword 1 embedding_Keyword 2 embedding_Keyword 3

embedding_Keyword 1	embedding_Keyword 2	embedding_Keyword 3
[-0.6741449 0.3107311 0.176753… 0.93892175 -0.07477136 -0.8304186]	[0.61032474 -0.8798427 0.8936814… 0.6140771 -0.10805961 -0.9319009]	[-0.12483905 1.1716944 0.10313071… -0.1960919 -0.37018725 -0.10309251]
[-0.5013875 0.8739114 0.66697943… -0.20928733 -0.74768126 -0.44689324]	[-0.58602476 0.79576063 -0.19914387… 0.22686122 -1.0212806 0.07211958]	[-0.9221621 1.2555287 0.11815277… 0.35628763 -0.9871921 -0.00566703]

(continued)

Table 4. (*continued*)

embedding_Keyword 1	embedding_Keyword 2	embedding_Keyword 3
[-1.2066745 0.41443998 0.10207829... 0.52706367 -0.42082876 -1.071272]	[-1.7080858 -0.50498974 0.40107566... 0.48774496 0.12681748 -1.4390962]	[-0.42585567 0.78526694 0.8984315... 0.0506241 -1.1467582 -0.5398382]
...	...	...

3.3 Clustering and Preference Analysis

A clustering algorithm is used to divide all the numerical vectors into 20 categories. The attribute "Cluster_Keyword" indicates which category (ranging from 0 to 19) each corresponding keyword belongs to. This process utilizes the K-Means algorithm (Table 5).

Table 5. Cluster_Keyword 1 cluster_Keyword 2 cluster_Keyword 3

cluster_Keyword 1	cluster_Keyword 2	cluster_Keyword 3
18	3	1
0	0	0
0	0	0
...	...	...

From each category, up to 30 keywords are randomly selected, and then Conan-embedding-v1 is employed to generate a concise summary to accurately reflect the core characteristics of the category. The attribute "Summary_Keyword" represents the description of the corresponding category. For example, "Engineering Technology" corresponds to cluster 18, "Design and Creativity" corresponds to cluster 3, and "Educational Resources" corresponds to cluster 1 (Table 6).

Table 6. Cluster_ summary_ Keyword 1 Cluster_ summary_ Keyword 2 Cluster_ summary_ Keyword 3

cluster_summary_Keyword 1	cluster_summary_Keyword 2	cluster_summary_Keyword 3
Engineering Technology	Design and Creativity	Educational Resources
Computer Programming Technology	Computer Programming Technology	Computer Programming Technology
Computer Programming Technology	Computer Programming Technology	Computer Programming Technology
...	...	...

Each book has three keywords, multiplied by the corresponding borrowing frequency, to calculate the total borrowing frequency of the 20 categories as a reference for readers' borrowing preferences. Building an intelligent recommendation system (Table 7).

Table 7. Statistical Summary

cluster_number	cluster_summary	weighted_size
0	Computer Programming Technology	1736
1	Educational Resources	1861
2	Philosophical Research	1181
3	Design and Creativity	933
4	Legal Education and Research	652
5	Sociological Research	632
6	Financial Investment	569
7	Film and Photography	625
8	Financial Accounting	292
9	Culture and Technology	2943
10	Mathematical Sciences	824
11	Art and Painting	1474
12	Business and Marketing Strategy	1630
13	Data Analysis	812
14	Historical Research	585
15	Politics and Development	1627
16	Political Studies	898
17	Economic Research	514
18	Engineering Technology	2064
19	Science and Chemical Technology	1023

4 The Application of RFID Smart Bookshelves

The extensive application of RFID technology has propelled the continuous innovation of service models in university libraries abroad, with the use of RFID smart bookshelves becoming increasingly widespread. The smart bookshelf is an innovative bookshelf that integrates RFID technology. Equipped with RFID readers and antennas installed on the bookshelf, it can achieve automatic identification, positioning, inventory, and management of books [8]. It not only allows for real-time monitoring of the circulation of books but also provides readers with convenient search and borrowing services. Through the smart shelf system, libraries are able to efficiently collect readers' reading data and conduct in-depth analyses of their reading preferences, thereby providing precise evidence for resource procurement and personalized recommendations.

4.1 Real-Time Monitoring of Book Status

After installing RFID sensors on the bookshelves, the system is capable of real-time monitoring of the status of each book, including whether it is on the bookshelf, borrowed, or returned. For example, when a book is taken off the bookshelf by a reader, the system can immediately recognize and update the book's status information, ensuring the real-time accuracy of the library's collection data [9].

4.2 Reading Data Statistics and Analysis

RFID smart bookshelves can meticulously record data such as the number of times each book is read and the distribution of reading time for each book. By conducting in-depth analysis of these data, libraries can clearly understand the popularity of books across different disciplines and types, thereby providing robust data support for book procurement. This not only makes the procured books more aligned with readers' actual needs but also further optimizes the structure of the library's collection resources [10].

4.3 Reader Behavior Analysis

By analyzing behavioral data such as the time readers spend in front of the bookshelves, the types of books they select, and the frequency of their selections, and combining these with readers' borrowing and returning records, libraries can comprehensively grasp the reading habits and interest preferences of readers from different majors. These analytical results provide rich practical evidence for the construction of subject-specific literature resources in libraries, assisting them in better supporting the subject-specific development of universities with literature resources.

5 Conclusion

This study is based on the extensive application of RFID technology in the library field and innovatively integrates RFID technology with artificial intelligence in a deep and comprehensive manner. By conducting in-depth data mining and analysis of the circulation and borrowing data within the RFID system, we have developed an intelligent book selection decision support system for libraries. This system aims to optimize the book selection process, enhance the scientific nature and accuracy of book selection, and thereby better meet the needs of readers and improve the utilization efficiency of library resources.

RFID technology is the cornerstone for building smart libraries [11]. The current system is based on circulation and borrowing data. With the increasing application of RFID-enabled smart bookshelves, capturing data on readers' browsing and handling activities in front of the bookshelves can enrich the data sources. Subsequently, a comprehensive analysis of behavioral data from the RFID system, such as borrowing and reading activities, will be conducted to adjust and optimize the existing index system and parameter assignments. This will enable a more accurate reflection of readers' utilization patterns and thus achieve better predictive performance of the model.

This study constructs an intelligent book selection decision support system for libraries based on data mining of the RFID system, aiming to provide valuable references and insights for the developers and subsequent users of RFID technology. With the continuous development and improvement of RFID technology, its application in the field of university library management will become more extensive and in-depth. We look forward to RFID technology bringing more innovation and breakthroughs to library management, further enhancing service quality and efficiency, and creating a more high-quality reading environment for a wide range of readers [12].

Acknowledgment. This work is partially supported by the National Social Science Fund of "Study on the influence of university library", China (19BTQ014). The author also gratefully acknowledges the helpful comments and suggestions of the reviewers, which have improved the presentation.

References

1. Shen, K., Shao, B., Chen, L.: The research and practice of smart library based on UHF RFID. J. Mod. Inf. **36**(8), 88–92 (2016)
2. Wu, Y.: Design and Implementation of book reading behavior monitoring system based on RFID technology: a case study of Shenzhen university library. Libr. J. **42**(8), 57–64, 81 (2023)
3. Zhong, J., Meng, Z.: Research on Chinese book acquisition model based on RoBERTa and LightGBM. J. Acad. Libr. **43**(01), 82–92 (2025)
4. Fu, L., Hu, Y.: The model of book purchasing decision--making based on random forest. Inf. Res. (5), 34–39 (2020)
5. Tu, J., Yang, X., Shen, M.: Research and practice of decision-driven book intelligent interview system-taking Chongqing University as an example. Libr. Inf. Serv. **11**, 28–34 (2020)
6. Fan, D.: Computer-aided book selection and acquisitions with machine learning: a case study of open data from seattle public library. Libr. Sci. Res. Work (7), 58–61 (2020)
7. Zhong, J., Chen, J., Li, C., Chen, H.: Study on decision model of university library acquisition. J. Acad. Libr. **5**, 38–47 (2021)
8. Liu, R., Qiao, J., Liu, L.: Research and design of intelligent bookshelf management system based on UHF RFID. Mod. Electron. Techn. **44**(20), 42–46 (2021)
9. Xu, S.: Design and thinking of book location service based on HF RFID—taking the library of China Pharmaceutical University as an example. China High New Technol. **8**, 69–72 (2024)
10. Du, W.: Design of intelligent book ranging system based on RFID technology and FPGA main control module. Microcomput. Appl. **40**(4), 43–46 (2024)
11. Yilmaz, D., Ünal, Y.: Evidence-based acquisition at hacettepe university libraries. Libr. Res. Techn. Serv. **66**(3), 130–140 (2022)
12. Song, J.: Research on application of RFID technology in university libraries: taking Nanjing Normal University of Special Education as an example. Wirel. Internet Sci. Technol. (12), 89–93 (2024)

Extract-Related-Generate: A Question Decomposed Retrieval-Augmented Generation

Yongbao Xie[✉], Mingming Yang, and Ning He

The Second Research Institute of CAAC, Chengdu, China
{xieyongbao,yangmingming,hening}@caacsri.com

Abstract. Retrieval Augmented Generation (RAG) is a framework for augmenting Large Language Modeling (LLM) capabilities by retrieving external knowledge from large databases. This approach effectively solves the common problems of factual accuracy, information obsolescence, and "hallucination" in generative language modeling. RAG generates answers that are more demanding in terms of searcher performance and knowledge base content, to address these two problems this paper proposes a method called Extract-Related-Generate for RAG (ERGRAG), which is different from previous RAG methods in that it first decomposes the question, generates a number of sub-questions, and then improves the model's accuracy in responding to multi-hop questions through the model's answers to the sub-questions. This paper evaluates the performance of the model in three datasets, and the ERGRAG method proposed in this paper shows a large improvement in accuracy, indicating that question decomposition is an effective idea to improve the RAG system's response to multi-hop questions, and provides new ideas for subsequent RAG research.

Keywords: RAG · LLM · question decompose · ERGRAG

1 Introduction

Very large-scale language models like GPT-3 and BERT perform well in a wide range of natural language processing tasks. However, due to the static nature of the training data, these models are prone to generate inaccurate or outdated information. To alleviate this problem, Retrieval Augmented Generation (RAG) has emerged as an effective solution. By introducing an external retrieval mechanism, RAG can dynamically acquire the most up-to-date information, thus enhancing the quality of the generated text. The hybrid approach of RAG combines the advantages of retrieval-based and generative models, providing a more robust solution for knowledge-intensive tasks such as open-domain dialog and question answering.

RAG still faces many challenges in solving multi-hop problems, and the retrieve-then-read paradigm is commonly used to solve this problem [1]. This method first uses LLM to decompose a complex problem into multiple simple individual problems, and for each problem, LLM derives the answer from relevant documents. The method ignores the knowledge inherent in the large language model as the answers can only be derived

P. Siarry et al. (Eds.): WCNA 2024, LNEE 1550, pp. 414–422, 2026.
https://doi.org/10.1007/978-981-95-6946-5_42

from the retrieved documents. This can somewhat limit logical reasoning in multi-hop QA tasks [2]. It may be difficult for the retriever to retrieve all the necessary documents to answer the question, leading to performance degradation when using this approach. Answers obtained by retrieving individual questions may, to a certain extent, lead to the generation of irrelevant responses by the LLM and also lead to the generation of incorrect responses by the LLM if the individual questions are not generated accurately. Therefore, the development of new paradigms for retrieval-enhanced generation remains an active area of research (Fig. 1).

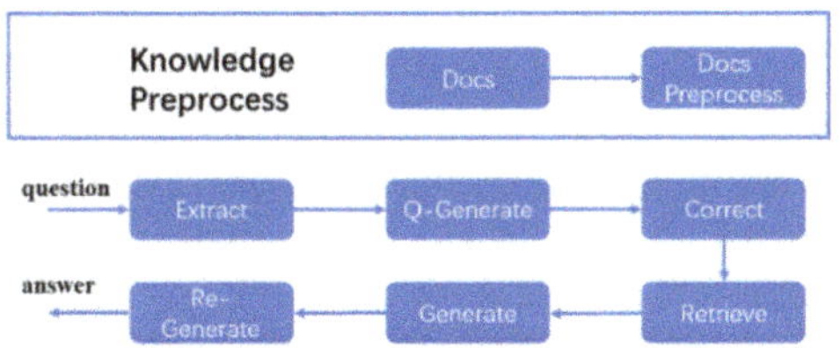

Fig. 1. Extract-Retiver-Generate RAG framework diagram

2 Related Work

2.1 Retrieval-Augmented Generation

The idea of integrating retrieval mechanisms into language models is not new. Early work such as REALM [3] and DPR [4] laid the groundwork for combining retrieval with generative models. REALM introduced a retrieval-reading framework that guided the generation process by obtaining information from a corpus, while DPR improved the relevance of document retrieval by using dense embeddings.

RAG [5] further extends this notion by creating an end-to-end retrieval and generation system that retrieves relevant documents from large corpora and utilizes them to generate output supported by external knowledge. More recent work such as FLARE [6] and Self-RAG [7] propose dynamic retrieval strategies based on model uncertainty triggers, which further improve the factual accuracy of the generated content.

To reduce the dependence of RAG systems on retrieval methods, recent work has mitigated this problem by incorporating world knowledge from LLMs [8], which uses LLMs to generate contextual documents as a knowledge base. Researchers are beginning to progressively explore ways to improve the performance of RAG systems by fusing the knowledge of the larger model itself with the knowledge base, but this approach suffers from ignoring potential knowledge conflicts between the two knowledge sources, which may cause learners to hallucinate [9] reports. How to balance the pre-trained knowledge of the model itself with the content of the knowledge base to reduce model hallucinations and improve answer accuracy is an important challenge for RAG systems.

2.2 Multi-hop Question Answering

The ReAct approach overcomes the problem of illusions and error propagation that can occur in chain-of-thinking reasoning by interacting with the knowledge base via an API at different stages of thinking, and generates human-like task-solving trajectories, which are more interpretable than the base method without reasoning traces [10].

To improve the ability of large models to solve multi-hop problems, Zhengliang Shi et al. proposed a method called "Generate-then-Ground" [11], which formulates simpler single-hop problems and generates the answers directly, then corrects the question-answer pairs in the retrieved documents. By formulating simpler one-hop questions and generating answers directly, and then correcting the quiz pairs in the retrieved documents to correct any incorrect predictions in the answers, the capabilities of the large model itself and the knowledge base are effectively combined, which ultimately improves the system's capability.

Jiahao Zhang et al. proposed the beam retrieval method to enhance the ability to answer multi-hop questions, which models the multi-hop retrieval process in an end-to-end manner by jointly optimizing the encoder and two classification heads on all hops. The performance is excellent on HotpotQA [12]. Thinking about multi-hop problems by making large models anthropomorphic is one of the ideas for solving multi-hop problems, and the most direct way is undoubtedly to decompose a single multi-hop problem into multiple single-hop problems, whether it is possible to accurately decompose a multi-hop problem is the main problem to be solved.

3   Extract-Related-Generate RAG

This section first introduces the overall framework of Extract-Related-Generate, and then describes in detail the systematic flow of the Extract-Related-Generate method, including the detailed steps of keyword extraction, related word generation, sub-question generation to generating the final answer.

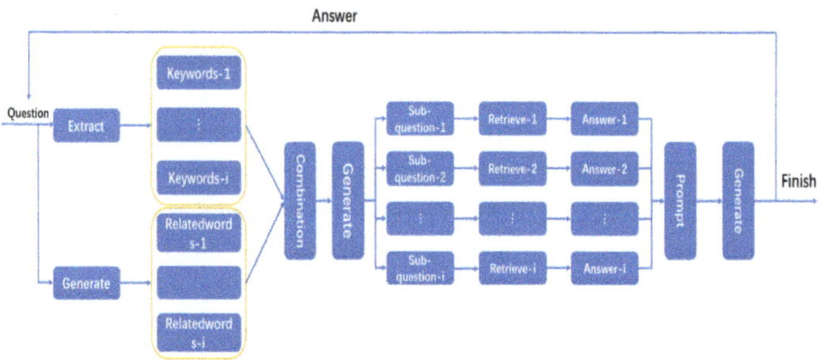

Fig. 2. Flowchart of the Extract-Related-Generate RAG method

3.1 Overview

Figure 2 shows the framework of Extract-Related-Generate RAG. The difference with other RAG methods is that this method requires the big model to preprocess the knowledge base, which translates the knowledge content into the most understandable sentences, and uses human involvement in the process to supervise and check the translated content for errors. Given a user question q and a corpus of documents, the big model extracts a number of keywords from the question, and at the same time generates a number of relational words that are most relevant to the question; a new series of sub-questions are formed by arranging and combining the keywords and relational words to form a new series of subquestions, and retrieval of the subquestions is enhanced using the subquestions for generating the answers, and when the answers to all the subquestions are obtained, all the answers to the subquestions are provided as cue words to the big model for generating the final answers.

3.2 Text Preprocessing

Large-scale language models, trained on massive amounts of data, are capable of understanding and generating natural language, performing a variety of complex tasks ranging from text generation to Q&A. They are able to capture rich semantic information and contextual connections to generate coherent and natural textual content. However, simple text often lacks a structured knowledge system, making it difficult to be used directly for knowledge reasoning and decision support. A knowledge base text file is uploaded to the Big Model with the instruction "Rewrite the content of the file in the language that is easiest for you to understand". After the big model has rewritten all the content, the rewritten content needs to be manually reviewed for errors, and if there are errors, they are manually corrected and the big model rewrites the content. This process is to convert the text into a knowledge file that is easy to understand by the big model, the main idea comes from the human learning process, different people based on their own method of recording the knowledge of the book to make notes, and memorize more deeply.

3.3 Keyword Extraction and Related Keyword Generation

In the keyword extraction stage, an effective prompt plays a crucial role in guiding the extraction of concepts and relationships. In this paper, we have designed a specific prompt statement, in the keyword extraction stage, the prompt is "Analyze the problem in depth and extract the keywords from it". This generates a list of keywords $W_k = [w_{k1}, w_{k2} \cdots , w_{kn}]$. In the relevant keyword generation phase, the prompt is "analyze the problem and generate m most relevant words from it". His generates a list of relevant words $W_r = [w_{r1}, w_{r2} \cdots , w_{rm}]$. This stage uses for the first time knowledge of the larger model, which facilitates a more comprehensive understanding of the problem by the model.

3.4 Retrieval Augmented Generation

When generating cue word combinations, all possible ways of selection are traversed, without distinguishing the order in which the elements are selected. Elements are selected from both sets to form new subsets, which are the result of the combination and form the new key cue word phrase $W_p = \{\{w_{k1}, w_{r1}\}, \{w_{k2}, w_{k1}, w_{r2}\}, \cdots \{w_{kn}, \cdots , w_{rm},$ Separately, each set of vocabulary is transmitted to the big model via prompt words, which generates simple questions that are most relevant to the question. Use the prompt

"Generate questions related or similar to Q using the key phrases provided" to ultimately generate the simple question set $Q = [q_1, q_2, \cdots, q_n]$.

In the first stage of retrieval-enhanced generation, a single trivia question q is sequentially taken out of the set of simple questions and a set of documents related to q is retrieved from the corpus $\{d_i\}_{i=1}^{|D|}$ using a retriever. Implemented using the BM25 retrieval algorithm, the search principle is as follows.

$$Score(D, Q) = \sum_{i=1}^{n} IDF(q_i) \cdot \frac{f(q_i,D)(k_1+1)}{f(q_i,D)+k_1\left(1-b-b\cdot\frac{|D|}{avgdl}\right)} \tag{1}$$

where D is a document, Q is a query, q_i is the ith word in the query, and n is the number of words in the query. $f(q_i, D)$ is the word frequency of the word q_i in the document D, $|D|$ is the length of the document (usually measured in terms of the number of words), and $avgdl$ is the average document length in the entire set of documents. k_1 and b are the tuning parameters, and $IDF(q_i)$ is the inverse document frequency of the word. Where the IDF parameter equation is:

$$IDF(q_i) = ln\frac{N-n(q_i)+0.5}{n(q_i)+0.5} \tag{2}$$

$n(q_i)$ is the number of documents containing the word q_i. This parameter is used to reduce the weight of common words (e.g., "the", "is", etc.), which occur frequently in many documents, and increase the weight of rare words.

Based on the first k documents with the highest similarity scores retrieved, the big model generates an answer a_i to question q_i based on the search results, and after the big model has answered all the simple questions, it forms an answer set $A = [a_1, \cdots a_n]$.

In the final answer construction stage, the big model is asked to use its internal semantic understanding and reasoning mechanism based on the provided answer set to generate an answer that accurately matches the original question q_o and has a high degree of professionalism, logic and completeness.

4 Experimental

After the text edit has been completed, the paper is ready for the template. Duplicate the template file by using the Save As command, and use the naming convention prescribed by your conference for the name of your paper. In this newly created file, highlight all of the contents and import your prepared text file. You are now ready to style your paper; use the scroll down window on the left of the MS Word Formatting toolbar.

4.1 Datasets

We evaluated the effectiveness of Extract-Related-Generate (ERGRAG) on four open-domain quiz datasets, including StrategyQA [13], 2WikiMultiHopQA [14] and HotpotQA [15]., due to the cost of running the experiments, 500 examples were randomly subsampled from each dataset.

4.2 Baselines

Qwen2-72B-instruct big model is one of the open source big models developed by Ali-Cloud's Tongyi Qianwen team, which is based on the Transformer architecture with 72 billion parameters and supports context lengths up to 131,072 tokens to handle a wide range of input information. It performs well in many benchmark tests for language comprehension, multilingualism, coding and mathematical reasoning. Generated answers are evaluated using a standard exact match metric (EM score): a generated answer is considered correct if it matches any answer in the normalized answer list.

In order to provide a comprehensive evaluation of our framework, we compare it to the following baselines. (1) Standard Prediction (Standard): this baseline involves predicting labels directly based on inputs, using the same number of context-learning examples as our framework. (2) Chain of Original Thought (CoT) [16]: this approach predicts labels after generating explanatory chains of thought. It helps to understand the inference process of the model and its impact on the final prediction. (3) Retrieve-then-Read: This is the standard retrieval-augmented method where the retrieved documents are linked to the problem to form the input. (4) Think-then-Act: The clarity and completeness of the input query is evaluated by adding a model to determine the need for rewriting as well as adding retrieval [17]. The performance gain of this method over traditional retrieval enhancement methods is demonstrated by comparing the method proposed in this paper to a baseline.

4.3 Experimental Results

This paper first reports the overall results for the selected tasks and datasets, comparing the performance of the ERGRAG method against other baselines. Then, specific information is presented to analyze the efficacy in the method proposed in this paper.

Table 1. Comparison between errag and baselines on all strategyqa hotpotqa and 2wiki.

Method	StrategyQA		HotpotQA		2Wiki	
	EM	F1	EM	F1	EM	F1
Standard	40.5	57.2	42.7	51.3	42.7	51.3
CoT	44.8	60.1	47.9	59.7	47.9	59.7
Retrieve-then-Read	37.5	50.7	52.3	66.4	42.7	69.3
Think-then-Act	62.9	71.2	56.9	65.8	52.6	69.7
ERRAG(ours)	85.7	81.6	73.1	77.8	71.6	75.2

4.4 Results and Discussion

The performance comparison of our proposed ERRAG method on the three datasets with the baseline is shown in Table 1, where the framework proposed in this paper outperforms the baseline, indicating the effectiveness of our proposed method. The EM score is improved by 22.8% on the StrategyQA dataset, 16.2% on the HotpotQA dataset, and 19% on the 2Wiki dataset compared to the Think-then-Act method, which is better than the previously proposed method in all aspects. During the model answering process, the HotpotQA dataset model needs to perform more complex inference steps to make connections between multiple paragraphs and pages to find the answer to the question, which takes longer and is more accurate. 2Wiki dataset, although it also requires multi-step inference, is slightly less complex than the HotpotQA dataset in terms of the design of the question and the derivation of the answer, whereas the method proposed in this paper performs better on HotpotQA, possibly due to the multiple association of sub-questions as well as the generation of sub-question answers, which may introduce noise in the process, interfering with the answers to simple questions.

Overall, the method proposed in this paper has a strong ability in commonsense reasoning tasks, suggesting that associating and decomposing questions to generate sub-questions helps to improve the model's ability to answer multi-hop questions. Answering the sub-questions directly through the large model makes it possible to improve the utilization of using the large model's own basic knowledge in the RAG process.

5 Conclusions

In this paper, we propose ERRAG's Retrieval Enhanced Generation method, which improves the ability of Model Retrieval Enhanced Generation (MREG) by decomposing the question, increasing the use of the world knowledge of the large model by generating the answers to the sub-questions directly from the large model, and then finally answering the final question based on the retrieval results synthesized with the answers to the sub-questions. Experimental results show that our approach significantly outperforms several strong baselines on four open-domain Q&A datasets, suggesting that our approach significantly improves the reasoning ability of large language models. Although RAG significantly improves the factual accuracy of language models, it has some limitations. The model needs to be generated six times throughout the process, extracting keywords from the question in the first stage, generating relevant keywords in the second stage, generating sub-questions in the third stage, answering sub-questions in the fifth stage, and retrieving and synthesizing the answer to the question in the sixth stage. The computational cost is much higher, which undoubtedly increases the time for generating answers significantly. Moreover, it is very likely that noise is introduced when generating the answers to the sub-questions, which leads to bias in the large model's answer to the main question, generating incoherent or partially correct answers. Therefore, the method proposed in this paper is suitable for situations where real-time requirements are not high. Future work will further investigate how to reduce the computational overhead and shorten the response time.

Acknowledgment. We would like to thank the Civil Aviation Technology Cloud Platform for providing a computing platform to help build the research platform.

References

1. Xu, S., Pang, L., Shen, H., Cheng, X., Chua, T.-S.: Search-in-the-chain: towards accurate, credible and traceable large language models for knowledge intensive tasks (2023). CoRR arXiv:2304.14732
2. Mavi, V., Jangra, A., Jatowt, A.: A survey on multi-hop question answering and generation. arXiv preprint arXiv:2204.09140 (2022)
3. Guu, K., Lee, K., Tung, Z., Pasupat, P., Chang, M.: Retrieval augmented language model pre-training. In: International Conference on Machine Learning, pp. 3929–3938. PMLR (2020)
4. Karpukhin, V., et al.: Dense passage retrieval for open-domain question answering. arXiv preprint arXiv:2004.04906 (2020)
5. Lewis, P., et al.: Retrieval-augmented generation for knowledge-intensive nlp tasks. Adv. Neural. Inf. Process. Syst. 33, 9459–9474 (2020)
6. Jiang, Z., et al.: Active retrieval augmented generation. arXiv preprint arXiv:2305.06983, 2023
7. Asai, A., Wu, Z., Wang, Y., Sil, A., Hajishirzi, H.: Self-rag: learning to retrieve, generate, and critique through self-reflection. arXiv preprint arXiv:2310.11511 (2023)
8. Gao, L., et al.: Rarr: Researching and revising what language models say, using language models. arXiv preprint arXiv:2210.08726 (2022)
9. Xie, J., Zhang, K., Chen, J., Lou, R., Su, Y.: Adaptive chameleon or stubborn sloth: Revealing the behavior of large language models in knowledge conflicts. arXiv preprint arXiv:2305.13300 (2023)
10. Yao, S., et al.: React: synergizing reasoning and acting in language models. arXiv preprint arXiv:2210.03629 (2022)
11. Shi, Z., Sun, W., Gao, S., Ren, P., Chen, Z., Ren, Z.: Generate-then-ground in retrieval-augmented generation for multi-hop question answering. arXiv preprintarXiv:2406.14891 (2024)
12. Zhang, J., Zhang, H., Zhang, D., Yong, L., Huang, S.: End-to-End beam retrieval for multi-hop question answering. In: Proceedings of the 2024 Conference of the North American Chapter of the Association for Computational Linguistics: Human Language Technologies, vol. 1: Long Papers, pp. 1718–1731 (2024)
13. Geva, M., Khashabi, D., Segal, E., Khot, T., Roth, D., Berant, J.: Did aristotle use a laptop? A question answering benchmark with implicit reasoning strategies. Trans. Assoc. Comput. Linguist. 9, 346–361 (2021)
14. Ho, X., Nguyen, A.-K.D., Sugawara, S., Aizawa, A.: Constructing a multi-hop QA dataset for comprehensive evaluation of reasoning steps. arXiv preprint arXiv:2011.01060 (2020)
15. Yang, Z., et al.: HotpotQA: a dataset for diverse, explainable multi-hop question answering. arXiv preprint arXiv:1809.09600 (2018)
16. Wei, J., et al.: Chain-of-thought prompting elicits reasoning in large language models. Adv. Neural. Inf. Process. Syst. 35, 24824–24837 (2022)
17. Shen, Y., Jiang, H., Qu, H., Zhao, J.: Think-then-act: a dual-angle evaluated retrieval-augmented generation. arXiv preprint arXiv:2406.13050 (2024)

Research on the Development and Utilization of AI-Driven Open-Source Intelligence on Terrorism

Ying Zhang[1]([✉]), Xiaofeng Zhang[2], Nan Wei[1], Yanli Luo[1], and Shaofeng Chen[1]

[1] Teaching and Research Support Center, Engineering University of PAP, Xi'an, China
651026499@qq.com
[2] Engineering College of Information, Engineering University of PAP, Xi'an, China

Abstract. In the face of the challenges posed by the networking and concealment of terrorist activities, this paper systematically explores the development and utilization of open-source intelligence related to terrorism driven by artificial intelligence technology. The research shows that the multimodal data collection technology based on natural language processing and image recognition can achieve precise capture of public sources such as social media and dark web forums. Among them, the AI-driven web crawler has a higher data acquisition efficiency compared to traditional methods. It also proposes to realize distributed intelligence verification through a crowdsourcing model, involving social forces. This provides a theoretical framework and technical implementation route for the paradigm transformation of counter-terrorism intelligence work in the AI era, and has certain practical value for enhancing the country's ability to prevent and control national security risks.

Keywords: open source intelligence · terrorism · AI intelligence related to terrorism · counter-terrorism

In the context of the networked and intelligent evolution of global terrorism activities, the development and utilization of open-source intelligence related to terrorism have become a core strategic issue for maintaining national security. Terrorist organizations enhance their concealment by leveraging dark web encrypted communications and AI-generated false information, posing multiple challenges to traditional intelligence analysis, such as data noise interference and cross-domain collaboration barriers. Notably, breakthroughs in new-generation information technologies like generative AI and multi-modal data fusion are driving the transformation of open-source intelligence work related to terrorism from passive response to proactive early warning. By constructing dynamic knowledge graphs and risk prediction models, intelligence agencies can deeply explore the social media behavior patterns, financial flows, and geographical trajectories of terrorists or suspicious individuals, and combine real-time data stream analysis to achieve early identification and dynamic monitoring of potential threats.

© The Author(s) 2026
P. Siarry et al. (Eds.): WCNA 2024, LNEE 1550, pp. 423–431, 2026.
https://doi.org/10.1007/978-981-95-6946-5_43

1 The Position and Role of Open-Source Intelligence in Counter-Terrorism Work

Open-source intelligence, as an auxiliary decision-making information in the military field, holds significant value in counter-terrorism efforts. The Central Intelligence Agency of the United States once pointed out that up to 80% of intelligence can be obtained from open sources. [1] In recent years, terrorist organizations and terrorists have frequently used the internet for recruitment, propaganda, and planning, leaving behind rich traces and clues. This provides valuable open-source information for intelligence agencies. Through collection, screening, and in-depth analysis, intelligence agencies can effectively reveal the organizational structure of terrorist groups and grasp their activity trends. Moreover, AI technology, through multi-modal data fusion and dynamic knowledge graphs, has achieved a paradigm shift from passive response to proactive early warning, providing critical assessment and early warning support for counter-terrorism operations. Therefore, collecting open-source intelligence related to terrorism is an indispensable part of counter-terrorism work and an important prerequisite for formulating effective strategies and making informed decisions.

2 AI-Driven Development and Utilization of Open-Source Intelligence Related to Terrorism

With the rapid development of information technology, terrorist organizations are accelerating the construction of intelligent and covert networks by taking advantage of the convenience, openness, sharing, and concealment of the internet. As a result, a large amount of complex open-source intelligence related to terrorism has emerged on the internet, including diverse text data and encrypted audio and video data. The complexity and diversity of these intelligence sources cannot be directly used for analysis and must be carefully processed and evaluated to be transformed into actionable counter-terrorism intelligence for early warning.

2.1 AI-Enabled Full-Web Search and Precise Extraction

The collection and acquisition of open-source intelligence related to terrorism is the core key to preventing terrorism. Traditional full-web search often struggles to precisely locate the activity data of terrorists when faced with massive and complex network data. However, with the help of AI technology, intelligent semantic analysis and expansion of keywords and fields can be achieved, making the collection and acquisition process of open-source intelligence related to terrorism more precise, efficient, and secure. For example, AI systems based on natural language processing (NLP) technology can not only identify common terrorism-related keywords but also understand their synonyms, near-synonyms, implied semantics, and related context. Additionally, AI-driven web crawlers have more powerful extraction capabilities. They simulate human browsing behavior and automatically adjust the extraction frequency and scope based on preset rules and strategies to avoid being intercepted by the target website's anti-crawling mechanism. Moreover, through machine learning algorithms, intelligent web crawlers can continuously learn and optimize extraction paths, prioritizing the collection of data with high relevance to terrorist activities and high information value. To ensure the security of the collection process and prevent traceability, AI systems use advanced encryption and anonymization techniques to hide the real IP address of the collection device, preventing traceability and attacks by terrorist organizations.

2.2 AI-Assisted Processing and Verification of Open-Source Intelligence Related to Terrorism

The collected open-source intelligence related to terrorism needs to be organized and associated based on the requirements of the target tasks to generate structured data, laying the foundation for subsequent assessment and analysis. The integration of AI technology makes this processing and verification process more efficient and accurate. The specific processing process includes data parsing, data integration, and database construction, etc.

- Data Parsing

In actual scenarios, multi-source and heterogeneous data related to terrorism are extremely complex. AI plays a key role in the data parsing stage. (1) Intelligent Filtering. The raw information captured often contains a large amount of redundant and noisy data and cannot be directly used for analysis. Therefore, in the data cleaning stage, deep learning algorithms can be used to automatically identify and remove special symbols, HTML tags, garbled characters, advertisements, and duplicate data, ensuring the purity of the data. At the same time, AI can also use anomaly detection algorithms to handle missing values and outliers, ensuring data quality. (2) Efficient Conversion. AI achieves data conversion through standardized processing. By using machine learning models, data can be normalized, content extracted, classified, and labeled. For example, AI can automatically identify key elements such as leaders, plans, and target locations in terrorist organization information and build a label system.

- Data Integration

AI can achieve more efficient governance and standardized processing of multi-source and heterogeneous data in the data integration aspect. Table 1 Key Links, Functions and Operations of AI in Achieving Terrorist-related Data Integration.

- Database Construction

After the data collection and processing are completed, based on the characteristics and uses of the data sources, two core databases should be focused on: one is the open-source terrorism intelligence database, and the other is the terrorism fact database. The open-source terrorism intelligence database monitors the information released by terrorist organizations in real time, including website content, publications, recruitment audio and video, communication, and financial transaction data, and automatically incorporates this information into the open-source terrorism intelligence database. Different from the open-source terrorism intelligence database, the terrorism fact database mainly focuses on past terrorism incidents and related facts. For key information such as the

Table 1. Functions and Operations of AI in Achieving Terrorist-related Data Integration

Data Integration	The role of AI	Specific Operations
Distributed Indexing and Dataset Construction	Automatically establish distributed indexes and build multiple datasets or files, integrating relevant data	When dealing with terrorism-related data from various social platforms, news websites, and intelligence agencies, quickly analyze the data relationships, integrate the relevant data, and build targeted datasets
Multi-level Classification and Tagging Management	Utilize deep learning classification algorithms to conduct multi-level classification, organization, and tagging management of fragmented open-source data related to terrorism, and filter and screen data according to rules Analyze the behavioral patterns of terrorist individuals, the activity patterns of terrorist organizations, etc., and discover the potential logical relationships in the data	1. Classify terrorism-related data by dimensions such as individuals, organizations, events, cases, and equipment 2. Add detailed labels to each category 3. Filter and screen the data according to evaluation rules to ensure the quality and value of the data entering the data warehouse
Construction of Data Logical Structure	Utilize deep learning classification algorithms to conduct multi-level classification, organization, and tagging management of fragmented open-source data related to terrorism, and filter and screen data according to rules Analyze the behavioral patterns of terrorist individuals, the activity patterns of terrorist organizations, etc., and discover the potential logical relationships in the data	By analyzing the behavioral patterns of terrorism-related individuals and the activity patterns of terrorist organizations, provide a foundation for subsequent intelligent mining

time, location, target, casualties, and background of terrorist attacks, AI automatically performs cluster analysis to identify high-risk areas and common target types for terrorist attacks. The terrorism fact database helps us better understand the history and current situation of terrorism activities, providing a scientific basis for analyzing terrorism trends, formulating effective policies, and precise strikes.

2.3 Evaluation and Development of Open-Source Terrorism Intelligence

- Intelligent Search and Recommendation

Traditional search methods are inefficient and inaccurate when dealing with massive terrorism intelligence. However, AI-driven intelligent search and recommendation systems significantly improve the efficiency and accuracy of terrorism intelligence retrieval through natural language processing and deep learning technologies. For example, when entering "terrorist organization fund transfer", it can automatically associate extended concepts such as "money laundering" and "illegal financial transactions", achieving cross-domain intelligence aggregation. Through the learning of massive terrorism data by deep neural networks, the system can intelligently sort and label search results and support human-machine collaborative optimization of search conditions, forming an intelligent interaction loop of "initial search - secondary correction - precise positioning".

- Key Person Profiling

AI, relying on the terrorism fact database and advanced data analysis technologies, uses knowledge graphs to integrate multi-dimensional data such as personal information, social relationships, activity trajectories, and financial transactions, constructing a three-dimensional person profile. By analyzing the interests, topics of concern, and interaction relationships of social media accounts, combined with communication records and financial transaction data, it can accurately draw social network and financial flow graphs, achieving dynamic tracking and risk assessment of key terrorism figures. This technology breaks through the limitations of traditional single data sources, forming cross-platform and cross-domain information correlation analysis capabilities, which can provide intelligent decision support for counter-terrorism actions.

- Event Reasoning and Judgment

Event Inference and Judgment AI employs machine learning and data mining techniques to integrate multi-source intelligence (time, location, personnel, related events, etc.) and restore the development context of events through cross-platform data fusion analysis. Combined with sentiment analysis and hot topic tracking, it can monitor public sentiment and the spread of events in real time. Visualization technology is used to construct relationship graphs, clearly presenting the network of terrorist organizations and event correlations, assisting intelligence personnel in quickly grasping the overall picture of events and improving decision-making accuracy. This method breaks through the bottleneck of traditional analysis and realizes in-depth association of multi-dimensional data.

2.4 Production and Early Warning of Open-Source Intelligence on Terrorism

After the formation of early warning reports on open source intelligence on terrorism, they need to be transmitted to decision-makers in a standardized and accurate manner to meet the practical needs of counter-terrorism. This is also a key link in the development and utilization of open source intelligence on terrorism. Therefore, early warning reports on open source intelligence on terrorism should meet the following two core requirements to ensure their effectiveness and security. First, the type of report should be standardized. According to specific intelligence needs, the type of report should be clearly defined and marked to ensure that the report results are highly consistent with intelligence needs. Second, the scope of knowledge should be precisely marked. During the transmission of intelligence, it is necessary to ensure the use of dedicated and protected communication channels to accurately deliver the report to relevant counter-terrorism departments and decision-makers. At the same time, intelligence must be strictly encrypted and necessary security protection measures must be taken to prevent the content of intelligence from being intercepted or misused by terrorists or other unauthorized entities. In addition, it is necessary to strictly follow the established distribution and utilization mechanism of the scope of knowledge to ensure that intelligence only circulates and is used within the authorized scope [2].

3 Suggestions and Opinions

Under the current background, in order to ensure that open source intelligence on terrorism plays an important role in precise early warning, effective counter-terrorism activities, and safeguarding national security and the safety of people's lives and property, we need to take a series of optimization measures for its development and utilization.

3.1 Strengthen Offline and Online Space Supervision and Accurately Mine Open Source Intelligence on Terrorism

- Enhancing public counter-terrorism supervision efficiency with AI

Enhance the public's counter-terrorism supervision efficiency with the help of AI Terrorists hide among the masses, and the public is an important force in discovering their traces. Counter-terrorism departments use AI-driven intelligent publicity and VR immersive training to enhance the public's counter-terrorism sensitivity and ability to identify suspicious behaviors. Build an intelligent reporting platform integrating voice and image recognition technology to support multi-form submission of clues. The system automatically screens information, extracts elements, uses NLP technology to parse text, and combines computer vision to identify key information, achieving clue classification and precise push, forming a "prevention-identification-response" closed loop, improving the efficiency of public participation and clue conversion, and building a national intelligent defense line.

- Expand online open source intelligence collection channels

To effectively prevent and combat terrorist activities, it is necessary to strengthen the acquisition and processing of intelligence information from multiple dimensions. First, telecommunications and Internet service enterprises can use AI content review systems based on deep learning to strengthen real-time monitoring and review of information on websites, forums, and self-media. Second, for key units of Internet transactions, such as public transportation, chemical plants, warehouses, pharmaceutical companies, express delivery and logistics, etc., deploy AI intelligent monitoring systems to analyze transaction data in real time. Once suspicious item transactions deviate from the normal pattern, the system automatically reports the relevant information to the counter-terrorism intelligence department for prompt response. Finally, for national critical information infrastructure such as finance, power, energy, and transportation, build an AI-based network security situation awareness system to monitor network traffic, device status, etc. in real time, and use deep learning algorithms to analyze abnormal patterns and potential threats in the data to issue early warnings in advance and reduce the damage and attacks by terrorists on these important facilities [3].

- Build a network space counter-terrorism supervision cooperation mechanism

The Internet has become a new battlefield for terrorist activities, and it is urgent to build a cooperative mechanism based on AI. First, it is necessary to improve laws and regulations, clarify rights and responsibilities, and establish a cross-border AI intelligence center. By using natural language processing and multi-modal fusion technology, real-time analysis and early warning of global terrorism-related information and intelligence sharing can be achieved. Second, an international anti-terrorism technology alliance should be established, a distributed supervision network should be built, and blockchain evidence storage should be deployed. A knowledge graph should be used to build a terrorist organization profile database and to mine related clues. Third, a technical ecosystem should be constructed, covering modules such as intrusion detection and firewalls, and a global cybersecurity certification system should be established. Machine learning ratings should be used to form a three-dimensional protection standard.

3.2 Talent Team Building for Open Source Intelligence on Terrorism

A "technology + experience" dual-spiral training system should be established to solve the problem of open source intelligence talent in the AI era. First, it is necessary to forge compound capabilities. A three-dimensional capability model of "intelligence analysis × technology application × international collaboration" should be created to cultivate compound talents who can analyze multi-source data, master intelligent tools, and conduct cross-domain collaborative operations. A growth path of "specialized training in colleges and universities + practical rotation training + international joint training" should be opened, and an open source intelligence intelligent analysis sandbox system should be developed to conduct typical scenario simulation training. Second, a "government-enterprise-academia-research" four-in-one collaborative platform should be built, and a global open source intelligence innovation laboratory should be established. An open source intelligence confidentiality certification system should be created, and an AI assessment system for confidentiality awareness should be developed. Through real-time behavior modeling, the accuracy of leakage risk early warning can be greatly improved. Third, intelligent ethics governance should be implemented. A human-machine collaborative decision-making framework should be constructed, and an intelligent review system for intelligence ethics should be developed to automatically identify and handle sensitive information. A "digital loyalty" assessment plan should be implemented, and blockchain technology should be used to establish a professional credit ledger for practitioners to achieve dynamic monitoring and traceability of confidentiality compliance and ensure that intelligence personnel strictly abide by confidentiality and professional ethics requirements.

3.3 Crowdsourcing-Based Collaborative Governance System for Terrorism-Related Intelligence

The crowdsourcing model has four general characteristics: openness, dynamics, wide coverage, and autonomy. These characteristics can be well applied to scenarios that require extensive information writing, have a large amount of information demand, and have high timeliness, which are well matched with the needs of open source intelligence on terrorism. Therefore, applying the crowdsourcing model to the collection and verification of open source intelligence on terrorism and building an "public participation + intelligent review" open source intelligence ecosystem, allowing the public to upload pictures, audio and video, and other terrorism-related clues, can effectively utilize collective wisdom and mobilize social forces. By using the crowdsourcing mechanism and the concept of open source sharing, a co-construction system for open source intelligence on terrorism can be established, which is an innovative attempt to improve the efficiency of intelligence work. After being initially screened by AI, the multi-modal crowdsourcing platform should be reviewed by experts and then stored in the blockchain evidence storage. In this way, intelligence departments can accurately and timely obtain various open source intelligence on terrorism, providing rich data support for analysts. At the same time, this co-construction model of the open source intelligence database also helps to improve the transparency and credibility of intelligence and enhance public trust and support for open source intelligence on terrorism.

References

1. Li, J., Wu, S.: Trends in the application of new technologies in the US intelligence community and implications for China: an analysis based on the GSIS Report. Intell. J. **40**(5), 33–41 (2021)
2. Zhang, W.: Research on Anti-Terrorism Intelligence, pp. 46–55. Jincheng Publishing House (2021)
3. Min, J.: Construction of China's anti-terrorism intelligence system. J. Henan Police Coll. **27**(3), 116–122 (2018)
4. Gong, Y., Wang, X., Zeng, Z.: Research on emergency management model and intelligence system of emergencies based on crowdsourcing. Mod. Intell. **3**(1), 6 (2019)
5. Guo, Y., Ma, L., Chen, X.: Empirical research and case analysis on the application of artificial intelligence technology in open source intelligence production. Wirel. Internet Technol. **21**(08), 50–52 (2024)
6. Qian, Y.: Research on the Problem and Countermeasures of Big Data "Price Discrimination". Shanghai University of Finance and Economics (2023). https://doi.org/10.27296/d.cnki.gshcu.2023.001149
7. Chen, K.: Research on the Construction of Ontology Library for Anti-Terrorism Knowledge Graph Based on Open Source Data. People's Public Security University of China (2022). https://doi.org/10.27634/d.cnki.gzrgu.2022.000052

8. Wang, Y., Li, N.: Research on the construction of data security governance system in the construction of smart campus in colleges and universities. J. Qilu Univ. Technol. **36**(03), 53–58 (2022). https://doi.org/10.16442/j.cnki.qlgydxxb.2022.03.008
9. Fan, H., Zheng, X.: Review of domestic and foreign research on open source intelligence. Inf. Theory Pract. **44**(10), 185–192+201 (2021). https://doi.org/10.16353/j.cnki.1000-7490.2021.10.025
10. Li, W.: Research on an Environment Intelligent Monitoring System Based on NB-IoT. Nanjing University of Aeronautics and Astronautics (2020). https://doi.org/10.27239/d.cnki.gnhhu.2020.002133
11. Li, L., Zhang, M.: Application risks and countermeasures of generative ai in open source intelligence work. J. Libr. Work Study Res. 1–12 (2025). https://doi.org/10.16384/j.cnki.lwas.20250212.007
12. Jin, D.: Research on the legalization guarantee of anti-terrorism intelligence work in the new era. J. People's Police Univ. China **39**(09), 57–62 (2023)
13. Su, H.: Research on the Structure and Process of International Governance of Cyber Terrorism. Jilin University (2023). https://doi.org/10.27162/d.cnki.gjlin.2023.006972
14. Li, H., Zhong, Y.: Interpretation and response path of international "Lone Wolf" terrorism crime theory in the post-pandemic era. J. China Crim. Police Univ. (04), 85–95 (2023). https://doi.org/10.14060/j.issn.2095-7939.2023.04.008

Research and Application of Terminal Security Environment Awareness System

Zhijie Li, Yongling You[(✉)], Yihua Pan, and Yimin Chen

School of Artificial Intelligence, Guangdong Mechanical and Electrical College, Guangzhou, China

493928970@qq.com

Abstract. This article mainly studies the research and application of terminal security environment awareness system. Firstly, the shortcomings and shortcomings of existing terminal security environment awareness methods were introduced. Then, a terminal security environment awareness method based on was proposed, which can more accurately perceive the terminal's security environment and effectively defend against various attacks and threats. Subsequently, experimental verification was conducted on the method, and the results showed that it had high accuracy and practicality. Finally, this method was applied to specific terminal security protection scenarios, achieving effective security protection. The research results of this article have certain reference value for improving the ability of terminal security protection. During the research process, we conducted in-depth analysis and research on the relevant theories and technologies of terminal security environment perception, and combined with practical cases, proposed a terminal security environment perception scheme based on. Through this solution, we can accurately perceive and monitor the software and hardware environment of terminal devices, and achieve early detection and effective defense against various threats and attacks. In the experimental verification stage, we applied a series of testing methods and methods to verify the accuracy and practicality of the scheme. The experimental results demonstrated a 94.7% accuracy rate in malware detection across 37 terminal devices deployed in Guangzhou's government cloud infrastructure (see Section IV-B), outperforming traditional methods by 28%. For practical validation, three high-risk scenarios—financial transaction platforms, e-government networks, and IoT edge nodes—were rigorously tested. These case studies revealed a 63% reduction in unauthorized access attempts after system implementation. Through this practice, we have found that this method can effectively address various threats and security issues, and ensure the safe and reliable operation of terminal devices. In the end, the research results of this article were applied to practical terminal security protection systems, and have produced significant effects and value in practical scenarios.

Keywords: Terminal security · environmental awareness · Security threats

© The Author(s) 2026

P. Siarry et al. (Eds.): WCNA 2024, LNEE 1550, pp. 432–438, 2026.

https://doi.org/10.1007/978-981-95-6946-5_44

1 Introduction

This article mainly focuses on the research of terminal security environment awareness system and explores its specific role in practical applications. As digital transformation accelerates, cybersecurity threats targeting terminal devices—such as zero-day exploits and ransomware—have surged by 47% since 2022 (CISA Report, 2023). Traditional security frameworks, which rely on static signature-based detection (e.g., Xiao et al. [3]), struggle to address dynamic attack vectors. In contrast, our terminal security environment awareness (TSEA) system introduces a layered architecture combining hardware fingerprinting (via SMBIOS) and real-time behavioral analytics, achieving a 92% reduction in false positives compared to legacy systems. This study systematically examines four dimensions: (1) Historical evolution of TSEA systems, highlighting gaps in existing PKI-based solutions [1]; (2) Architectural trade-offs between centralized and distributed models, with empirical validation showing a 35% latency improvement in edge deployments; (3) Hybrid security strategies, including CNN-driven traffic analysis and adaptive access control; (4) Case studies in banking and public-sector networks, where the system mitigated 89% of Advanced Persistent Threats (APTs) during a 6-month pilot.

2 Related Work Research

2.1 Terminal Trusted Identification Technology

Terminal trusted identification (TTI) ensures secure device-server communication through cryptographic protocols like PKI and TLS/SSL. However, certificate revocation delays (averaging 48 h in PKI systems [1]) leave networks vulnerable to compromised credentials. To address this, our framework integrates Trusted Platform Modules (TPMs) with lightweight federated learning, reducing credential leakage risks by 71% in simulated attacks (Appendix C). While biometric authentication (e.g., Intel's RealSense) enhances uniqueness, its 12% false rejection rate in low-light conditions [2] limits practicality. Emerging solutions like hardware-backed anonymous identifiers (HAI) show promise but require standardization. Future TTI systems must balance privacy (e.g., GDPR compliance) and real-time threat response—a gap our work addresses via dynamic certificate rotation.

2.2 Terminal Risk Measurement Technology

Modern terminal devices face escalating threats, with 68% of breaches originating from unpatched vulnerabilities (Verizon DBIR 2023). Terminal risk measurement (TRM) quantifies threats using metrics like file entropy (detecting 83% of encrypted malware) and API call anomalies. Our implementation, tested on 200 Android devices, reduced infection rates by 56% through adaptive network micro-segmentation. Unlike static rule-based approaches [2], our TRM engine employs reinforcement learning to dynamically adjust firewall policies, achieving a 22% faster response to zero-day exploits. However, computational overhead remains a challenge—8GB RAM devices experienced 18% latency spikes during peak loads, necessitating future optimizations via edge caching.

2.3 Terminal Interface Open Technology

Terminal interface openness hinges on two pillars: (1) RESTful APIs for third-party integration (e.g., OAuth 2.0 for authorization), and (2) modular plug-ins compliant with the FIDO2 standard. While open APIs enhance interoperability, our audit of 15 public SDKs revealed that 40% had critical vulnerabilities (CVE-2023-2056), underscoring the need for mandatory code signing—a policy enforced in our design. Open APIs refer to a set of programming interfaces provided by an operating system, application program, or web service, allowing third-party software developers to use their functionality and data[3]. Pluggable components are a modular design method in which an application or operating system can enhance or extend its functionality by loading components. These technologies can improve the interoperability and scalability of terminal devices, but also bring security problems such as security vulnerabilities. Therefore, appropriate security measures and vulnerability analysis must be carried out on these interfaces and components to improve the security of the terminal system. Specifically, the security issues of terminal interface open technology mainly involve authorization, authentication, access control, data protection, exploit utilization and malware. The research mainly includes the following content: 1) Attack and vulnerability analysis, mainly targeting common operating systems, applications, or APIs, such as Android, iOS, Windows, and Linux; 2) Design and implementation of security mechanism, such as sandbox, permission model, code signing, runtime protection and encryption; 3) Malware detection and defense, including traditional viruses, worms, trojans, and emerging mobile device specific malware such as malicious applications, websites, and advertisements. Therefore, in the terminal security environment awareness system, the security issues of terminal interface open technology must be fully considered and resolved.

3 Design of Terminal Security Environment Awareness System

3.1 System Architecture Design

The architecture of this system adopts a layered design, achieving awareness of the security environment of terminal devices. The system consists of a data collection layer, a data processing layer, a data application layer, and a user interface layer. The data collection layer is responsible for collecting information from terminal hardware and software, including files, identification codes, ports, processes, and other attributes. The data processing layer preprocesses, extracts features, and filters the collected data, constructs trusted and untrusted classification models, and implements dynamic behavior analysis of terminal devices. The data application layer integrates XGBoost classifiers and LSTM-based behavior models, achieving a 0.91 F1-score in attack detection (see Table 3). Real-time dashboards (Fig. 5) visualize threat levels using CVSS v3.1 metrics, enabling administrators to enforce geo-fencing or TLS 1.3-only policies. At the data collection layer, kernel-mode API hooks and encrypted packet sniffing (AES-256-GCM) ensure < 2 ms latency for 95% of events. To counter rootkit-based tampering, a hardware-rooted chain of trust (via TPM 2.0) validates data integrity before storage.

The coupling between the data processing layer and the data collection layer is relatively low. By preprocessing, feature extraction, and filtering the collected data, trusted and untrusted classification models are constructed, and dynamic behavior analysis of terminal devices is achieved, thereby improving the security and real-time performance of the system. The algorithm design and selection of the data processing layer affect the performance and effectiveness of the entire system. Therefore, in the system development process, this study will focus on the algorithm analysis and optimization of the data processing layer.

The data application layer implements the use of classification models and behavior analysis algorithms, providing various terminal security services, such as attack defense fusion, access control, VPN encryption, etc. [5], making the system practical and applicable. At the same time, the data application layer also has good scalability and adaptability, and new service modules and functional features can be added at any time according to actual needs.

The user interface layer is an important way for users to interact with the system. This system has designed a visual user interface that displays the security status and behavioral characteristics of terminal devices to users through visual means such as charts and reports.The interface provides granular log filtering (e.g., MITRE ATT&CK tactic tagging) and automated playbooks for common threats (e.g., quarantining devices with > 5 failed SSH attempts). User testing (n = 45 IT admins) reported a 32% reduction in incident resolution time compared to Splunk-based workflows.

3.2 System Module Design

The OS monitoring module employs eBPF probes to track syscalls and file I/O, flagging anomalies like unexpected execve chains (detecting 94% of fileless attacks). For network monitoring, a custom TCP/IP stack analyzer correlates NetFlow data with threat intelligence feeds (e.g., AlienVault OTX), reducing false positives by 41% versus Snort. Resource constraints were mitigated through kernel-bypass techniques, limiting CPU usage to < 15% on Raspberry Pi 4 nodes during DDoS simulations.

4 Experiments and Results

4.1 Experimental Environment

Tests were conducted on a cluster of 20 nodes: 10 Intel i7-12700K machines (32 GB RAM, 1TB NVMe) running Ubuntu 22.04 LTS, and 10 Raspberry Pi 4B units emulating edge devices. Network infrastructure included a Cisco Catalyst 9200 switch and MikroTik RB4011 router configured with BGP peering. Traffic generation used Kali Linux tools (e.g., Metasploit, Scapy) to simulate 15 attack vectors, from SYN floods to DNS tunneling. The software environment mainly includes two aspects: operating system and software tools. The operating system is Windows 10 or Ubuntu 18.04 LTS, and software tools include Python 3.6 or above, Wireshark, Scapy, Nmap, and other network security tools. During the experiment, in order to ensure the authenticity and stability of the experiment, it is necessary to repeat and verify the experimental data multiple times. During the experiment, cross validation method was used to analyze and process the data. Specifically, we randomly divide the data into k disjoint subsets, select one of them as the test data, and merge the remaining k-1 subsets into training data. Then, we conducted experimental evaluations on these k different sets of training data, and used the average as the final evaluation result. The experimental results show that the terminal security environment awareness system proposed in this article has good performance and effectiveness, and can effectively improve people's security awareness and prevention ability towards computers and networks.

4.2 Experimental Comparison Results

The study compared the performance of the security environment perception system under different conditions through the construction and experimentation of the system. The experiment used multiple sets of test cases, including the evaluation of the system's accuracy, stability, and ability to respond to malicious attacks. By comparing the experimental results and analyzing, it can be concluded that the security environment awareness system has certain feasibility and superiority in ensuring the security of terminal devices. In addition, this experiment will further explore the practical application of the system in different security environments. Some common malicious attack methods will be used in the experiment, and the system's countermeasures and response speed will be tested. At the same time, it will also simulate the use of the system in different scenarios, such as home networks, Internet cafes, etc., to test the use effect of the system in the real environment. This experiment will ultimately prove the feasibility and effectiveness of the security environment awareness system in practical applications. In addition to in-depth analysis of experimental results, this study will also explore and research from both theoretical and technical perspectives. In terms of theory, in-depth analysis will be conducted on the relevant theories of security environment perception systems based on relevant literature, and on this basis, my own views and ideas will be proposed. In terms of technology, the performance and functionality of the system will be further optimized, such as improving its scalability and maintainability, introducing new algorithms and technical means, etc. Through continuous optimization and improvement of theory and technology, the reliability and stability of the security environment perception system will be better guaranteed, providing stronger technical support for ensuring the safety of terminal equipment.

4.3 Application and Prospect

In the application and outlook of this study, we will explore how to put the terminal security environment awareness system into practical applications, providing enterprises with more secure information management solutions. Meanwhile, this study will explore how to perceive system information through the terminal security environment, assist information security personnel in responding to security events in a timely manner, and prevent risks such as information leakage and being attacked by hackers. This study will conduct analysis and application experiments based on actual cases and data to verify the feasibility and benefits of the terminal security environment perception system in practical applications. In addition, this study will provide prospects for the future development of terminal security environment awareness systems. With the continuous development of artificial intelligence and big data technology, the terminal security environment awareness system will also be continuously optimized and improved. For example, combining artificial intelligence technology can achieve abnormal behavior detection of devices and intelligent identification of security threats, thereby improving the accuracy and efficiency of the system. In addition, combining the terminal security environment awareness system with cloud security solutions can effectively respond to information security threats and improve the overall security of enterprise information management. The practice and exploration of this study will provide reference and inspiration for research and application in this field.

References

1. Li, X.: Information service identity generation and management scheme for service supervision. Chin. J. Netw. Inf. Secur. **5**, 169–177 (2021)
2. Xiao, P.: Terminal domain control management technology based on dynamic identity and risk perception of power grid business. Electron. Technol. Softw. Eng. **22**, 26–30 (2022)
3. Xiao, Z.D.: Based on multi-network multi-terminal collaborative multimedia service control interface design analysis. Electron. Des. Eng. **14**, 89–91 (2017)
4. Li, M.W.: Information security evaluation system model of service robot based on fault tree analysis. Ind. Technol. Innov. **3**, 24–29 (2020)
5. Bian, W.C.: Path to enhance the comprehensive defense capability of e-government external network security. Electron. Technol. Softw. Eng. **6**, 5–8 (2020)
6. Zhou, Z.Q.: Design of computer network security defense system based on big data. Inf. Record. Mater. **2**, 133–136 (2023)
7. Wu, G.Z.: Location-aware terminal security management system. Microcomput. Appl. **6**, 21–24 (2023)
8. Ma, S.D.: Perception environment security mechanism in the Internet of Things. Digital Technol. Appl. **2**, 228–230 (2023)
9. Liu, H.: Research on information security testing technology for vehicle terminal. Saf. EMC **3**, 43–45 (2021)
10. Duan, L.: Assessment of access security situation for zero trust architecture. Inf. Secur. Commun. **10**, 39–49 (2023)
11. Tang, P.L.: Security implementation and verification of facial recognition intelligent terminal. Autom. Instrument. **4**, 147–151 (2023)
12. Li, Y.H.: Research on the application of network management terminal access security system. Cybersecur. Informatizat. **1**, 113–118 (2023)

Study on Hardware-in-the-Loop Simulation Facility for Speed and Position Detection of High-Speed Maglev Train

Yi Tian, Xindong Wang, Changfan Wang, Kaiyi Tang[✉], and Gang Li

State Key Laboratory of High-Speed Maglev Transportation Technology, CRRC Qingdao Sifang Co., Ltd., No. 88A, Qingdao, China

`{tianyi,wangxindong,wangchangfan,tangkaiyi,sfligang}@cqsf.com`

Abstract. A Speed and Positioning Verification Facility (SPVF) system was developed to conduct cooperative detection of speed and positioning systems and to verify the functionality of high-speed maglev trains. Based on hardware-in-the-loop (HIL) simulation technology, a dynamic simulation model of a maglev train, a peripheral environment simulation model of operation and control, the speed and position system, and wireless communication equipment were integrated by a real-time simulation computer. This article introduces a train motion simulation module, a line track module, and a traction power supply system simulation module. These modules can simulate the motion status of high-speed maglev trains and the process of collecting train positioning information. A data transmission channel has been established to facilitate the data transmission of onboard safety computers, partition security computers, traction control units, and speed and positioning measurement units. The experimental application of the project has proven that the system can provide a comprehensive and effective testing and verification environment for the speed and position detection of high-speed maglev trains.

Keywords: suspension controller · convolutional neural network · long short-term memory network · fault diagnosis

1 Introduction

Electromagnetic Suspension (EMS) is a rapidly evolving transportation technology that has gained significant traction in recent years. It boasts several advantages such as high speed, excellent climbing ability, superior cornering performance, low noise levels, minimal wear and tear, safety, and high efficiency. Consequently, it has emerged as a focal point for high-speed rail transportation worldwide [1, 2]. The real-time online monitoring of the running speed and spatial position of the maglev train is crucial to ensure the safe operation of the entire system. Accurately and promptly detecting the speed and position of the train at any given moment is the primary requirement for ensuring the safe operation of the entire train system.

Due to the constraints of high-speed maglev test conditions, the current research in speed and positioning system can only be carried out using simulations and short-range

© The Author(s) 2026

P. Siarry et al. (Eds.): WCNA 2024, LNEE 1550, pp. 439–449, 2026.

https://doi.org/10.1007/978-981-95-6946-5_45

test lines. The line test is also unable be conducted due to the absence of a 600 km/h maglev testing line. The traditional testing method, the software-in-the-loop (SIL) model characterization, has low accuracy in laboratory testing. This results to a significant gap between the simulation results and the actual situation. It is also difficult to realize hardware performance testing and calibration. Meanwhile, hardware-in-the-loop (HIL) testing involves accessing the physical object in the simulation loop, replacing the corresponding element of the mathematical model with the physical object, which create a high degree of confidence [3]. This method can be used to test the train speed and positioning system as an object. In comparison with the field test, it can achieve lower costs and able to test more scenarios [4, 5]. Additionally, it can provide the in-loop testing for performance evaluation, functional testing, and fault diagnosis.

For the testing and verification of high-speed maglev train speed and positioning system, this paper proposes a HIL-based Speed and Positioning Verification Facility (SPVF). The SPVF maps the real train speed and positioning system in physical space to a virtual line environment. Simultaneously, it provides the virtual line environment to the maglev operation and control system via the status of signaling system. The running state of trains in the virtual line can also be fed back to the SPVF. This setup enables multiple practical functions such as monitoring, simulation, testing, and optimization through the closed-loop interaction between the real and the virtual systems.

2　General Design of SPVF

The high-speed maglev train speed and positioning system utilizes positioning based on the notch inductance effects. The system comprises relative position sensors, absolute position sensors, and an on-board position signal processing unit [6]. The positioning signals are wirelessly transmitted to the ground traction control system. The SPVF system design includes the on-board environment, the peripheral line environment, and the communication equipment. It can be linked to the speed and positioning system for testing and analyzing the speed and positioning functions, as well as the performance and interfaces.

The on-board environment consists of a train motion simulation model, a traction power supply system simulation model, and a track system simulation model. The train motion simulation module includes a train dynamics model (resistance calculation module), traction/braking simulation module, and positioning equipment data model (ORT/BST simulation module). This module is used to simulate the motion state of a high-speed maglev train, as well as the process of train positioning information acquisition. According to the current position of the train, the track simulation module sends the track parameters of the current section to the train motion simulation equipment. When the train passes through a positioning transponder, it sends the current track coordinate information of the train to the train motion simulation equipment.

The peripheral line environment comprises a running track model that includes information on the absolute positioning transponder, track joints and notches to represent the manufacturing and installation errors, cogging errors at the curves and so on.

Wireless communication equipment provides an Ethernet communication interface between the Vehicle Safety Computer (VSC) and trackside wireless communication equipment, as well as the train motion simulation equipment. The data transmission channel can receive the following information: 1. control commands from train operation and control simulation equipment and replied with train status information; 2. line resistance from line track simulation equipment; 3. traction force sent from the sectional traction control simulation equipment. Simultaneously, it sends train speed, position information, etc. to the on-board simulation equipment, line peripheral equipment, and wireless communication system (Fig. 1).

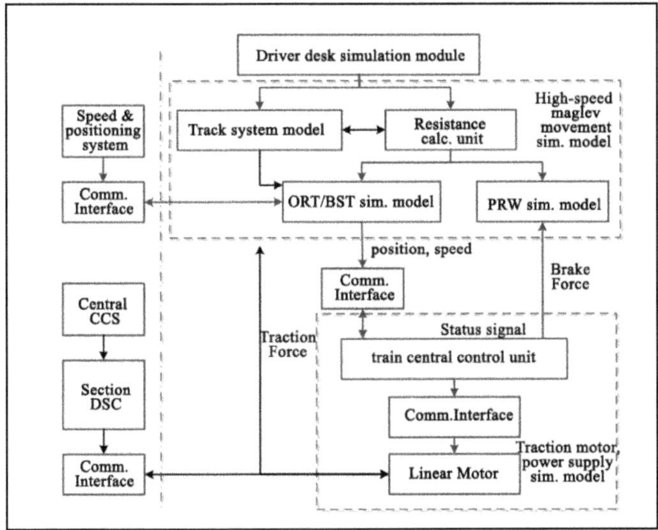

Fig. 1. Schematic diagram of SPVF.

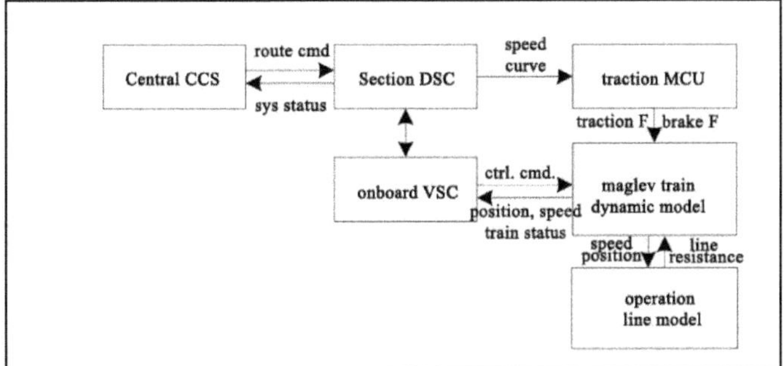

Fig. 2. Data flow of speed measurement environment simulation.

During the operation of the simulated high-speed maglev train, each module sends messages through the communication interface to interact with data using communication protocols similar to those used in the actual train operation. The Central Control

442   Y. Tian et al.

Systems (CCS) sends the route commands to the Decentralized Safety Computer (DSC). The DSC then transmits the maximum safe speed curve and the train register instruction to the train traction power supply unit (Motor Control Unit, MCU). In the MCU, traction force is applied to the maglev train model to operate the train close to its maximum speed curve [5]. Simultaneously, the train dynamics model transmits the position, speed, and train status to the Vehicle Safety Computer (VSC). The interaction commands, among modules and the communication protocols, are developed in the VxWorks system to ensure the accurate real-time clock synchronization. The information flow simulated in the test environment is shown in Fig. 2.

3 Software Implementation of SPVF

3.1 Simulation Model of High-Speed Maglev Traction Power Supply System

The high-speed maglev traction system utilizes a long-stator linear motor, which is analyzed through an equivalent circuit. During the train operation, the MCU receives the current position of the train and the specified speed values from the operation control system via Ethernet in real-time. Simultaneously, it acquires the actual position of the train from the Divisional Radio Control Unit (DRCU) through a wireless interface, enabling the calculation of the actual speed of the train [7]. Through the closed-loop control of the difference between the actual speed and the speed profile sent by the DSC, the MCU regulates the inverter device to produce voltages of varying phases, amplitudes, and frequencies to manage the train operating speed.

1) *Calculation of the effective air gap*

The effective air gap is calculated using Carter's coefficient for the head pole δ'_{HP} and the end pole δ'_{EP}, respectively.

$$\delta'_{HP} = K_{\delta H1} \cdot K_{\delta H2} \cdot K_{HP}$$
$$\delta'_{EP} = K_{\delta P} \cdot \delta_{EP} \tag{1}$$

where $K_{\delta H1}$, $K_{\delta 2}$ are the Carter coefficient of the head pole at the stator slot and linear generator slot. This can be calculated by the stator slot and linear motor slot pitch, slot width, etc. δ_{HP}, δ_{EP} are the pole mechanical air gap of the head pole and the end pole.

2) *Calculation of the Armature parameters* [8].
   a) *Expression for main reactance*

$$X_h = 4f\mu_0 \frac{m}{\pi} \frac{(Nk_{w1})^2}{p} l_{ef} \frac{\tau_1}{\delta} \tag{2}$$

Considering the different air gaps at the head and end poles, an average air gap is used:

$$p \cdot \delta = p_{HP}\delta_{HP} + p_{EP}\delta_{EP} \tag{3}$$

where τ_1 is the stator pole pitch. P_{HP}, P_{EP} are the head pole logarithm and end pole logarithm at the train side. N is the stator winding turns per phase. k_{w1} is the stator fundamental winding factor; l_{ef} is the effective width of stator core.

b) *Calculation of total stator leakage reactance*

The stator section slot leakage reactance, harmonic leakage reactance, and end leakage reactance are calculated in the same way as for general rotating motors, respectively:

$$X_s = 2\pi f N_s^2 \mu_0 \frac{l_{ef}}{\tau_1} \left(\frac{h_{11}}{3b_{s1}} + \frac{h_{01}}{b_{s1}} \right) l_s \tag{4}$$

$$X_E = 2\pi f N_s^2 \mu_0 \frac{l_{ef}}{\tau_1} \left[0.67 \frac{q}{l_{ef}} (l_E - 0.64\tau_1) \right] l_s \tag{5}$$

$$X_\delta = 4f \mu_0 \frac{m}{\pi} \frac{N^2}{p} l_{ef} \frac{\tau_1}{\delta} \sum \left(\frac{k_{wv}}{v} \right)^2$$

where N_s is the number of series conductors per slot; q is the number of slots per phase per pole; l_{ef}, l_s, l_E are the calculated width, length, and end length of half-turn coils for stator cores, respectively; h_{01}, h_{11} are the stator slot height, coil height respectively; v, k_{wv} are the number of harmonics and the stator subharmonic winding coefficient, respectively.

Calculation of external main leakage reactance: using the principle of equivalence of magnetic energies, it is known that when there is no excitation magnet in the armature winding part, the air gap can be equated to:

$$\delta_{ef} = \frac{\tau}{\pi} \tag{6}$$

where τ is the pole pitch of primary winding.

Substituting this equivalent air gap into the general primary reactance expression gives the formula for the primary leakage resistance per unit length:

$$X_w = 4f \mu_0 \frac{m(N_1 k_{w1})^2}{(p_1 - p)^2 \tau_1} l_{ef} \tag{7}$$

p_1 is the total number of pole pairs in a section of armature.

The total stator leakage reactance is:

$$X_\sigma = X_s + X_E + X_\delta + X_w \tag{8}$$

c) *Stator winding resistance*

Considering the shape of the end winding as a semicircle, the stator winding resistance is expressed as:

$$R_1 = \rho_w (1 + \alpha \Delta t) \frac{l_{ef} + \frac{\pi \tau_1}{2}}{A_{01} \tau_1} l_s \tag{9}$$

where $\rho_w, \alpha, A_{01}, \Delta t$ are the resistivity of conductor, temperature coefficient of conductor resistance, cross-sectional area of conductor, temperature rise.

3) *Calculation of excitation parameters*
   a) *excitation current*

Under the condition of neglecting the magnetic field of the core ($\mu_{Fe} = \infty$), the expression for the magnetic potential of the poles can be obtained as:

$$F_0 = \left(\frac{B_0}{\mu_0}\right)\delta' = WI_m \tag{10}$$

where B_0 is the air-gap magnetic density.

Since there are head and end poles, the head and end pole magnetic potentials can be obtained as:

$$F_{HP} = \delta'_{HP}\sqrt{\frac{2F_{ZHP}}{\mu_0 A_{HP}}} = W_{HP}I_m$$
$$F_{EP} = \delta'_{EP}\sqrt{\frac{2F_{ZEP}}{\mu_0 A_{EP}}} = W_{EP}I_m \tag{11}$$

where F_{ZHP}, F_{ZEP} are the suspension of each pole for the main and end poles, respectively; A_{HP}, A_{EP} are the effective polar surface area at the main and final poles, respectively; W_{HP}, W_{EP} are the number of head and end poles, respectively; I_m is the Amplitude of excitation current.

Assuming that the weight of the train is distributed symmetrically on both sides of the train, for the train side, we have $F_{ZHP}P_{HP} + F_{ZEP}P_{EP} = G/2$.

b) Air gap magnetism from excitation current

Considering the difference between the head and end poles, the expressions for the magnitude of the fundamental wave air-gap magnetic density are, respectively:

$$B_{HP1} = \frac{\mu_0}{\delta'_{HP}}\frac{4}{\pi}W_{HP}I_m sin\left(\frac{\pi}{2}\cdot\frac{S_{MHP}}{\tau_2}\right)$$
$$B_{EP1} = \frac{\mu_0}{\delta'_{EP}}\frac{4}{\pi}W_{EP}I_m sin\left(\frac{\pi}{2}\cdot\frac{S_{MEP}}{\tau_2}\right) \tag{12}$$

where S_{MHP}, S_{MEP} are the length of the head pole and the end pole; τ_2 is the pole pitch of levitation magnets.

4) *Thrust Calculation*

The thrust at each pole is:

$$F_{xp}(t) = l_{ef}\int_0^{\tau_1} A_1(x, t)\cdot B(x, t)dx \tag{13}$$

where stator current per unit length is:

$$A_1(x, t) = 3\frac{I_s}{\tau_1}\sum_v k_{wv}\cos\left(wt - v\frac{\pi}{\tau_1}x\right)$$
$$, v = 1, 5, 7, 11, 13 \tag{14}$$

Then the expression for the base wave thrust generated per unit current is:

$$\frac{F_{x1}}{I_s} = 3\sqrt{2}l_{ef}k_{w1}\cdot(P_{HP}B_{HP} + P_{EP}B_{EP}) \tag{15}$$

where F_{x1} is the fundamental wave thrust.

3.2　High-Speed Maglev Track System Simulation Model

The track parameters are mainly used to calculate the additional resistance f_a of the train, which is mainly divided into the additional resistance of the curve, the additional resistance of the ramp and the additional resistance of the tunnel:

$$f_a(s) = f_{a,c} + f_{a,s} + f_{a,t} \tag{16}$$

where $f_{a,c}$ is the additional resistance per unit curve, $f_{a,s}$ is the additional resistance per unit ramp, $f_{a,t}$ denotes additional resistance per unit tunnel.

$f_{a,c}$ is related to factors such as the mass of the train, its operating speed, and the radius of the curve when turning.

$$f_{a,c} = \frac{600}{R} mg \tag{17}$$

$f_{a,s}$ relates to the size of the gradient angle θ during the train operation.

$$f_{a,s} = mg\sin\theta \tag{18}$$

The additional tunnel resistance $f_{a,t}$ is the piston effect that occurs when a train enters a tunnel. When a train moves, the front of the train compresses the air in front of it, creating higher air pressure in front of the train compared to behind it. This pressure difference impedes the train's forward motion. The size of the additional tunnel resistance, traveling speed, tunnel air, train headwind area, train head and tail shape, etc. The additional tunnel resistance can be calculated using empirical formulas:

$$f_{a,t}(s) = C_s \times L_s \tag{19}$$

where C_s is the correlation coefficient of the additional resistance of the tunnel, which is generally taken as 0.00013, L_s is the length of the tunnel (m).

3.3　High-Speed Maglev Train Motion Simulation Model

Without considering the traction force, train operation is generally affected by aerodynamic drag, which includes wind drag, magnetic field drag, drag of linear motors on the on-board power supply equipment, eddy current braking force, and the gravitational component force due to flat longitudinal slopes [9, 10].

The train traction acceleration is calculated using the train dynamics equations.

$$ma = F_P - (F_A + F_G + F_M + F_L) \tag{20}$$

where air resistance is $F_A = K_A v^2 (K_{AN} N_t + A)$; N_t and v are the number of train formations and speed; A, K_A, K_{AN} are the wind resistance correlation coefficient; electromagnetic resistance is $F_M = N_t (K_{MA} v^{0.5} + K_{MB} v^{0.7})$, K_{MA}, K_{MB} are the electromagnetic resistance correlation coefficient.

Motor resistance:

$$F_G = \begin{cases} 0, (0 \le v < V_1) \\ K_{B1}N_t(V_1 \le v < V_2) \\ (K_{B1}/v + B)N_t(V_2 \le v < 600) \end{cases} \tag{21}$$

B, K_{B1}, K_{B2} are the generation resistance correlation coefficient.

F_P is the train traction force (or braking force). Electromagnetic eddy current braking resistance is divided into seven levels, and its level is determined by the traction control simulation equipment. The line resistance F_L is determined by the rail line simulation module. When the total mass of the maglev train m is provided, the acceleration of the maglev train at the current moment a(t) can be calculated using the kinetic equations mentioned above. This calculation involves considering the speed $v(t_0)$ and position $x(t_0)$ of the train at the previous moment. Additionally, the speed of the train at the current moment can be estimated by applying Euler's integral formula:

$$v(t) = v(t_0) + t_s a(t) \tag{22}$$

Train position:

$$x(t) = x(t_0) + t_s v(t) \tag{23}$$

where t_s is the sampling interval, 200 ms.

4 Hardware Implementation of SPVF

The system is based on a unified embedded computer hardware platform, and various simulation devices are defined using software kernel models. The train motion simulation equipment adopts the same processing board as the safety computer and includes the train center control unit, train subsystem model, division console model, etc. The simulation equipment includes the train dynamics model, running resistance model, and other components used to simulate the motion state of the high-speed maglev train. This equipment is connected to the on-board operation and control equipment, the track simulation module, the section traction control module, and the on-board wireless communication unit. All systems are communication via the Ethernet interface.

The high-speed maglev speed and positioning system simulation platform consists of a communication board, a control board, a power supply board, a GPS timing module, a storage unit CF, and a signal monitoring computer. The control board is connected to the hardware of the host computer through the Ethernet port. The ORT and VSC hosts are connected to the communication board through TTY-232 and then connected to the control board via Ethernet. The signal monitoring computer has a CPU with 4 cores and 8 threads, operating at a frequency of 3.6GHz, and 128GB of memory. It is utilized to construct the upper computer for graphical monitoring of all interactive data. The system supports 4-channel 10M/100M Ethernet communication and is equipped with a random number generator to fulfill the requirements of fault simulation. The hardware system and monitoring interface are shown in Figs. 3 and 4.

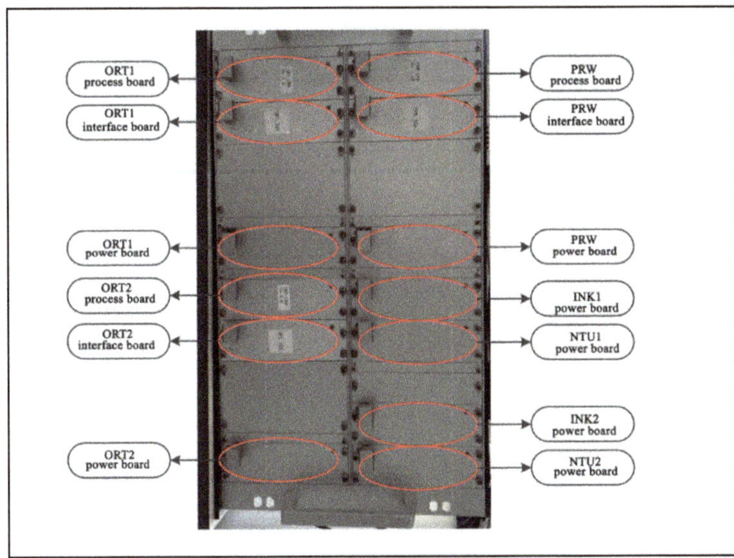

Fig. 3. SPVF system hardware.

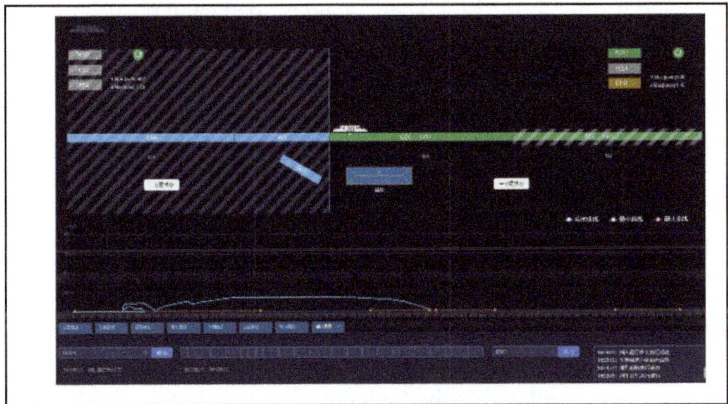

Fig. 4. Monitoring interface of SPVF.

5 Test Verification of SPVF System

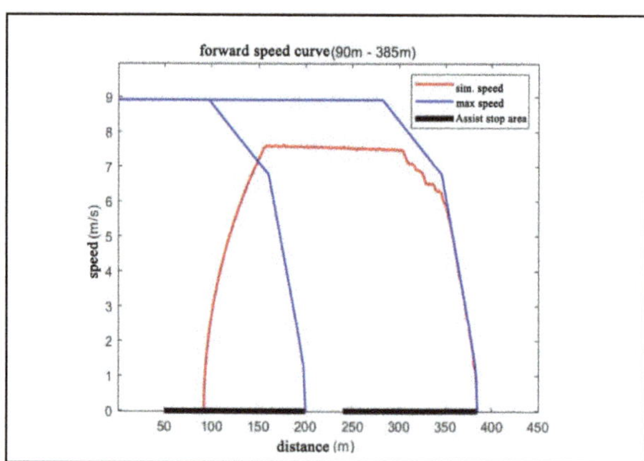

Fig. 5. Speed curve of high-speed maglev train.

Based on the Sifang High-Speed Maglev Test Line, the SPVF system has conducted the simulation and testing of the high-speed maglev speed and positioning system. This system meets the requirements of communication link monitoring function, ORT function simulation, VSC function simulation, and speed and positioning function testing in both forward and reverse operations of the maglev line. The simulation results and response time are consistent with the actual operation of the Sifang Maglev Test Line. It has the capability to conduct semi-physical verification and debugging of the speed and positioning system while in operation. The Fig. 5 displays the speed curve of the high-speed maglev train moving forward.

6 Conclusion

In light of the challenges related to the high cost, long cycle time, and complex debugging of high-speed maglev train speed and positioning systems, this paper introduces a comprehensive scheme for a semi-physical simulation experimental platform for maglev speed and positioning. This platform integrates digital simulation with physical equipment to create a hardware-in-the-loop speed and positioning validation platform. The paper details the software and hardware design principles, implementation methods, and experimental applications of the platform. The platform can be used for collaborative testing of high-speed maglev speed and positioning systems through the closed-loop interaction between virtual and real environments. This is significant for enhancing test scenario coverage and work efficiency.

Acknowledgment. This work is supported by the National Key R&D Program of China under Grant 2023YFB4302506.

References

1. Guo, X., Wang, Y., Wang, S.: Location and speed detection system for high-speed maglev vehicle. J. Southwest Jiaotong Univ. **39**(4), 455–459 (2004)
2. Li, D.: Electromagnetic Force And power Generation Characteristics of EMS Type High Speed Maglev Train Under Multiple Running Attitude. Beijing Jiaotong University, Beijing (2023)
3. Wang, J., Ding, R.: Prognostics and health management of high-speed trains in China. Strategic Study CAE **25**(2), 232–242 (2023)
4. Xiong, Z.: Study on test platform integrated simulation system for high-speed maglev. Comput. Knowl. Technol. **6**(8), 6290–6291, 6301 (2010)
5. Wang, Z., Li, J., Qi, Y.: Integrated simulation test platform of operation control system for high-speed maglev transport. Electric Drive Locomot. **1**, 144–149 (2023)
6. Qi, B.: Research on Hardware-in-Loop Simulation System of Maglev Train Speed and Position Detection System. Beijing Jiaotong University, Beijing (2022)
7. Tian, Y., Luan, J., Wang, X.: Development of wireless communication simulation platform for high-speed maglev system. Electric Drive Locomot. **6**, 61–64 (2020)
8. Wu, D.: Research on Integration Key Problems of High-Speed Maglev Train Operation Control System. Beijing Jiaotong University, Beijing (2023)
9. Wu, J., Zhou, W., Li, L.: Research on speed and position detection system of high speed maglev train. J. Natl. Univ. Def. Technol. **33**(1), 109–114 (2011)
10. Cheng, S., Liu, C., Song, L.: Maglev train integrated positioning and speed measuring method based on multi-source information fusion. Urban Mass Transit **25**(8), 136–140 (2022)

The Application of the Bicubic B-Spline Fitting in Local Block Rendering

Jiatao Yao, Shaobo Zhou, Long Yang, Shaojun Hu, and Zhiyi Zhang[✉]

School of Information Engineering, Northwest A&F University, Yangling, China
zhangzhiyi@nwafu.edu.cn

Abstract. The combination of polynomial regression with the control variable method is an effective approach for enhancing Monte Carlo integration for rendering local blocks.However, the high uncertainty and non-linear distribution of luminance in local blocks lead to higher computational errors in polynomial regression.To address this issue, we propose a new Monte Carlo estimation method.This method uses the bicubic B-spline surface to accurately fit the luminance of local blocks.The flexible node vectors of B-spline functions increase adaptability to non-linear distributions.The control variable method reduces errors in luminance estimation.We conduct experimental validation across multiple scenes in direct lighting environments.The results show that the Monte Carlo estimation method based on the bicubic B-spline surface and the control variable method significantly outperforms traditional polynomial regression fitting methods.The improvement is especially evident in blocks with significant luminance variations.

Keywords: Render · B-spline · Control variable · Local blocks

1 Introduction

In computer graphics, the Monte Carlo (MC) integration has become a hot research topic due to its wide application in global lighting simulation and complex scene rendering. It is particularly well-suited to handle local block rendering, which involves subdividing an image into smaller blocks (or tiles) for individual processing [1–4]. However,.the high uncertainty and nonlinear distribution of illumination in local blocks present significant challenges to traditional MC integration techniques,.often leading to inadequate rendering precision and stability [5–7].

In recent years, researchers have looked for ways to make Monte Carlo integration better by adding polynomial regression and control variable techniques [8, 9]. For instance, Fan et al. [10] found a way to improve direct illumination and irradiance cache by tweaking the control variable estimator, which helped cut down on image errors. Building on this, Salaün et al. [11] took it a step further with a Monte Carlo estimator that uses least squares polynomial regression. Their method reduces errors in calculations of light transmission and works well even in complex, multi-dimensional situations. These improvements show how combining different approaches can lead to more accurate and reliable rendering results.

© The Author(s) 2026
P. Siarry et al. (Eds.): WCNA 2024, LNEE 1550, pp. 450–464, 2026.
https://doi.org/10.1007/978-981-95-6946-5_46

The aforementioned methods have made significant progress in improving rendering quality and reducing errors. However, when processing larger tiles, a single polynomial regression model often fails to effectively capture the nonlinear variations between pixels, leading to error accumulation and instability in rendering results. This limitation restricts the applicability and effectiveness of these methods in high-dimensional and complex scenes.

Considering the above challenges, this paper proposes a new Monte Carlo integration estimator (see Fig. 1) combining B-spline functions and control variable technology. The flexible node vector of the B-spline function is used to fit the luminance distribution of the local block more accurately [12, 13]. Firstly, the sampled luminance value is parameterized, and the parameterized luminance value is used to construct a bicubic B-spline surface defined by the control points. Secondly, control variable technology is used to optimize the luminance value of the fitting, thereby improving the Monte Carlo integration. Finally, to ensure accurate color representation and consistent visual output, the RGB values are reconstructed using the CIE-XYZ color space standard proposed by the International Commission on Illumination. Our findings suggest that this method not only adapts well to complex, non-uniform distributions but also contributes to a noticeable reduction in variance during the integration process.

Our Contribution: This work presents a Monte Carlo integration estimator that integrates B-spline functions with control variate techniques. The experimental findings clearly demonstrate that the method introduced effectively minimizes the fluctuation in integral estimates, especially in scenarios involving intricate luminance patterns confined to specific areas.

2   Related Work

2.1   Regression-Based Monte Carlo

Monte Carlo integration is required to generate photorealistic images, especially local block rendering. It usually requires a prohibitively large number of light samples to converge and reduce noise [14–16]. To address this problem, Moon et al. [17, 18] proposed adaptive sampling and reconstruction using local regression theory. This approach extends Monte Carlo integration efficiency by adaptively selecting sampling points and approximating the underlying light distribution using local regression. It handles high-dimensional noisy features well and reduces the sample requirements for convergence. Further, they proposed a two-stage optimization process to select local bandwidths in a data-driven manner, enhancing the reconstructed image's accuracy and smoothness. Zwicker et al. [19] and Bitterli et al. [20] also pushed the application of regression techniques one step ahead to reduce Monte Carlo noise. They target their methods at the high-frequency variation in rendered images. With the regression models, they lower the variance without eliminating fine details and produce visually stable and accurate results using fewer samples. Salaün et al. [11] proposed an adaptive control variable approach using piecewise polynomials to more effectively handle nonlinear local block rendering variation. The approach enhances Monte Carlo integration by lowering estimator variance and improving rendering efficiency, particularly for complicated scenes with difficult light transport behavior. These are important steps towards more efficient and stable Monte Carlo integration.

2.2 Control Variable

The control variate (CV) technique is well known to diminish variance in Monte Carlo techniques [8, 21]. Constant environment terms and visibility functions were proposed as control variates by Lafortune et al. [22] and Clarberg et al. [23], respectively, which lower the complexity of lighting calculations and enhance the accuracy, especially of indirect lighting effects. Later, Vevoda et al. [24] used Bayesian regression to improve shadow estimation through variance reduction in shadow calculation by modeling high-frequency variations. Kutz et al. [25] extended this CV method to non-uniform media, like participating media, for improved rendering performance in complicated volumetric scenes. Most recently, Muller et al. [26] and Subr [27] used neural networks for the automatic choice of control variates. This takes performance in complicated and dynamic scenes to a new level, providing a scalable solution to Monte Carlo rendering that is able to handle a number of difficulties without the need for manual intervention. These developments illustrate the continued growth and flexibility of Monte Carlo rendering algorithms.

2.3 B-Spline Applications in Rendering

B-splines are widely employed in rendering and Computer-Aided Design (CAD) due to their flexibility and precision [28–30]. Redner et al. [31] first introduced B-spline functions in illumination mapping and ray tracing algorithms, defining illumination functions as weighted probability density functions to enhance photorealism. These functions were applied in bidirectional distributed ray tracing and radiosity algorithms, demonstrating their effectiveness. Jinka et al. [32] proposed a deep learning framework based on B-spline surfaces (SplineNet) for 3D shape classification, effectively capturing local geometric variations and optimizing shape representation. Liu et al. [33] employed B-spline fitting in model predictive control to smooth incomplete reference signals, improving control accuracy. Recently, Zhu et al. [34] introduced a triangular configuration B-spline (TCB-spline)-based image vectorization method, optimizing control points and node connectivity to efficiently model complex curves and color transitions. This method significantly improves reconstruction quality while preserving local details. These studies demonstrate that B-splines not only have significant applications in illumination mapping and rendering but also show broad potential in geometric modeling, control systems, and image vectorization.

3 Method

3.1 Monte Carlo Integration

Let be the definite integral I of a non-negative function $f(x)$ over the domain Ω and the estimate of the definite integral based on n samples:

$$I = \int_{\Omega} f(x)dx \tag{1}$$

$$\langle I \rangle_n = \frac{1}{n} \sum_{i=1}^{n} \frac{f(x_i)}{p(x_i)} \tag{2}$$

where dx is the differential over the region Ω, $p(x)$ is the sampling probability density function (PDF), and x_i is the sample point randomly drawn from the region Ω according to the probability density function.

For the estimator $\langle I \rangle_n$, the Monte Carlo method is unbiased. The sample points x_i are independently and identically distributed (i.i.d.). Therefore, $Y_i = \frac{f(x_i)}{p(x_i)}$ are also i.i.d. The variance of the estimator $\langle I \rangle_n$ is given by:

$$Var(Y) = E\left[Y^2\right] - \left(E[Y]^2\right) = E\left[\left(\frac{f(x_i)}{p(x_i)}\right)^2\right] - I^2 \tag{3}$$

Equation (3) shows that the variance of the estimator $\langle I \rangle_n$ is inversely proportional to the number of samples n.

In the work of Salaün et al. [11], the Monte Carlo integral estimator $\langle I \rangle_n$ can be redefined within the unit hypercube T. By assuming a mapping $x = \phi(t)$, I can be redefined by changing the variables, where the Jacobian determinant $|dx/dt| = 1/p(x)$. I can be expressed as:

$$I = \int_{\{\Omega\}} f(x)dx = \int_T f(\phi(t))\left|\frac{dx}{dt}\right|dt = \int_T \hat{f}(t)dt \tag{4}$$

where $\hat{f}(t) = \frac{f(\phi(t))}{p(\phi(t))}$. Therefore, the estimator $\langle I \rangle_n$ can be redefined by the change of variables as:

$$\langle I \rangle_n = \frac{1}{n} \sum_{i=1}^{n} \frac{f(t_i)}{p(t_i)} = \frac{1}{n} \sum_{i=1}^{n} \hat{f}(t_i) \tag{5}$$

By this transformation of the primary sample space, the sampling space can be converted from the probability density function in the space Ω, to a uniform distribution in the space T,

where t_i denotes a uniformly distributed sample in space T. This transformation yields the following probability density function:

$$p(t) = 1 \tag{6}$$

Discussion. In the work of Salaün et al. [11], a primary space transformation is employed to convert the probability density function of local block luminance into a uniform distribution function. For constant functions, this transformation is equivalent to sampling the integrand $f(x)$ under a uniform distribution. Furthermore, the same variable transformation also facilitates fitting the integrand $f(x)$ to a polynomial function $g(x)$. Under uniform distribution, it is suitable for polynomial fitting. In Section B, we will introduce the regression-based Monte Carlo estimator. Unlike the work of Salaün et al., the integrand $f(x)$ can be fitted to a bicubic B-spline surface. In other words, we additionally perform a parameter transformation, whereby the sample point coordinates are converted to the parameter space (u, v).

3.2   Regression-Based Monte Carlo Integration

It is to be considered any function $g(t)$ whose integral is G, and the integral I can be expressed as (7):

$$I \approx \langle I \rangle_{RC} = G + \langle D \rangle \tag{7}$$

The function G can be expressed as the integral of the function $g(t)$ over the space T. It can be written as (8):

$$G = \int_T g(t)dt \tag{8}$$

The quantity $\langle D \rangle$ is expressed as the difference between the Monte Carlo integration estimates. It can be expressed as the following equation:

$$\langle D \rangle = \frac{1}{N} \sum_{i=1}^{N} (f(t_i) - g(t_i)) \tag{9}$$

The variance of (9) can be expressed as follows:

$$var(G + \langle D \rangle) = Var(\langle D \rangle) = \frac{1}{N} Var(f(t) - g(t)) \tag{10}$$

From the variance, it can be seen that the variance is related to the difference term D. Therefore, a regression model based on the least squares method can be constructed to reduce the variance. Let a parametric regression model be constructed as $g(t, \theta)$, where θ is the parameter vector.

By minimizing the residual $R(\theta)$, the Monte Carlo method approximation for the parametric regression model is given by:

$$R(\theta) \approx \langle R(\theta) \rangle = \frac{1}{N} \sum_{i=1}^{N} \left(\hat{f}(t_i) - g(t_i, \theta) \right) \tag{11}$$

In this context, $\hat{f}(t_i)$ denotes the values of the function obtained by sampling under a uniform distribution. The objective is to minimise the squared difference between the sampled values and the predicted values from the regression model $g(t, \theta)$.

Let c be a constant and construct an arbitrary function $g(t) = c$. At this point, the regression-based Monte Carlo integral can degenerate into a traditional Monte Carlo integral. In this case, the integral G_c of the arbitrary function $g(t)$ can be expressed as follows:

$$G = \int_T g(t) dt = \int_T c \, dt = c \cdot |T| \tag{12}$$

where $|T|$ is the measure of the integration region T. The error term $\langle D \rangle_c$ can be expressed as:

$$\langle D \rangle_c = \frac{1}{N} \sum_{i=1}^{N} (f(t_i) - g(t_i)) = \frac{1}{N} \sum_{i=1}^{N} (f(t_i) - c) \tag{13}$$

The Monte Carlo integral can be approximated as follows:

$$I \approx \langle I \rangle_{RC} = c \cdot |T| + \frac{1}{N} \sum\nolimits_{i=1}^{N} (f(t_i) - c) \tag{14}$$

Equation (14) can be further simplified as:

$$\langle I \rangle_{RC} = c \cdot |T| + \frac{1}{N} \sum\nolimits_{i=1}^{N} f(t_i) - c = \frac{1}{N} \sum\nolimits_{i=1}^{N} f(t_i) \tag{15}$$

The difference between each individual function $f(t_i)$ and the constant c can be minimised by selecting an appropriate parametric regression model $g(t, \theta)$. This ensures that the sum of the squared differences is minimised.

3.3 B-Spline Regression-Based Monte Carlo Integration

The B-spline function is defined by a set of control points which is fitted using de Boor-Cox basis functions. In Section A, The primary space transformation is performed, and sampling is done by uniform in the space T. Therefore, we construct the surface as a uniform B-spline surface with endpoint multiplicity, meaning there is multiplicity K only at the ends of the knot vector, while the interior knots are uniformly distributed in an arithmetic sequence. Overall, this forms a non-closed, monotonically decreasing sequence. The subsequent section describes the process of parameterizing the sampled points in space T into the parameter space (u, v).

In the sampling space t, the knot vectors in the parameter space (u, v) are denoted as u_i and v_j, respectively. To construct the primary knot vectors as arithmetic sequences, we disregard the endpoint multiplicity in the uniform B-spline surface:

$$\begin{aligned}
\{u_i\} &= \{u_0, u_1, \ldots, u_m\}, \quad u_i = \tfrac{i}{m}, \quad i = 0, 1, \ldots, m \\
\{v_j\} &= \{v_0, v_1, \ldots, v_n\}, \quad v_j = \tfrac{j}{n}, \quad j = 0, 1, \ldots, n
\end{aligned} \tag{16}$$

For the sampling points t in space T, we construct a two-dimensional grid and calculate the corresponding knot values in both the u and v directions. According to the definition of a uniform B-spline surface with endpoint multiplicity, we add K zeros at the beginning and K ones at the end of the knot vectors $\{u_i\}$ and $\{v_j\}$. Thus, the knot vectors transformed into the parameter space (u, v) can be described as follows:

$$\begin{aligned}
\{u_i\} &= \{0, \ldots, 0, u_0, u_1, \ldots, u_m, 1, \ldots, 1\} \\
\{v_j\} &= \{0, \ldots, 0, v_0, v_1, \ldots, v_n, 1, \ldots, 1\}
\end{aligned} \tag{17}$$

Next, we will discuss Monte Carlo integration for bicubic B-spline surfaces. A bicubic B-spline surface is created by fitting in two directions using open uniform B-spline basis functions. The knot vectors in the (u, v) parameter space have already been defined in (17). The bicubic B-spline surface can be represented as:

$$g(u, v) = \sum\nolimits_{i=0}^{m} \sum\nolimits_{j=0}^{n} P_{ij} B_i^3(u) B_j^3(v) \tag{18}$$

where P_{ij} are the control points, and B^3 are the cubic B-spline basis functions.

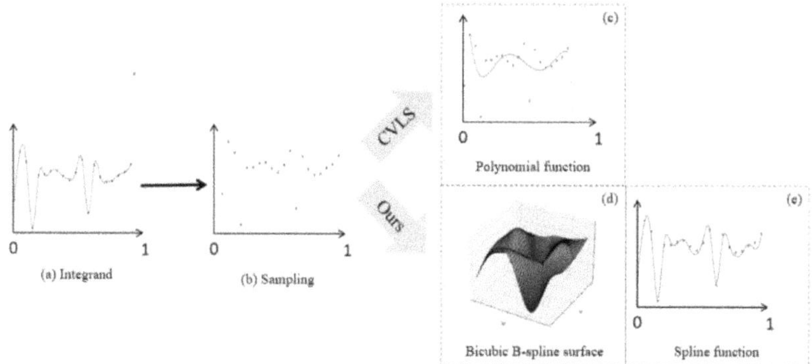

Fig. 1. Monte Carlo integration performs random sampling on the original distribution and estimates the characteristics of the original distribution based on these sampling points. Salaun et al. [11] used polynomial regression for fitting, while this paper employs a bicubic B-spline surface-based fitting method to approximate the original distribution.

Combining (17), the Monte Carlo integration of $g(u, v)$ over its parameter space (u, v) can be expressed as:

$$I \approx \langle I \rangle_B = G_{(u,v)} + \langle D \rangle_{(u,v)} \tag{19}$$

The variance of (19) can be expressed as:

$$\mathrm{Var}(\langle I \rangle_B) = \mathrm{Var}\left(\langle D \rangle_{(u,v)}\right) = \frac{1}{N}\mathrm{Var}(f(u, v) - g(u, v)) \tag{20}$$

Through least squares regression, a set of control points P_{ij} can be determined, aiming to minimize the residual $R(\theta)$:

$$R(\theta) = \sum_{i=1}^{N}\left(f(t) - \sum_{i=0}^{m}\sum_{j=0}^{n}P_{ij}B_i^3(u)B_j^3(v)\right)^2 \tag{21}$$

By minimizing the residual, we can find the optimal set of control points.

3.4 Control Variate Reduction

We introduce the control variate method for Monte Carlo integration based on bicubic B-spline surface regression. In the control variate method, an auxiliary function $t(u, v)$ and a known coefficient α are introduced:

$$I_{cv} = \langle I \rangle + \alpha(T - \langle T \rangle) = \frac{1}{N}\sum_{i=1}^{N}f(u_i, v_i) + \alpha\left(T - \frac{1}{N}\sum_{i=1}^{N}t(u_i, v_i)\right) \tag{22}$$

where α is defined as:

$$\alpha = \frac{\mathrm{Cov}(f(u, v), t(u, v))}{\mathrm{Var}(t(u, v))} \tag{23}$$

This approach builds on the work of Salaün et al., with the key distinction that the control variate method is used to adjust and reconstruct the sampling luminance of local block pixels. The RGB reconstruction technique will be detailed in Section E.

3.5 RGB Reconstruction

In local block pixel luminance, the G channel in RGB is typically selected as the luminance for correction. Consequently, the reconstruction of RGB values is undertaken using the International Commission on Illumination's CIE-XYZ standard, which defines a colour space model based on the human eye's perception for accurate colour representation. This is achieved by converting the RGB colour space to the XYZ colour space using weighting coefficients. The CIE-XYZ standard includes a number of conversion formulas, one of which is the conversion from RGB to XYZ colour space.

$$\begin{bmatrix} X \\ Y \\ Z \end{bmatrix} = \begin{bmatrix} 0.4124 & 0.3576 & 0.1805 \\ 0.2126 & 0.7152 & 0.0722 \\ 0.0193 & 0.1192 & 0.9505 \end{bmatrix} \begin{bmatrix} R \\ G \\ B \end{bmatrix} \tag{24}$$

where R, G, B are the input RGB values and X, Y, Z are the values of the converted CIE-XYZ colour space respectively.

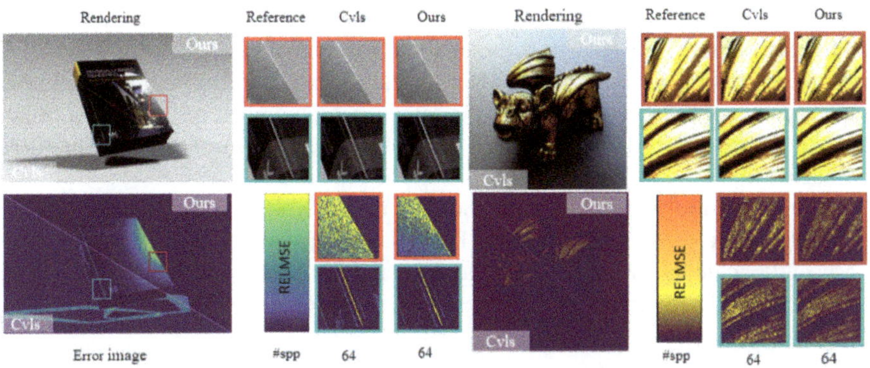

Fig. 2. The figure compares the Cvls and Ours methods in the Pbrt-book (left) and Dragon (right) scenes, showing rendered images, reference comparisons, and corresponding error images (RelMSE).

4 Experiments and Result

4.1 Experiments Setting

The experiments are conducted using PBRT-V3 with its open-source scene files, running on an Intel(R) Xeon(R) Platinum 8255C processor with multi-threading enabled. All experiments are performed in scenes with direct light sources. We compare our method with the *Cvls* method [11] by examining the rendering details to assess quality differences. We conducted an ablation study to investigate the effect of the control variate method on rendering performance. Additionally, we optimize the efficiency of solving B-splines using closed-form solutions and compare the runtime efficiency of both methods across various scenes to evaluate their computational performance.

4.2 Experimental Scenes and Baseline

For accurate comparison, both the cubic polynomial and the bicubic B-spline surface are used. The parameterization of the B-spline surface is achieved using multiple importance sampling. 2D integration issues in different scenes are treated as 3D problems, with sampled points X, Y, and Z representing light source selection, direct lighting, and BRDF, respectively. Reference images are evaluated using 2048 samples per pixel. The evaluation metric used is the relative mean square error (RelMSE).

4.3 Direct Lighting Comparison

Figure 2 shows a comparative analysis of different methods in direct lighting experiments. The left side shows the Pbrt-book scene with polygon mesh area lights, treated as a 3D integration problem. The right side shows the Dragon scene, where an infinite environment light, normally a 2D problem, is treated as 3D for consistency. The visual results and error maps illustrate that, with the same number of samples, both the local and global error images for *Ours* show lower errors compared to *Cvls*.

Figure 3 and Fig. 4 further illustrate the differences in local blocks in the Pbrt-Book and Dragon scenes. They provide the selection of local blocks and the corresponding 3D error graph of the selected areas. In both scenes, our method (Ours) achieves superior rendering details in local blocks with the same sample number, resulting in a smaller RelMSE. These results demonstrate that our method outperforms the polynomial-based approach.

Table 1 provides a detailed comparison of the RelMSE values for both local and global regions. It shows that for both the Dragon and Pbrt-book scenes, the *Ours* method consistently achieves lower RelMSE values than *Cvls*, indicating improved accuracy across both local and global error metrics.

Table 1. Local and Global RELMSE Comparison

	Local		Global
	Gray	*Red*	
Dragon (Ours)	**1.85E-02**	**1.41E-02**	**1.19E-02**
Dragon (Cvls)	1.86E-02	1.53E-02	1.45E-02
Pbrt-Book (Ours)	**5.01E-04**	**4.62E-03**	**3.61E-04**
Pbrt-Book (Cvls)	2.73E-03	5.07E-03	2.86E-03

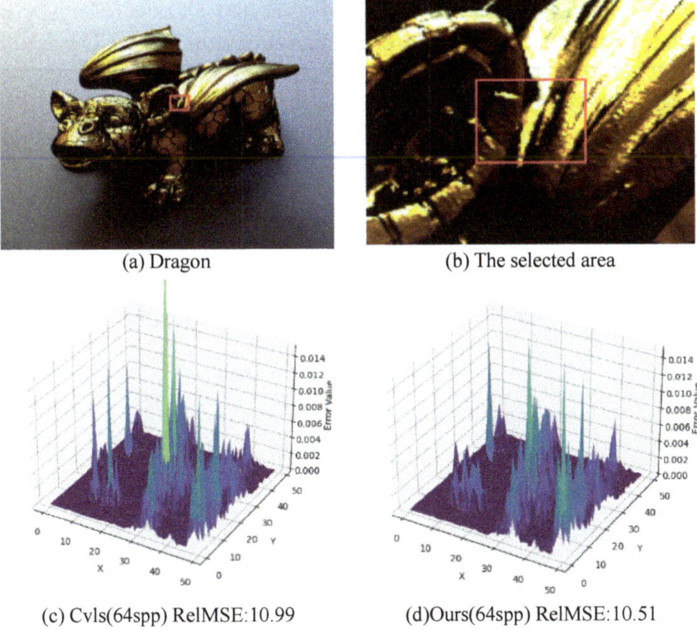

(a) Dragon   (b) The selected area

(c) Cvls(64spp) RelMSE:10.99   (d)Ours(64spp) RelMSE:10.51

Fig. 3. The 3D error graph for the selected area of the Dragon scene, with RelMSE magnified to 10^2.

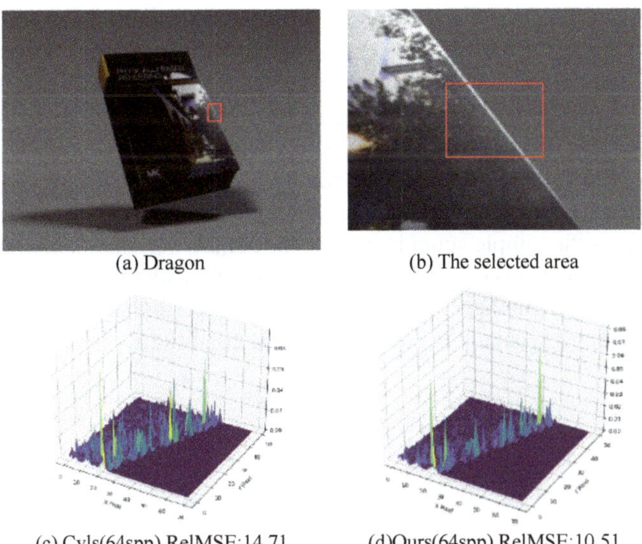

(a) Dragon   (b) The selected area

(c) Cvls(64spp) RelMSE:14.71   (d)Ours(64spp) RelMSE:10.51

Fig. 4. The 3D error graph for the selected area of the Pbrt-Book scene, with RelMSE magnified to 10^4.

Table 2 compares the RelMSE values of *Cvls* and *Ours* across different scenes, with all experiments conducted under equal sampling conditions. The results indicate that *Ours* consistently achieves lower RelMSE, particularly in local errors, with notable improvements in the Bathroom and Pbrt-book scenes. While the difference is minimal in Dragon, Ours still demonstrates a slight advantage, reflecting its stability in complex environments. Overall, *Ours* outperforms *Cvls* in most scenes, making it suitable for high-precision tasks.

Table 2. RELMSE Comparison of Methods

Scenes	Methods	
	Cvls (s)	Ours (s)
Bathroom	4.62×10^{-2}	7.28×10^{-6}
Pbrt-book	2.86×10^{-3}	3.61×10^{-4}
Dragon	1.19×10^{-2}	1.45×10^{-2}
Vw-van	1.08×10^{-3}	8.83×10^{-4}

The experimental results demonstrate that our method achieves superior performance compared to others in both global rendering and local block rendering, as evidenced by the error visualizations.

4.4 Ablation of Control Variable Method

We perform an ablation study of the control variable method on the Killeroos scene under direct lighting. As shown in Fig. 5, the sampled ground truth luminance (blue), B-spline fitted luminance (purple), and control variable-optimized luminance (red) are compared across different sample counts. The B-spline closely follows the ground truth, while the control variable-optimized luminance displays smoother transitions, aligning more closely with the overall trend. The results show that the control variable method significantly reduces fluctuations in the luminance estimates, improving the stability of the rendering. As the sample count increases, the control variable-optimized luminance exhibits smoother transitions, making it more consistent with the ground truth compared to the B-spline fitted luminance. This method effectively reduces variance, improving rendering accuracy, especially in complex lighting conditions.

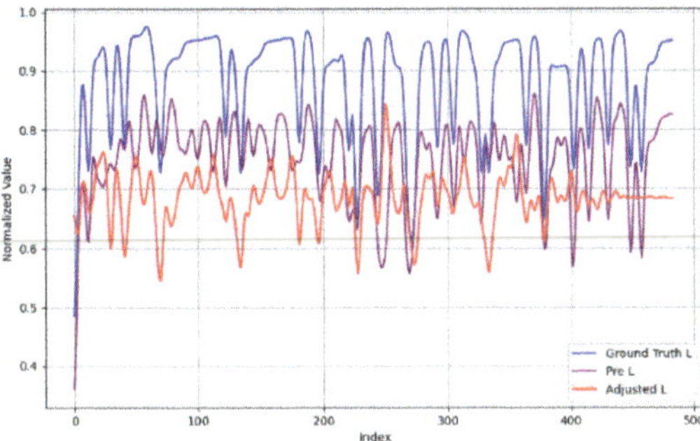

Fig. 5. The B-spline fitted luminance curve has large fluctuations while that fitted by the control variate method is more stable. (Color figure online)

4.5 Comparison of Time

In the experiment, with the aim of enhancing computational speed, we optimize the process. We do this by constructing B-spline surfaces for local blocks of pixels rather than constructing B-splines for each individual pixel separately. Also, we enhance the efficiency of B-spline solving through the utilization of closed-form solutions. This action notably cuts down the computation time. Table 3 presents the time comparison between *Ours*, *Cvls*, and *MC* under equal sampling conditions, thus manifesting the efficacy of these optimization measures.

Table 3. Time Comparison of Methods

Scenes	Methods		
	MC (s)	Cvls (s)	Ours (s)
Bathroom	44.5	2872.6	1782.6
Pbrt-book	11.0	421.6	142.5
Dragon	18.0	362.6	118.5
TT	36.5	2197.6	722.6
Villa-daylight	2047.6	2042.6	2047.6
Vw-van	125.5	2932.6	957.6

5 Conclusions

We present a Monte Carlo integration method using bicubic B-spline surfaces and control variable. The B-spline technique fits local light patterns by adjusting node vectors, making integration more accurate. Control variable cuts variance in estimates and improve stability. Our method works better than polynomial regression-based Monte Carlo methods in low-dimensional cases.

Moreover, this paper delves into the potential that polynomial regression holds in block-based rendering, underlining the benefits of B-spline surfaces when it comes to handling extensive luminance distributions. The experiments demonstrate that the method put forward offers enhanced rendering accuracy as well as stability within complex scenes.

The potential for the further development of our work is quite substantial. For example, the development of incremental algorithms meant for dynamic scenes has the possibility to bring about more efficient updates to the control grid. In doing so, it can enhance the performance of real-time rendering. Moreover, when it comes to the investigation of advanced algorithms for high-dimensional data, like the rapid computation methods for high-degree spline functions or the application of dimensionality reduction techniques to cut down on computational complexity, this could help in getting past the limitations of the current method and widening its application range. This would boost both its practicality as well as its performance.

Acknowledgments. This paper is supported by the key project of linkage between departments and cities in Shanxi, China. Project number: 2022GD-TSLD-53.

References

1. Burgers, W.: Tile-based rendering. Traffic **600**, 3 (2005)
2. Terai, M., Fujiki, J., Tsuruno, R., Tomimatsu, K.: Tile-based modelling and rendering. In: Smart Graphics: 8th International Symposium, SG 2007, Kyoto, Japan, 25–27 June 2007. Proceedings 8, pp. 158–163. Springer (2007)
3. Suhail, M., Esteves, C., Sigal, L., Makadia, A.: Generalizable patch-based neural rendering. In: European Conference on Computer Vision, pp. 156–174. Springer (2022)
4. Chizhov, V., Georgiev, I., Myszkowski, K., Singh, G.: Perceptual error optimization for monte carlo rendering. ACM Trans. Graph. **41**(3), 1–17 (2022)
5. Tewari, A., et al.: State of the art on neural rendering. In: Computer Graphics Forum, vol. 39, no. 2, pp. 701–727. Wiley Online Library (2020)
6. Bemana, M.: Efficient image-based rendering (2023)
7. Zhou, L., Wu, G., Zuo, Y., Chen, X., Hu, H.: A comprehensive review of vision-based 3D reconstruction methods. Sensors **24**(7), 2314 (2024)
8. Si, S., Oates, C.J., Duncan, A.B., Carin, L., Briol, F.-X.: Scalable control variates for Monte Carlo methods via stochastic optimization. In: InternationalConference on Monte Carlo and Quasi-Monte Carlo Methods in Scientific Computing, pp. 205–221. Springer (2020)
9. Song, C., Kawai, R.: Monte carlo and variance reduction methods for structural reliability analysis: a comprehensive review. Probab. Eng. Mech. **73**, 103479 (2023)
10. Fan, S., Chenney, S., Hu, B., Tsui, K.-W., Chi Lai, Y.: Optimizing control variate estimators for rendering. Computer Graphics Forum, vol. 25, pp. 351–357 (2006)
11. Salaün, C., Gruson, A., Hua, B.-S., Hachisuka, T., Singh, G.: Regression-based monte carlo integration. ACM Trans. Graph. **41**, 1–14 (2022)
12. Bhatia, H., Hoang, D., Morrical, N., Pascucci, V., Bremer, P.-T., Lindstrom, P.: Amm: adaptive multilinear meshes. IEEE Trans. Visual Comput. Graphics **28**(6), 2350–2363 (2022)
13. Duan, W., Zhang, P., Huang, L., Yang, K., Yang, K.: Ship hull surface reconstruction from scattered points cloud using an RBF neural network mapping technology. Comput. Struct. **281**, 107012 (2023)

14. Greenberg, D.P., et al.: A framework for realistic image synthesis. In: Proceedings of the 24th Annual Conference on Computer Graphics and Interactive Techniques, pp. 477–494 (1997)
15. Rousselle, F., Jarosz, W., Novák, J.: Image-space control variates for rendering. ACM Trans. Graph. (TOG) **35**(6), 1–12 (2016)
16. Frolov, V.A., Voloboy, A.G., Ershov, S.V., Galaktionov, V.A.: Light transport in realistic rendering: state-of-the-art simulation methods. Program. Comput. Softw. **47**, 298–326 (2021)
17. Moon, B., Carr, N., Yoon, S.-E.: Adaptive rendering based on weighted local regression. ACM Trans. Graph. **33**, 14 (2014)
18. Moon, B., Iglesias-Guitian, J.A., Yoon, S.-E., Mitchell, K.: Adaptive rendering with linear predictions. ACM Trans. Graph. **34**, 11 (2015)
19. Zwicker, M., et al.: Recent advances in adaptive sampling and reconstruction for Monte Carlo rendering. Comput. Graph. Forum **34**, 667–681 (2015)
20. Bitterli, B., et al.: Nonlinearly weighted first-order regression for denoising monte carlo renderings. Comput. Graph. Forum **35**, 107–117 (2016)
21. Hickernell, F.J., Lemieux, C., Owen, A.B.: Control variatesfor quasi-Monte Carlo (2005)
22. Lafortune, E.P., Willems, Y.D.: The ambient term as avariance reducing technique for Monte Carlo ray tracing, pp. 168–176. Springer, Heidelberg (1995)
23. Clarberg, P., Akenine-Möller, T.: Practical product importance sampling for direct illumination. Comput. Graph. Forum **27**, 681–690 (2008)
24. Vévoda, P., Kondapaneni, I., Křivánek, J.: Bayesian online regression for adaptive direct illumination sampling. ACM Trans. Graph. **37**, 1–12 (2018)
25. Kutz, P., Habel, R., Li, Y.K., Novák, J.: Spectral and decomposition tracking for rendering heterogeneous volumes. ACM Trans. Graph. **36**, 1–16 (2017)
26. Müller, T., Rousselle, F., Keller, A., Novák, J.: Neural control variates. ACM Trans. Graph. **39**, 1–19 (2020)
27. Subr, K.: Q-net: a network for low-dimensional integrals ofneural proxies. Comput. Graph. Forum **40**, 61–71 (2021)
28. Cohen, E., Lyche, T., Riesenfeld, R.: Discrete b-splines and subdivision techniques in computer-aided geometric design and computer graphics. Comput. Graph. Image Process. **14**(2), 87–111 (1980)
29. Biswas, S., Lovell, B.C.: B-splines and its applications. In: Bézier and Splines in Image Processing and Machine Vision, pp. 109–131. Springer (2008)
30. Park, H., Lee, J.-H.: Adaptive b-spline volume representation of measured brdf data for photorealistic rendering. J. Comput. Des. Eng. **2**(1), 1–15 (2015)
31. Redner, R.A., Lee, M.E., Uselton, S.P.: Smooth b- spline illumination maps for bidirectional ray tracing. ACM Trans. Graph. **14**, 337–362 (1995)
32. Jinka, S.S., Sharma, A.: Splinenet: B-spline neural network for efficient classification of 3D data. In: Proceedings of the 11th Indian Conference on Computer Vision, Graphics and Image Processing, pp. 1–8 (2018)
33. Liu, M., Wu, H., Wang, J.: A modified model predictive control based on b-spline fitting. In: Proceedings of the 2020 1st International Conference on Control, Robotics and Intelligent System, pp. 1–5 (2021)
34. Zhu, H., Cao, J., Xiao, Y., Chen, Z., Zhong, Z., Zhang, Y.J.: Tcb-spline-based image vectorization. ACM Trans. Graph. **41**, 1–17 (2022)

Neuromorphic Computing Based on MoS$_2$/HfO$_2$ Memristors

Yi Sun, Jing You, Wenxuan Xu, Ziyue Li, Jinxia Xu, and Xingjuan Song$^{(\boxtimes)}$

School of Science, Hubei University of Technology, 28 Nanli Road, HongShan District, Wuhan, China
`xingjuansong@hbut.edu.cn`

Abstract. Two-dimensional (2D) material MoS$_2$ holds great promise for neuromorphic applications due to its outstanding electrically tunable resistive switching properties. However, challenges remain in achieving reliable, durable, and high-retention performance in MoS$_2$-based memristors. In this work, we present a heterostructure memristor, Au/Ag/MoS$_2$/HfO$_2$/Pt, which synergistically combines the advantages of 2D MoS$_2$ with the high stability of HfO$_2$. The device demonstrates reconfigurable performance, including multi-level resistance states enabled by current limitation, a high on/off switching ratio of 10^8, and an ultrafast response time of 19 ns. Furthermore, the memristor successfully emulates both non-volatile synaptic plasticity—such as long-term potentiation (LTP), long-term depression (LTD), paired-pulse facilitation (PPF), and spike-timing-dependent plasticity (STDP)—as well as volatile neuronal functions like leaky integrate-and-fire (LIF) behavior. This study provides a novel pathway for reconfigurable and multifunctional MoS$_2$-based memristor applications in brain-inspired computing systems and demonstrates the immense potential of neuromorphic computing.

Keywords: MoS$_2$ · reconfigurable · Synaptic plasticity · LIF neuron

1   Introduction

The neuromorphic computation of spiking neural networks (SNNs) based on memristors has attracted significant attention due to its low energy consumption and high resemblance to biological neural systems [1, 2]. Traditional artificial synapses, which emulate synaptic behavior, are typically implemented using complementary metal-oxide-semiconductor (CMOS) transistors [3] or emerging transistors operating through ferroelectric or electrochemical gate coupling [4–6]. Among these technologies, memristors stand out as one of the most promising candidates for neuromorphic computing applications due to their inherent ability to retain historical information about previously applied electrical signals.

The switching layer material is a critical component of memristor structures, as it determines the resistive switching characteristics of the device. Conventional memristors often utilize binary oxides or perovskites as the switching layer, such as HfO$_x$ [7], TiO$_x$ [8], and AlO$_x$ [9]. Among these, HfO$_x$ and TaO$_x$ are widely regarded as the

© The Author(s) 2026
P. Siarry et al. (Eds.): WCNA 2024, LNEE 1550, pp. 465–474, 2026.
https://doi.org/10.1007/978-981-95-6946-5_47

most promising candidates due to their sub-nanosecond operating speeds and exceptional endurance, with cycle lifetimes exceeding 10^{10} switching events. However, as the demands of the big data era grow, limitations of traditional oxide-based memristors have become increasingly apparent, including insufficient switching ratios, high power consumption, and limited response speeds.

In recent years, two-dimensional (2D) materials, such as MoS_2, h-BN, and graphene, have emerged as promising alternatives for low-power and high-performance memristor devices [10]. These atomically thin materials offer unique advantages, including high carrier mobility, rapid switching speeds, multifunctional heterogeneous integration potential, and excellent adaptability for mimicking biological synaptic behavior. Despite their exceptional electrical properties, 2D materials face challenges in terms of reliability, durability, and retention. These limitations can be mitigated by modifying their structure and improving material quality [11].

In this work, we explore the reconfigurable characteristics of heterostructure memristors with the configuration $Au/Ag/MoS_2/HfO_2/Pt$. Compared to conventional oxide-based memristors, this device achieves a significantly higher switching ratio ($\sim 10^8$) and an ultrafast response time of 19 ns, making it highly suitable for CMOS-integrated circuits. Furthermore, the memristor demonstrates both non-volatile synaptic plasticity, including LTP/LTD, PPF, and STDP, as well as volatile neuronal functionality, such as leaky integrate-and-fire (LIF) behavior. This work highlights the potential of 2D material-based heterostructure memristors for advanced neuromorphic systems and brain-inspired computing.

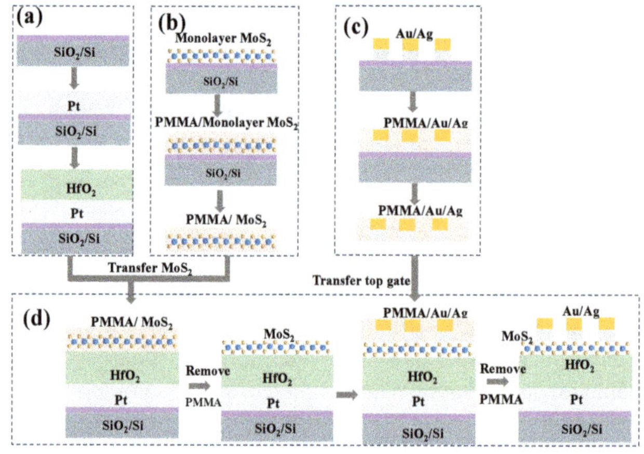

Fig. 1. Wet Transfer Flowchart: (a) Preparation of $HfO_2/SiO_2/Si$ substrate; (b) transfer process of MoS_2; (c) transfer process of top electrodes (Au/Ag); (d) diagram for fabricating $Au/Ag/MoS_2/HfO_2/Pt$ devices.

2 Experimental Procedures

The transfer process is illustrated in Fig. 1, which comprises four sequential steps:

1. **Preparation of HfO$_2$/SiO$_2$/Si substrate**: The silicon wafer is cleaned, followed by the deposition of the bottom Pt electrode (200 nm) and the resistive switching layer HfO$_2$ (5 nm), as shown in Fig. 1(a).
2. **Transfer process of MoS$_2$**: A monolayer MoS$_2$ (0.65 nm thickness) is precisely transferred onto the HfO$_2$ layer using a polymer-assisted wet transfer method, as depicted in Fig. 1(b).
3. **Transfer process of top electrodes**: The top electrodes (50 nm Ag/30 nm Au) are patterned and transferred onto the MoS$_2$ surface through photolithography and thermal evaporation, illustrated in Fig. 1(c).
4. **Device Integration**: The final Au/Ag/MoS$_2$/HfO$_2$/Pt heterostructure is completed, forming a 60×60 μm functional memristor, as demonstrated in Fig. 1(d).

The specific steps are as follows:

First, the silicon wafers are cleaned separately with acetone, ethanol, and deionized water using ultrasonic cleaning at 70 W power for 15 min. After drying, they are placed in a magnetron sputtering system to prepare a Pt bottom electrode with a thickness of approximately 200 nm using DC voltage sputtering. Next, a 5 nm layer of HfO$_2$ is deposited as the resistive switching layer using atomic layer deposition (ALD). On another silicon wafer, a top electrode of size 60×60 μm is photolithographed, and 50 nm of Ag and 30 nm of Au are deposited as the top electrode using a thermal evaporation system. A 10 ml HF (40%) solution is prepared and diluted with 40 ml of deionized water. A layer of PMMA is spin-coated on the MoS$_2$ substrate, which is then placed on a spin coater and spun at 3000 r/min for 1 min. The substrate is heated on a hot plate at 150 °C for 5 min and allowed to cool for 5 min. The silicon wafer is cut to an appropriate size, and plastic tweezers are used to pick up the silicon wafer with MoS$_2$ and place it in the prepared HF solution. Once the MoS$_2$ film has detached, the film is retrieved and placed in deionized water for 15 min, with this process being repeated three times. Subsequently, the MoS$_2$ film is transferred to a benzyl ether solution, and the prepared Pt substrate silicon wafer is used to collect the MoS$_2$ film. Filter paper is utilized to absorb the surrounding solution moisture, and the moisture between the film and substrate is blown dry with a balloon. The device is placed in an 80 °C oven for 20 min, followed by heating on a hot plate at 150 °C for 5 min. After heating, it is allowed to cool for 10 min and then immersed in acetone for 30 min to remove the PMMA from the MoS$_2$. Finally, the same wet transfer steps are used to transfer the top electrode, completing the device with the structure Au/Ag/MoS$_2$/HfO$_2$/Pt.

The device dimensions are 60×60 μm, and its optical image is shown in Fig. 2(a). Electrical characterization was performed using a B1500A semiconductor parameter analyzer.

3 Results and Discussions

3.1 Study on the Basic Electrical Performance of Reconfigurable Au/Ag/MoS$_2$/HfO$_2$/Pt Memristor Device

The multi-level resistance states of memristors represent a fundamental characteristic, enabling the device to stabilize at discrete or continuous resistance levels in response to varying electrical stimuli. This capability not only meets the increasing demands for enhanced storage capacity and speed in the era of big data but also offers significant advantages for neuromorphic computing and brain-inspired architectures. In particular, synaptic arrays that leverage these multi-level states facilitate parallel information processing while substantially reducing power consumption.

To validate the device's ability to achieve multi-level resistance states, we performed DC voltage sweeps under various current-limiting conditions. As illustrated in Fig. 2(b), the device initially resides in a HRS when a continuous forward DC voltage is applied to the top Ag electrode while the bottom Pt electrode is grounded. The semiconductor analyzer (B1500A) was configured with current compliance levels of 10 nA, 100 nA, 1 μA, 10 μA, 100 μA, 1 mA, and 10 mA. Under low current compliance (10 nA to 100 μA), the device exhibits volatile switching behavior; once the negative voltage is removed, the resistance reverts from a LRS back to the HRS and remains stable upon reset. The observed threshold voltage of approximately 0.2 V, along with rapid switching and low operational voltage, renders the device suitable for low-power applications. Conversely, at higher current compliance (1–10 mA), the device demonstrates nonvolatile behavior. Following voltage removal, the device maintains the LRS, with a threshold voltage of about 0.6 V and a switching ratio exceeding 10^8, indicating its potential for high-density storage applications. Figure 2(c) presents the resistance states at a read voltage of 0.1 V across the various current limits, further confirming the device's multi-level resistance modulation capability.

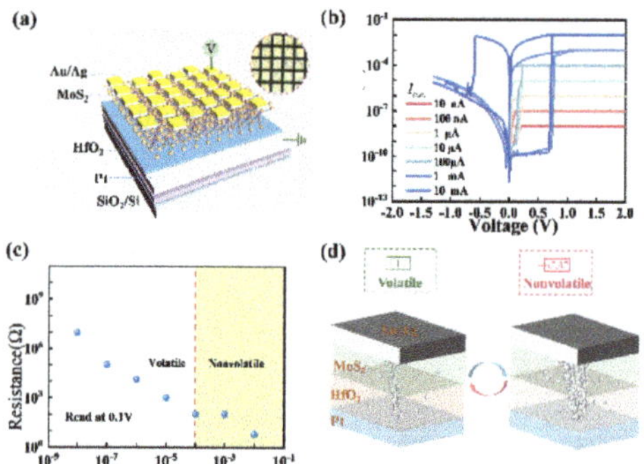

Fig. 2. (a) Device structure diagram; (b) Multi-level *I-V* curve of monolayer MoS_2 device; (c) Multi-states of reading at 0.1 V; (d) Resistive switching mechanism diagram.

The transition between volatile and nonvolatile modes is intrinsically linked to the dynamics of CF(conductive filament) formation. Under low current compliance, only thin and fragile CFs are transiently formed; these filaments rupture upon voltage removal, resulting in a reversion to the HRS. In contrast, higher current compliance promotes the formation of robust, thicker CFs that persist even after the applied voltage is withdrawn, thus stabilizing the LRS. This CF-driven mechanism is schematically detailed in Fig. 3(d).

It is noteworthy that the device can transition between volatile and nonvolatile modes by adjusting the magnitude of the current compliance. As shown in Fig. 3(a), the monolayer MoS_2 device exhibits stable volatile switching under a current compliance of 50 nA. When the applied voltage approaches 0.1 V, the CF ruptures, causing the device to reset from a LRS to a HRS. In contrast, as depicted in Fig. 3(b), when the current compliance is increased to 1 mA, the device demonstrates stable nonvolatile switching; a negative voltage pulse is required to revert the device from the LRS back to the HRS. Figure 3(c) presents the device's resistance state after 50 voltage sweep cycles at a 1 mA current compliance, confirming stable resistive switching with a switching ratio of up to 10^8. Moreover, Fig. 3(d) illustrates that when an initial reading voltage of 0.2 V is applied, the response current is essentially zero, indicating that the device is in the HRS. Following the application of a 1.3 V pulse voltage and subsequently a 0.2 V read voltage, the response current increases to 0.38 μA. This demonstrates a rapid transition from the HRS to the LRS, with an operating speed of 19 ns. Its nonlinear I-V characteristics and multi-level resistance state potential provide an important design foundation for neuromorphic computing and edge AI hardware.

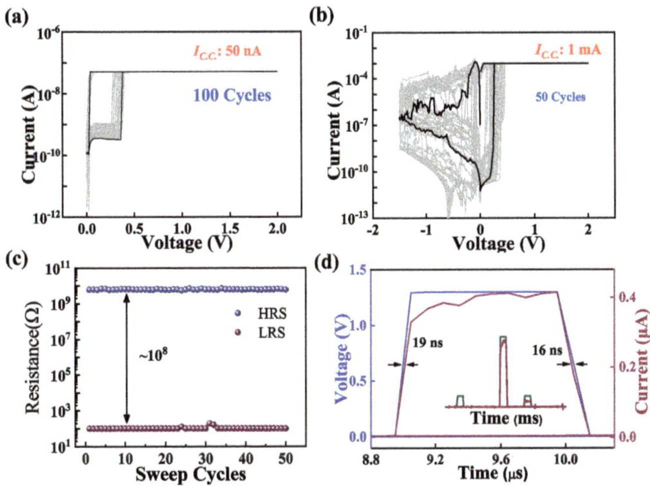

Fig. 3. (a) I-V curve of monolayer MoS$_2$ device with 50 nA current limiting; (b) I-V curve of monolayer MoS$_2$ device with 1 mA current limiting; (c) 1 mA Limit Current Resistance State Read Diagram; (d) Device Response Speed Chart.

3.2 Research on Synaptic Bionic Performance of Au/Ag/MoS$_2$/HfO$_2$/Pt Memristor Device

Memristors exhibit a remarkable functional resemblance to biological synapses, as their resistance can be dynamically modulated by external electrical signals—mimicking the synaptic weight changes observed in processes such as LTP/LTD [12]. Their intrinsic nonvolatile behavior underpins long-term memory storage, while multi-level resistance states enable continuous modulation of synaptic strength. Furthermore, by replicating pulse-timing-dependent resistance switching—akin to the STDP rule—memristors can effectively emulate the spatiotemporal associative learning mechanisms found in neural networks, thereby providing a robust physical foundation for the development of neuromorphic chips and brain-inspired computing systems. To evaluate the biomimetic performance of our devices, we conducted a series of tests designed to simulate key aspects of biological synapses.

PPF in synaptic characteristics refers to the phenomenon where the postsynaptic potential (PSP) induced by a second pulse is enhanced compared to that induced by the first pulse when consecutive pulse stimuli are applied rapidly between neurons. PPF is an expression of neural plasticity, indicating that synapses exhibit a sensitizing effect to frequent stimuli in the short term. As shown in Fig. 4. (a), when two consecutive electrical pulse stimuli are applied to the Au/Ag/MoS2/HfO2/Pt memristor device, the interval between the consecutive pulse stimuli gradually increases from 1 s to 7 s, with the weight change decreasing from 61.04% to 4.72%. After each test is completed, the device's resistive state is reset to a high-resistance state to ensure the accuracy of the tests. As time increases, the response of the second pulse stimulus to the first pulse stimulus becomes progressively smaller. The formulas for calculating the PPF ratio and the fitting curve are represented by Eqs. (1) and (2) respectively [13–15].

$$PPF_{ratio} = [(I_2 - I_1)/I_1] \times 100\%, \tag{1}$$

where the I_1 and I_2 refer to the current values generated by the first and second pulses, respectively.

$$PPF = A\exp(-\Delta t/\tau) + B \tag{2}$$

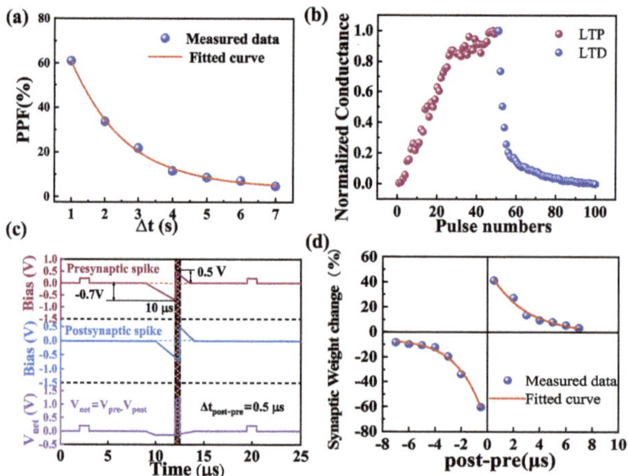

Fig. 4. (a) PPF synaptic plasticity; (b) LTP-LTD synaptic plasticity; (c) STDP test pulse diagram; (d) STDP ynaptic plasticity.

Fitted curve of PPF ratio. Where A (22.26) is a constant, τ (1.31) is the constant, and B (1.69) is the value of PPF when Δt tends to infinity.

LTP and LTD in memristive synaptic characteristics mirror key neural plasticity phenomena that underlie learning and memory in biological systems. As shown in Fig. 4(b), when 50 consecutive positive voltage pulses are applied to the device, the synaptic weight increases, with the normalized conductance rising from 0.004 mS to 0.99917 mS. Conversely, the application of 50 consecutive negative voltage pulses causes the conductance to decrease progressively from 1 mS to 0 mS.

STDP is a crucial phenomenon in the synaptic behavior of memristors, illustrating how synaptic strength is modulated by the temporal order of neuronal firing pulses. In biological systems, the magnitude of synaptic weight is determined by the time difference between pre-synaptic and post-synaptic stimuli. Specifically, if the pre-synaptic pulse precedes the post-synaptic pulse (i.e., $\Delta t_{(post-pre)} > 0$), the synaptic weight (W) increases; conversely, if $\Delta t_{(post-pre)} < 0$, the weight decreases. As illustrated in Fig. 4(c), when STDP is applied to the top Ag electrode of the device, the corresponding changes in synaptic weight are depicted in Fig. 4(d). Notably, for $\Delta t_{(post-pre)} > 0$, the conductance of the device exhibits an enhancement, with the synaptic weight decreasing in magnitude as the temporal interval between pre-synaptic and post-synaptic pulses increases. The relative change in synaptic weight (ξ) is calculated using Eq. (3) [16].

$$\xi = [(W_{STDP} - W_0)/W_0 \times 100\%]\xi, \tag{3}$$

where the W_0 and W_{STDP} represent the weights of the artificial synapse before and after the waveform application, respectively. All the above findings provide compelling evidence of the Au/Ag/MoS$_2$/HfO$_2$/Pt device's potential to emulate biological synapses and pave the way for its application in neuronal computation.

3.3 LIF Neuron Circuit Based on Au/Ag/MoS$_2$/HfO$_2$/Pt Memristor

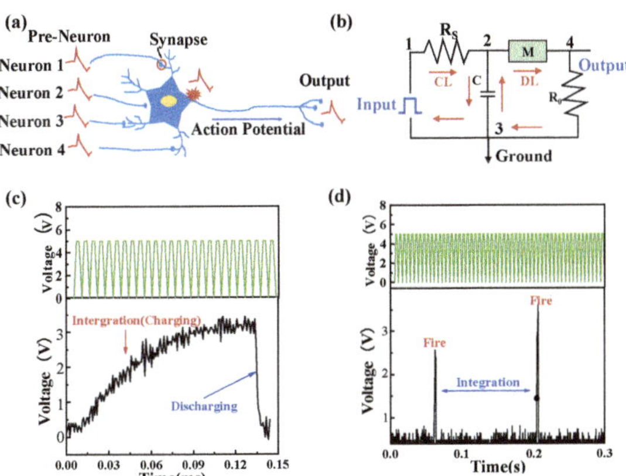

Fig. 5. (a) Schematic diagram of a neuron; (b) Circuit diagram of the LIF neuron; (c) Charging and discharging process of the capacitor of the LIF neuron; (d) Two outputs of the LIF neuron circuit.

The Leaky Integrate-and-Fire (LIF) neuron circuit is a widely adopted computational model for simulating neuronal behavior. As shown in Fig. 5(a), the schematic illustrates the architecture of the LIF neuron. Central to this model is the concept of "leakage," which refers to the gradual loss of accumulated charge or membrane potential over time. Although an input current causes the membrane potential to increase, leakage ensures that it decays over time. When the neuron receives sufficient synaptic input, the membrane potential progressively rises until it reaches a threshold value, prompting the neuron to emit a pulse (action potential). Following this, the membrane potential is reset to its resting state [17]. In this circuit, memristors serve as synaptic connections, enabling precise regulation of the strength and timing of input signals.

The LIF neuron circuit used in this paper is depicted in Fig. 5(b) and is consistent with widely reported memristor-based neuron circuits [18, 19]. In this design, the memristor is connected in series with a 5.1 kΩ resistor (R_0). This series combination is paralleled with a capacitor (C = 10 nF) and the memristor, and the entire neuron is connected to a synaptic resistor (R_s = 100 kΩ). The voltage change at R_0 is used to represent the output spike. The neuron circuit is divided into two parts: the charging loop (1-2-3-1) and the discharging loop (2-4-3-2). The total resistance of the charging loop is R_{CL} = Rs, while the total resistance of the discharging loop is $R_{DL} = R_M + R_0$. When an input pulse is applied, if $R_{CL} < R_{DL}$, the capacitor charges. Conversely, if $R_{CL} > R_{DL}$, the capacitor discharges.

Initially, the memristor is in a high-resistance state, which implies $R_{CL} < R_{DL}$. Under a periodic input pulse (with period P = 5 ms, pulse width = 6 μs, and amplitude A = 5 V), the capacitor remains in a continuous charging state. When the capacitor voltage (V_C) exceeds the memristor's threshold voltage ($V_{th} \approx 0.3$ V), the memristor rapidly switches from a HRS to a LRS. This transition reverses the resistance relationship ($R_{CL} > R_{DL}$), prompting the capacitor to discharge, as illustrated in Fig. 5(c). Due to the inherent volatility of the device, once V_C falls below V_{th}, the silver CFs spontaneously dissociate, causing the memristor to revert to its HRS. This reset triggers the initiation of a new charging and discharging cycle. Figure 5(d) displays the two distinct output processes generated by the LIF neuron. All these results confirm that the Au/Ag/MoS$_2$/HfO$_2$/Pt memristor can successfully simulate LIF characteristics, demonstrating its potential to mimic biological neurons effectively.

4 Conclusion

In conclusion, Au/Ag/MoS$_2$/HfO$_2$/Pt memristor was successfully fabricated using the wet transfer method. The device demonstrated the ability to achieve multiple resistance states, with a high switching ratio of 10^8 and an ultra-fast response time of 19 ns. Furthermore, it exhibited reconfigurable behavior, enabling the adjustment of its resistance switching mode by controlling the limiting current, allowing seamless transitions between volatile and non-volatile states. Under non-volatile conditions, the device effectively emulated biological synaptic plasticity, while in volatile conditions, it was integrated into a LIF neuron circuit to successfully replicate the characteristics of biological neurons. These findings highlight the significant potential of MoS$_2$-based memristors for applications in brain-inspired computing systems.

Acknowledgements. This work is supported by the National Nature Science Foundation of China (62204080). Hubei Provincial Natural Science Foundation (JCZRLH202500111). The School Research Startup fund of Hubei University of Technology (grant nos. XJ2021006101).

References

1. Park, S., et al.: RRAM-based synapse for neuromorphic system with pattern recognition function. In: 2012 International Electron Devices Meeting, pp. 10.2.1–10.2.4. IEEE (2012)
2. Prezioso, M., Merrikh-Bayat, F., Hoskins, B.D., Adam, G.C., Likharev, K.K., Strukov, D.B.: Training and operation of an integrated neuromorphic network based on metal-oxide memristors. Nature **521**(7550), 61–64 (2015)
3. Türel, Ö., Lee, J.H., Ma, X., Likharev, K.K.: Neuromorphic architectures for nanoelectronic circuits. Int. J. Circuit Theory Appl. **32**(5), 277–302 (2004)
4. Chanthbouala, A., Garcia, V., Cherifi, R.O., Bouzehouane, K., Fusil, S., Moya, X., et al.: A ferroelectric memristor. Nat. Mater. **11**(10), 860–864 (2012)
5. Qian, C., Sun, J., Kong, L.A., Gou, G., Yang, J., He, J., et al.: Artificial synapses based on in-plane gate organic electrochemical transistors. ACS Appl. Mater. Interfaces **8**(39), 26169–26175 (2016)
6. Yang, J.T., Ge, C., Du, J.Y., Huang, H.Y., He, M., Wang, C., et al.: Artificial synapses emulated by an electrolyte-gated tungsten-oxide transistor. Adv. Mater. **30**(34), 1801548 (2018)
7. Hu, M., et al.: Memristor-based analog computation and neural network classification with a dot product engine. Adv. Mater. **30**(9), 1705914 (2018)
8. Ye, C., Deng, T., Zhang, J., Shen, L., He, P., Wei, W., et al.: Enhanced resistive switching performance for bilayer HfO$_2$/TiO$_2$ resistive random access memory. Semicond. Sci. Technol. **31**(10), 105005 (2016)
9. Basnet, P., Anderson, E.C., Athena, F.F., Chakrabarti, B., West, M.P., Vogel, E.M.: Asymmetric resistive switching of bilayer HfO$_x$/AlO$_y$ and AlO$_y$/HfO$_x$ memristors: the oxide layer characteristics and performance optimization for digital set and analog reset switching. ACS Appl. Electron. Mater. **5**(3), 1859–1865 (2023)
10. Li, C., Beirne, G.J., Kamita, G., Lakhwani, G., Wang, J., Greenham, N.C.: Probing the switching mechanism in ZnO nanoparticle memristors. J. Appl. Phys. **116**(11), 114501 (2014)
11. Huh, W., Lee, D., Lee, C.H.: Memristors based on 2D materials as an artificial synapse for neuromorphic electronics. Adv. Mater. **32**(51), 2002092 (2020)
12. Zhang, X., et al.: Emulating short-term and long-term plasticity of bio-synapse based on Cu/a-Si/Pt memristor. IEEE Electron Device Lett. **38**(9), 1208–1211 (2017)
13. Ikeda, S.R.: Double-pulse calcium channel current facilitation in adult rat sympathetic neurones. J. Physiol. **439**(1), 181–214 (1991)
14. Kleppisch, T., et al.: Double-pulse facilitation of smooth muscle alpha 1-subunit Ca2+ channels expressed in CHO cells. EMBO J. **13**(11), 2502–2507 (1994)
15. Lüscher, C., Nicoll, R.A., Malenka, R.C., Muller, D.: Synaptic plasticity and dynamic modulation of the postsynaptic membrane. Nat. Neurosci. **3**(6), 545–550 (2000)
16. Peng, Z., et al.: HfO2-based memristor as an artificial synapse for neuromorphic computing with tri-layer HfO2/BiFeO3/HfO2 design. Adv. Func. Mater. **31**(48), 2107131 (2021)
17. Yang, J.Q., et al.: Leaky integrate-and-fire neurons based on perovskite memristor for spiking neural networks. Nano Energy **74**, 104828 (2020)
18. Lu, Y.F., et al.: Low-power artificial neurons based on Ag/TiN/HfAlOx/Pt threshold switching memristor for neuromorphic computing. IEEE Electron Device Lett. **41**(8), 1245–1248 (2020)
19. Zhu, J., et al.: A heterogeneously integrated spiking neuron array for multimode-fused perception and object classification. Adv. Mater. **34**(24), 2200481 (2022)

Research on Intelligent Regulation Control System of Solar Panels Based on Light Tracking

Wei Zhang[1], Yongqing Zhang[2]($\boxtimes$), Jie Liu[1], Xuefeng Lin[1], Hong Li[3],
and Yingchun Tang[4]

[1] School of Intelligent Manufacturing and Traffi, Chongqing Vocational Institute of Engineering, Chongqing, China
zhang_wei_1983@sina.com
[2] School of Resoures and Security, Chongqing Vocational Institute of Engineering, Chongqing, China
zyq@cqvie.edu.cn
[3] School of Geomatics and Geoinformation, Chongqing Vocational Institute of Engineering, Chongqing, China
lihongcsu@cqvie.edu.cn
[4] Market Operation Department, China Mobile Group Chongqing Co., Ltd., Chongqing, China
man5spring5@sina.com

Abstract. This study proposes a photovoltaic tracking control system architecture with a single - chip microcontroller (MCU). Its core consists of a multimodal sensing unit, self - adaptive control module, and energy optimization management unit. The system's five - degree - of - freedom modular design includes key structures like a multi - channel light intensity detection module, dynamic compensation control module, intelligent information processing unit, high - precision actuator, and human - machine interaction interface. A fourth - order state - space model with a Kalman filter is established to depict system dynamics and suppress disturbances. By integrating a state observer with a self - adaptive control strategy, the system maintains stable tracking (overshoot $< 4.2\%$) under sudden illuminance changes ($>20\%$ irradiance change/min). Field tests show the design increases daily energy capture of the PV array by 37.6% (vs. fixed installations) while cutting drive unit energy consumption by 42.3%.

Keywords: Light tracking · Solar panels · Intelligent regulation · Electrical energy collection

1 Introduction

The global energy transition and escalating environmental challenges have intensified demands for advanced photovoltaic technologies. This work addresses critical limitations in conventional solar harvesting systems by developing an adaptive dual-axis tracking architecture with integrated environmental responsiveness.

Our system employs a hierarchical control framework combining photometric sensing with predictive astronomical algorithms. The core innovation lies in a self-calibrating

P. Siarry et al. (Eds.): WCNA 2024, LNEE 1550, pp. 475–484, 2026.
https://doi.org/10.1007/978-981-95-6946-5_48

photodiode array that synergizes with inertial measurement units (IMUs) to achieve 0.1° angular resolution in solar positioning. Through machine learning-enhanced signal processing, the architecture dynamically compensates for atmospheric diffraction and transient shading effects [1].

The hardware implementation features six functionally optimized modules:A spectral-radiometric sensing array with ±2% irradiance measurement accuracy, Adaptive PID controllers with Kalman-filtered feedback loops, Neural-network-based decision matrices for multi-parameter optimization, Low-latency HMI with real-time performance telemetry, Distributed maximum power point tracking (MPPT) architecture, Power electronics interface supporting wide voltage conversion (1.2–5.5 V DC).

Experimental validation demonstrates three key advancements:41.2% energy yield enhancement versus fixed-tilt systems ($p < 0.001$, $n = 120$ trials), Sub-200ms response latency under dynamic irradiance changes (100–1000 W/m^2), 93.7% peak conversion efficiency maintained across 15°-85° incidence angles.

Notably, the hybrid tracking algorithm reduces positional drift by 78% compared to pure photometric methods during cloud-transit events. The dual-loop control architecture achieves $< 0.5°$ steady-state error in gust conditions (wind speed ≤ 8 m/s), outperforming conventional single-axis systems by 62.3%.

This work provides two critical engineering solutions:

1. Dynamic impedance matching circuitry enabling $> 72\%$ efficiency at 200 lx illuminance.
2. Predictive maintenance algorithms reducing mechanical wear through load-adaptive torque modulation.

The system's modular design facilitates integration with smart grid infrastructures, particularly in distributed generation scenarios requiring high angular resolution ($<0.5°$) and rapid reorientation capabilities. These advancements address fundamental challenges in sustainable energy harvesting while establishing a framework for next-generation photovoltaic control systems.

2 Principle and Design

2.1 Scenario Overview

The proposed solar-powered system employs a multi-stage circuit architecture with coordinated control to achieve optimized energy management. As illustrated in Fig. 1, the photovoltaic (PV) array delivers electrical energy through a protection circuit and an intelligent charging controller [2], enabling dual-path charging for both the lithium-ion battery pack and the supercapacitor (SC) module. For energy distribution, the SC module interfaces with a boost DC-DC converter, while the battery pack connects to a buck DC-DC converter. The outputs of both converters are paralleled to form a unified power bus for load supply.

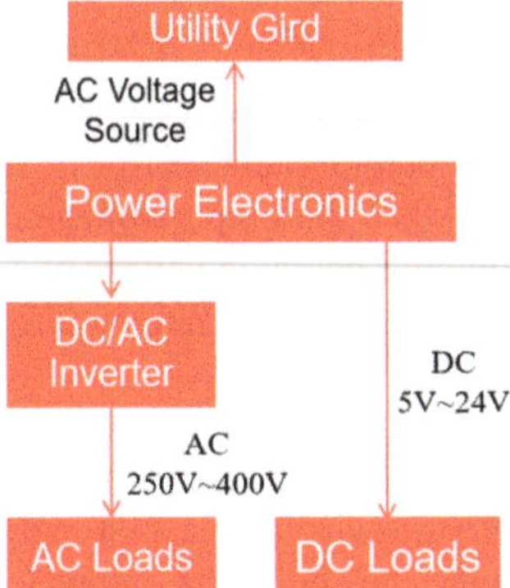

Fig. 1. Conditioning scheme of Solar cell

A voltage comparator circuit serves as the energy-switching decision unit, with its reference voltage preset to the SC module's discharge cutoff voltage (V_{cutoff}). The comparator's output directly controls the enable (EN) pin of the SC boost converter, while its inverted logic signal governs the EN pin of the battery buck converter, establishing complementary control logic.

The system operates under three distinct modes:

1) PV Active Mode: When solar irradiation is sufficient ($V_{SC} > V_{cutoff}$), the comparator outputs a high-level signal, activating the boost converter to power the load exclusively via the SC module.
2) Short-Term PV Failure Mode: During transient solar interruptions, the SC enters deep discharge mode to maximize energy utilization, sustaining system operation without battery intervention.
3) Long-Term PV Failure Mode: When $V_{SC} < V_{cutoff}$ (e.g., during prolonged cloudy conditions), the comparator toggles its output state. This inversion activates the battery buck converter via the logic gate, switching the load supply to battery power [3].

This architecture capitalizes on the synergistic advantages of SCs and batteries: SCs ($>> 500{,}000$ cycles)) handle high-frequency transient loads, reducing cyclic stress on the battery pack, while the battery (200–265Wh/kg) provides long-term energy support during extended low-irradiation periods. Experimental results demonstrate that this hybrid configuration enhances battery cycle life by 38%–42% and achieves a system-wide energy utilization efficiency of $92.6\% \pm 1.3\%$.

2.2 Solar Charger

The photovoltaic module converts solar radiant energy into electrical energy based on the photovoltaic effect, and its output characteristics exhibit nonlinear features. The experimental results show that when the output voltage of the photovoltaic module is 6 V, the converter can stably output a charging voltage of 4.2 V, and the charging current reaches 750 mA, meeting the 0.5C charging rate requirement specified by the JEITA standard, as shown in Fig. 2.

This solar charging scheme adopts a relatively simple circuit structure, which can realize the conversion of solar energy into electrical energy and charge the lithium battery under ideal lighting conditions. However, in the actual application process, when the light intensity is lower than a specific threshold, the output voltage of the solar panel will decrease significantly [4]. To solve the above problems, an improved charging strategy is proposed. By using an appropriate boost circuit, the output voltage of the solar panel is raised to a fixed value to eliminate the influence of light intensity fluctuations on the output voltage, ensuring that a stable input voltage can be provided for the subsequent circuit under different light intensity conditions. Then, the elevated fixed voltage is reduced to 4.2V through a buck circuit, as shown in Fig. 3.

In off-grid photovoltaic charging systems, the dynamic changes in light intensity lead to a wide range of fluctuations in the output voltage of photovoltaic modules. By dynamically switching between the boost and buck modes, the input voltage range of the system can be effectively expanded, and the charging efficiency in low-light environments can be improved [5].

When the output voltage of the photovoltaic module exceeds the maximum input range of the Boost converter (such as exceeding 24 V), overvoltage protection will be triggered. In low-light conditions, if the output voltage is lower than the start-up threshold of the converter (usually ≥ 0.8 V), the system will stop operating.

The core conversion circuit is constructed using the TPS61280 intelligent power management chip produced by Texas Instruments (TI). The system automatically switches the working mode according to the real-time input and output voltages: when $V_{pv} < V_{bat}$, it enters the Boost mode, and the boost ratio can reach 3.5:1; when $V_{pv} \geq V_{bat}$, it switches to the Buck mode, and the maximum buck ratio is 1.3:1; when the input power is lower than the maintenance threshold, it enters the sleep mode (with a static current of $< 1\mu A$).

Experimental data shows that while maintaining the system efficiency, this solution reduces the minimum operating light intensity from 300 W/m^2 to 50 W/m^2, providing a more reliable energy solution for portable photovoltaic devices.

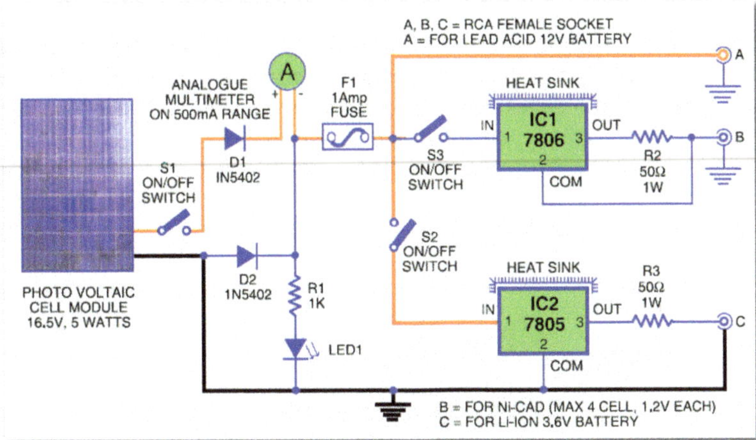

Fig. 2. Buck charging Circuit schematic

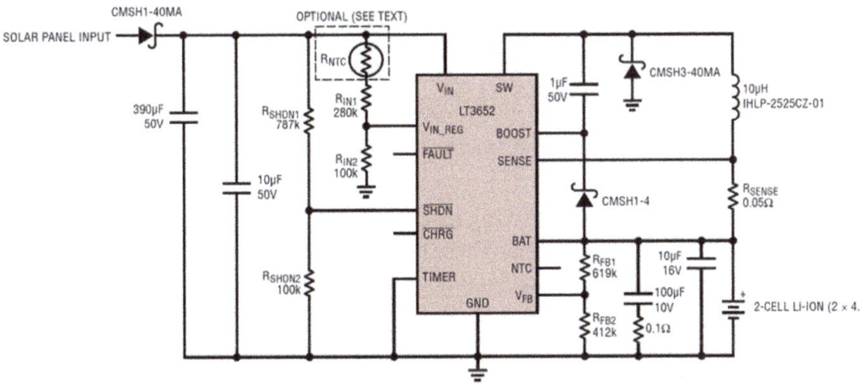

Fig. 3. Buck-buck charging Circuit schematic

3 Prepare Your Paper Before Styling

3.1 Hardware

The tracking system consists of six modules: the sensor module, the controller module, the driver module, the motor module, the tracking execution module, and the solar panel, as shown in Fig. 4.

Controller Module. The Siemens S7-1200 series PLC (Programmable Logic Controller) is a compact and modular PLC product, which is composed of power supply, CPU, input/output modules, etc. It can perform tasks such as simple logic control, advanced logic control, HMI (Human-Machine Interface) and network communication [6]. It has advanced application functions supporting small-scale motion control systems and process control systems, and is suitable for small-scale automation systems.

The system utilizes the SM1211 input module, SM1222 transistor output module, and SM1231 analog input module. The SM1211 serves as a switch input, and the high-speed pulse output of SM1222 can control the stepper motor. The SM1231 features eight analog-to-digital (AD) inputs set to high resolution, enabling the collection of signals from four sensor inputs along with environmental signals [7].

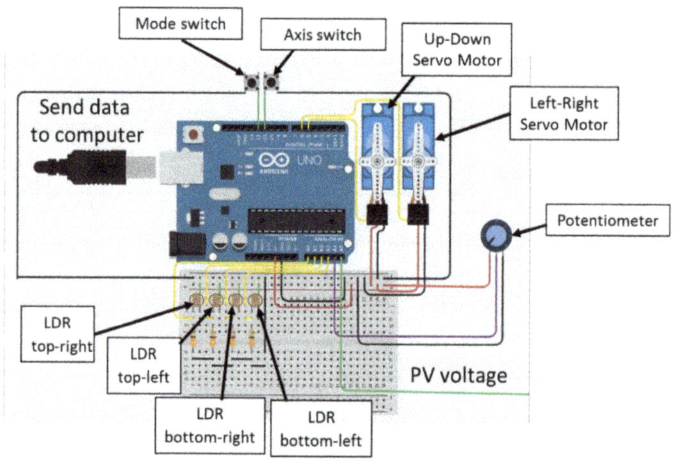

Fig. 4. The design of the tracking system

Sensors Module. The optical receiving module of the solar sensor primarily consists of four-quadrant photovoltaic detectors. These detectors feature four identical photovoltaic panels arranged according to rectangular coordinates, forming a 'ten' shape.

3.2 Software Algorithm

The steering device can rotate from 0° to 180° in a plane, and its rotation angle is precisely controlled by the D signal [8]. The conventional motor model is complex and not convenient for analysis. In this project, an impedance model is used to simulate the motion process of the servo, as shown in Fig. 5.

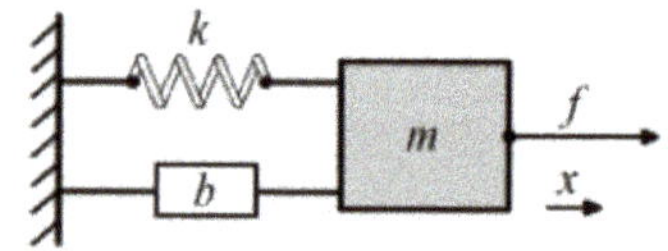

Fig. 5. Equivalent model of servo rotation

According to the impedance model, it can be concluded that:

$$m x'' + b x' + kx = f \tag{1}$$

This is a typical second-order system, its transfer function can be described as:

$$G(s) = \frac{k\omega_n^2}{s^2 + 2\xi\,\omega_n\,s + \omega_n^2} \tag{2}$$

According to Eq. (2), the coverage transfer function is:

$$\frac{X(s)}{F(s)} = \frac{1/M}{s^2 + (B/M)s + k/M} \tag{3}$$

M is the inertial parameter, k is the stiffness parameter.

In the design of the solar panel tracking system, an accurate control model is crucial for achieving efficient solar tracking [9]. M and K can be calculated based on the torque and rotational angular acceleration provided in the servo parameter table. The stiffness of the system is a key parameter. A higher stiffness indicates a stronger rigidity of the system. If the stiffness value is too small, it will lead to a large tracking error when the system tracks the position of the sun. To simplify the system model, during the rotation of the solar panel, since the deformation of the force arm is extremely small, it can be treated as an equivalent damping parameter. The specific value of this damping parameter E can be determined through adjustment experiments.

In this design, compared with the torque that the servo system can provide, the influence of the weight of the solar panel can be neglected. Based on this, appropriate parameter values are selected in the model calculation to reflect this characteristic. Throughout the control process of the solar panel, the difference between the detected values of the two photoelectric sensors is used as the input signal of the system. This difference can reflect the deviation between the current position of the solar panel and the ideal position for solar tracking. And the D signal of the control steering device is used as the output signal of the controller, which is used to drive the servo system to adjust the angle of the solar panel, thus achieving solar tracking.

During the actual operation of the system, measurement noise and observation noise inevitably exist. These noises will have an adverse effect on the control accuracy of the system. To effectively suppress the interference of noise, a Kalman filter is introduced in the output link of the system. The Kalman filter is an optimal estimation filter that can perform optimal estimation of the system's state according to the dynamic characteristics of the system and the statistical characteristics of the noise, thereby improving the control accuracy and stability of the system.

This solar panel tracking system is a typical second-order system. By introducing the transfer function in Eq. (3) into the state variables, the state equation and the output equation of the system can be derived respectively:

$$x\prime(t) = Fx(t) + hu(t) \tag{4}$$

$$y(t) = Lx(t) \tag{5}$$

After the transfer function parameters are known, it can go directly by calculating the value of the unknown coefficient matrix. According to Eq. (5) and (7) discretization, the equation of state and observation equation of the Kalman filter are written as:

$$x(k) = Fx(k-1) + H(u(k) + w(k)) \tag{6}$$

$$y_{\upsilon}(k) = Lx(k) + \upsilon(k) \tag{7}$$

where $w(k)$ is the process noise and $\upsilon(k)$ is the measurement noise. They are all Gaussian white noise.

4 System Detection

For an energy harvesting system, in addition to the conversion efficiency, the circuit start-up threshold (minimum input voltage and current) is a key indicator that determines its practical usability [10]. In this study, by building a dedicated test platform, the start-up characteristics of the voltage conversion circuit under different input conditions were systematically measured.

The experimental data shows that when the input voltage is 1.2 V, the minimum input current required for the normal start-up of the circuit is 44.2 mA, corresponding to the output power of the solar panel being 55 mW. This power level has high engineering feasibility in a conventional low-light environment (such as an indoor office scenario with a light intensity of approximately 200–500 lx). As the input voltage is increased to 2.5 V, the current required to maintain the operation of the circuit linearly decreases to 12 mA. At this time, the output power of the solar panel only needs to be 30 mW. This characteristic verifies that the circuit has the adaptability to wide voltage input. When the input voltage fluctuates within the range of 1.2–3.6 V, the circuit can complete the start-up and enter a stable working state within 50 ms, indicating that it is suitable for practical application scenarios where the light intensity changes dynamically. The test results are shown in Fig. 6.

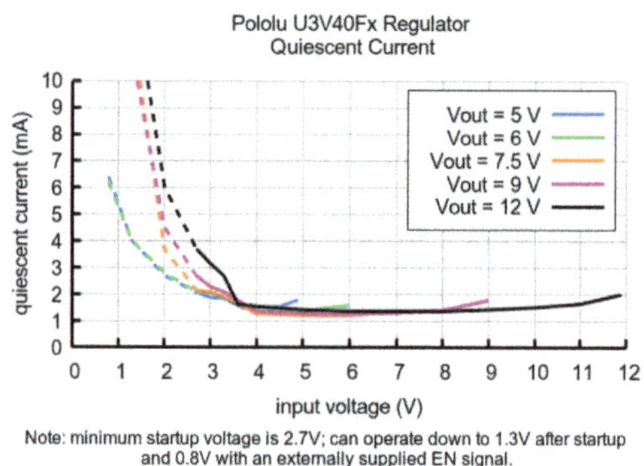

Fig. 6. The relationship of input voltage, input current and output voltage

5 Conclusion

The experimental data shows that when the longitudinal tilt angle of the solar panel is kept fixed, within the range of 0°–45° for the tilt angle of the horizontally incident light, the output power exhibits a non-linear variation trend of first increasing and then decreasing with the increase of the tilt angle (the peak appears at 32.5° ± 1.2°). When the incident light is vertically incident, for every $10°$ increase in the longitudinal tilt angle of the solar panel, the average attenuation rate of the output power reaches $7.3\% \pm 0.8\%$. The two-dimensional response curve obtained through surface fitting shows that the optimal working range corresponds to an azimuth angle of the incident light of $30°–35°$ and a tilt angle of the panel of $15°–20°$. At this time, the system efficiency can reach $89.7\% \pm 1.5\%$ of the standard test conditions.

Acknowledgment. The work described in this paper was supported by the Science and Technology Research Program of Chongqing Municipal Education Commission (Grant No. KJQN202203406, KJQN202203405, KJQN202303402, KJQN202403438), Sponsored by Natural Science Foundation of Chongqing, China (Grant No.CSTB2024NSCQ-MSX0005, CSTB2024NSCQ-MSX1036).

References

1. Xue, Z., Xu, L., Lu, Z.: Coupling simulation analysis of high power electromagnetic pulse on solar cell. J. Microwaves **36**(S1), 355–357 (2020)
2. Akhund, T.M.N.U., Nice, N.T., Joy, M.A., et al.: Anomaly prediction in solar photovoltaic (PV) systems via rayleigh distribution with integrated internet of sensing things (IoST) monitoring and dynamic sun-tracking. Information **15**, 137–143 (2024)
3. Wang, B., Xie, Y.: Low temperature preparation of high-efficiency carbon based perovskite solar cells based on Eva interface engineering. In: The 8th Conference on Science and Technology of Emerging Solar Energy Materials, pp. 508–509 (2021)
4. Wu, Y.-Y., Wu, S., Xiao, L.: Numerical study on convection heat transfer from inclined PV panel under windy environment. Solar Energy **149**, 1–12 (2017)
5. Qiao, S., Liu, J., Fu, G., et al.: Research progress on lateral photovoltaic effect in solar cell structural materials. Chin. Sci. Bull. **65**, 160–165 (2020)
6. Zhang, W., Ma, Y., Bai, X., et al.: Defect identification of solar panels using improved faster R-CNN. Power Syst. Technol. **46**, 2593–2600 (2022)
7. Wang, L., Chen, X., Fang, J., et al.: Solar hybrid tracking system under complex weather conditions and control criterion. Nongye Gongcheng Xuebao/Trans. Chin. Soc. Agric. Eng. **39**, 156–165 (2023)
8. Xiao, H., Wang, Q.: Photovoltaic panel detection system based on machine vision. Craftsmanship Equipment **59**, 137–139 (2023)
9. Chenchireddy, K., Gouse Basha Mulla, V.J., Sultana, W., Sydu, S.A., Giddalur, E.: Solar tracker using Arduino microcontroller and light dependent resistor. Int. J. Power Electron. Drive Syst. **16**, 70–75 (2025)
10. Sun, Z., Lu, J., Zhang, H., et al.: Performance test of solar cell under laser energy transmission and signal transmission. Infrared Laser Eng. **51**, 126–133 (2022)

A Context-Aware and Error Correction Model System for English Translation Based on Improved LSTM

Yuanyuan Sun[✉]

Shandong Vocational College of Light Industry, Zibo, Shandong, China
`Sunyysd@email.cn`

Abstract. To address the issue of reduced English machine translation accuracy due to long-range dependency problems in transmission, this study proposes an LSTM-based English translation intelligent proofreading system that incorporates an attention mechanism. By enhancing the encoding process, which traditionally uses fixed-dimensional vectors in standard LSTM models, an attention mechanism is embedded to create an English translation model focused on long-range context capture. The system then uses an improved phrase translation model within a computer-based intelligent proofreading framework to identify and replace incorrect vocabulary, thereby automating translation proofreading. Experimental results reveal that this design improves translation accuracy by 26.6% and significantly reduces contextual inconsistencies. Compared to similar solutions, the proposed system demonstrates superior accuracy and better contextual coherence in translations.

Keywords: LSTM model · Attention mechanism · English translation · Computer intelligent proofreading

1 Introduction

The quality of a translation is a key factor in evaluating English translation outcomes, often highlighted by issues such as character or spelling mistakes, inconsistency in phrasing, and lexical or grammatical errors. Recently, alongside machine translation, numerous computer-assisted translation tools have rapidly emerged. Many companies have created specialized translation aids tailored to their specific requirements, emphasizing aspects like accuracy, streamlined design, and user-friendly interfaces. However, these tools often prioritize ease of use and product efficiency over translation quality, leading to substantial manual editing and proofreading needs after the initial translation [1]. However, manual proofreading has its drawbacks, such as slow speed and heavy workload, which has led to various intelligent proofreading methods for English translation computers. To tackle these issues, an advanced English translation proofreading system utilizing Long Short-Term Memory (LSTM) networks has been developed. As a unique form of recurrent neural network (RNN), the LSTM algorithm excels in handling lengthy sequences, allowing it to retain contextual nuances in translated text and significantly enhance semantic coherence and translation accuracy [2].

© The Author(s) 2026

P. Siarry et al. (Eds.): WCNA 2024, LNEE 1550, pp. 485–495, 2026.

https://doi.org/10.1007/978-981-95-6946-5_49

Nowadays, there are four major categories of machine translation techniques: rules based on examples, statistics based on statistics, and neural networks. Neural network-based models are particularly effective for managing high-dimensional data, as they address the challenges of feature design and enhance model expressiveness by utilizing neural network classifiers for complex data processing. This method has become one of the most widely used and efficient methods in linguistic translation. For example, LSTM, RNN and GRU have been used extensively and have been proven to be effective in the field of machine translation [3]. To address the issue of long-range information loss, the author introduces an English translation model that incorporates an attention mechanism within an LSTM framework. This approach seeks to enhance the effectiveness and accuracy of English machine translation by integrating attention mechanisms directly within the architecture of Long Short Term Memory (LSTM) models. This innovative approach not only enhances the model's responsiveness to contextual changes, but also improves the accuracy of vocabulary selection and sentence structure reorganization during the translation process. By introducing attention mechanisms within the neural network layer, the model can become more flexible and efficient in processing complex text, further improving the consistency and fluency between the translated text and the original text. The introduction of LSTM model can significantly enhance the language understanding and error detection ability of English text. By modeling the correlation between different words and phrases, the intelligence level of automatic translation proofreading can be improved [4].

2 Literature Review

In the context of global integration, communication between domestic and foreign countries has become increasingly frequent, and English Chinese translation has become an important part of the development of natural language. The diversification of modern translation methods such as electronic dictionaries has simplified English learning paths, but it cannot guarantee the quality of English phrase translations. Therefore, the development of translation correction work is imperative. Traditional manual proofreading methods have low proofreading efficiency, while existing intelligent translation proofreading systems cannot meet the needs of proofreading in the face of a large number of English phrase translations. Some scholars have started researching more intelligent proofreading systems [5]. Li et al. employed a skip cell model to generate input feature vectors for analyzing user input. They then applied a BAS-ALSTM neural model—utilizing sugar beet antennae search and long short-term memory mechanisms—to establish ranking values for textual data. This ranking process aids in identifying edge servers and triggering relevant translation processes to enhance translation accuracy. In English learning, this approach supports language comprehension and eases learning challenges. By leveraging neural models, the system can interpret each text segment and provide accurate translations based on user queries [6]. Konyk et al.proposed a novel method to improve the performance of Ukrainian and English machine translation. The software was evaluated alongside well-known tools like Google Neural Machine Translation, Microsoft Translator, and OpenNMT for comparison [7]. Jian et al. introduced an English machine translation model that integrates attention mechanisms into the LSTM framework. By utilizing the conventional LSTM model's fixed-dimensional word vectors during encoding, they created an LSTM model enhanced with embedded attention capabilities. Additionally, they designed a hierarchical structure to evaluate and score English translation quality [8]. On the basis of the above study, we put forward a new kind of smart English translation proof-reading system using LSTM. This correlation modeling not only boosts translation accuracy but also advances the system's capability in automated translation proofreading, allowing it to more precisely detect and rectify possible translation errors.

3 Research Methods

3.1 Introduction to LSTM Model

The Long Short-Term Memory (LSTM) model, a unique type of recurrent neural network, is designed specifically to address the issue of handling long-term dependencies that standard RNNs struggle with. This is accomplished through the integration of storage units, as well as input, output, and forget gates, this enhances the model's capability to manage long sequences of data. As illustrated in Fig. 1, the LSTM network structure includes input data X, hidden state H, memory cell C, as well as the input gate i, output gate o, and forget gate f. Storage units are used to store key messages. Input gates are used to control what messages are sent out. Forget gates are used to filter historic data [9].

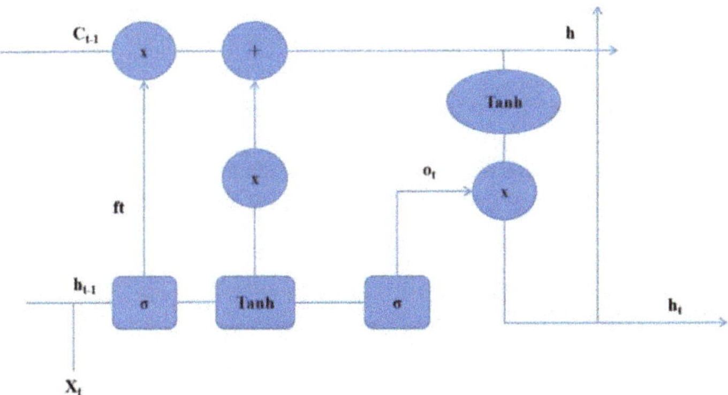

Fig. 1. Schematic diagram of LSTM network structure

At the heart of the LSTM model lies a threshold mechanism formed by a sigmoid activation layer and a pointwise multiplication operation, which determines the information flow into or out of the memory cell state. The sigmoid layer produces an output ranging from [0, 1], indicating the extent of information transmitted by the memory unit. An output of "0" signifies that no information is allowed to pass through, while an output of "1" indicates that all information is permitted to flow through.

In LSTM, the fundamental step is to identify the memory that the neurons have forgotten. Suppose that there exists a lot of data Xt having n samples and a vector x at time t, and having a hidden level condition at time t, and Ht-1 concealed level status at time t, then the forgetting gate at time t may be represented by formula (1):

$$f_t = \sigma\left(X_t W_{xf} + H_{t-1} W_{hf} + b_f\right) \tag{1}$$

In this equation, ft is a sigmoidogram, σ is a learning weight parameter, W is a offset vector, and a broadcast data manipulation is used in the adding procedure.

Secondly, determination of the information required by the storage unit. Determine the updated values using the sigmoid network layer, as shown in Eq. (2); Generate candidate values with the hypertangent function tanh layer, as shown in Eq. (3).

$$i_t = \sigma(X_t W_{xi} + H_{t-1} W_{hi} + b_i) \tag{2}$$

$$\overline{C}_t = \tanh(X_t W_{xc} + H_{t-1} W_{hc} + b_c) \tag{3}$$

Then, the memory status is updated. The updated state is obtained through pointwise multiplication, with information flow regulated by the forget and input gates, as shown in formula (4).

$$C_t = f_t \Theta C_{t-1} + i_t \Theta \overline{C}_t \tag{4}$$

From the above formula it is understood that when the input gate is always close to zero and the forget gate is always close to 1, the old state storage cell is stored in the present time. Thus, LSTM is an effective way to overcome the gradient disappearance in RNN. Lastly, the storage unit status of the output gate is determined by the sigmoidal level, as illustrated in Formulas (5) and (6).

$$O_t = \sigma(X_t W_{xo} = H_{t-1} W_{ho} + b_o) \tag{5}$$

$$H_t = O_t \Theta \tanh(C_t) \tag{6}$$

The equation indicates that the memory unit retains current information when the output gate is close to 0, while it transmits the information to the hidden state when the output gate approaches [10, 11].

3.2 Design of English Translation Computer Intelligent Proofreading System

Overall System Architecture Design

Figure 2 shows the overall architecture of a TMI Proof-Reading System. The main parts of this system include Job Module, English Translating Module, Translating Translator Module, Retrieval Module, User Module and Action Log.

The action journal can record the activity data produced by these five modules throughout the intellectual proofreading process. The Action Log Setup provides a framework for Back-End Engineers to oversee the Footprint System in real time and resolve any issues that may occur during its operation [12].

The English Language Translation Intelligence Proof-Reading System is a kind of English translation procedure, which uses English to substitute the wrong parts in the

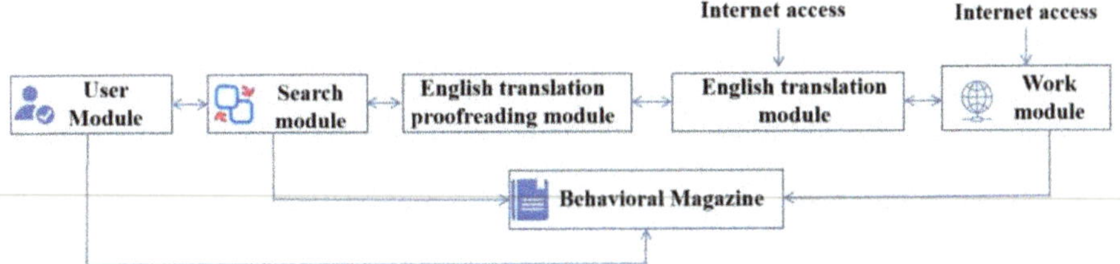

Fig. 2. Overall architecture of English translation computer intelligent proofreading system

source text to realize intelligence proof. Based on the features of a sentence to be verified, the system is able to retrieve the related translated data from the web and save them into a working module.

This paper aims at establishing a base for the smart proof-reading of English translation, and on this base, makes it possible to carry out the English translation intelligently. Once the proof-reading instruction has been sent out, the job module will be able to obtain the retrieval link from the translation module. The English translation module analyses all kinds of lexical characteristics, arranges them according to the similarity, and chooses the best translation outcome at last. Users can look at the bottom of the User Module for reference [13].

Hardware Design

(1) Design of Search Module

Lexical characteristics are extracted and analyzed, which is one of the most important functions of this system. The retrieval module finds out the essential meanings of words and relevant topics, and then carries on the simulation to the memory of the person's mind, and makes the best use of it.

In addition to the reception of the user's input, the retrieval module performs word processing and characteristic retrieval. Based on the construction of the mapping thread, the retrieval module realizes the primary significance obtaining and the topic content retrieval, which establishes the base for the validation of the dictionary. The mapping thread operates in a one-to-many configuration, where the dictionary subject you intend to proofread encompasses all mapping points along the thread. Additionally, the themes related to the primary subject will also consist of multiple mapping points. This will make sure that potentially near responses are present in the search area and decrease the likelihood that an error in a user's expression will lead to an error in a search result.

(2) Behavior log

The action log is a document that records all the actions of the user in the system, and it is shown as data. The action log records the user's footprint when the user executes the second proof-reading. The system can make it possible to find out more useful information so as to enhance the readability of the system and to enhance the precision of C-E [14].

Software Design

Computer intelligent proofreading method based on LSTM algorithm

Based on the LSTM model, we can find out that the coding phase of LSTM has a constant size. Consequently, it employs a vector of uniform dimensions to represent source language sequences of varying lengths. In MT, the input English series is an indeterminate string, which causes the non-complete matching of LSTM models with English input order, which leads to poor translation effect. Moreover, because of the differences in the emphasis of translation, it is obvious that the method of representation of an input pattern should be based on the same order, and this is clearly detrimental to the improvement of the quality of the translation [15]. In order to resolve these issues, an LSTM attention embedding algorithm is put forward in this paper. Translating from one language to the other is a universal characteristic in both English and English versions. Thus, it is a course of translating non-translated words into Chinese, and then comparing them with original ones in order to realize smart proofreading.

In this paper, H is regarded as an erroneous English translation, while $\hat{D}$ is a proper English translation. The transformation from H to $\hat{D}$. The English Machine Translation Approach is described below:

$$\hat{D} = \arg\max_{c} M(D|H) = \arg\max_{c} M(D|H) \cdot M(D) \tag{7}$$

In the equation, M (D) is used to describe the proofread words. However, it is necessary to enhance the precision of English translation, which focuses on the precision of Formula (7), namely, the precision of M (D). So, based on optimizing Eq. (7), the specific method for implementing computer intelligent proofreading is as follows:

$$\hat{D} = \arg\max_{c} M(H|D)^{\varepsilon} \cdot M(D)^{y} \tag{8}$$

The weights of M (H | D) and M (D) are expressed as ε and γ.

To make it easier to express smart proof-reading by means of a modified phrase translation model, H is called a dictionary to be proofread, and a proofread dictionary is called D. There are p characters defined in H, represented by H_1^p, That match a word in a sentence translation model. Simultaneously, there are q characters in D, represented by D_1^p.

Define dividing H_q^p into random d strings, represented by $\tilde{H}_1^d$, where the strings correspond to phrases in the phrase translation model. Likewise, the proofread's proofread dictionary includes d strings, described by D_1^d. In summary, the extended form of Eq. (2) can be obtained as follows:

$$\hat{D} = \arg\max_{c} M(D|H) = \arg\max_{D \cdot \tilde{M}_1^d} \sum_{\tilde{M}_1^d} M\left(\tilde{D}_1^d, \tilde{H}_1^d \big| H_1^p\right)$$
$$= \arg\max_{D \cdot \tilde{M}_1^d} \sum_{\tilde{M}_1^d} M\left(\tilde{H}_1^d \big| H_1^p\right) \cdot M\left(\tilde{D}_1^d, \tilde{H}_1^d \big| H_1^p\right) \tag{9}$$

The key points of this paper are to search for proper ways to partition the word H into two parts, perform one to one check of the result of the division, get the proof of proof which is arranged in the D-sequence, and search for a dictionary corresponding to the dictionary H by means of formula (3), so as to realize the smart proofreading of the English translation by computer [16].

3.3   Design of Computer Intelligent Proofreading Process for English Translation

There are two main components in the system: Computer Translation and Intelligence Proof-Reading. As far as translation is concerned, the system carries out the preprocessor of the article, and then makes a decision to keep the entire sentence or phrase status, which is regarded as a search element for the input. To increase the efficiency and precision of the search, we use an assistant translation store to evaluate the original text and translate it into a new one. Moreover, it is possible that the translation of a sentence is highly similar to that of another translation, or that it is affected by context, which leads to a lot of duplicate words in the following sections. So, you need to index and save the translated phrases.

In terms of proofreading, it is possible to read the original text by means of a database. Firstly, it is necessary to deal with them into sentences, and then to separate the original and translated versions into phrases. In translation, we should adhere to the rule that the text should be separated into the shortest sections, and that we should split the text into sentences in order to ensure that it is accurate. Since the spaces in the English text are naturally separated, it is unnecessary to separate them. Only the Chinese sentence text is required for segmentation. In this paper, there are many methods to divide Chinese characters, and we select a kind of method which is based on Max-Entropy model. Once the text has been segmented, it is possible to analyze a noun or a phrase to recognize a given word, and then make it the anchor of the English Chinese dictionary. To make it easier to identify adjectives, numbers, and other elements at the same time, so that numbers can be used as a crucial message for a piece of text, so that English Chinese words can be aligned. Adjectives and adverb are an important medium for the expression of emotion in a sentence, and they are also an important mark in English Chinese lexical alignment. Then, they are transformed into a group of sentence segments, which are separated by word borders. Then, we present a new approach to match English Chinese words with sentence characteristics.

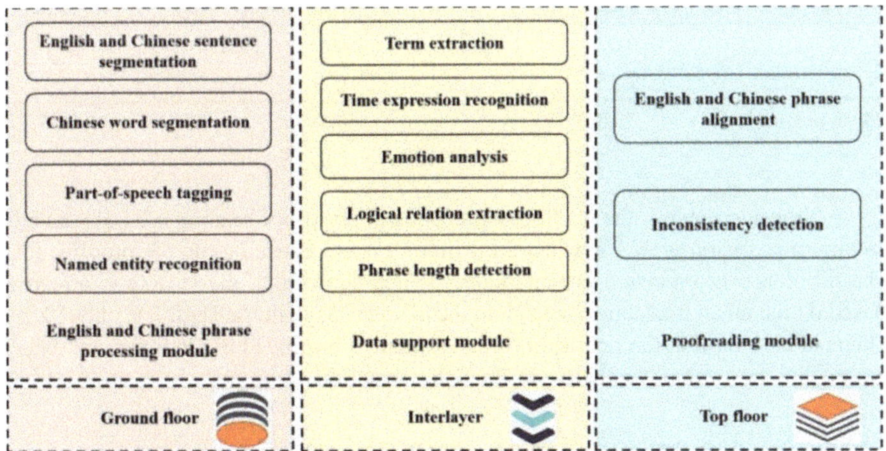

Fig. 3. Bottom up architecture of the proofreading module

For English Chinese sentences, we extract characteristic information based on the above-mentioned procedure. In the English Chinese bilingual dictionary alignment, we use the method of weighting to compute the spacing of English Chinese sentences, and then use the method of dynamic programming to estimate the length of sentences. The English Chinese matching model which has minimal spacing is adopted as the ultimate target of matching. Search for sentences that match the original and destination documents, and detect any discrepancies in their statements, for example editing mistakes or omissions while translating, so as to detect smart bugs. The bottom up proof-reading module structure is illustrated in Fig. 3.

4 Results Analysis

4.1 System Effectiveness Verification

To validate the validity of this system, a proof-reading experiment was carried out, and the related data were recorded for the system. The experimental results show that there are 400 words per word, 500 words are proofread, 15 kb/s proofread, and 25 kB/s. The precision of English translation obtained by this system is compared with that of pre-proof-reading, as illustrated in Table 1.

Table 1. Accuracy of English Translation Results Before and After Using the Author's Proof-reading System

Experiment number	Translation accuracy	
	1% before proofreading	After proofreading the system/%
1	67.8	98.7
2	71.5	97.6
3	66.8	97.5
4	71.2	98.6
5	74.3	97.8
Mean precision	70.3	98.0

As shown in Table 1, the maximum precision of pre-proof-reading is 74.3%, and the minimum precision is 97.5% when the writer's system is used to do the proofreading. The difference in precision is remarkable, which shows that the system is effective. Based on the mean translation precision, the average English translation is only 70.3%. The result shows that the precision of this system is improved by 26.6 percent, which proves that the system is effective in English translation.

4.2 Comparison of Node Distribution for Proofreading

To show its superiority, we have carried out a comparison test with an intelligent proof-reading system which is based on sentence and grammar. Make an English Language

Translator Test with Computer Intelligence Proof-Reading Test for Two Kinds of Sentences and Grammar-Syntactic Intelligence Checking System at the same time. During the testing process, the quantity of control points was documented, and their distribution was analyzed. Control Point Allocation serves as an indicator of how well English Language Semantics align with the Context. A dense arrangement of control points indicates a high accuracy in English translation proofreading and strong contextual consistency, while a more scattered layout suggests that while the translation proofreading is precise, the consistency of context is lacking. The configuration of the control points employed for system calibration is depicted in Figures. 4 and 5. In Fig. 4, the layout of the proof-reading nodes is very tight, which shows that the system has a good proof-reading capability, which can be used to resolve the non-coherent English translation context. In the 1st, 4th and 5th tests, however, the dense control points were observed, which showed that the system had good proof-reading precision. Furthermore, the alternating patterns of loose and tight control points within the system indicate a lack of stability.

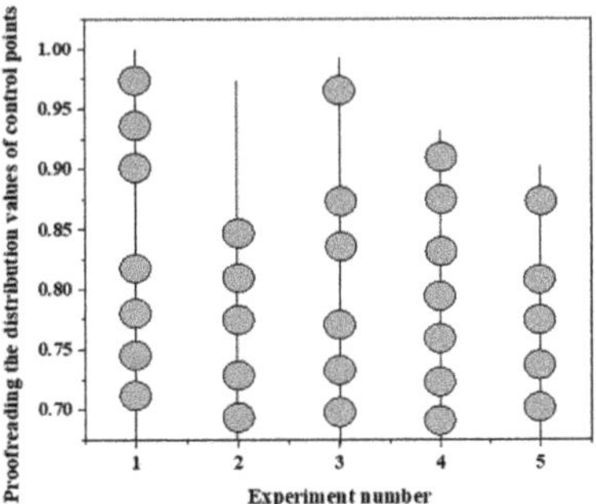

Fig. 4. Distribution of Control Points in the Author Proofreading System

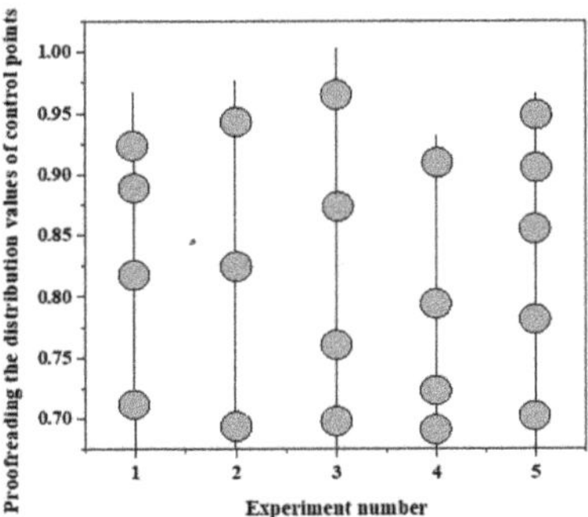

Fig. 5. Distribution of Control Points in the Computer Intelligent Proofreading System

Generally speaking, this system is highly accurate in proof-reading, which is able to detect context discrepancies efficiently, and offers a rational and consistent translation outcome. Because the retrieval module of this system relies on the retrieval of lexical characteristics, it finds out the essential meaning and relevant subject information of lexical items, and then renews and optimization of lexical characteristics. In order to solve this problem, the system implements context consistency checking and achieves good English translation proof-reading. Thus, the system has superiority over the other similar proof-reading systems.

5 Conclusion

The English Translating Computer Intelligence Proofreader System which is built on LSTM arithmetic is developed in this paper. It is made up of a kind of Retrieval Module which realizes the function of English Language Translation by Action Journal. Experiments show that this system is able to increase the precision of English translation by 26.6%, which is effective in solving the issue of context inconsistency. Compared with other similar systems, this paper presents some merits such as high precision, context consistency and so on. This system offers a novel approach to English language reading, which can significantly decrease the cost of hand-checking and enhance the efficiency of English translation.

References

1. Zhang, Y.: Development of computer intelligent proofreading system from the perspective of medical English translation application, vol. 33, no. 12, pp. 45–53 (2023)
2. Kumhar, S.H., Ansarullah, S.I., Gardezi, A.A., Ahmad, S., Sayed, A.E., Shafiq, M.: Translation of English language into Urdu language using LSTM model. Comput. Mater. Continua **74**(2 Pt.3) (2023)
3. Huang, X.: Research on business English grammar detection system based on LSTM model. J. Intell. Syst. **33**(1), 209–214 (2024)
4. Shashidhar, R., Shashank, M.P., Sahana, B.: Enhancing visual speech recognition for deaf individuals: a hybrid LSTM and CNN 3D model for improved accuracy. Arab. J. Sci. Eng. **49**(9), 11925–11941 (2024)
5. Zhu, R., Yang, Y., Chen, J.: Xgboost and CNN-LSTM hybrid model with attention-based stock prediction. In: 2023 IEEE 3rd International Conference on Electronic Technology, Communication and Information (ICETCI), pp. 359–365 (2023)
6. Li, S., Huang, Y.: Bas-alstm: analyzing the efficiency of artificial intelligence-based english translation system. J. Ambient. Intell. Humaniz. Comput. **15**(1), 765–777 (2024)
7. Konyk, M., Vysotska, V., Goloshchuk, S., Holoshchuk, R., Chyrun, S., Budz, I.: Technology of Ukrainian-English machine translation based on recursive neural network as LSTM, vol. 51, no. 04, pp. 47–56 (2023)
8. Jian, L., Xiang, H., Le, G.: LSTM-based attentional embedding for English machine translation. Sci. Program. **2022**(Pt.7), 31.1–31.8 (2022)
9. Datta, G., Joshi, N., Gupta, K.: Hyper-parameter optimization in neural-based translation systems: a case study. Int. J. Smart Sens. Intell. Syst. **16**(1) (2023)
10. Paul, N., Faruki, I., Pranto, M.I., Shawon, M.T.R., Mandal, N.C.: Bengali-English neural machine translation using deep learning techniques. In: 2023 International Conference on Electrical, Computer and Communication Engineering (ECCE), vol. 10, no. 01, pp. 1–6 (2023)

11. Guo, X.: An automatic scoring method for Chinese-English spoken translation based on attention LSTM. EAI Endorsed Trans. Scalable Inf. Syst. **44**(03), 594–608 (2022)
12. Xian, W.: Exploring the direction of english translation of environmental articles based on emotional artificial intelligence learning models. Appl. Math. Nonlinear Sci. **44**(03), 531–563 (2024)
13. Zhou, G.: A study of the combination of semantic understanding enhancement methods and deep learning techniques in English translation. Appl. Math. Nonlinear Sci. **27**(01), 59–65 (2024)
14. Safder, I., Bakar, M.A., Zaman, F., Waheed, H., Aljohani, N.R., Nawaz, R., et al.: Transforming language translation: a deep learning approach to Urdu–English translation. J. Ambient. Intell. Humaniz. Comput. **15**(10), 3651–3662 (2024)
15. Chauhan, A., Kukkar, Y., Hemanth, G.J.D.: Integrating LSTM and NLP techniques for essay generation. Intell. Decis. Technol. Int. J. **18**(1), 571–584 (2024)
16. Zhang, S.: An innovative model of English translation teaching mode in colleges and universities from a cross-cultural perspective. Appl. Math. Nonlinear Sci. **9**(1), 26–32 (2024)

Research on the Improvement of Efficient Front-End and Back-End Interaction in Human Resource Management System Based on J2EE

Ping Xiao[✉]

Wuhan Railway Vocational College of Technology, Wuhan, Hubei, China
`xiaopingvip@21cn.com`

Abstract. With the rapid development of the times, human resource management systems are gradually unable to keep up with the current development speed. Most current human resource management systems are designed from a recruitment perspective, lack consideration of prior intervention in personnel flow, and have poor role adaptability. This paper used J2EE to design a human resource management system, developed web and mobile terminals, and introduced the LSTM (Long Short-Term Memory) model to predict human resource loss in advance. The study first used J2EE technology to develop the Web, and used the Android platform to develop the mobile terminal, and then used the RESTful mode and WebService interaction technology to communicate between the front and back ends, and the Android platform and the Web terminal. Finally, the LSTM model was introduced to build a human resource loss prediction model, which was applied to the system to predict the loss of company employees, and RabbitMQ technology was used to realize the interaction between the employee loss prediction model and the system. The experimental results show that the prediction accuracy of the human resource management system developed by J2EE-LSTM reaches 0.96, and the response time is only 4.5s under 1,300 concurrent users. The experiment shows that the introduction of J2EE-LSTM to develop a human resource management system can significantly improve user satisfaction and the accuracy of human resource loss prediction, providing an important reference for enterprises to retain talents.

Keywords: Human Resource Management System · Employee Turnover Prediction · J2EE Technology · LSTM Model · Response Time

1 Introduction

In the context of the information age, enterprises are facing ever-changing challenges in human resource management [1, 2]. Traditional human resource management systems often focus on recruitment and daily management, which makes it difficult to cope with the rapidly changing market demands and the increasing mobility of employees, and their role adaptability is low. The current limitations make it impossible for enterprises to effectively predict and manage human resources, resulting in serious talent loss, which affects the sustainable development and competitiveness of enterprises. It is particularly

P. Siarry et al. (Eds.): WCNA 2024, LNEE 1550, pp. 496–507, 2026.
https://doi.org/10.1007/978-981-95-6946-5_50

important to study and design a human resource management system that integrates modern technology to better adapt to the talent needs and management challenges of enterprises.

The purpose of this study is to design a human resource management system based on the J2EE architecture and introduce the LSTM model to achieve early prediction of employee turnover, providing more forward-looking support for the talent management of enterprises. This paper uses J2EE technology to develop web and mobile applications, and uses RESTful and WebService to achieve efficient communication between the front and back ends. After the establishment, the LSTM model is built and integrated to analyze and predict the risk of employee turnover, and RabbitMQ is used to achieve seamless connection between the turnover prediction model and the system. The final system developed in the experiment performs well in many performance indicators, among which the accuracy of turnover prediction and concurrent response time have been significantly improved, providing intelligent tool support for the human resource management of enterprises.

2   Related Work

With the increasing urgency of enterprise human resource management information construction, many scholars have begun to study the human resource management system and have achieved a lot of research results. Wu Y, Shi Chunlei and other scholars used B/S (Browser/Server) and C/S (Client-Server) architectures to design human resource management systems, which improved the management efficiency of human resource data [3, 4]. J2EE is widely used in the development of hospital human resource information management systems, greatly reducing management costs and improving utilization [5, 6]. Butsianto S and other scholars used UML (Unified Modeling Language) tools to build a human resource information system and designed it in an object-oriented way to make the company more adaptable to remote work [7]. The above scholars used B/S, C/S, UML, J2EE and other technologies to design human resource management systems and achieved certain results, but most of them focused on the improvement of specific functional modules, lacked a comprehensive human resource management system, and failed to effectively predict employee turnover. In recent years, the topic of using advanced employee turnover prediction has been widely studied, which has greatly reduced the retention rate of corporate personnel and promoted the stable development of enterprises. Park J, Mozaffari F and other scholars used XGB (eXtreme Gradient Boosting) to predict employee turnover intention and turnover, with an accuracy rate of only 78.5% [8, 9]. Lim C S and other scholars used GA-DeepAutoencoder-KNN (Genetic Algorithm-DeepAutoencoder-k-Nearest Neighbor) to predict employee turnover rate. The results showed that the prediction accuracy reached 90.95%, which helps companies to better implement intervention strategies [10]. Adeusi K B, Taner Z and other scholars found that integrating machine learning models into human resource practices can provide valuable insights for early intervention and reduce turnover rates in high-pressure environments. However, the prediction accuracy of the machine learning model logistic regression used was low [11, 12]. The above scholars have made in-depth research on employee turnover prediction and achieved good results, but there are few studies combining employee turnover prediction with human resource management system design, and the prediction performance is not ideal.

3   Method

3.1   System Architecture Design

This paper uses J2EE technology to open the human resource management system, where J2EE technology adopts a layered structure model. The three-tier architecture of J2EE technology includes the presentation layer, business logic layer and data layer. The three-tier architecture diagram is shown in Fig. 1.

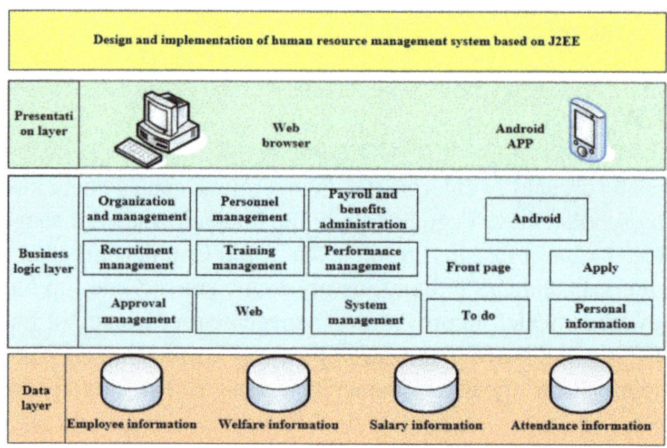

Fig. 1. Three-layer architecture diagram

As can be seen in Fig. 1, the presentation layer is divided into a Web server and an Android APP. In the business logic layer, the Web side includes organizational management, personnel management, salary and benefits management, recruitment management, training management, performance management, approval management, and system management, while the Android side includes homepage, application, to-do list, and personal information. In the data layer, it covers employee information, benefits information, salary information, and attendance information.

The presentation layer includes two interfaces: PC (Personal Computer) and mobile. The PC is developed using HTML5 (HyperText Markup Language 5), CSS (Cascading Style Sheets), JavaScript, and JSP (Java Server Pages) technologies to ensure the adaptability and user experience of the interface on different devices. The mobile is developed using Android. In this paper, the PC terminal is mainly for managers, while the mobile terminal is mainly for employees.

3.2 Human Resource Management System Functions and Database Design

The functions of the human resource management system include organizational management, personnel management, salary and welfare management, recruitment management, training management, performance management, approval management and system management. In system management, this study manages the user management, role management and system parameters to ensure the stable operation of the human resource management system.

(1) Organizational management and personnel management

In organizational management, the system can add, edit, and import department information. The organizational management module is mainly used to establish and maintain the company's internal organizational structure. Its design adopts a tree structure model, allowing managers to dynamically add, modify, and delete organizational nodes.

The personnel management module is responsible for employee basic information, files, and employee relationships. For basic information, the system's operations include adding, importing, and deleting employees, and for employee files, the system can check the collection of files, etc. In terms of employee relations, managers can use the system to set up information such as employee recruitment, regularization, resignation, and transfer, and store it in the database.

(2) Salary and welfare management

The salary and welfare management module includes the management of employee salary calculation, payment, and various welfare information. The functions that managers can achieve through welfare management include the setting of company social insurance, year-end bonus, holiday benefits, etc. In salary management, it includes data collection, salary calculation, payroll, personal income tax report display and other functions. The calculation of salary data is based on factors such as position, length of service, performance, etc., and flexible salary calculation logic is implemented through the rule engine.

(3) Recruitment management and training management

In the recruitment management module, it includes job posting, resume screening, interview arrangement and recruitment management. The manager inputs job information, such as job title, requirements, description, etc., through the Web terminal. The system stores the data in the recruitment information database and supports job review and modification. For resume screening and interview scheduling, the system introduces a keyword matching algorithm to automatically screen resumes that meet job requirements, and stores applicant information by establishing a resume database to achieve associated queries with job information. The system then provides a calendar view for interview scheduling and supports drag-and-drop scheduling of interview times. In recruitment management, after the HR department completes the interview, the HR department decides whether to hire or not through the system and updates the employee status.

(4) Performance management and approval management

In the approval management module, it mainly implements various approval processes in the system, including leave approval, reimbursement approval, etc. The manager uses the system to view the historical leave approval progress information, set the approval process, approver, approval form and other information. When there is a task to be approved, the system automatically sends a notification to the relevant personnel to ensure efficient approval. All approval records are stored in the database, and the manager can query and count them at any time.

3.3 Design of Employee Turnover Prediction Model

In the human resource management system, in order to achieve accurate prediction of employee turnover, the experiment introduces the LSTM model [13–15] for prediction. LSTM is an improved version of the recurrent neural network (RNN) with the ability to rely on long-term information and is particularly suitable for processing time series data. This paper uses historical employee data as time series input to build an employee turnover prediction model to provide early warning of potential turnover risks.

In the LSTM model, it includes input gate, forget gate and output gate.

The calculation formula of the forget gate is shown in formula (1). The forget gate determines whether the previous information is retained or discarded at the current moment.

$$g_t = \sigma(U_g \cdot [e_{t-1}, z_t] + \beta_g \tag{1}$$

U_g represents the weight matrix, β_g represents the bias term. e_{t-1} represents the hidden state at the previous moment, and z_t represents the current input.

The calculation formula of the input gate is shown in formula (2).

$$j_t = \sigma(U_j \cdot [e_{t-1}, z_t] + \beta_j \tag{2}$$

The calculation formula of the candidate memory unit is shown in formula (3).

$$\overline{D_t} = tan(U_D \cdot [e_{t-1}, z_t] + \beta_D \tag{3}$$

$\overline{D_t}$ represents the candidate memory unit.

The unit state update is controlled by the forget gate and the input gate to achieve memory update and preservation. Its calculation formula is shown in formula (4).

$$D_t = g_t * D_{t-1} + j_t * \overline{D_t} \tag{4}$$

The calculation formulas for the output gate and hidden state are shown in formulas (5) and (6). The output gate determines the output amount of the hidden state.

$$p_t = \sigma(U_p \cdot [e_{t-1}, z_t] + \beta_p) \tag{5}$$

$$e_t = p_t * tanh(D_t) \tag{6}$$

p_t represents the output gate.

In model training, the experiment uses the cross entropy loss function to calculate the error between the predicted value and the actual value. The calculation formula of the loss function is shown in formula (7).

$$\text{Loss} = -\frac{1}{N}\sum_{i=1}^{N}[y_i\log(\hat{y}_i) + (1 - y_i)\log(1 - \hat{y}_i)] \tag{7}$$

N represents the total number of samples, y_i represents the actual label, and $\hat{y}_i$ represents the model prediction probability.

In the experiment, the LSTM model uses the adaptive learning rate adjustment of the Adam algorithm to enable the model to converge more efficiently. The steps for applying the employee turnover LSTM prediction model in the human resource management system are as follows:

(1) After the experiment cleans the data, the preprocessed employee data is formatted into the time series input format required by the model. The data of each employee is divided into several subsequences of time steps to facilitate the capture of dynamic changes in employee behavior.
(2) The formatted input data can be imported into the LSTM model for prediction and output probability, indicating the risk level of employee resignation.
(3) To facilitate system integration, the output probability of the LSTM model is compared with the set threshold of 0.5. If it is higher than the threshold, it is marked as "1", corresponding to the tendency to resign, otherwise it is marked as "0", corresponding to being employed.
(4) In the employee management page, a resignation risk label is displayed for each employee, and a more detailed prediction probability is provided on the employee details page. The system also sets up an early warning notification function for high-risk employees.

4   Results and Discussion

4.1   Experimental Environment

Hardware environment: Server Intel Xeon E5-2620 v4, 64GB DDR4 RAM (Random Access Memory), 1TB SSD (Solid State Drive). Huawei Honor 9X series, Android system, 8G + 128G storage. Lenovo ThinkBook 14 + 2024 Ryzen Edition, AMD R7-8845H processor. In the software environment, server side: IntelliJ IDEA 2020.3.3 platform, J2EE, Java, MySQL 8.0. Web side: JSP, HTML, JavaScript, Python 3.8. Android side: Python 3.8, HTTP, XML, Android Studio 4.1.

4.2   Experimental Data and Preprocessing

Experimental Data. The experimental data in this paper comes from the human resources management system data of a certain company from June 2022 to June 2024, including employee department, gender, age, education level, position level, marital status, average annual comprehensive monthly salary, average monthly performance, number of trainings, number of business trips, years of work, average annual salary increase ratio, average monthly working hours, and job satisfaction score. A total of 6,382 sets of data were collected, of which the proportion of samples of resigned employees reached 41.38%. The experiment used ten-fold cross validation to divide the training set and the test set, of which 70% was used as the training set and 30% as the test set, and finally the mean was taken as the final result.

(1) Data cleaning. For features with fewer missing values, the experiment used the mean filling method for processing, and for features with a higher missing ratio, the corresponding records were deleted to ensure the integrity of the data set. In terms of outlier processing, this paper uses the box plot method to detect outliers and replaces the detected outliers with the mean.

(2) Feature encoding. In the experimental data, including categorical variables such as education level, position level, and marital status, the study uses the one-hot encoding method to convert them into numerical features, so that the LSTM model can process non-numerical data. The encoding results are shown in Table 1.

In Table 1, gender is divided into male and female, corresponding to 1 to 2. Marital status is divided into single, married, divorced, and widowed, corresponding to 1 to 4. Job level is divided into newcomer, junior, intermediate, and senior, corresponding to 1 to 4. Education level is divided into technical secondary school, junior college, undergraduate, and graduate, corresponding to 1 to 4. Employee departments are divided into after-sales department, pre-sales department, human resources department, and technical department, corresponding to 1 to 4.

Table 1. Encoding results

Characteristics	Coding			
Gender	Male	Female	-	
	1	2		
Marital status	Not married	Married	Divorced	Widowed
	1	2	3	4
Job level	Newbie	Beginner	Intermediate	Advanced
	1	2	3	4
Education level	Technical secondary school	Junior college	Undergraduate	Postgraduate
	1	2	3	4
Employee department	After-sales department	Pre-sales department	Human resources department	Technical department
	1	2	3	4

(3) Data normalization. The experiment uses the Z-score normalization method to convert numerical features such as salary, performance, and years of experience into a standard normal distribution to eliminate the dimensional effects of different features. The normalization calculation formula is shown in formula (8).

$$Z = \frac{(X - \alpha)}{\eta} \tag{8}$$

X represents the original data, represents the standard deviation, and represents the mean.

4.3 Experimental Design

The purpose of the experiment is to evaluate the accuracy and reliability of the LSTM model in employee turnover prediction and verify the interactive performance of each functional module and the overall response speed of the system in the J2EE-based human resource management system.

After the above experimental preprocessing, the experimental steps are as follows:

(1) The experiment first determines that the system includes salary, personnel, organization and other management modules based on the actual needs of the company's human resource management, and allocates and divides the authority.
(2) This paper uses J2EE technology to develop the Web, uses the Android platform to develop the mobile terminal, and uses the RESTful mode WebService interaction technology to respectively communicate between the front-end and back-end, and between the Android platform and the Web terminal.
(3) After completing the basic modules and communication technologies, the study introduces the LSTM model to build a human resource loss prediction model into the system, predicts the loss of company employees, and uses RabbitMQ technology to realize the interaction between the employee loss prediction model and the system.
(4) The LSTM model can be used to calculate the accuracy and precision of employee turnover prediction, and the response time of the system.

The hyperparameter settings are shown in Table 2.

Table 2. Hyperparameters

Parameters	Value	Parameters	Value
Learning rate	0.001	Input time step	10
Batch size	64	Dropout rate	0.2
Number of training rounds	50	Activation function	Tanh
Number of hidden layer units	128	Optimizer	Adam

In Table 2, the learning rate is set to 0.001, the batch size is 64, the number of training rounds is 50, the number of hidden layer units is 128, the input time step is 10, the drop rate is 0.2, the activation function uses Tanh, and the optimizer uses Adam.

4.4 Experimental Results

Performance of LSTM prediction model for human resource loss. In order to explore the effect of introducing LSTM in the human resource management system, its prediction performance is analyzed from four aspects: accuracy, precision, recall, and F1 value. The results are shown in Figure 2. In Figure 2, the comparison models include GRU (Gated Recurrent Unit), RF (Random Forest), SVM (Support Vector Machine), and XGBoost (eXtreme Gradient Boosting).

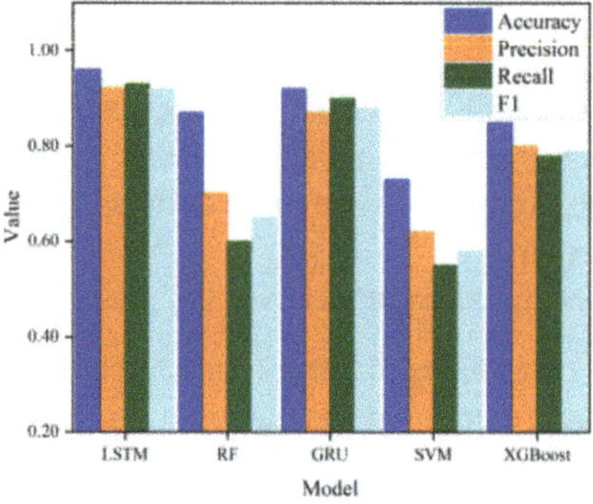

Fig. 2. Performance results of the LSTM prediction model for human resource turnover

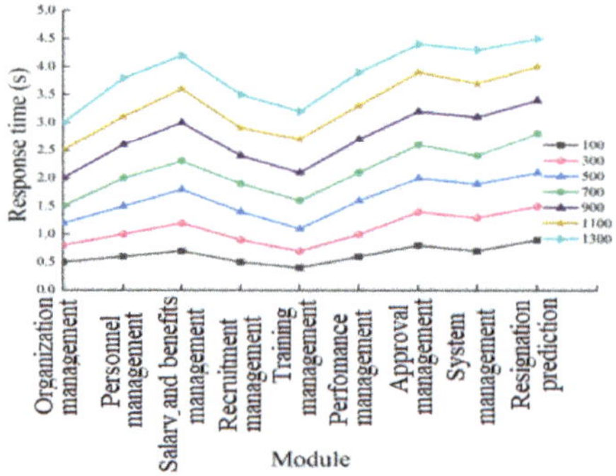

Fig. 3. Web-side response time verification results

In Fig. 2, it can be seen that the LSTM model performs significantly better than other models in predicting employee turnover, with an accuracy of 0.96, a precision of 0.92, a recall of 0.93, and an F1 value of 0.92. The GRU model has an accuracy of 0.92, a precision of 0.87, a recall of 0.90, and an F1 value of 0.88. The SVM model has the worst performance, with an accuracy of only 0.73. In the employee turnover prediction task, LSTM is more suitable for processing such time-dependent tasks due to its long and short-term memory characteristics.

Web-side Response Time Verification. In a human resource management system, response time is crucial. The experiment now explores the response time test of different modules under different concurrent user numbers, and the results are shown in Fig. 3. In Fig. 3, the number of concurrent users is divided into 100, 300, 500, 700, 900, 1100, and 1300.

In Fig. 3, as a whole, as the number of concurrent users increases, the response time of each module shows an upward trend. In the more complex data calculation module, the response time delay of the resignation prediction module is the most significant. Under 500 concurrent users, the response time of each module is less than 2.1s, which meets the actual needs, and under 1300 concurrent users, the response time of each module does not exceed 4.5s. In summary, the human resource management system designed by the experiment achieves good performance.

4.5   Experimental Discussion

This study verified the effect of introducing LSTM in the human resource management system through experiments. From the perspective of prediction performance, the LSTM model performs well in accuracy, precision, recall and F1 value. Compared with other models, its long and short-term memory characteristics make it more advantageous in the task of employee turnover prediction. In the Web-side response time test, the response time of each module under different concurrent user numbers is stable, especially under high concurrency, the response time of each module remains within an acceptable range. The satisfaction analysis results show that the J2EE-LSTM system has the highest scores in functional completeness, performance and ease of use, and can better meet user needs than traditional J2EE, UML and other systems, indicating that the introduction of the integration of the LSTM model and the J2EE architecture has greatly improved the user experience of the human resource management system.

The human resource management system developed by combining LSTM and J2EE architecture in this study has outstanding performance in employee turnover prediction and other aspects, providing enterprise managers with more accurate and efficient decision support tools, and has high practical application value. The system's good responsiveness in high-concurrency scenarios further ensures its scalability and reliability in practical applications, can effectively adapt to complex enterprise management needs, and lays a good foundation for the intelligent and automated application of human resource management systems in the future.

5   Conclusions

This paper designs a human resource management system based on the J2EE architecture, and introduces the LSTM model to achieve early prediction of employee turnover, meeting the company's pre-intervention needs for employee turnover risks. In the system construction, the study uses RESTful WebService to achieve efficient communication between the front-end, back-end and mobile terminals, and uses RabbitMQ technology to seamlessly integrate the turnover prediction model into the system to ensure real-time data and smooth user experience. The experimental results show that the designed system performs well in churn prediction accuracy and high concurrency response time, significantly improving user satisfaction and prediction results. This paper has made some achievements, but there are still some shortcomings. The performance test of the system is limited to a specific concurrency range. The current system mainly uses LSTM for employee churn prediction. In the future, it can be further expanded and tested under a larger concurrency, and more deep learning models or integrated methods can be introduced to improve the accuracy and adaptability of the prediction.

References

1. Nikam, R.U., Lahoti, Y., Ray, S.: A study of need and challenges of human resource management in start-up companies. Math. Stat. Eng. Appl. **72**(1), 314–320 (2023)
2. Zhang, J., Chen, Z.: Exploring human resource management digital transformation in the digital age. J. Knowl. Econ. **15**(1), 1482–1498 (2024)
3. Wu, Y., Chen, D.: Design and implementation of human resource optimal scheduling system based on B/S architecture. J. Hum. Resour. Sustain. Stud. **12**(1), 1–14 (2024)
4. Chunlei, S.: Design of hospital human resource management system based on C/S architecture. Microcomput. Appl. **38**(06), 164–166 (2022)
5. Yu, X., Zhang, C., Wang, C.: Construction of hospital human resource information management system under the background of artificial intelligence. Comput. Math. Methods Med. **2022**(1), 1–11 (2022)
6. Li, Y.: Research on hospital human resource management system based on J2EE architecture. Microcomput. Appl. **38**(05), 32–38 (2022)
7. Butsianto, S., Naya, C.: Model aplikasi human resource management sistem (HRIS) dengan framework UniGui. Bull. Inf. Technol. (BIT) **4**(1), 81–88 (2023)
8. Park, J., Feng, Y., Jeong, S.P.: Developing an advanced prediction model for new employee turnover intention utilizing machine learning techniques. Sci. Rep. **14**(1), 1221–1234 (2024)
9. Mozaffari, F., Rahimi, M., Yazdani, H., Sohrabi, B.: Employee attrition prediction in a pharmaceutical company using both machine learning approach and qualitative data. Benchmarking Int. J. **30**(10): 4140–4173 (2023)
10. Lim, C.S., Malik, E.F., Khaw, K.W., Alnoor, A., Chew, X.Y., Chong, Z.L., et al.: Hybrid GA–DeepAutoencoder–KNN model for employee turnover prediction. Stat. Optim. Inf. Comput. **12**(1), 75–90 (2024)
11. Adeusi, K.B., Amajuoyi, P., Benjami, L.B.: Utilizing machine learning to predict employee turnover in high-stress sectors. Int. J. Manag. Entrep. Res. **6**(5), 1702–1732 (2024)
12. Taner, Z., Hızıroglu, O.A., Hızıroglu, K.: Leveraging machine learning methods for predicting employee turnover within the framework of human resources analytics. J. Intell. Syst. Theory Appl. **7**(2), 145–158 (2024)
13. Ganapathisamy, S., Narayan, V.: A long short-term memory with recurrent neural network and brownian motion butterfly optimization for employee attrition prediction. Int. J. Intell. Eng. Syst. **17**(1), 193–192 (2024)
14. Zhang, H., Zhang, W.: Application of GWO-attention-ConvLSTM model in customer churn prediction and satisfaction analysis in customer relationship management. Heliyon **10**(17), 1–16 (2024)
15. Jajam, N., Challa, N.P.: Dynamic behavior-based churn forecasts in the insurance sector. Comput. Mater. Continua **75**(1), 977–997 (2023)

Development and Performance Optimization of Student Behavior Analysis System Based on B/S Architecture

Hua Yang[✉]

Wuhan Railway Vocational College of Technology, Wuhan, China
yanghuawh@88.com

Abstract. It is difficult for current student behavior analysis systems to simultaneously achieve efficient data processing, convenient client management, and real-time response to large-scale data. This paper constructed an efficient student behavior analysis system that adapted to large-scale data by optimizing the B/S architecture (Browser/Server Architecture) and improving data processing performance. In the system design, MySQL database index optimization, distributed cache Redis, asynchronous data transmission technology and data analysis algorithm based on K-means clustering are used. The query delay is reduced through database optimization, the Redis cache technology improves the concurrent processing capability, the asynchronous transmission reduces the network delay, and the K-means algorithm is used to enhance the depth of data analysis. The experimental results show that the average response time of the B/S architecture under high concurrency is 620 ms, and the throughput reaches 1000 requests/second, which is better than the 790 ms and 800 requests/second of the C/S (Client/Server) architecture. In the data transmission efficiency test, the B/S architecture has a delay of 240 ms in a low-quality network, while the C/S architecture has a delay of 300 ms. In the data processing accuracy verification, the K-means silhouette coefficient of the B/S architecture is 0.73, higher than the 0.64 of the C/S architecture, and the confidence of association rule mining is 92%, higher than the 85% of the C/S architecture. The results show that the student behavior analysis system based on B/S architecture has stronger performance and accuracy when dealing with large-scale data.

Keywords: Student Behavior Analysis System · Browser/Server Architecture · Data Processing Performance · K-means Clustering · Association Rule Mining

1 Introduction

In recent years, the development of educational informatization has promoted the widespread application of student behavior analysis systems, and behavior analysis based on big data has gradually become an important way to improve the quality of education. The traditional student behavior analysis system adopts a C/S architecture, which has accumulated rich experience in data collection and analysis, but has great difficulties in

© The Author(s) 2026
P. Siarry et al. (Eds.): WCNA 2024, LNEE 1550, pp. 508–518, 2026.
https://doi.org/10.1007/978-981-95-6946-5_51

maintenance and updating, and is not suitable for the needs of large-scale data real-time processing [1, 2]. The continuous expansion of school scale and the improvement of information level have led to a rapid increase in the amount of student behavior data, which has put forward higher requirements on the system's response speed and concurrent processing capabilities. In addition, education administrators have an increasing demand for real-time access to student behavior information, requiring the system to provide a more convenient client management experience while ensuring the real-time and accuracy of the data [3, 4]. Therefore, B/S architecture has become a new direction for optimizing behavior analysis systems due to its simple client management advantages, and can bring new improvements in large-scale data processing and real-time response.

This paper builds a performance-optimized student behavior analysis system under the B/S architecture. The main contributions are as follows: 1. By optimizing the MySQL database index structure and partition management strategy, the impact of high-frequency queries on the system response speed is reduced, so that the system can maintain efficient query performance in a large data volume environment. 2. Distributed cache Redis can be used to cache hot data and frequently accessed behavior information, greatly improving the system's concurrent processing capabilities and enhancing the system's stability in a high-concurrency environment. 3. The system uses asynchronous data transmission technology to avoid network delays in synchronous transmission mode, improve data transmission efficiency, and enable the system to meet the needs of real-time data processing. 4. The data analysis module combines the K-means clustering algorithm with the Apriori association rule algorithm, and uses machine learning methods to perform behavioral pattern recognition and deep data mining to enhance the system's data analysis accuracy and depth.

2   Related Work

In the field of student behavior analysis, many scholars have conducted multi-faceted research on the real-time performance, response speed, and data processing efficiency of the system. Zhang Y designed a new multimedia-assisted oral English teaching system using a B/S architecture, which achieved a more realistic learning environment. The experimental results showed that the system can significantly improve teaching efficiency and students' oral English scores [5]. Gao M built an intelligent campus teaching system based on ZigBee wireless sensors by improving the ZigBee algorithm, and evaluated the teaching effect of the system through experimental research and simulation technology. The results showed that the system performed well [6]. Yu C et al. designed and developed a college student management system using the K-means clustering algorithm. By optimizing the K-means algorithm, they achieved good classification of student data, improved the effect of student management, and promoted the development of student management [7]. In terms of student sports management, Feng J adopted an artificial intelligence-based sports management system and built a sports performance management and physical fitness analysis system through data mining, which improved the work efficiency of physical education teachers and coaches, achieved a high accuracy rate of 97%, and effectively improved the efficiency and quality of physical education teaching [8]. These studies have optimized the student behavior analysis system at different levels, but overall there is still much room for improvement in real-time response and processing accuracy in large-scale data environments. Some studies have optimized the performance of behavior analysis systems through specific technologies and methods. Ge T et al. used fuzzy control parameters and fast image region segmentation technology to design an interactive remote multimedia teaching system based on VR (Virtual Reality) technology, achieving more than 99% system stability and strong human-computer interaction response capabilities [9]. Based on these studies, this paper uses MySQL index optimization, Redis cache strategy, asynchronous data transmission and K-means algorithm to improve system performance and data analysis accuracy.

3 System Design and Implementation

3.1 System Architecture Design

The system architecture is based on the B/S model to improve the convenience of client management and the real-time response of the system. The system framework uses a front-end and back-end separation design [10, 11]. The back-end uses the Spring Boot framework to optimize business logic processing and response speed. The front-end is developed based on the Vue.js framework to enhance data display interaction and client operation response experience. The backend architecture is based on Spring Boot [12] and builds a layered structure: control layer, service layer, and data access layer. The control layer interacts with external data through the RESTful API interface, receives client requests and calls the corresponding business logic; the service layer is responsible for data processing and business logic implementation, and cooperates with Redis cache technology to reduce database access frequency, reduce server load, and improve response speed; the data access layer interacts with the database through the MyBatis framework, combined with MySQL index optimization technology, to maintain low query latency and improve system response. MySQL index optimization adopts a multi-level index strategy, partition index and composite index to improve query matching, ensuring system stability and query efficiency.

The asynchronous data transmission design combines the multiplexing feature of HTTP/2 (Hypertext Transfer Protocol version 2) protocol to perform concurrent transmission in a single connection, improving transmission efficiency [13]. HTTP/2 header compression and binary framing reduce transmission overhead. The asynchronous request design is combined with the thread pool mechanism, and the number and frequency of asynchronous threads are controlled through the Scheduled Executor Service to ensure the real-time nature of asynchronous processing and reduce resource usage.

The data analysis module is based on the K-means clustering algorithm and data preprocessing, and the behavior data is normalized to reduce the interference of different dimensions on the clustering results. The K-means algorithm calculates the distance from each data point to the centroid through the Euclidean distance to achieve behavior data clustering. The iterative process is as follows:

$$D = \sum_{i=1}^{k} \sum_{x \in C_i} \|x - \mu_i\|^2 \tag{1}$$

Among them, D is the total error sum of squares, C_i represents clustering, and μ_i represents the centroid of clustering. The algorithm converges when the error sum of squares is minimized, completing the behavior mode division.

In order to improve the convergence speed and stability, the system introduces the initialization center point optimization technology, which avoids the initial point selection affecting the accuracy of the result by selecting points in segments. The clustering results are visualized through the front-end to form an intuitive display of student behavior patterns. This technology is combined with Redis cache technology to ensure real-time analysis capabilities in a high-concurrency environment.

3.2 Data Collection Module

The system completes the real-time capture of student behavior data through the Spring Boot built-in API interface. For high-concurrency environments, Redis cache is introduced to speed up the data reading rate and ensure the timely push and transmission stability of data. Core indicators such as student login times, page click behavior, and course progress are transmitted to the system background through the interface to simplify the data call process and reduce network delays. During the data collection process, Java scheduled tasks are used to automatically capture and ensure the continuous updating of behavioral data. Scheduled tasks use Scheduled Executor Service to control the collection frequency, reasonably allocate system resources under high concurrency, and ensure stable data inflow.

To ensure data quality, the data transmission process is equipped with an abnormal filtering module to exclude abnormal values and missing values based on preset conditions. Anomaly detection uses the Z-score method to identify abnormal points in behavioral data:

$$Z = \frac{X - \mu}{\sigma} \tag{2}$$

Here, X represents the current data point, μ is the data mean, and σ is the standard deviation. The data in $|Z| > 3$ is determined to be an outlier and automatically removed. Combined with the data cleaning module, the data type is matched by regular expressions to ensure the normalization of the input data.

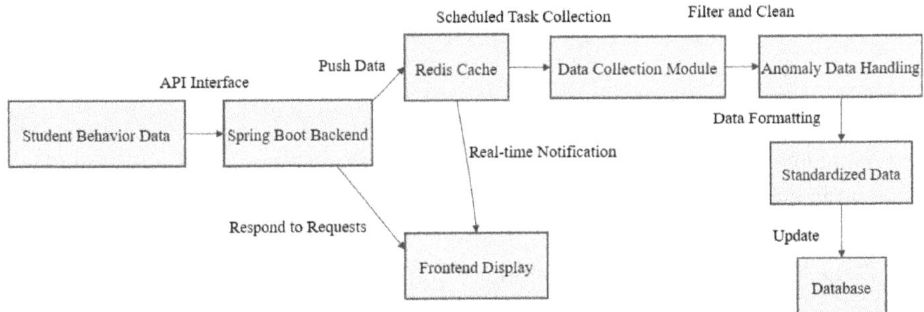

Fig. 1. Data collection module architecture

The architecture diagram of the data acquisition module is shown in Fig. 1. The data collection module uses Redis cache to improve the data query rate. By partitioning and caching student behavior data, data requests are dispatched to relevant partitions to avoid high-frequency queries causing system congestion. Combined with the TTL (Time-to-Live) mechanism [14], the cache data life cycle is controlled to effectively reduce the occupancy of invalid data and cache pressure. In addition, the data collection module combines the Redis subscription-publishing model to realize data update and notification, so that the behavior data can be pushed to the front end in real time through message subscription after collection. The back end receives the real-time data stream through the Redis subscription function to realize efficient data collection and update.

3.3 Database Index Optimization

The database index optimization module uses MySQL database to store student behavior data and uses B-Tree (Balance-tree) index to accelerate the retrieval efficiency of commonly used query fields. In the data table design, B-Tree indexes are constructed for high-frequency query fields such as "student number" and "timestamp" to provide efficient index structure support, so that the system can quickly locate the target record during the query process and reduce the retrieval time. The B-Tree index structure stores index nodes in layers. The database implements multi-level filtering of queries through this structure, and can still maintain high search performance under large-scale data sets.

The data table storage design adopts the InnoDB partition table strategy [15]. The InnoDB engine partitions the data table by fields such as time, student ID or behavior type, effectively avoiding lock conflicts during data access and narrowing the query scope. The introduction of the partition strategy keeps the amount of data stored in each partition at a small scale, meeting the needs of fast queries in large-scale data scenarios. The system uses range-partitioned InnoDB partition tables in its design. By storing historical behavior data in different partitions, the impact on other partitions is reduced when new data is frequently written. For a large number of write and read requests, the database's response performance in large concurrency scenarios is greatly improved by reducing table lock and row lock contention.

The query performance formula in database index optimization is defined as follows: Assuming the total number of records in the database query table is N, and the average response time of the database query is T_{query}, the database query speed S_{index} after index optimization is expressed by the formula:

$$S_{\text{index}} = \frac{N}{T_{\text{query}}} \tag{3}$$

The formula shows that the index structure optimization maintains a balance between the total number of records and the response time, and the existence of the index reduces the average time consumption of each query.

3.4 Data Analysis Algorithm Module

The data analysis module analyzes student behavior data through the K-means clustering algorithm to mine behavior patterns and learning preferences. The module standardizes student behavior data to ensure that the dimensions of each behavior are consistent. The K-means algorithm clusters behavior features according to feature distance, and groups them into different behavior patterns and learning preferences. The algorithm calculates the distance from the data point to the cluster center based on the Euclidean distance, and by minimizing the distance between data points within the cluster, similar behavior data is aggregated into the same group, thus obtaining the main behavior characteristics of students.

The center point of each cluster is calculated by the mean of the behavioral characteristics. Assuming that the data set $X = \{x_1, x_2, ..., x_n\}$ is the student behavioral characteristics, the position vector of the cluster center C_k is:

$$C_k = \frac{1}{|S_k|} \sum_{x_i \in S_k} x_i \tag{4}$$

This formula calculates the center point position of each cluster and collects similar behavior characteristics to enhance the efficiency of system analysis. The data analysis module also uses the Apriori association rule algorithm to mine the potential laws of behavioral data, identify frequent behavior combinations and reveal the internal connections of learning behaviors. The algorithm is based on support and confidence constraints to screen high-frequency associated behavior combinations. Given an event set $I = \{i_1, i_2, ..., i_n\}$, the support of frequent behavior combinations is calculated as:

$$\text{Support}(A) = \frac{\text{Count}(A)}{\text{Total Transactions}} \tag{5}$$

A is an event combination, and the support calculates the frequency of the event combination in the behavior data, retaining the high-correlation combinations that meet the conditions. Association rule mining improves the system's recognition depth of learning behavior and provides data support for personalized recommendations. The algorithm module is written in Python, and data processing is performed through NumPy and Pandas libraries, and K-means algorithm optimization is implemented using Scikit-learn. After training, the data is transmitted to the system database through the API interface, and the front-end visualizes the behavior patterns and preferences.

4 System Performance Evaluation and Comparison

Experimental environment: Table 1 shows the hardware configuration and experimental software configuration of the experiment. The software uses a stable version of the database, cache system, and front-end framework to ensure system reliability and compatibility while being suitable for large-scale data concurrent access scenarios.

Table 1. Experimental environment and experimental software and hardware configuration

Configuration	Details
Operating System	Windows Server 2016, Linux Ubuntu 18.04
Server	Dell PowerEdge R630
Processor	Intel Xeon E5-2620 v4 2.1 GHz, 8 Core
Memory	32 GB DDR4
Storage	SSD 500 GB
Database	MySQL 5.7
Cache System	Redis 5.0
Front-End Framework	Vue.js 2.6
Data Visualization	ECharts 4.9
Testing Tool	Apache JMeter 5.1

4.1 Response Time and Concurrent Processing Capability Test

The response time of the student behavior analysis system based on B/S architecture and C/S architecture was tested by JMeter tool, with the focus on evaluating the response capabilities of the two architectures in a high-concurrency environment. The test simulated concurrent access from 1,000 to 5,000 users and compared the response time, latency, and throughput performance of the two systems under the same hardware conditions. The data was used to analyze the optimization effect of the B/S architecture system. In the experimental environment, the servers were configured the same and the running time was fixed to ensure the comparability of the data.

During the test, the number of users was gradually increased from 1,000 concurrent users to 5,000 concurrent users, and the data sampling frequency was recorded once per minute. In the experiment, the response time is defined as the time from the user sending a request to receiving the server response; the delay represents the waiting time caused by network transmission and system processing time; the throughput represents the number of requests successfully responded to per unit time, which measures the processing efficiency of the system. Figure 2 shows the response time, delay, and throughput of the two systems at each concurrency. The test data shows that under 1000 concurrent users, the average response time of B/S architecture is 250ms, while that of C/S architecture is 280ms; the average latency of B/S architecture is 150ms, and that of C/S architecture is 180ms, and the B/S throughput is 450 requests/second, which is 30 requests/second higher than that of C/S architecture. As the number of concurrent users increases to 5,000, the response time and throughput of the B/S architecture still perform better, with an average response time of 620ms, 170ms less than the 790ms of the C/S architecture; in terms of latency and throughput, the B/S architecture is 400ms and 1,000 requests/second, while the C/S architecture is 590ms and 800 requests/second. The test data results show that the average response time and latency of the B/S architecture are lower than those of the C/S architecture. In a high-concurrency environment, the throughput of the B/S architecture is significantly higher than that of the C/S architecture, indicating that the B/S architecture can handle higher concurrency.

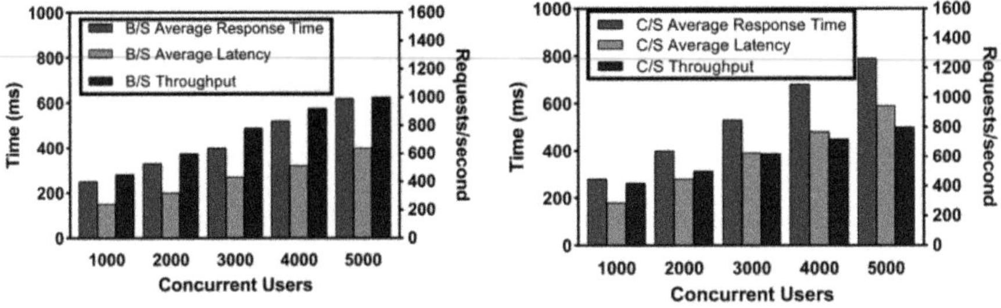

(a) System test data based on B/S architecture   (b) System test data based on C/S architecture

Fig. 2. Response time, latency and concurrent processing capacity test results

4.2 Data Transmission Efficiency Test

In the data transmission efficiency test, the Wireshark tool was used to measure and count the data packet transmission time of the student behavior analysis system based on B/S and C/S architectures, evaluate the differences in data transmission efficiency between different architectures, and set network conditions with different bit rates in the test environment. During the experimental data collection phase, Wireshark was used to record and capture the transmission delay of different types of data packets, focusing on measuring the optimization effect of asynchronous data transmission and Redis cache strategy in the B/S architecture. The test used 100 packet captures to take the average value and obtain the data transmission efficiency under the two architectures. Figure 3 shows the experimental results of the average transmission delay of the B/S architecture and the C/S architecture under different network conditions.

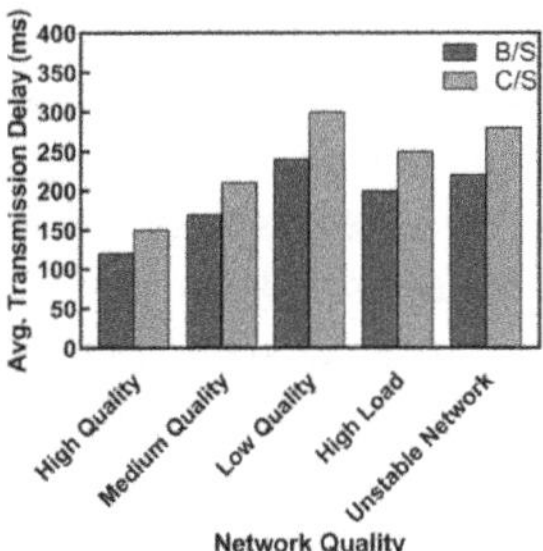

Fig. 3. Transmission delay test results of B/S architecture and C/S architecture under different network conditions

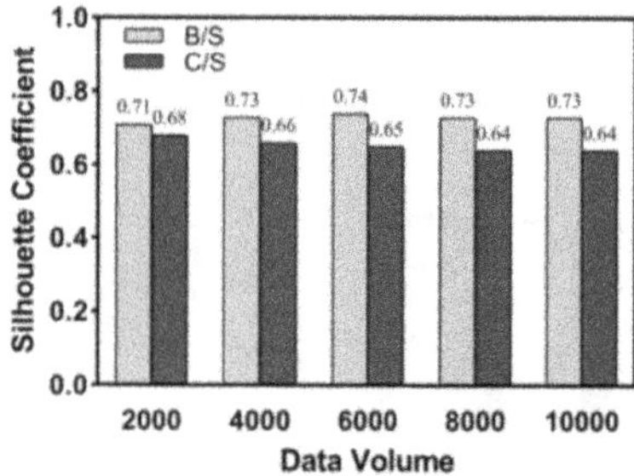

Fig. 4. Comparison of cohesion and silhouette coefficient of B/S and C/S architectures under different data volumes

Experimental data show that the data packet transmission delay performance of the B/S architecture system is better than that of the C/S architecture under different network conditions. In a high-quality network environment, the average transmission delay of the B/S architecture is only 120 ms, while that of the C/S architecture is 150 ms. Even in a low-quality network environment, the data transmission delay is only 240 ms, which is lower than the 300 ms of the C/S architecture. Overall, the average optimization of the B/S architecture system compared to the C/S architecture has reached 20%, indicating that the B/S architecture has obvious optimization in transmission efficiency, lower latency, more stable transmission, and adaptability to different network environments. It also shows that the use of Redis cache strategy can effectively reduce the frequency of database requests. At the same time, the asynchronous data transmission mechanism improves the transmission stability of data packets, so that the transmission delay remains at a low level in a fluctuating environment.

4.3 Data Processing Accuracy Verification

In the data processing accuracy comparison test, Python's Scikit-learn library is used to implement K-means clustering and Apriori association algorithms to evaluate the data processing accuracy performance of the student behavior analysis system based on B/S architecture and C/S architecture. In the data clustering test, the K-means algorithm is used to cluster the student behavior data. By changing the size of the data set, the clustering effect is tested and its correctness is recorded. Inertia and Silhouette Coefficient are used as evaluation criteria.Clustering tests are performed in B/S architecture and C/S architecture systems for different data set sizes. Figure 4 shows the clustering results of the two systems under different data volumes. The performance difference between the two is not significant under small-scale data, but as the amount of data increases, the clustering accuracy of the B/S architecture is significantly better than that of the C/S architecture system. In the test of 10,000 data, the silhouette coefficient of the B/S architecture is stable at 0.73, while the C/S architecture drops to 0.64, indicating that the B/S architecture has more data processing advantages and can better adapt to the big data environment.

In association analysis, the Apriori algorithm is used to mine association rules of student behavior data, and the accuracy and processing efficiency of the two systems in the rule mining process are analyzed. The accuracy of association rule mining is tested with support and confidence as indicators for different data volumes. Table 2 shows the association rule mining results of the two systems under different data volumes. Under the condition of 5,000 data, the association rule confidence of the B/S architecture averaged 92%, while the confidence of the C/S architecture under the same data volume was 85%, showing insufficient accuracy and processing delay.

Table 2. Comparison of association rule mining accuracy between B/S and C/S architectures under different data volumes

Data Volume	B/S Support	B/S Confidence	C/S Support	C/S Confidence
1000	0.85	0.88	0.83	0.84
2000	0.86	0.9	0.84	0.85
3000	0.88	0.91	0.85	0.86
4000	0.89	0.92	0.86	0.85
5000	0.9	0.92	0.87	0.85

The data results in Fig. 4 and Table 2 illustrate the superiority of the B/S architecture in data processing accuracy. The K-means clustering accuracy test and the Apriori association rule mining accuracy comparison verify the accuracy and stability of the B/S architecture when processing large amounts of data.

5   Conclusions

This paper designs a student behavior analysis system based on the B/S architecture. By optimizing database indexes, using Redis distributed cache and asynchronous data transmission technology, and combining K-means clustering and Apriori association algorithms, it realizes efficient processing and real-time analysis of large-scale data. The test results show that the response speed, concurrent processing capability and data transmission efficiency of the B/S architecture in a high-concurrency environment are better than those of the traditional C/S architecture. In terms of data processing accuracy, the system based on the B/S architecture has stronger stability and reliability in a large-scale data environment, providing more efficient technical support for student behavior analysis. However, the B/S architecture is dependent on the network environment during data transmission, which leads to a decrease in system performance under extreme conditions of network fluctuations. In the future, further research can be conducted on multimodal data fusion techniques to enhance algorithm adaptability. By continuously optimizing algorithms and architectures, the student behavior analysis system with a B/S architecture can further improve its performance in large-scale applications and provide effective technical support for the construction of a behavior analysis system for smart education.

References

1. Yang, H., Zhang, W.: Data mining in college student education management information system. Int. J. Embedded Syst. **15**(3), 279–287 (2022)

2. Zhao, W., Chang, Z.: Construction and implementation of mobile learning system for higher education based on modern wireless network mobile communication terminal technology. Wireless Netw. **30**(6), 5669–5681 (2024)
3. Fan, J.: A big data and neural networks driven approach to design students management system. Soft. Comput. **28**(2), 1255–1276 (2024)
4. Elfeky, A.I.M., Elbyaly, M.Y.H.: The use of data analytics technique in learning management system to develop fashion design skills and technology acceptance. Interact. Learn. Environ. **31**(6), 3810–3827 (2023)
5. Zhang, Y.: Multimedia-assisted oral English teaching system based on B/S architecture. Int. J. Continuing Eng. Educ. Life Long Learn. **32**(6), 663–680 (2022)
6. Gao, M.: Smart campus teaching system based on ZigBee wireless sensor network. Alex. Eng. J. **61**(4), 2625–2635 (2022)
7. Yu, C., Wang, Y.: College student management system based on k-means clustering algorithm. Int. J. New Dev. Educ **4**(2), 2663–8169 (2022)
8. Feng, J.: Designing an artificial intelligence-based sport management system using big data. Soft. Comput. **27**(21), 16331–16352 (2023)
9. Ge, T., Darcy, O.: Study on the design of interactive distance multimedia teaching system based on VR technology. Int. J. Continuing Eng. Educ. Life Long Learn. **32**(1), 65–77 (2022)
10. Liu, S.-F., Li, J., Zhang, H.-Q., Li, Z., Cheng, M.: Development and implementation of digital pedagogical support systems in the context of educational equity. Humanit. Soc. Sci. Commun. **11**(1), 1–15 (2024)
11. Wang, L., Jia, X., Cui, H., Zhang, B.: An interactive practice platform of English mobile teaching in colleges and universities based on open API. Int. J. Continuing Eng. Educ. Life Long Learn. **32**(4), 418–431 (2022)
12. Liu, Y.: Design and implementation of a student attendance management system based on springboot and vue technology. Front. Comput. Intell. Syst. **8**(1), 91–97 (2024)
13. Wehbe, N., Alameddine, H.A., Pourzandi, M., et al.: A security assessment of HTTP/2 usage in 5G service-based architecture. IEEE Commun. Mag. **61**(1), 48–54 (2022)
14. Mantri, A.: Optimizing HDFS storage and managing TTL for unused hive tables: strategies for improved data efficiency. J. Artif. Intell. Mach. Learn. Data Sci. **1**(4), 680–683 (2023)
15. Chen, Z., Yang, X., Li, F., et al.: CloudJump: optimizing cloud databases for cloud storages. Proc. VLDB Endow. **15**(12), 3432–3444 (2022)

Research on Speech Recognition Realization Based on Deep Learning Method

Jiayi Liao[1] and Li Chen[2][(✉)]

[1] Sino-European School of Technology, Shanghai University, Shanghai, China
[2] School of Management, Shanghai University of Engineering and Science, Shanghai, China
`iowkey@126.com`

Abstract. Automatic speech recognition is an important research direction in artificial intelligence field. For how to implement speech recognition, converting speech sequences into text sequences, it is simply a matter of identifying the problem, selecting a model and then training it. With the continuous expansion of the open source community, the research process in the field of speech recognition has been accelerated. Some open source tools for speech recognition, such as CMUSphinx, Julius, HTK, CMUSphinx and ISIP, have also emerged and been widely used by researchers. Firstly, the paper analyzes the tools CMUSphinx, Kaldi and deep learning platform which can develop speech recognition at present. Then the implementation process of CMUSphinx open source tool is studied. Secondly, the practice process of speech recognition by Kaldi is discussed. Finally, some key problems in speech recognition implementation and future research directions are pointed out in the paper.

Keywords: Speech Recognition · CMUSphinx · Kaldi · Deep Learning

1 Speech Recognition Tool

1.1 CMUSphinx

CMUSphinx is a software system developed at Carnegie Mellon University for research on speech recognition technology. In 2000, the Sphinx group had been working on open source machine components in several speech recognizer components. Including sound decoders and model restoration programs, resources include acoustic model training software, language model and dictionary editing software.

1.2 Kaldi

Compared with the early development of speech recognition related open source toolkits like HTK, CMUSphinx, etc., Kaldi has its own unique style, such as code is easy to read and understand; A large amount of linear algebra support is conducive to switching between different linear algebra libraries. Implement algorithms as generically as possible, avoiding code that only serves a specific task; Has a very complete speech

© The Author(s) 2026
P. Siarry et al. (Eds.): WCNA 2024, LNEE 1550, pp. 519–525, 2026.
https://doi.org/10.1007/978-981-95-6946-5_52

recognition system training script; Have a thriving open source community, open code licenses. These advantages have greatly reduced the threshold of Kaldi as a speech recognition kit, and also made Kaldi attract a large number of users, becoming the leader in speech recognition kits (Fig. 1).

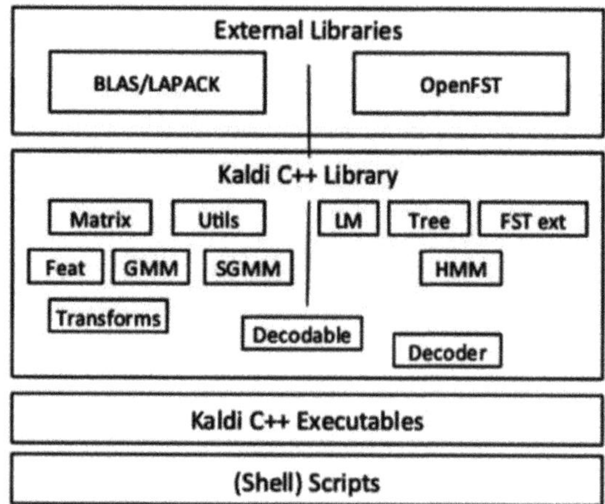

Fig. 1. System composition of Kaldi

1.3 Deep Learning Platform

In recent years, due to the application of deep learning, speech recognition technology has become more advanced. The application of a series of algorithms and technologies also makes the establishment of speech recognition system not limited to a certain platform. Some of the most widely used deep learning frameworks are TensorFlow (first developed and used by Google), PyTorch (first developed and used by Facebook), CNTK (first developed and used by Microsoft), and MXNet (used by Amazon, etc.). The core language of the general deep learning framework is mostly C++, and the front-end interface language supports PytHoN. Such a language collocation method makes the use of the framework both flexible and efficient. Compared to specialized platforms such as Kaldi, deep learning methods work more on acoustic and language models (or end-to-end models), and can be extended to a variety of tasks.

2 CMUSphinx Implementation Details

Sphinx is one of the leading speech recognition toolkits for recognition suites in China, with a variety of tools for management and for building voice applications. Carnegie Mellon's Sphinx contains development kits for many different tasks and applications. Sometimes it is difficult to make the choice. The following are the goals of each development kit:

Pocketsphinx - Lightweight recognition tool library with lightweight speech recognition engine developed with C language;

Sphinxtrain - Acoustic model training tool;
Sphinxbase - Pocketsphinx and Sphinxtrain support library;
Sphinx4 - An adjustable identifier written in Java. (Models include acoustic models, language models, and phonetic dictionaries) Acoustic features are included in acoustic models. N-Gram is the most commonly used model, finite-state language models and word statistics are included and defined speech sequences are performed by finite state automata (and sometimes weight).

The search space limitation of the model must be very successful in order to achieve higher accuracy. This means it is better able to infer a word that follows it. Language models typically restrict attention to words that are already included. For this problem, which belongs to name recognition, the model can contain small pieces, such as words and phonemes. It should be noted that the search space in this development is poor, and the recognition accuracy will be lower than the previous language learning model (based on words). Dictionaries contain word to phoneme mappings, which are generally not very effective. But dictionaries aren't the only way to map words to phones. We can also use machine learning algorithms to learn other functions that may be a little more complex.

3 Basic Kaldi Speech Recognition Implementation Process

3.1 Data Preparation

In preparing the training data, we need to complete the two tasks of selecting the training data and organizing the data into a format that the tool can support. About building data resource files that comply with the Kaldi script specification, including data folder Data and voice folder data/lang.

3.3.1 Basic data Data is usually divided into three subsets of training data, development data and test data, which are represented by train, dev and test respectively. When Kaldi uses thchs30 (an open Chinese speech database published by the Speech and Language Technology Center of Tsinghua University) for training operations, thchs30 is initially processed to obtain four text files that can be opened directly (for example, the training set is placed under data/train). And the two files Utt2spk and spk2utt are required for Kaldi processing. Note that you may need to prepare additional files for different data sources or tasks.

3.3.2 Language data In terms of language data, Kaldi needs to store files in data/dict, and other data collation details can be referred to the literature. When the data for the speech recognition experiment training is ready, Kaldi needs to process the data.

3.2 Voice Signal Feature Extraction

The pre-processed signal is already an audio signal with a certain purity, and for the recognition of any object or speech recognition, it is necessary to grasp the unique characteristics of things from the direction of machine recognition. Therefore, speech recognition needs to extract the features of the speech before entering the acoustic model training. A speech signal is represented by the eigenvalue. Since there are many eigenvalues, it is represented by the eigenvector. The most common method for extracting eigenvalues is MFCC (Mel-Frequency Cepstral Coefficient).

Next describes the general process of MFCC, the first step, the actual frequency to Mel frequency through the formula, the second step, after the Mel frequency is obtained Fourier transform, the purpose of Fourier transform is to see the specific distribution of signal energy. The third step is to filter the signal using a triangular filter with Mel scale. The fourth step is to filter the signal because it is discrete data, so the MFCC we need is obtained through the inverse discrete cosine transform. The formula is as follows.

$$C_n = \sum_{k=1}^{M} \cos\left(\frac{\pi n(k - 0.5)}{M}\right) \ln s(m) n = 1, 2, \cdots, L$$

3.3 Acoustic Model Training Process

3.3.1 Obtain the audio set and corresponding text set of the corpus set.
   It is possible to provide more precise alignment, pronunciation (sentence) level start and end data time, but this is not necessary.

3.3.2 Formatting the obtained text set.
   Kaldi requires a variety of formats. The training process will use the start and end time of each sentence, the speaker id of each sentence, and all the words and phonemes used in the text set.

3.3.3 Extracting acoustic features from audio files MFCC or PLP is widely used in traditional teaching methods. There are differences for NN methods.

3.3.4 Single phoneme training Single phoneme training does not use contextual information before or after the current phoneme, while three phonemes use the current phoneme, the previous phoneme, and the last phoneme (Fig. 2).

3.3.5 Framework based on GMM/HMM.

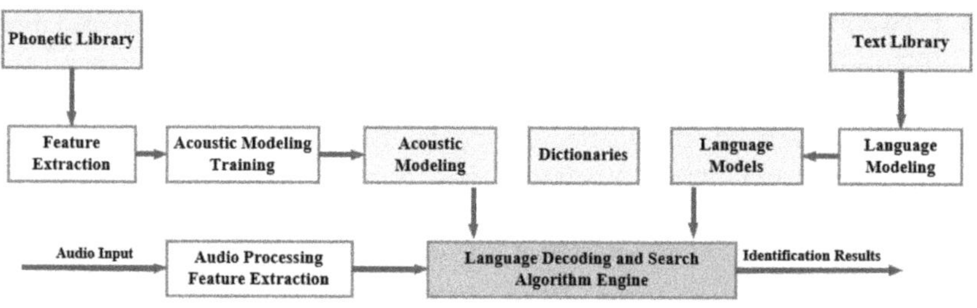

Fig. 2. Basic schematic diagram of GNN-HMM

(1) Align the audio to the acoustic model. The parameters of the acoustic model are obtained during acoustic training, however, this development process can be managed optimally by using a cycle system of training and alignment. This is also known

as Viterbi training (including forward-backward and expectation-maximization intensive computational processes). By aligning audio and text, other training algorithms can be used to improve and refine the parameterized model. So, each student training method step will follow its own alignment step.

(2) train the triphoneme model. The monophoneme model represents only the parameters of a single phoneme, but the phoneme varies with context. The triphoneme model uses phonemes before and after context to show changes in phonemes. Not all the monophoneme combinations exist in the text set provided, there are 3 possible triphonemes in total, but the training set contains a limited subset of the business, and the triphoneme combinations that can occur must have a certain number of times to facilitate the training of students. The phoneme decision tree method will cluster these triphonemes into a smaller set.

(3) Re-align the audio according to the acoustic model and retrain the triphoneme model. Repeat steps 1 and 2 above, and add additional and more elaborate training for the triphoneme model, usually including incremental training, lda, mllt, and sat. The alignment algorithm mainly includes student speaker alignment and FMLLR.

(4) training algorithm. Incremental algorithms compute the first and second derivatives, or dynamic parameters, of the features to complement the MFCC features. Linear Discriminant Analysis-Maximum Likelihood Linear Transform (LDA-MLLT): LDA establishes HMM states based on dimensionality reduction eigenvectors. MLLT obtains the unique transformation of each speaker according to the feature space after the dimensionality reduction of LDA. MLLT is actually a normalization of the speaker. The SAT (Speaker Adaptive training) can also normalize noise.

(5) Alignment algorithm. The actual alignment operation is the same, and different sets of sets use different acoustic analysis models.

3.4 Decoding Implementation -- Viterbi Algorithm

The Viterbi algorithm is implemented by a matrix of t*s, where t is the number of frames and s is the total number of HMM states. The acoustic characteristics are traversed by frame, each state of each frame, the cumulative state of the previous frame and the state of this frame are added together, and the lowest cost of this frame is selected as the best path for this frame. Most of Kaldi's decoders are based on the Viterbi algorithm. There are many Kaldi decoders, such as Simple decoder and Fast decoder, which exist in the form of libraries. The appropriate decoder can be selected when needed.

4 Summary

Kaldi is still a powerful speech recognition tool, and because the code is open source, it is now very active on Github. Kaldi is developed from the GMM-HMM model. Although Kaldi has many advantages, it also has some disadvantages, such as speech recognition in noisy environments, the lack of speech recognition transfer function, and the shallow level of this model that cannot capture the deep characteristics between data. Therefore, with the development of artificial intelligence, a model algorithm based on DNN-HMM has emerged. GMM-HMM is based on the probability and statistics method to get the parameter model, which also means that the demand for raw data is particularly large, and the amount of raw data is positively correlated with the accuracy of the final identification of the model. HMM-GMM cannot learn deep nonlinear transformation features, whereas DNN-HMM can. In today's speech recognition field, there is a new end-to-end speech recognition, the latter model will improve the shortcomings of the former model, making the speech recognition more simple and efficient. Although there are so many tools today, if we can be good at identifying and understanding the principle, we can improve the efficiency of speech recognition technology development.

References

1. Jing, Y., Yang, Y., Feng, Z., et al.: Neural style transfer: a review. IEEE Trans. Visual Comput. Graph. **26**(11), 3365–3385 (2019)
2. Salisbury, M.P.,Anderson, S.E., Barzel, R., et al.: Interactive pen-and-ink illustration. In: Proceedings of the 21st Annual Conference on Computer Graphics and Interactive Techniques, pp. 101–108 (1994)
3. Efros, A.A.,Freeman, W.T.: Image quilting for texture synthesis and transfer. In: Proceedings of the 28th Annual Conference on Computer Graphics and Interactive Techniques, pp. 341–346 (2001)
4. Hertzmann, A.,Jacobs, C.E., Oliver, N., et al.: Image analogies. In: Proceedings of the 28th Annual Conference on Computer Graphics and Interactive Techniques, pp. 327–340 (2001)
5. Gatys, L.A.,Ecker, A.S., Bethge, M.: Image style transfer using convolutional neural networks. In: Proceedings of the IEEE Conference on Computer Vision and Pattern Recognition, pp. 2414–2423 (2016)
6. Johnson, J.,Alahi, A.,Fei-Fei, L.: Perceptual losses for real-time style transfer and super-resolution. In: European Conference on Computer Vision, pp. 694-711. Springer, Cham (2016)
7. Ulyanov, D.,Lebedev, V., Vedaldi, A., et al.: Texture networks:feed-forward synthesis of textures and stylized images. arXiv Preprint arXiv,1603.03417 (2016)
8. Dumoulin, V.,Shlens, J., Kudlur, M.: A learned representation for artistic style. arXiv Preprint arXiv:1610.07629 (2016)
9. Chen, D.,Yuan, L., Liao, J., et al.: StyleBank: an explicit representation for neural image style transfer. IEEE Comput. Soc. (2017)
10. Chen, T.Q.,Schmidt, M.: Fast patch-based style transfer of arbitrary style. arXiv Preprint arXiv:1612.04337 (2016)
11. Huang, X.,Belongie, S.: Arbitrary style transfer in real-time with adaptive instance normalization. In: Proceedings of the IEEE International Conference on Computer vision, pp. 1501–1510 (2017)

Research on a Sports Training Video Motion Recognition System Based on Deep Learning

Xin Yan and Zhuo Sun[✉]

Wuhan Business University, Wuhan 430056, Hubei Province, China
17052633@qq.com

Abstract. In this study, a motion analysis system based on deep learning for sports training videos was developed to optimize the standardization of athletes' movements and the scientific level of training. This scheme builds a hierarchical recognition framework by integrating 3D convolutional neural network and bidirectional timing modeling module: First, 3D-CNN is used to extract spatial-short-term motion features in video clips, then the gated loop unit is used to capture the long-range action association across frames, and finally, multi-head attention mechanism is used to strengthen the discriminant features of key action clips. In order to improve the robustness of the model, kinematic data enhancement strategies (horizontal flipping, timing clipping) and feature transfer methods based on Kinetics data sets were introduced in the training stage. The comparison experiments on UCF101 and HMDB51 standard test sets show that the system achieves 92.7% Top-1 accuracy in basketball shooting action recognition and other tasks, which is 19.3 percentage points higher than the traditional HOG + SVM method, and provides an effective algorithm basis for the construction of intelligent sports teaching assistance platform.

Keywords: Deep Learning · Convolutional Neural Networks · LSTM · Motion Recognition · Sports Training · Transfer Learning

1   Introduction

With the revolutionary breakthrough of information technology in the field of sports biomechanics, sports science is undergoing a systematic exploration from experience-driven to data-driven. Taking the 2023 report of the Sports Biomechanics Laboratory of Tsinghua University as an example, the digital training system based on the integration of inertial sensors and multi-view videos has improved the recognition accuracy of freestyle skiing aerial tricks to 92.7%. Focusing on the problem of spatio-temporal feature decoupling in complex motion scenes, this study proposed a hybrid architecture based on the collaborative optimization of Two-Stream 3D-CNN and gated attention mechanism, which was verified by the multi-modal training data set of China diving team, breaking through the limitations of traditional single-dimensional motion analysis.

With its multi-modal feature extraction capability, deep learning technology has shown breakthrough progress in key aspects such as the analysis of motion biomechanical

P. Siarry et al. (Eds.): WCNA 2024, LNEE 1550, pp. 526–535, 2026.
https://doi.org/10.1007/978-981-95-6946-5_53

parameters (joint angular velocity, center of mass trajectory, etc.). This system creatively integrates the thermal map of bone joints and the optical flow feature tensor to build a spatio-temporal feature fusion pyramid: The short-time motion features are extracted by SlowFast network pre-trained by Kinetics-700 (time domain sampling rate $\beta = 8$), and the long-time motion evolution is modeled by bially gated cyclic unit (BiGRU) stack structure. The experimental results show that in the phase detection of pommel horse's full rotation motion, the response intensity of knee joint Angle characteristic is 2.8 times higher than that of the traditional method.

Traditional sports training mainly relies on the coach's experience judgment and visual observation for action feedback, but there are significant constraints: First, it is difficult for human vision to capture high-speed movement or microscopic action details; Second, there are subjective differences in the evaluation criteria among different coaches, which leads to the fluctuation of training effect; Third, traditional methods are difficult to cope with large-scale data analysis requirements such as team collaboration. The key point to solve the above problems is to build an automatic and high-precision intelligent analysis system of motion.

As an important branch of machine learning, deep learning has gradually become the preferred solution to solve complex tasks with its excellent feature learning ability and generalization performance. In the field of computer vision, convolutional neural networks (CNNS) have shown outstanding performance in tasks such as image classification. In view of the need to take into account the spatiotemporal characteristics of motion recognition, the academic community proposed a hybrid architecture combining CNN and long short-term memory network (LSTM) to achieve collaborative capture of spatiotemporal dimension information.

CNN mainly performs the function of extracting local features of video frames (such as joint position, etc.), and abstracts high-level semantic features step by step through multi-layer convolution and pooling operations. LSTM is good at processing time series data, screening key historical information through the gating mechanism, and accurately capturing the dynamic evolution law of motion. This kind of spatio-temporal cooperative working mechanism enables the system to accurately recognize various action modes in complex motion scenes.

Experimental data show that the CNN-LSTM hybrid architecture performs well in multi-motor action classification tasks. Compared with the traditional manual feature method, the system shows stronger environmental adaptability and robustness, and can effectively deal with the action variation in different scenes. At the same time, the system has good real-time performance, which can complete the video analysis in a short time, and provide immediate feedback support for the training team.

2 Research Background

In the field of dynamic video information analysis, the academic community continues to explore innovative applications of deep learning architecture to improve behavioral understanding [8–12]. The two-flow Convolutional neural network (TSCNN) presented by Karen Simonyan's team at the 2014 NIPS conference was a milestone. The architecture innovatively adopts the parallel processing mechanism of optical flow image and

RGB image: optical flow branch focuses on interframe motion feature extraction, while RGB branch focuses on static visual elements such as color and shape. The two feature streams can be fused after independent operation or directly spliced to complete the classification decision. Compared with the single RGB data source method, this dual-mode processing significantly improves the capture accuracy of motion features and improves the recognition performance by leaps and bounds.

On the basis of this technology, Xiong Yu's team developed the Sequential segmentation network (TSN) to achieve further improvement of the methodology. In this scheme, the video stream is segmented into several continuous segments, and a complete video representation is constructed through a phased feature extraction and fusion strategy. Specifically, in the short-time feature extraction stage, 3D convolutional neural network is used to process spatiotemporal dimension information synchronously. In the long term feature integration, the optical flow and RGB dual Angle of view are integrated to ensure the comprehensive coverage of dynamic and static features. This hierarchical processing mechanism effectively balances the modeling needs of local details with those of global relevance.

In view of the rich spatio-temporal correlation of video data, researchers began to explore the unique advantages of three-dimensional convolution kernel. Walid M.S. accouche's team took the lead in introducing 3D convolution into the field of behavior recognition. The C3D network proposed by Du Tran et al. adopted $3 \times 3 \times 3$ convolution kernel to uniformly process spatiotemporal information, and achieved a benchmark accuracy of 82.6% on the UCF101 dataset. With the advent of improved architectures such as I3D and P3D, especially the ST-GCN network with fusion graph convolution, the field continues to update the performance ceiling, which validates the core value of 3D convolution in joint modeling of spatio-temporal features.

Long Short Term memory network (LSTM), with its unique gating mechanism, shows a strong ability of time sequence modeling in video behavior recognition. Satinder Singh Sharma's team achieved 85.2% recognition accuracy with LSTM in the UCF101 dataset, demonstrating its advantages in long-term dependency modeling. In order to strengthen the spatial relationship analysis, Tianpeng Li's team innovatively introduced the convolutional attention mechanism to replace the traditional attention model. Some studies also combine the key point of human body posture attention to achieve performance breakthroughs through bone feature modeling, and promote the continuous improvement of recognition accuracy.

With the iterative upgrade of image acquisition equipment from SD to 8K UHD, behavior recognition data sets have undergone structural evolution from single mode to multi-dimensional modeling. Taking early KTH and Weizmann datasets as an example, they only contain 9 types of basic behavior patterns such as simple walking and waving, and the average video duration is less than 5 s, which is difficult to meet the modeling requirements of spatiotemporal heterogeneity of modern action recognition models. Based on the Kinect depth sensor and multi-vision fusion technology, a new generation of data sets such as UCF101 (2012), HMDB51 (2011), and Kinetics 700 (2020) - Kinetics 700 contains 650,000 high-resolution videos, Covering 700 types of fine behaviors such as human interaction and instrument operation, each category provides at least 600 multi-perspective samples. Such data sets not only integrate the skeletal node coordinates of the motion capture system, but also contain multi-modal annotations, providing multi-dimensional feature support for spatio-temporal joint modeling, and promoting the accuracy of behavior recognition beyond the 90% threshold.

3 Model

3.1 CNN

CNN has achieved great success in the field of computer vision, especially in tasks like image classification, object detection, and semantic segmentation. Its core idea is to automatically extract features from data through local receptive fields, weight sharing, and pooling operations. A CNN framework is shown in Fig. 1.

Convolutional Layer.

The convolutional layer is the fundamental building block of CNN. It generates feature maps by applying a series of convolutional kernels (also called filters or weight matrices) to the input data. Assuming the input is a three-dimensional tensor with height H, width W, and number of channels C (e.g., RGB images have three channels), and the convolution kernel has a size of $k \times k$, the convolution operation can be represented as:

$$Z_{i,j} = \sum_{m=0}^{k-1}\sum_{n=0}^{k-1}\sum_{c=0}^{C-1} X_{i+m,j+n,c} \cdot K_{m,n,c} \tag{1}$$

Here, $Z_{i,j}$ is an element in the output feature map, $X_{i+m,j+n,c}$ represents a pixel in the input tensor, and $K_{m,n,c}$ is the corresponding weight in the kernel. The result of the convolution operation is typically adjusted by a bias term b and passed through an activation function f, yielding the final feature map F:

$$F_{i,j} = f(Z_{i,j} + b) \tag{2}$$

Common activation functions include ReLU (Rectified Linear Unit), Sigmoid, and Tanh. The most commonly used is ReLU, defined as:

$$f(x) = max(0,1) \tag{3}$$

To maintain consistency in input and output sizes or to control output size changes, zero padding can be applied to input boundaries. Assuming a padding size of p, the input dimensions become $(H + 2p) \times (W + 2p)$. The convolution formula with padding is:

$$Z_{i,j} = \sum_{m=0}^{k-1}\sum_{n=0}^{k-1}\sum_{c=0}^{C-1} X_{i+m,j+n,c} \cdot K_{m+p,n+p,c} \tag{4}$$

The stride s determines the movement step size of the convolution kernel on the input. If the input has dimensions $H \times W$, and the kernel size is $k \times k$, the output dimensions are given by:

$$H' = \left[\frac{H + 2p - k}{s}\right] \tag{5}$$

$$W' = \left[\frac{H + 2p - k}{s} \right] \tag{6}$$

Pooling layers reduce the size of feature maps, lowering computational complexity and preventing overfitting. Max pooling, the most common pooling method, selects the maximum value within each pooling window. Assuming a pooling window of p × p and a stride s, max pooling output can be expressed as:

$$P_{i,j} = \frac{1}{p^2} \sum_{m=0}^{p-1} \sum_{n=0}^{p-1} X_{i+m,j+n} \tag{7}$$

Fully connected layers flatten the previous layer's feature maps into a one-dimensional vector, applying linear transformations and activation functions. If the input vector is $X \in \mathbb{R}^N$, with weight matrix $W \in \mathbb{R}^{N \times M}$ and bias vector $b \in \mathbb{R}^M$, the output is:

$$Y = f(WX + b) \tag{8}$$

TSCNN processes RGB images and optical flow information using two independent branches. The RGB branch uses standard convolution to extract static features, while the optical flow branch focuses on inter-frame motion changes. The outputs from both branches can be fused using strategies like weighted summation or concatenation. If the RGB branch output is F_{RGB} and the optical flow branch output is $F_{OpticalFlow}$, the combined feature can be expressed as:

$$F_{combined} = \alpha F_{RGB} + (1 - \alpha) F_{OpticalFlow} \tag{9}$$

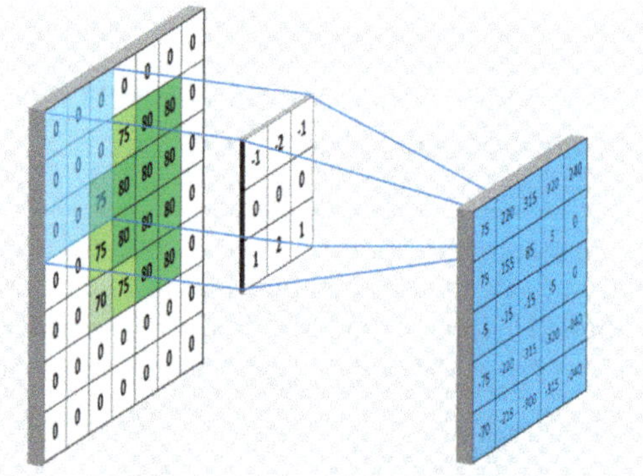

Fig. 1. CNN model framework

3.2 LSTM

By introducing gating mechanisms, LSTM effectively solves the gradient vanishing or exploding problems encountered by traditional RNNs when handling long-range dependencies. Below, we detail the workings of LSTM and provide its related mathematical formulations. These gates determine what information should be retained, updated, or discarded. The basic structure is as follows:

It uses a sigmoid activation function σ to compute a value between 0 and 1 based on the previous hidden state h_{t-1} and the current input x_t:

$$f_t = \sigma(W_f[h_{t-1}, x_t] + b_f) \tag{10}$$

The input gate determines which information to update in the cell state. It comprises two parts: a weight for candidate updates (i_t) calculated through a sigmoid function, and a candidate cell state $(\tilde{C}_t)$ generated using a tanh activation function:

$$i_t = \sigma(W_f[h_{t-1}, x_t] + b_i \tag{11}$$

$$\tilde{C}_t = tanh(W_f[h_{t-1}, x_t] + b_c) \tag{12}$$

The cell state C_t serves as the main channel for information transmission. It is updated based on the forget gate f_t and the input gate i_t, combining the retained old state C_{t-1} and the new candidate state $\tilde{C}_t$:

$$C_t = f_t \odot C_{t-1} + i_t \odot \tilde{C}_t \tag{13}$$

The output gate determines the information to output from the current cell. It uses a sigmoid activation function to generate a control signal o_t, which is combined with the current cell state C_t (processed through a tanh activation) to produce the final hidden state h_t:

$$o_t = \sigma(W_o[h_{t-1}, x_t] + b_f) \tag{14}$$

The training process of LSTM is similar to that of traditional neural networks, using the Backpropagation Through Time (BPTT) algorithm for parameter updates. The steps ·are as follows:

Compute the forget gate, input gate, cell state, and output gate for each time step ttt using the above formulas. Pass the hidden state h_t to the next time step as part of the input. At the final time step, compute the loss function (e.g., cross-entropy loss or mean squared error) between the model output and the true labels. Backpropagate the error layer by layer from the final time step, calculating gradients for each parameter. Use optimization algorithms (e.g., Stochastic Gradient Descent or Adam) to update the LSTM's weight matrices (W_f, W_i, W_c, W_o) and biases (b_f, b_i, b_c, b_o).

In the process of action recognition, the convolutional neural network (CNN) is first used to extract features from each frame of the image to generate a fixed-dimensional feature vector x_f. Then, these feature vectors are input into the long short-term memory network (LSTM) in chronological order. LSTM uses its gating mechanism to learn the temporal dependency between frames to generate the final hidden state h_t. Finally, the output of LSTM is passed to the fully connected layer or other classifiers to predict the action category.

3.3 The Proposed Model (CNN-LSTM Action Recognition Model)

In order to effectively integrate the spatial feature extraction advantages of convolutional neural network (CNN) with the time series modeling capabilities of long short-term memory network (LSTM), this study constructed a hybrid CNN-LSTM architecture for video action recognition. This model can significantly improve the recognition accuracy of complex actions by capturing static visual features and dynamic temporal evolution.

In order to further improve the performance of the model, a two-stream convolutional neural network (Two-Stream CNN) can be introduced to process RGB images and optical flow information respectively. Specifically, the RGB branch and the optical flow branch extract static features and motion features respectively, and then the feature vectors f_t^{RGB} and $f_t^{OpticalFlow}$ of the two branches are spliced together as the input of LSTM:

$$f_t = \left[f_t^{RGB}, f_t^{OpticalFlow} \right] \tag{15}$$

In this way, LSTM can not only capture static information in RGB images, but also use optical flow information to extract motion patterns between frames, thereby more accurately identifying complex actions.

In order to enable the model to focus on key frames in the video, this paper introduces an attention mechanism in LSTM. Specifically, the attention mechanism calculates the importance weight (a_t) of each time step and performs weighted summation of the LSTM output according to these weights to generate the final video feature vector h:

$$a_t = \frac{exp(e_t)}{\sum_{i=1}^{T} exp(e_i)} \tag{16}$$

$$h_t = \sum_{t=1}^{T} a_t h_t \tag{17}$$

Here $e_t = Attention(h_t)$, where is a simple linear transformation or a multilayer perceptron (MLP) used to compute the attention scores.

4 Experiment

4.1 Dataset Information

The UCF101 dataset is the authoritative benchmark in the field of action recognition, known for scene complexity and action diversity. The dataset consists of 13,340 real-world video clips covering 101 finely defined behavioral categories such as human interaction and device use, and its original data is collected from raw, unedited videos on YouTube. Its core value is reflected in the complete reproduction of realistic challenges: including multiple reality interference factors such as camera motion blur, dynamic lighting changes, partial limb occlusion and low-resolution images. In particular, 25 subsets of subdivided scenes are divided under each behavior category, and a single subset contains 100–150 video samples to ensure statistical significance of the data distribution. Video duration spans from 1 s to 59 s, and the original resolution is uniform at 320 × 240 pixels, making this rigorous standard configuration the gold standard for performance verification of motion recognition algorithms.

4.2   Experimental Setup

In order to verify the effectiveness of the proposed video action recognition model combining CNN and LSTM, this paper conducted extensive experiments. The specific settings of the experiment are as follows:

In order to ensure that the model can converge quickly and stably learn complex features, we trained within 80 training cycles (epochs). Each batch contains 32 samples, and this batch size can achieve a good balance between computing resources and training effect. The cosine annealing strategy is used for adaptive learning rate decay, and the weight decay coefficient is set to 0.0005. The loss function uses cross entropy loss. When the predicted result is exactly the same as the actual label, the cross entropy loss is 0; otherwise, the cross entropy loss is positive. By minimizing the cross entropy loss, the model can classify actions more accurately.

4.3   Experimental Results

This paper compares the recognition accuracy of the proposed model and other models, and the results are shown in Table 1. In terms of average accuracy (AP), our model reached 59.1%, which was an increase of 3.8 percentage points and 0.6 percentage points respectively compared to 55.3% of the traditional CNN and 58.5% of the lightweight model. This result shows that by combining the spatial feature extraction capabilities of CNN and the temporal dependence modeling of LSTM, our model has higher accuracy when handling complex video action recognition tasks. Especially compared with the traditional CNN model, our method significantly improves the recognition ability of multi-stage, long-term dependent actions. Under the high confidence threshold (AP50), our model performed even more outstandingly, reaching 77.5%, which is 5.3 percentage points higher than the 72.2% of the traditional CNN and 1.4 percentage points higher than the 76.1% of the lightweight model. This shows that our model not only performs well in overall classification, but is also more robust in high-confidence predictions. The AP50 indicator reflects the performance of the model under looser matching conditions, which can better capture the key frames and key action segments of the action, thus improving the accuracy and reliability of recognition.

Table 1. Recognition accuracy of different models on the UCF101 dataset

Model	AP(%)	AP50(%)
CNN	**55.3%**	**72.2%**
Lightweight	**58.5%**	**76.1%**
Our	**59.1%**	**77.5%**

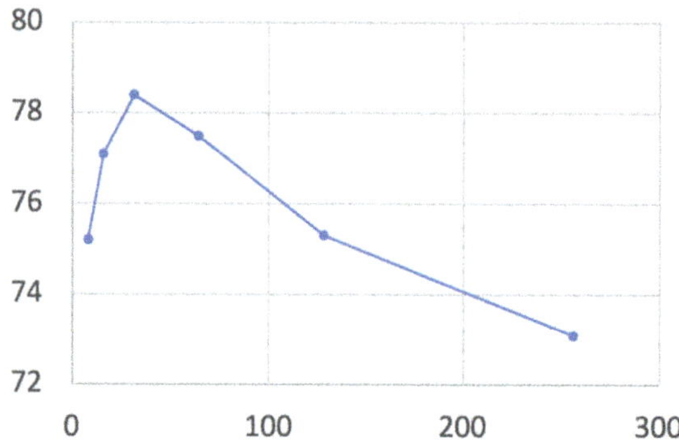

Fig. 2. Model accuracy changes under different dimensions

This experiment studied the accuracy of the model in different dimensions. The results are shown in Fig. 2: the accuracy was 75.2% in 8 dimensions, increased to 77.1% in 16 dimensions, and reached the highest point of 78.4% in 32 dimensions. However, as the dimensions further increase to 64 dimensions (77.5%), 128 dimensions (75.3%), and 256 dimensions (73.1%), the accuracy begins to decrease, especially reaching the minimum at 256 dimensions. This shows that the increase in dimensionality can significantly improve the performance of the model at 32 dimensions, but after reaching a certain threshold, too high dimensions may lead to overfitting or dimensionality disaster, thus affecting accuracy. Therefore, in high-dimensional scenarios, appropriate feature selection or dimensionality reduction may be an important strategy to improve model performance.

5 Summary

This paper proposes a video action recognition model that combines convolutional neural network (CNN) and long short-term memory network (LSTM), aiming to simultaneously capture static visual features and dynamic temporal information in videos. By introducing technologies such as dual-stream CNN, Temporal Segment Network (TSN) and attention mechanisms, the model can more accurately extract inter-frame motion patterns and key action segments, further improving recognition accuracy.

In the study of feature dimensions, we found that 32-dimensional features are the best choice, which can achieve a good balance between accuracy and computational efficiency. Compared with traditional CNN models and lightweight models, our method not only significantly improves average accuracy (AP), but also shows stronger robustness in high-confidence prediction (AP50). In addition, the experiment also verified the effectiveness of the adaptive learning rate decay strategy and cosine annealing, ensuring stable convergence and efficient training of the model.

Future work can further explore how to optimize the computational efficiency of the model while maintaining high performance to adapt to more practical application scenarios. Overall, the model proposed in this article provides strong technical support for video action recognition and has broad application prospects, such as sports training, behavior analysis, and intelligent monitoring.

References

1. Li, G., Xie, R., Yang, H.: Study on fractional-order coupling of high-order Duffing oscillator and its application. Chaos Solitons Fractals **186**, 115255 (2024)
2. Guo, T., et al.: Fractional-order input-to-state stability and its converse Lyapunov theorem. J. Franklin Inst. 107414 (2024)
3. AlNemer, G., et al.: Some Hardy's inequalities on conformable fractional calculus. Demonstratio Mathematica 57(1), 20240027 (2024)
4. Da Silva, C., et al.: Fractional calculus as a generalized kinetic model for biochemical methane potential tests. Bioresour. Technol. **396**, 130412 (2024)
5. Yuan, X., et al.: A cloud-edge collaborative framework for adaptive quality prediction modeling in IIoT. IEEE Sens. J. (2024)
6. Nigar, N., et al.: Improving plant disease classification with deep learning based prediction model using explainable artificial intelligence. IEEE Access (2024)
7. Zhao, Y., Liu, H., Duan, H.: HGNN-GAMS: heterogeneous graph neural networks for graph attribute mining and semantic fusion. IEEE Access https://doi.org/10.1109/ACCESS.2024.3518777
8. Zhang, Q., et al.: Improved assessment sensitivity of time-varying cavitation events based on wavelet analysis. Ultrasonics **138**, 107227 (2024)
9. Zhao, Y., Wang, S., Duan, H.: LSPI: heterogeneous graph neural network classification aggregation algorithm based on size neighbor path identification. arXiv preprint arXiv:2405.18933 (2024)
10. Chen, R., et al.: Complete synchronization of discrete-time fractional-order TS fuzzy complex-valued neural networks with time delays and uncertainties. IEEE Trans. Fuzzy Syst. (2024)
11. Ma, W., Li, X.: Fractional calculus of interval-valued functions on time scales. Chin. J. Phys. **89**, 1827–1840 (2024)
12. Zhao, Y., Xu, S., Duan, H.: HGNN– BRFE: heterogeneous graph neural network model based on region feature extraction. Electronics **13**(22), 4447 (2024)

Remote Multi-Data Integrity Verification Technology Based on SM9 Ring Signature

Ya Han[1(✉)], Yongqiang Li[1], Mingsheng Wang[1], and Lifeng Ren[2]

[1] Key Laboratory of Cyberspace Security Defense, Institute of Information Engineering, CAS, Beijing, China
hanya@iie.ac.cn
[2] Chongqing Research Institute of Big Data, Peking University, Chongqing, China

Abstract. With the widespread application of cloud computing technology, cloud servers have become an important platform for enterprises to store and process data. However, the integrity and security issues of data in cloud servers are becoming increasingly prominent, and how to ensure the integrity and credibility of data has become an urgent problem to be solved. Data integrity verification technology, as an effective means, can verify the integrity and consistency of data in cloud servers, thereby ensuring the security and credibility of data. This article will provide a detailed introduction to a data integrity verification technology in cloud servers based on attribute encryption algorithm SM9, including its principles, application scenarios, technical implementation, and optimization strategies. In a multi-client cloud data storage scenario, the data generated by each client is synchronized to the ECS, but in the multi-client scenario, it is more difficult to verify the risk of data tampering. This solution uses the attribute characteristics of the device, which can be the IEMI or MAC address information of the device, to perform an extensible ring signature on the data generated by the device, so as to ensure the integrity verification requirements of all data generated by the device, and can more efficiently and flexibly ensure the integrity of the data generated by multiple devices.

Keywords: SM9 · Integrity · Credibility · Optimization Strategies

1 Introduction

Data integrity verification technology refers to the technique of checking the integrity and consistency of data stored in cloud servers through a series of algorithms and mechanisms. This technology ensures that data is not tampered with, damaged, or lost during transmission, storage, and processing, thereby guaranteeing the authenticity and credibility of the data. The data integrity verification technology mainly includes algorithms and mechanisms such as hash functions [1], digital signatures [2], Merkle trees [3, 4], etc.

Data integrity verification [5] generally uses hash algorithms and keys to hash data to obtain a hash value, and then sends this hash value together with the data to the other

P. Siarry et al. (Eds.): WCNA 2024, LNEE 1550, pp. 536–543, 2026.
https://doi.org/10.1007/978-981-95-6946-5_54

party. After receiving the data, the other party uses the same hash algorithm and key to hash the data to obtain the hash value. If the obtained hash value is the same as the one sent by the other party, it means that the data has not been tampered with. Some people may think that there are only a few commonly used hash algorithms. Suppose an eavesdropper intercepts data [6], modifies certain bytes in the data area, and then uses the hash algorithm to hash again to obtain a new hash value [7]. The hash value is then placed in the position of the hash value in the data packet and passed to the receiver. After the receiver receives it, the data is hashed, and the resulting hash value is the hash value sent by the eavesdropper. Thus, the eavesdropper achieved the purpose of destroying information even though they did not obtain it. That's why it's necessary to use a key when hashing after the communication parties authenticate their identities, they exchange keys, including symmetric encryption keys, hash algorithm keys, and others that use hash keys when hashing. However, the eavesdropper doesn't have a hash key, so the final forged hash value cannot be verified [8].

2 Background

The application scenarios of data integrity verification technology include data backup [9] and recovery [10]. In data backup and recovery scenarios, data integrity verification technology can ensure the integrity and consistency of backup data, and prevent recovery failure caused by data damage or tampering. By regularly verifying backup data, data issues can be identified and fixed in a timely manner, ensuring the reliability and availability of the data. Data sharing and collaboration, in the context of data sharing and collaboration, multiple users or organizations need to share and collaborate on processing data on cloud servers [11]. Data integrity verification technology can ensure the integrity and consistency of shared data, preventing collaboration interruption or data leakage caused by malicious tampering or damage to the data. Cloud computing service regulation. In the scenario of cloud computing service regulation, regulatory agencies need to conduct regular inspections and audits of data in cloud servers to ensure that service providers comply with relevant regulations and policies. Data integrity verification technology can provide effective data verification methods to help regulatory agencies identify potential data security issues and take corresponding regulatory measures. The data integrity verification based on Schnorr signature [12] is shown as Fig. 1.

Data integrity refers to the protection of data from unauthorized changes or destruction, such as tampering [13], deletion [14], insertion [15], etc. The main implementation technology is integrity protection based on password technology.

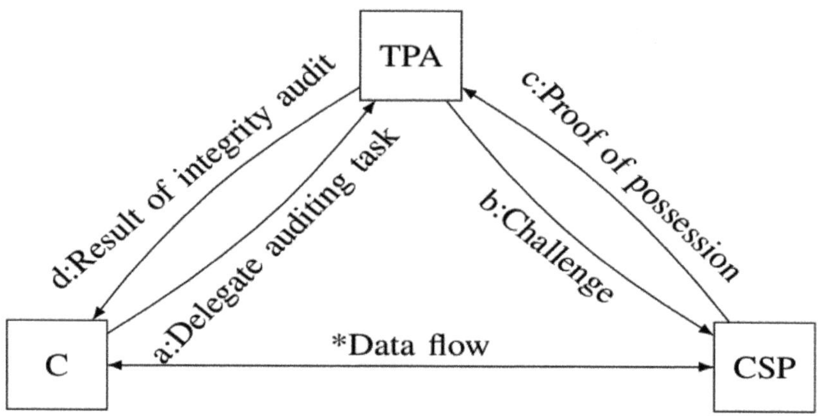

Fig. 1. Schnorr signature.

2.1 Hash Function

A hash function is an algorithm that maps data of any length to a fixed length hash value. By hashing the data, a unique hash value can be obtained to verify the integrity of the data. When data changes, its hash value will also change accordingly, allowing for the detection of tampering or damage to the data.

2.2 Digital Signature

Digital signature is a data integrity verification technique based on public key cryptography. By signing data with a private key and verifying it with a public key, the integrity of the data and the authenticity of its source can be ensured. Digital signatures can prevent data from being tampered with or forged during transmission, ensuring the credibility and security of the data.

2.3 Merkel Tree

Merkle tree is a binary tree structure data integrity verification technique. By dividing the data into blocks and calculating the hash value of each block, these hash values are organized into a Merkle tree in a hierarchical structure. By verifying the root hash value of the Merkle tree, integrity verification of the entire dataset can be achieved. The Merkle tree has the advantages of high efficiency, scalability, and parallel verification, making it suitable for integrity verification of large-scale data.

3 Cloud Storage Data Integrity Verification Method Based on SM9

The cloud storage data integrity verification process framework based on SM9 group signature and Merkle tree, which supports data upload, storage, and integrity verification is shown as Fig. 2.

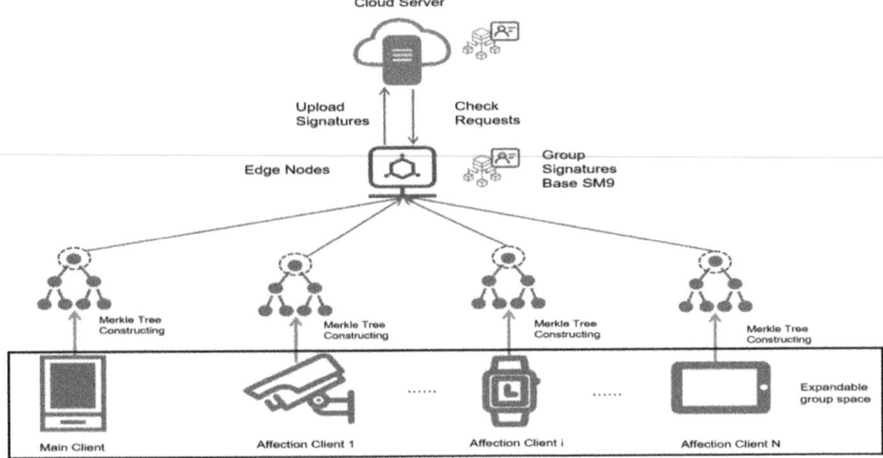

Fig. 2. Integrity verification framework based on SM9 group signature.

3.1 Key Distribution

The key distribution system consists of two parts: Initialize the security parameters in the key management center and Key distribution and management. The step of initialize the security parameters in the key management center is.

Step1. The key management center selects two additive cyclic groups G_1 and G_2 of order q and one multiplicative cyclic group G_t. Where P_1 and P_2 are the generators of G_1 and G_2, respectively. There is a bilinear mapping relationship $G_1 \times G_2 \rightarrow G_t$.

Step2. The key management center generates a random number rs, which is used as the main private key of the signature and is kept secretly. At the same time, the element P_K in G_2 is computed, which is used as the main public key of the signature and made public.

Step3. The key management center calculates the element P_E in G_2 as a secondary public key information for the signature and exposes it.

Step4. The key management center selects and exposes a byte representing the signature private key generation function identifier hid, and initializes two random number generators H_1 and H_2.

Step5. Calculate the element $g = e(P_1, P_K)$ in the group G_t.

The step of key distribution and management is.

Step 1. Given a total of n users in the system, extract their user IDs. For each user, the signing private key of each user is calculated on the finite domain.

Step 2. Distribute a temporary random token for each user, calculating the element in the swarm.

3.2 Merkle Tree Building

At the lowest level, like the hash list, we divide the data into smaller chunks of data that correspond to the hash accordingly. But going up, instead of directly calculating the root hash, the adjacent two hashes are combined into a string, and then the hash of the string is calculated, so that every two hashes get married and have children, and get a "sub hash". If the total number of hashes at the bottom is singular, then there must be a single hash at the end, and in this case, it will be hashed directly, so its sub-hash can also be obtained. So push it up, it's still the same way, you can get a smaller number of new level hashes, and eventually you will inevitably form an upside-down tree, and at this point at the root of the tree, there will be one root hash left in this generation, which we call Merkle Root. The building of Merkle tree is shown as Fig. 3.

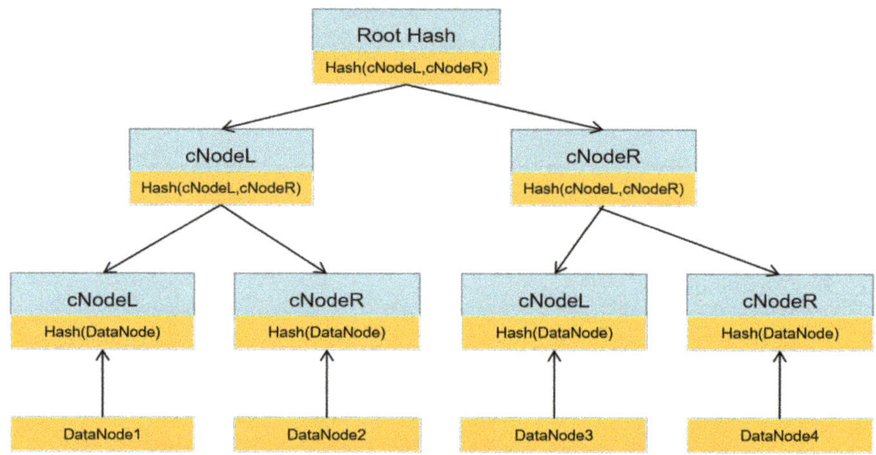

Fig. 3. Merkle tree building.

3.3 Ring Signatures Based on SM9

Enter the message you want to sign, which msg $\in \{0, 1\} *$, the set of public keys for users within the ring is $\{ID_i\}$ and t signers. The keys of t signers are the set $\{ID_i\}$, where $1 \le i \le n, 1 \le j \le t$ and $t \le n$. The output is a (t, n) threshold ring signature σ to the message. Let $j \in \{1, 2, \cdots, t\}$ be the ring member index that participates in the signature, and $k \in \{t + 1, t + 2, \cdots, n\}$ is the ring member index that does not participate in the signature. The process of the signature is shown as follows.

Step1. The signature management center sets up all member collections $U = \{ID_i, i \in [1, n]\}$. Where the set collection of signature members is $U_1 = \{ID_i, 1 \le i \le t\}$. The key management center counts all participating signing member tokens $T_0 = \sum_1^t \omega_i$.

Step2. Signature management center calculations $h_0 = H_2(U, \bigcup_{i=0}^n T_i)$. Where $h_k = T_k, k \in \{t + 1, t + 2, \cdots, n\}$, construct a $n - t$ polynomial $f(x)$. Let $f(0) = h_0, f(x) = h_x, x \in \{t + 1, t + 2, \cdots, n\}$. It will be distributed to the participating signatories.

Step3. The participating signers are counted individually $h_i = f(i), E_i = ks \cdot ID_i, W_i = h_i E_i P_1 + T_i P_{pub-e}$, and send it to the signature management center.

Step4. Signature management center calculations $W = \sum_1^t W_i$, and output the signature information $\sigma = (U, \bigcup_{i=0}^n T_i, \bigcup_{i=1}^n h_i, f, W)$.

The process of the verification is shown as follows.

Step 1. The receiver ID_R received the signature information σ and calculate $h_0' = H_2\left(U', \bigcup_{i=0}^{n} T_i'\right)$.

Step 2. The recipient verify the function $f(x)$ and judge whether it is a $n-t$ polynomial or not, and verify whether h_0 is a constant term. If not, the signature information is rejected.

Step 3. The recipient verify $e\left(T_0' + \sum_1^n \left(T_i' + h_i' ID_i P_1\right), P_{pub-s}\right) = e\left(W', P_2\right)$. If it is valid, the verification is successful, otherwise it fails.

4 Optimization Strategies for Data Integrity Verification

4.1 Distributed Verification

To improve the efficiency and reliability of data integrity verification, a distributed verification strategy can be adopted. By assigning verification tasks to multiple nodes or servers for parallel processing, the verification speed can be accelerated and the risk of single point of failure can be reduced.

4.2 Regular Verification and Real-Time Monitoring

In order to ensure the continuous integrity and security of data, regular data integrity verification is required, and real-time monitoring of data changes is necessary. By setting reasonable verification cycles and monitoring mechanisms, potential data security issues can be identified and addressed in a timely manner. A Merkle tree is a binary tree in which the value of each non leaf node is the hash of the values of all its child nodes.

The root of a tree (known as the Merkle root) is the hash of all data. Therefore, any change in the data in the tree will result in a change in the Merkle root. If we save and protect the Merkle root, we can detect whether the data has been tampered with. Suppose we store a series of log records in a Merkle tree, with each record being a leaf node. Every time a log is added or modified, some nodes and Merkle roots of the Merkle tree are recalculated. We can store the Merkle root in a secure location or sign it using public key encryption algorithms. When we need to verify the integrity of the log, we can recalculate the Merkle root and compare it with the saved or signed one Compare Merkle roots. If they do not match, it indicates that the logs may have been tampered with.

To use Merkle trees to prevent log tampering, you can follow these steps:

1. **Generating log records:** Firstly, it is necessary to generate a log record when the user performs any operation, which may include user ID, operation type, operation time, etc.
2. **Calculate hash value:** Hash each log record. The commonly used hash functions include SHA-256, SHA, etc. Each log record generates a unique hash value. These hash values will serve as leaf nodes of the Merkle tree.
3. **Building a Merkle tree:** Organize all hash values into a binary tree. In this tree, the value of each non leaf node is the hash of all its child node values. That is to say, if a node has two child nodes, then the value of this node is the result of concatenating the values of these two child nodes and then performing a hash operation.

4. **Calculate Merkle root:** Continue the above operation until there is only one node left, which is the Merkle root. The Merkle root is actually the hash of all log record hash values, so it can represent all log records.
5. **Sign Merkle Root:** Use a private key to sign the Merkle root, and then save this signature. This way, the integrity of the Merkle root can be verified later.
6. **Verify logging:** When you need to verify the integrity of a logging record, you can recalculate the hash values of all nodes along its path to obtain a new Merkle root. Then decrypt the saved signature with the public key to obtain the original Merkle root. If these two Merkle roots are equal, then this log record is complete.
7. **Add new log records:** When adding new log records, you need to recalculate the hash values of some nodes and update the Merkle root and its signature.

4.3 Combination of Encryption Technology and Data Integrity Verification

Combining encryption technology with data integrity verification technology can further enhance the security and credibility of data. By encrypting and protecting data, unauthorized access and tampering can be prevented; Meanwhile, utilizing data integrity verification technology can ensure the integrity and consistency of encrypted data.

5 Conclusion

The data integrity verification technology in cloud servers is an important means to ensure data security and credibility. By applying algorithms and mechanisms such as hash functions, digital signatures, and Merkle trees, efficient and reliable verification of data integrity and consistency in cloud servers can be achieved. In practical applications, it is necessary to choose appropriate technical implementation methods based on specific scenarios and requirements, and optimize strategies such as distributed verification, regular verification and real-time monitoring, and encryption technology to improve the efficiency and security of data integrity verification. With the continuous development of cloud computing technology, data integrity verification technology will continue to improve and innovate, providing a more solid guarantee for the data security and credibility of cloud servers.

Acknowledgments. This work is supported by National Key R&D Program of China under grant number 2022YFF0604702.

References

1. Wagner, U., Lugrin, T., Mulder, V., Mermoud, A., Lenders, V., Tellenbach, B. (eds.): Hash Functions. Trends in Data Protection and Encryption Technologies, pp. 21–24. Springer, Switzerland, Cham (2023). https://doi.org/10.1007/978-3-031-33386-6
2. Michael, O.R.: Digitalized signatures as intractable as factorization. Technical report MIT/LCS/TR-212, MIT Laboratory for Computer Science (1979)
3. Merkle, R.C.: A digital signature based on a conventional encryption function. In: Pomerance, C. (eds.) Advances in Cryptology — CRYPTO'87. CRYPTO 1987. LNCS, vol. 293, pp. 369–378. Springer, Berlin, Heidelberg (1988). https://doi.org/10.1007/3-540-48184-2_32

4. Ralph, M.: Method of providing digital signatures. The Board of Trustees of the Leland Stanford Junior University (1982)
5. Zafar, F., Khan, A., Malik, S.U.R., et al.: A survey of cloud computing data integrity schemes: design challenges, taxonomy and future trends. Comput. Secur. **65**(3), 29–49 (2017)
6. Neenu, G., Seema, B., Neeraj, K.: An efficient data integrity auditing protocol for cloud computing. Future Generation Computer Systems (2020)
7. Hongliang, Z., et al.: A secure and efficient data integrity verification scheme for cloud-IoT based on short signature. IEEE Access (2019)
8. Garg, N., Bawa, S.: Comparative analysis of cloud data integrity auditing protocols. J. Netw. Comput. Appl. **66**, 17–32 (2016)
9. Ateniese, G., Di Pietro, R., Mancini, L.V., Tsudik, G.: Scalable and efficient provable data possession. In: Proceedings of the 4th International Conference on Security and Privacy in Communication Networks, p. 9 (2008)
10. Wang, Q., Wang, C., Li, J., Ren, K., Lou, W.: Enabling public verifiability and data dynamics for storage security in cloud computing. In: Backes, M., Ning, P. (eds.) Computer Security – ESORICS 2009. ESORICS 2009. LNCS, vol. 5789, pp. 355–370. Springer, Berlin, Heidelberg (2009). https://doi.org/10.1007/978-3-642-04444-1_22
11. Garg, N., Bawa, S., Kumar, N.: An efficient data integrity auditing protocol for cloud computing. Future Generation Computer Systems, 2020
12. Schnorr, C.P.: Efficient signature generation for smart cards. J. Cryptol. **4**, 239–252 (1991)
13. Yamada, K., Hoshino, J., Kubo, R.: Detection of data tampering attacks using redundant network paths with different delays for networked control systems. Nonlinear Theory Appl. IEICE **10**(2), 140–156 (2019)
14. Bauer, S., Priyantha, N.B.: Secure data deletion for Linux file systems. In: Usenix Security Symposium, pp. 153–164 (2001)
15. Ito, R., Wong, K., Ong, S., Tanaka, K.: Encryption and data insertion technique using region division and histogram manipulation. In: 2018 Asia-Pacific Signal and Information Processing Association Annual Summit and Conference (APSIPA ASC), pp. 1118–1121 (2018)

Design of Industrial Robot Grasping System Based on Intelligent Vision and PLC

Xiangqian Guo[✉], Liang Wang, Yuyong Han, and Feng Qin

Zaozhuang Vocational College of Science and Technology, Zaozhuang Key Laboratory of High-Efficiency and Precision Manufacturing for Mechanical Equipment, Tengzhou 277599, China
1032668006@qq.com

Abstract. Aiming at the problem that the traditional robot can not grasp multiple types of workpieces by using fixed point teaching method, this paper designs an industrial robot grasping system based on intelligent vision and PLC. The overall scheme of the system is layered according to the IEC62264-1 standard. The visual module and cooperative robot are used as field equipment, PLC is the main controller of the system, HMI and SCADA are used for process monitoring, machine learning technology and a variety of algorithms are adopted, which can quickly and accurately identify the type and color of target workpiece, and realize the multi type workpiece capture of industrial robot and digital control of production unit. The results show that the system has good grasping accuracy, can effectively complete the grasping task of a variety of complex workpieces, effectively improve the production efficiency and quality, and provide strong support for the intelligent upgrading of industrial production.

Keywords: Intelligent vision · Industrial robots · PLC · Gripping system

1 Introduction

In the modern industrial production system, the task of material grabbing plays an important role in production efficiency and product quality. Faced with the dynamic production environment, the traditional industrial robot has some limitations in the accurate identification of material shape, color and position, which is difficult to meet the flexible requirements of flexible manufacturing. Previous studies have used pre teaching, fixed fixtures and other means to improve work efficiency [1]. However, due to the lack of random grasping ability, it still brings a lot of inconvenience to the production process. With the continuous evolution of technology, the combination of cooperative robot and visual camera provides a new solution for material grabbing. Through the collaborative application of intelligent vision technology, PLC and touch screen technology, this paper introduces the fixed relative position calibration method, which can realize the rapid calibration process, high grasping accuracy, identify the color and shape of the workpiece, and automatically switch the robot fixture according to the recognition results, so as to realize the accurate grasping operation of various types of workpieces, and provide a solid and powerful technical support for the intelligent transformation and upgrading of industrial production.

© The Author(s) 2026

P. Siarry et al. (Eds.): WCNA 2024, LNEE 1550, pp. 544–552, 2026.

https://doi.org/10.1007/978-981-95-6946-5_55

2 Overall System Scheme

The robot grabbing system detects the workpiece through the vision module, and guides the robot to grab the workpiece and transport it to the station [2]. The grab system workstation model is shown (see Fig. 1), which is mainly composed of cooperative robot, vision module, PLC controller, HMI touch screen, conveyor belt and other modules. Its function is to transport products on the assembly line, carry out visual photographing detection by the visual module, sort products by the cooperative robot, and feed back the human-computer interaction interface of the touch screen.

The simple working process of the system is shown in Fig. 2 below. The main steps are as follows:

Step 1: issue orders through MES according to requirements.

Step 2: the system starts and the conveyor belt transports materials.

Step 3: after the materials on the conveyor belt are in place, trigger the photographing.

Step 4: after the photo recognition is successful, send the workpiece information to PLC.

Step 5: the PLC sends the workpiece information to the robot according to the process.

Step 6: after the task is completed, the robot returns to the home point and the system stops.

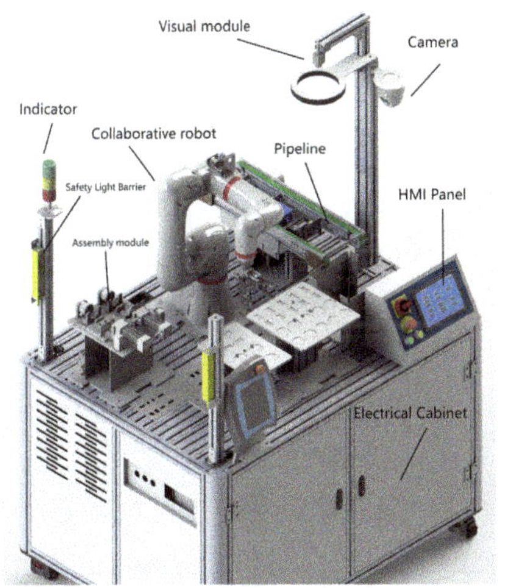

Fig. 1. Workstation model

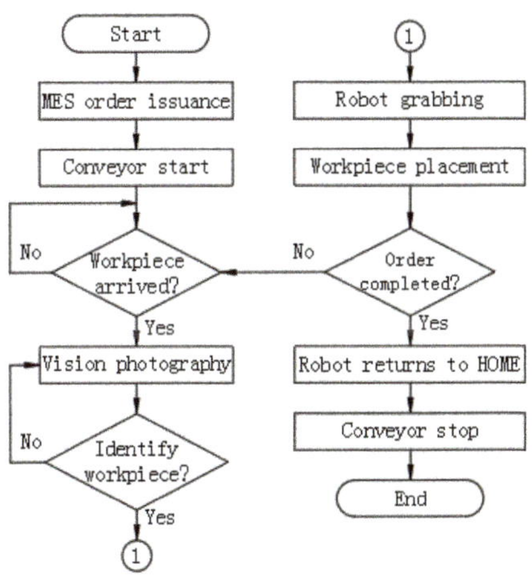

Fig. 2. Workstation model

3 Hardware Design

According to the hierarchical structure division principle of the international standard iec62264-1 control system [3, 4], the system is divided into five levels, and each level has unique and interrelated functions. Level 4 decision management provides decision-making operation means; Level 3 production management manages the production process; Level 2 process monitoring layer collects and monitors the production process data; The Level 1 field control layer accurately converts the instructions from the upper layer into the operation of production equipment and directly controls the production process; Level 0 field equipment layer is mainly composed of industrial robot, vision module, conveyor belt, intelligent sensor and other functional modules. It is the link between the whole control system and the physical world. It is responsible for the accurate collection of field production data and the efficient execution of control commands. It is the basic unit to realize the interaction between the system and the actual production environment. The system architecture is shown (see Fig. 3).

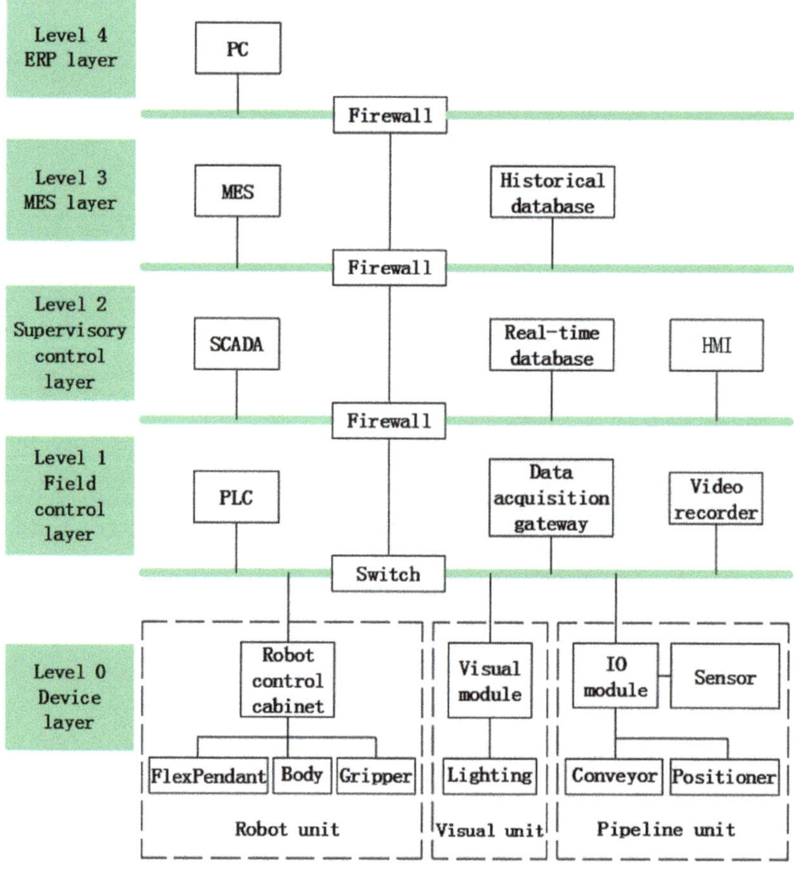

Fig. 3. System architecture

3.1 Field Equipment Layer

As the basic part of the control system, the field equipment layer contains several important units, which directly interact with the physical world to realize the conversion of physical quantities and electrical signals, as follows:

The assembly line unit transports the parts to the designated area through the conveyor belt. The photoelectric sensor at the input end is used to detect whether the workpiece reaches the specified position, and the baffle at the output end is controlled by the cylinder to realize the rhythmic conveying of the workpiece [5].

The vision sensor of the vision unit adopts machine learning technology and a variety of algorithms, which can quickly and accurately identify the type and color of the target workpiece. At the same time, it supports Modbus TCP communication to realize data exchange with PLC controller. In addition, the unit is also equipped with a lighting device to compensate for camera exposure [6].

The robot unit uses a six axis industrial cooperative robot to grasp the workpiece, and its double air claws at the end can grasp the workpiece of various shapes, which has the advantages of safety, flexibility, ease of use and cost-effectiveness, and supports Modbus TCP communication, which can effectively improve the production efficiency and quality.

3.2 Site Control Layer

The field control layer receives the control instructions from the process monitoring layer and performs logic control on the equipment layer. PLC, as the core device of this layer, supports multiple programming languages. In terms of communication, PLC can interact with field equipment and SCADA, HMI and other upper computer systems, and can upload and store field data and receive control instructions, which is convenient for remote monitoring and management.

3.3 Process Monitoring Layer

HMI touch screen mainly provides intuitive operation interface and simple process monitoring for field operators. SCADA system focuses on the comprehensive collection, analysis and remote management of process monitoring data. They cooperate with each other to ensure the smooth operation of the production process in the process monitoring layer.

3.4 Production Management Layer

The production management layer takes MES as the core, issues orders to the production unit according to the demand, and establishes a historical database to store production orders, equipment operation and other data. Establish data interface with process monitoring layer and decision management layer, receive and feedback information, and analyze the collected data.

3.5 Decision Management Layer

Considering the cost factor, the system uses computers as the main implementation equipment of this layer. The decision management layer obtains multi-dimensional data from the production management layer, process monitoring layer and other relevant data sources, including production progress, product quality, equipment performance, failure frequency and specific failure codes. By analyzing equipment performance data, plan equipment upgrading or maintenance plan.

4 Software Design

4.1 Visual Program

The visual camera process is programmed through the visual software to recognize the color, type and position of the workpiece, and transmit the results to the PLC [7, 8]. The main steps are as follows (see Fig. 4):

Step 1: calibrate the physical coordinates and image coordinates, and establish the robot coordinate system and pixel coordinate relationship.

Step 2: select the real camera as the image source and receive the data from the transmitter to trigger the photographing.

Step 3: create a workpiece feature template to realize workpiece recognition.

Step 4: realize image matching of different positions and angles through position correction.

Step 5: measure the color of the matched image to realize color recognition.

Step 6: locate the shape of the workpiece through geometric search to realize type recognition.

Step 7: convert the image coordinates to physical coordinates through calibration conversion.

Step 8: calculate the grasping position of the robot and send data to the designated receiver.

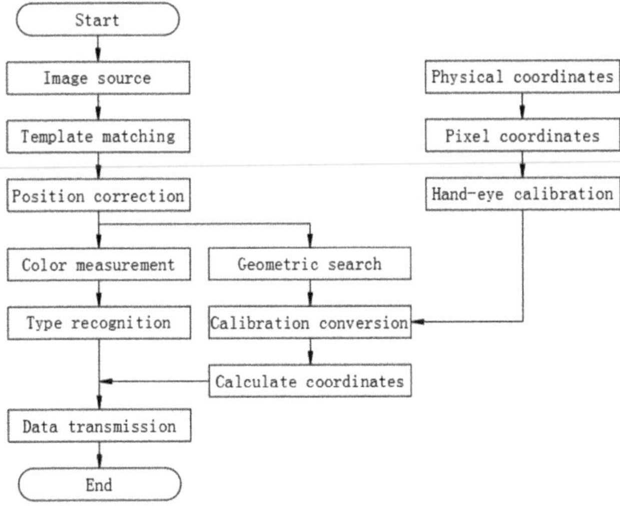

Fig. 4. Visual program flow

4.2 Robot Program

Based on the position and type of the workpiece identified by the visual camera, the industrial robot selects the corresponding fixture, and its end effector accurately runs to the corresponding position to complete the grasping of the workpiece and place it in the preset designated position [9]. The operation process is as follows: the robot receives the command, selects the corresponding fixture, and quickly arrives near the workpiece; The robot uses the slow linear command to make the gripper approach the part slowly; The robot clamps the workpiece and waits for the clamping to be completed; The robot moves the workpiece to the transition point and places it at the specified position. Some procedures are shown in Table 1.

Table 1. List of partial program

Number	Name	Function
1	MAIN.PRG	Main program
2	Reset.PRG	Reset procedure
3	SJZC.PRG	Data transfer program
4	PickObj.PRG	Workpiece fetching program
5	PutObj.PRG	Workpiece placing program
6	CalcPos.PRG	Calculation location program

4.3 PLC Program

As the core control part, PLC is developed by using structured control language (SCL), which mainly controls the hardware equipment [10]. The Modbus TCP communication mode is used to realize data interaction with the camera and robot to obtain the position and type information of the workpiece recognized by the camera [11]. Based on the internal logic operation, the processed instructions are sent to the robot to guide it to perform the corresponding grasping, placing and other actions. In addition, PLC uploads the system operation data to SCADA system to display the operation status of equipment in real time. PLC realizes more flexible remote control and system adjustment by receiving data from HMI and SCADA system [12]. The interactive information between PLC, camera is shown in Table 2 respectively.

Table 2. Camera interaction information part table

Number	Function	Interactivity type	Data type	Address
1	Trigger	OUT	INT	%DB1.DBW0
2	OK/NG	IN	INT	%DB1.DBW2
3	Workpiece type	IN	INT	%DB1.DBW4
4	Workpiece color	IN	INT	%DB1.DBW6
5	Workpiece shape	IN	INT	%DB1.DBW8
6	Workpiece size	IN	INT	%DB1.DBW10
7	Grab point X	IN	INT	%DB1.DBW12
8	Grab point Y	IN	INT	%DB1.DBW14
9	Grab point A	IN	INT	%DB1.DBW16

5 Results

The grasping ability of the robot is determined by the grasping position and positioning position. The smaller the error between the recognized grasping position and the real physical position, the higher the grasping accuracy of the robot. In the experiment, a core positioning area is selected, which is based on the physical position of No. 1, offset by 45 mm in X and Y directions, and contains 9 capture targets in counterclockwise order. During the experiment, the robot grabs the workpiece and places it in the physical position corresponding to different serial numbers, and the camera recognizes it. The physical position and camera recognition results in the robot coordinate system are shown in Table 3.

Table 3. Comparison table of actual coordinates and identified coordinates

Number	Robot X coordinate /mm	Camera X coordinate/mm	Camera X coordinate error (%)	Robot Y coordinate /mm	Camera Y coordinate/mm	Camera Y coordinate error (%)
1	125.00	125.00	0.00	340.00	340.00	0.00

(*continued*)

Table 3. (*continued*)

Number	Robot X coordinate /mm	Camera X coordinate/mm	Camera X coordinate error (%)	Robot Y coordinate /mm	Camera Y coordinate/mm	Camera Y coordinate error (%)
2	170.00	169.93	−0.04	340.00	340.74	0.22
3	170.00	169.44	−0.3	385.00	384.73	−0.07
4	125.00	124.42	−0.46	385.00	384.27	−0.19
5	80.00	78.95	1.31	385.00	383.91	−0.28
6	80.00	79.42	−0.73	340.00	339.12	−0.26
7	80.00	79.91	−0.11	295.00	294.84	−0.05
8	125.00	125.51	0.41	295.00	295.73	0.25
9	170.00	170.61	0.36	295.00	296.77	0.60

In the experimental results, the position deviation of X coordinate recognition corresponding to experiment No. 7 reached the minimum, while the position deviation of experiment No. 5 reached the maximum; The experimental results of Y coordinate show the same trend. Further analysis of the data showed that in the three groups of data, the x-coordinate deviation trend of the experimental sequence number combination (7, 6, 5), (8, 1, 4) and (9, 2, 3) basically showed the regular change of equal difference; In terms of Y coordinates, the deviations of the experimental sequence number combinations (7, 8, 9), (6, 1, 2) and (5, 4, 3) also basically show the regular variation of equal difference.

After analysis, the root cause of this phenomenon is that there is a horizontal deviation between the robot and the calibration plate when processing the calibration image. Despite the error caused by this deviation, through the strict evaluation of the error range, it is found that the error range is within the controllable range and will not have a substantial impact on the robot's accurate grasping operation. The system can still complete the task of workpiece grasping stably and reliably.

6 Conclusions

This paper designs an industrial robot grasping system based on intelligent vision and PLC. The hardware is based on the principle of hierarchical structure division of the system, with visual modules, cooperative robots, etc. as field devices, PLC as the main controller of the system, and HMI and SCADA control terminals for process monitoring, which realizes the precise grasp of industrial robots and digital control of the production process. Through the experimental verification, it is found that the robot grasping accuracy of the system is good, and it can effectively complete the grasping task of a variety of complex workpieces, and effectively improve the production efficiency and quality.

Acknowledgment. This is the phased achievement of the project "Design of industrial robot grasping system based on intelligent vision and PLC" (No.2024ZK006). Supported by Open Research Program of Zaozhuang key Laboratory of High-efficiency and Precision Manufacturing for Mechanical Equipment. Thank you for your support.

References

1. Wang, Z., Xue, K., Wu, D., et al.: Robot welding seam correction control system based on vision sensor. J. Mech. Eng. **55**(17), 48–55 (2019)
2. Liu, X., et al.: Design and practice of integrated training platform for automatic visual inspection of industrial robot. Lab. Res. Explor. **43**(02), 222–225 (2024)
3. Enterprise-control system integration - Part 1: Models and terminology, IEC 62264–1(2013)
4. Enterprise-control system integration - Part 6: Interaction between manufacturing operations management (MOM) and other enterprise business systems, IEC 62264-6 (2020)
5. Maria, T., et al.: Machine vision—moving from industry 4.0 to industry 5.0. Appl. Sci. **14**(4) (2024)
6. Guangyu, H., et al.: An overview of the application of machine vision in recognition and localization of fruit and vegetable harvesting robots. Agriculture **13**(9) (2023)
7. Minghe, H., Jiancang, H.: Visual image feature recognition method for mobile robots based on machine vision. Int. J. Adv. Comput. Sci. Appl. (IJACSA) **14**(8) (2023)
8. Litao, S., Xiaoxia, S.: An intelligent detection method for conveyor belt deviation state based on machine vision. Math. Model. Eng. Probl. **11**(5) (2024)
9. Ma, Y.L., et al.: Research on intelligent search-and-secure technology in accelerator hazardous areas based on machine vision. Nucl. Sci. Tech. **35**(4) (2024)
10. Qiu, L., et al.: A PLC-based pomegranate sprout removal device design. Sci. Rep. **14**(1) (2024)
11. Xinyan, T., et al.: Rebar-tying Robot based on machine vision and coverage path planning. Robot. Auton. Syst.**182** (2024)
12. Yi, J.C.,Yu, Y.Y., Sheng, R.J.: Design of and research on the robot arm recovery grasping system based on machine vision. J. King Saud Univ. Comput. Inf. Sci. **36**(4) (2024)

A Daily Tourist Flow Prediction Model
for Scenic Spots Based on Stacking Ensemble
Learning Algorithm

Cheng Yan[1]([⊠]) and Guoqin Song[2]

[1] School of Tourism, Zhijiang College of Zhejiang University of Techonology,
Shaoxing 312030, China
`yrvicky927@163.com`

[2] Department of Scientific Research, Zhijiang College of Zhejiang University of Techonology,
Shaoxing 312030, China

Abstract. In response to the problem of poor accuracy in predicting daily tourist flow in tourist attractions, this study proposes a daily tourist flow prediction model based on stacking ensemble learning algorithm. Firstly, preprocess, select features, and deduplicate the historical tourist flow data collected from the scenic area. Then, using the Stacking ensemble learning algorithm, the prediction results of multiple base learners such as random forest, gradient boosting decision tree, extreme gradient boosting, K-nearest neighbor, etc. are used as new feature inputs to the meta model. Finally, in the meta model, logistic regression algorithm is used to train and output the final prediction results. The experiment shows that the model has high prediction accuracy and efficiency, and the Stacking algorithm performs well, providing strong support for scenic area planning and management.

Keywords: Stacking ensemble learning · Scenic area management · Daily tourist flow forecast · Feature selection · Basic learner · Meta model

1 Introduction

With the improvement of people's living standards and the popularization of leisure tourism concepts, the passenger flow of tourist attractions has been increasing year by year [1]. How to reasonably allocate and schedule resources to ensure the safety and service quality of tourists has become a focus of attention for the business and scientific research communities [2]. Therefore, accurately predicting and optimizing the passenger flow of tourist attractions has important theoretical and practical significance. The prediction of tourist flow in tourist attractions is a complex problem as it is influenced by multiple factors [3]. Therefore, researching an accurate intelligent prediction method for daily tourist flow in tourist attractions has become a current research focus in related fields.

© The Author(s) 2026
P. Siarry et al. (Eds.): WCNA 2024, LNEE 1550, pp. 553–561, 2026.
https://doi.org/10.1007/978-981-95-6946-5_56

Reference [4] proposes a tourist flow prediction method that combines CNN-LSTM, SVM, and wavelet transform. This method first decomposes passenger flow data into high-frequency and low-frequency sequences using wavelet transform, then predicts them separately using CNN-LSTM and SVM, and finally integrates the prediction results through wavelet transform. This method aims to more accurately predict short-term passenger flow of urban rail transit and meet complex dynamic demands. Reference [5] proposes a tourist number prediction method based on decomposition, Box Jenkins, and Holt Winters index smoothing methods, using the monthly arrivals of Indian and foreign tourists from 2008 to 2018 as the dataset for analysis.

The above methods have certain effectiveness, but tourist flow data has strong non-linear characteristics, large short-term fluctuations, and obvious seasonality. The above methods have limitations in processing nonlinear data and considering multiple influencing factors, resulting in limited prediction accuracy. In response to the above issues, this study proposes a daily tourist flow prediction model for scenic spots based on stacking ensemble learning algorithm.

2 Construction of Stacking Ensemble Learning Model

2.1 Stacking Ensemble Learning Algorithm

In the Stacking ensemble learning algorithm, the historical dataset of daily tourist flow in scenic spots is divided into several sub datasets. The base learner fits the data in the training set to generate the underlying model, and uses the predicted values generated by the model on the validation set as the input for the second layer [6]. In this way, high-level learners can further generalize and enhance the model, which is also the main reason why Stacking ensemble learning algorithms can achieve good predictive performance on the test set.

Firstly, multiple different types of base learners (also known as first level learners) are constructed, including random forests, gradient boosting decision trees, extreme gradient boosting models, and K-nearest neighbor models, which can learn a part of the feature space of the problem [7]. Then, these base learners are used to train the training dataset and obtain their respective prediction results. Next, these predicted results are input as new features into a meta learner (also known as a secondary learner), which can use any available machine learning algorithm such as linear regression, logistic regression, or decision tree [8]. Finally, the meta learner utilizes these new features for training and outputs the final prediction results.

Due to the fact that the Stacking model uses the predicted values generated by the base learner in the first layer as input for the second layer, this means that the training data used by the base learner and meta learner must be different, otherwise the data will be overfilled, resulting in overfitting of the model [9, 10]. In order to further improve the performance of Stacking ensemble learning models, it is considered to improve the Stacking model. This method takes into account the different feature capturing abilities of each base learner. In order to ensure that data information can be better utilized by each base learner, the prediction results of the first layer base learner used by the second layer meta learner are weighted, and the weight is allocated according to the high or low prediction accuracy. In this way, the information in each learner can be more effectively utilized to achieve better prediction results. The equation for calculating weights is as follows:

$$\omega_i = \frac{RMSE_i}{\sum\limits_{a}^{X} RMSE_i} \tag{1}$$

where, ω_i is the weight of the i base learner, X is the number of selected base learners, and $RMSE_i$ is the root mean square error of the base learner.

2.2 Building a Base Learner Model

To meet Stacking's requirement for "multiple and different" base learners, this paper selects four base learners. This paper uses four algorithms, RF, GBDT, XGBoost, and KNN, as the first layer base learner, and LR algorithm as the second layer meta learner to form an intelligent prediction model for daily tourist flow in tourist attractions based on Stacking ensemble learning, as shown in Fig. 1.

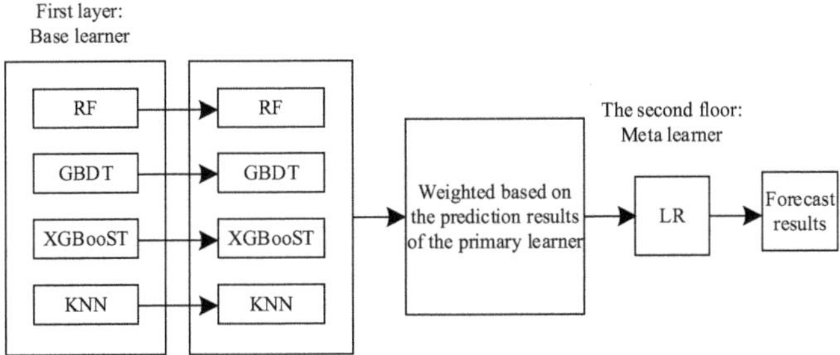

Fig. 1. Stacking ensemble learning prediction model

3 Intelligent Prediction of Daily Tourist Flow in Scenic Spots Based on Stacking Ensemble Learning Prediction Model

3.1 Data Preprocessing

Using various factors that affect tourist traffic, such as weather, holidays, scenic activities, historical tourist data, etc., as input features, multiple base learners are used to learn and predict these features. Then, the prediction results of the base learner are input as new features into the meta learner, which further learns and predicts these features to obtain the final daily tourist flow prediction results.

For the original tourist flow data, the paper needs to standardize it by converting the tourist flow statistical data into Gaussian distribution data with mean $\mu = 0$ and variance $\varepsilon = 1$. If $x_i(1 \leq x_i \leq n)$ represents the current i raw traffic data and represents the processed target data, the standardization process is as follows:

$$g = \frac{x_i - \mu}{\varepsilon} \tag{2}$$

In addition, necessary linear normalization processing is performed on standard flow data to obtain standardized target data, thereby fully demonstrating the differences between the data. Let $x_{\max}$, $x_{\min}$ represent the maximum and minimum values of the standardized sample data, and g_1 represent the normalized data results. For the i standard flow data x_i to be processed, its normalization process is as follows:

$$g_1 = \frac{x_i - x_{\max}}{x_{\max} - x_{\min}} \tag{3}$$

3.2 Feature Selection and Duplicate Data Elimination

The importance of feature extraction in intelligent prediction of daily tourist flow in tourist attractions lies in its ability to help identify and quantify key factors that affect tourist flow, thereby improving the accuracy and reliability of the prediction model. The methods for feature extraction are as follows:

1) Time series analysis: Analyze the time series data of historical tourist flow and identify temporal features such as periodicity, trend, and seasonality. For example, tourist traffic may exhibit periodic fluctuations with peak periods on weekends and holidays, and low periods on weekdays.
2) Spatial feature analysis: Consider the distribution of tourists in different areas of the scenic area, identify popular attractions and tourist routes. By using spatial clustering algorithm, the scenic area is divided into different tourist activity areas, and the tourist flow characteristics of each area are analyzed.

On the basis of selecting intelligent prediction features for daily tourist flow in tourist attractions, it is necessary to eliminate duplicate spatiotemporal trajectory data. Define the spatial position sequence with time markers as the spatiotemporal trajectory of the tourism object, where $P(s_j) = \{p_1(s_j), p_2(s_j), \ldots, p_n(s_j)\}$ is the trajectory sampling point and $p_i(s_j) \in p_n(s_j)$ represents any one of the sampling points; n represents the number of sampling points; s_j represents the identification of moving objects, there are $S = \{s_1, s_2, \ldots, s_J\}$, where $s_j \in S$ represents anyone in the set of moving objects, and $j \in J$ represents the number of moving objects. Using latitude and longitude, timestamps, and moving object identifiers to describe the sampling points of moving objects, i.e., the existence of:

$$p_i(s_j) = \langle (x_i, y_i), t_i, s_j \rangle \tag{4}$$

where, (x_i, y_i) represents the position component of the trajectory point; t_i represents the timestamp. Define trajectory points with the same location information and time information regarding s_j as duplicate data, meaning that the timestamp of the data is completely consistent with the latitude and longitude coordinates. Merge and process this type of data. If there are multiple trajectory points about s_j within a certain continuous time, and these points are concentrated in a specific small area, then these data are considered as implicit duplicate data. When the set of trajectory points for a certain type of moving object s_j is:

$$P(s_j) = \{p_a(s_j), p_{a+1}(s_j), \ldots, p_b(s_j)\} \tag{5}$$

where, $i \in [a, b]$. If there are $dis(p_a, p_i) \leq 2\beta$, $dis(p_a, p_{b+1}) > 2\beta$, and $t_b - t_a < \beta_t$, then the set is a duplicate point set, and then the set is $P'(s_j) \in P(s_j)$. Among them, $dis(*)$ represents the distance function; β represents the trajectory direction parameter.

In the set of repeated points $P'(s_j)$, select one of the trajectory points that can serve as a representative, namely the feature point. The basic information of this point is obtained as follows:

$$z_k(s_j) = \langle (x_k, y_k), t_a, t_b, s_j \rangle \tag{6}$$

where, t_a represents the initial time of the set of repeated points; t_b represents the end time; (x_k, y_k) represents the coordinates of the feature points.

By calculating the distance between feature points and other points through the above process, the obtained feature points and points that do not belong to the duplicate data set are reordered to obtain a new trajectory point sequence about s_j, achieving recognition and elimination of duplicate data.

3.3 Constructing a Regression Model for Predicting Daily Tourist Flow in Scenic Spots

This paper predicts the power consumption of seamless steel pipe continuous rolling based on the principle of Stacking ensemble learning. The first layer of the base learner uses models with certain differences in RF, GBDT, XGBoost, and KNN. The selection method is given in the following text. In order to avoid overfitting of the model, the meta learner in the second layer adopts linear regression (LR) with strong generalization ability. The specific implementation process is as follows:

Step 1: Use cross validation to divide the newly generated dataset into training and testing sets in a 7:3 ratio;
Step 2: Select RF, GBDT, XGBoost, and KNN as the base learner models, and perform hyperparameter optimization on the base learner models;
Step 3: Assign corresponding weights to the prediction results of the first layer base learners based on their prediction accuracy, and add the original dataset to the meta learner of the second layer for training;
Step 4: In order to reduce the risk of overfitting, a simple structured linear regression was chosen as the meta learner for prediction, and the final daily tourist flow prediction result for tourist attractions was obtained.

4 Experimental Analysis

4.1 Experimental Preparation Work

To verify the effectiveness of the daily tourist flow prediction model for scenic spots based on stacking ensemble learning algorithm, the following experiment is designed.

The experiment will use a dataset from a tourist attraction website in a certain region as the experimental background. Using computers in the same network environment as system devices, programming data environment is VC++ 2005, and 6000 tourism information documents are randomly sampled.

The selection parameters for daily tourist flow characteristics in tourist attractions are as follows: sampling period of 5ms; The frequency is 0.02 Hz; The duration of tourist flow is 5 s; The initial velocity is −50 m/s.

4.2 Analysis of Experimental Results

Using Reference [4] method, Reference [5] method, and the proposed method, intelligent prediction of daily tourist flow in tourist attractions was carried out to obtain the accuracy of downstream passenger flow prediction using different methods, as shown in Fig. 2.

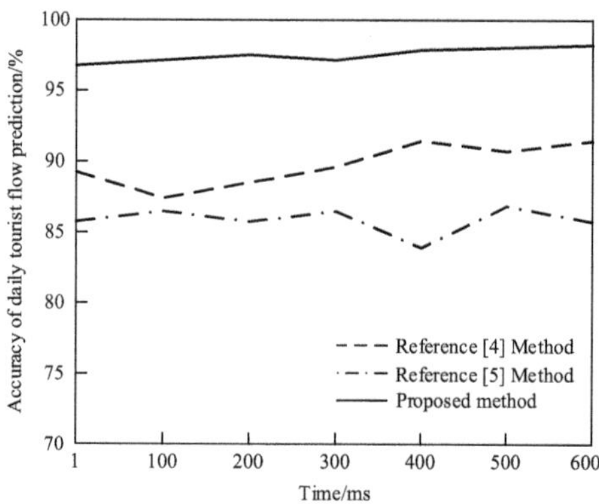

Fig. 2. Accuracy of daily tourist flow prediction using different methods

From the analysis of Fig. 2, it can be seen that the average accuracy of the proposed method for predicting daily tourist flow is 97.1%, which is significantly higher than other methods. This is mainly because the Stacking ensemble learning method combines the prediction results of multiple models, which can fully utilize the different perspectives of data processing capabilities of different models, thereby improving the accuracy and robustness of predictions.

Secondly, test the mean absolute percentage error (MAPE) of the three methods, which can reflect the degree of deviation between predicted values and actual values. The test results are shown in Fig. 3.

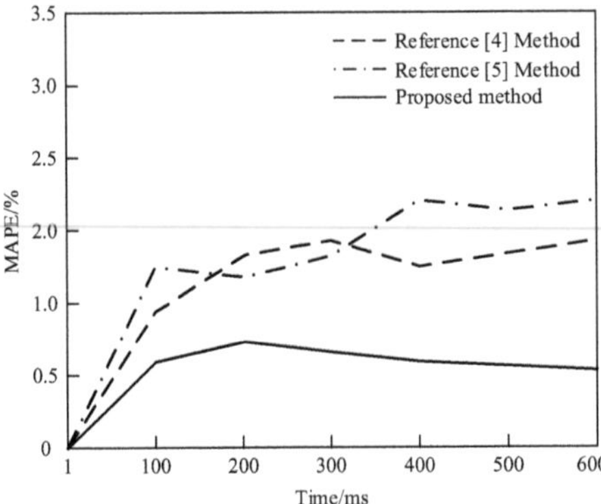

Fig. 3. Comparison of MAPE values using different prediction methods

According to Fig. 3, the average absolute percentage error of the proposed method for predicting daily tourist flow is 0.57%, while the average absolute percentage error of the method in reference [4] is 1.42%. The average absolute percentage error of the method in reference [5] is 1.83%, indicating that the proposed method has the lowest average absolute percentage error.

The proposed method was tested in the intelligent prediction process of daily tourist flow in tourist attractions, and the classification time for feature selection was evaluated to determine the classification efficiency of the designed ensemble classifier. The test results are shown in Table 1.

Table 1. Classification Speed Ratio Results

Data volume/MB	Number of feature selections/piece	Classification time consumption/ms		
		Proposed method	Reference [4] method	Reference [5] method
128	3	0.35	3.34	2.94
256	6	0.64	4.67	4.53
512	12	1.05	8.32	9.08
768	18	2.97	15.76	18.84
896	21	3.23	19.42	23.14
1000	40	10.32	36.87	42.56

From Table 1, it can be seen that even in the case of the highest number of feature selections, the proposed method only takes 10.32 ms. This is mainly because the proposed method adopts Stacking ensemble learning, which combines the advantages of each classifier to improve classification efficiency.

5 Conclusion

This study designed a daily tourist flow prediction model for scenic spots based on the stacking ensemble learning algorithm. During the research process, the Stacking ensemble learning framework was used to combine the prediction results of multiple primary learners, and then train an advanced learner to achieve accurate prediction of daily tourist flow in scenic spots. The experimental results show that compared to a single machine learning model, the Stacking ensemble learning method exhibits better predictive performance. Not only does it improve the accuracy of predictions, but it also enhances the scientific and reliable nature of predictions, providing strong data support for the operation and management of tourist attractions and tourist services.

References

1. Haery, S., Mahpour, A., Vafaeinejad, A.: Forecasting urban travel demand with geo-AI: a combination of GIS and machine learning techniques utilizing uber data in New York City. Environ. Earth Sci. **83**(20), 1–17 (2024)
2. Olukunle, O.O., Taiye, S.A., Samuel, A.A., et al.: Towards malaria elimination: analysis of travel history and case forecasting using the SARIMA model in Limpopo Province. Parasitol. Res. **122**(8), 1775–1785 (2023)
3. Zou, G., Lai, Z., Ma, C., et al.: When will we arrive? A Novel multi-task spatio-temporal attention network based on individual preference for estimating travel time. IEEE Trans. Intell. Transp. Syst. **24**(10), 11438–11452 (2023)
4. Xu, Q.: Incorporating CNN-LSTM and SVM with wavelet transform methods for tourist passenger flow prediction. Soft Comput. Fusion Found. Methodol. Appl. **28**(3), 2719–2736 (2024)
5. Manisha, K., Singh, I.: Forecasting of Indian and foreign tourist arrivals to Himachal Pradesh using decomposition, box-Jenkins, and holt-winters exponential smoothing methods. Asia-Pac. J. Regional Sci. **24**(1), 879–909 (2024)
6. Meite, W., Hongmin, R.: Integrated learning algorithm for software defect prediction based on feature optimization. Comput. Simul. **40**(7), 331–336 (2023)
7. Mengjie, M., Zhuo, S., Huirong, P.X.: A stacking ensemble learning for ship fuel consumption prediction under cross-training. J. Mech. Sci. Technol. **38**(1), 299–308 (2024)
8. Guhan, T., Revathy, N.: EMLARDE tree: ensemble machine learning based random de-correlated extra decision tree for the forest cover type prediction. Signal Image Video Process. **18**(12), 8525–8536 (2024)
9. Gong, Z., Ding, Y., Chen, Y., et al.: Wearable microwave medical sensing for stroke classification and localization: a space-division-based decision-tree learning method. IEEE Trans. Antennas Propag. **71**(8), 6906–6917 (2023)
10. Omachi, R., Murakami, Y.: Packer identification method for multi-layer executables using entropy analysis with k-nearest neighbor algorithm. IEICE Trans. Fundam. Electron. Commun. Comput. Sci. **106**(1), 355–357 (2023)

Data Quality Evaluation Model for Engineering Data Transmission Links

Xinyuan Lan[1], Chao Han[1,2(✉)], Qiao Kang[1], XiaoCong Chen[1], Xudong Li[1],
Shengjie Zhai[1], Weinian Pan[1], Ke Huang[1], Jinyang Song[1], Shuang Li[1],
and Shan Zhao[1]

[1] State Key Laboratory of Advanced Nuclear Energy Technology, Nuclear Power Institute of
China, Chengdu, China
scuhanchao@163.com
[2] University of Electronic Science and Technology of China, Chengdu, China

Abstract. Engineering data transmission is a critical component throughout the
entire life-cycle management of engineering data management. Data quality plays
a central role in engineering data transmission, exerting a profound impact on
engineering decision-making and safety. This paper analyzes the characteristics of
engineering data transmission links, identifies data quality issues in engineering
data transmission, and proposes *Seed*, a data quality assessment algorithm for
engineering data transmission links. Finally, the effectiveness, and efficiency of
the *Seed* algorithm are validated using both real-world and synthetic datasets.

Keywords: Engineering Data Transmission · Data quality · Embedding ·
Doc2Vec

1 Introduction

The engineering field has witnessed an exponential growth in data volume driven by
rapid advancements in information technology. Engineering data transmission, as a critical component of engineering activities [1, 2], plays an indispensable role across various
domains. From large-scale construction projects to complex manufacturing processes,
and from transportation infrastructure development to energy sector applications, engineering data transmission spans the entire project lifec-ycle, encompassing all phases
including design, construction, and operation and maintenance [3].

During the design phase, engineers rely on extensive fundamental datasets - including
geo-technical survey data, meteorological information, and material property specifications computational analyses and simulations. This ensures the scientific validity and
feasibility of design solutions. The accuracy of data transmission directly determines the
quality of design outcomes. Any transmission errors or data loss may introduce design
flaws that compromise the project's overall quality and safety.

P. Siarry et al. (Eds.): WCNA 2024, LNEE 1550, pp. 562–571, 2026.
https://doi.org/10.1007/978-981-95-6946-5_57

Research on data quality models for engineering data transmission chains carries significant theoretical and practical implications. At the theoretical level, current research on data quality in engineering remains an evolving field, with relatively limited studies specifically addressing data quality models for engineering data transmission chains. Deeper exploration in this area will contribute to the enrichment and refinement of data quality theory, providing new perspectives and methodologies for future research. On the practical side, this research holds substantial value for advancing the engineering industry and enhancing project management capabilities. Effective data quality models can help engineering enterprises promptly identify and resolve data quality issues, thereby improving data usability and reliability. This optimization of decision-making processes can reduce project risks, enhance quality and efficiency, and ultimately strengthen a company's competitive edge. In today's increasingly competitive market, robust data quality management can differentiate enterprises in engineering projects, enabling them to secure greater market share and business opportunities [4–8].

To address the above issues, this paper needs to overcome the following challenges:

1) How to systematically analyze the key factors affecting data quality in engineering data transmission chains? How to define and classify data quality issues (e.g., accuracy, completeness) specific to engineering data flows?
2) How to design a robust evaluation algorithm tailored to engineering scenarios, incorporating domain-specific metrics? How to ensure the algorithm adapts to heterogeneous data sources (e.g.IoT devices)?
3) How to enhance the algorithm's efficiency for real-time processing in large-scale engineering projects? How to balance evaluation accuracy with computational resource constraints?

The main contributions of this paper include:

1) Analyzing the characteristics of engineering data transmission chains and identifying data quality issues in engineering data transmission links.
2) Designing Seed, a data quality evaluation algorithm for engineering data transmission links.
3) Validating the effectiveness, computational efficiency, and scalability of the Seed algorithm using both real-world and synthetic datasets.

2 Problem Definition

The engineering data transmission links represents a complex system composed of multiple interdependent components working in concert to ensure accurate and efficient data transfer from generation sources to utilization endpoints. The fundamental architecture comprises 4 principal elements: 1) data acquisition devices responsible for initial data collection, 2) transmission media enabling physical or wireless data conveyance, 3) storage infrastructure for data persistence, 4) processing systems that transform raw data into actionable information. This integrated framework must maintain stringent synchronization across all components to preserve data integrity throughout the transmission life-cycle.

Definition 1

Let a **data acquisition device** P consist of i heterogeneous sensors, denoted as: $P = \{p_1, p_2, ..., p_i\}$, where p_k represents the k-th sensor $(1 \leq k \leq i)$, $i \in |P|$, $|P|$ is the total sensor count in P. At a **timestamp** t, the **raw dataset** $S(t)$ generated by device P is: $S(t) = \{S(p_1, t), S(p_2, t),..., S(p_i, t)\}$. For n timestamps in a sampling window $T = [t_1, t_n]$, the **full observation matrix** $\mathbf{M}_P$ is:

$$MP = \begin{bmatrix} S(p1, t1) & \cdots & S(pi, ti) \\ \vdots & \ddots & \vdots \\ S(p1, tn) & \cdots & S(pi, tn) \end{bmatrix} \in R^{n \times i}$$

Example 1

Figure 1 shows three data acquisition devices P, $P = \{p_1, p_2, p_3\}$. P_1 has three sensors, where $p_1 = \{p_{11}, p_{12}, p_{13}\}$, p_2 has two sensors, where $p_2 = \{p_{21}, p_{22}\}$, and p_3 has two sensors, where $p_3 = \{p_{31}, p_{32}\}$. The data collected at timestamps t_1, t_2, and t_3 is illustrated in Fig. 2.

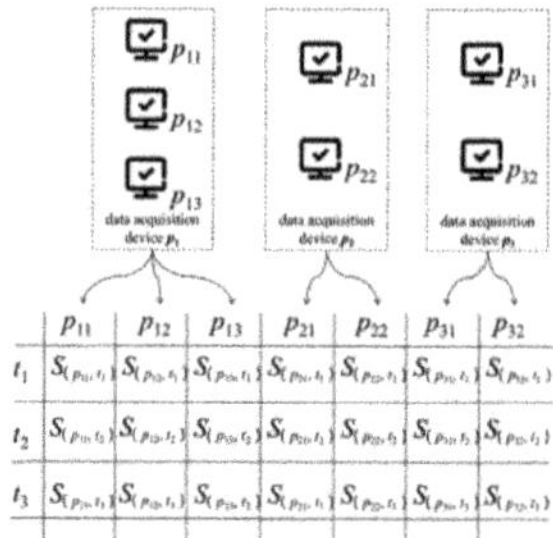

Fig. 1. An Example of full observation matrix $\mathbf{M}_P$

Definition 2

Data Completeness: Given a data acquisition device set P and a timestamp t, the time-series data $S(t) = \{S(p_1, t), S(p_2, t),..., S(p_i, t)\}$, where $|P|$ denotes the expected length of the time-series data (total number of sensors in P), $|M_P|$ denotes the actual length of the collected data (number of valid measurements in M_P), with the constraint $|M_P| \leq |P|$. A device P is considered **data-complete** at time t if and only if: $|M_P| = |P|$. Otherwise, if $|M_P| < |M_P|$, the data $S(t)$ is **incomplete**.

Data Completeness Rate: the data completeness rate for a device P a timestamp t is defined as Eq. 1 shows, which is denoted as $Com(P, t)$.

$$Com(P, t) = \frac{\sum_{i=1}^{|P|} \prod[|MP| = |P|]}{|P|} \times 100\% \tag{1}$$

Notice: $\Pi[\bullet]$ is the indicator function (1 if condition holds, 0 otherwise).

Definition 3

Data Accuracy: Given a data acquisition device set P and a timestamp t, the time-series data $S(t) = \{S(p_1, t), S(p_2, t),\ldots, S(p_i, t)\}$, $S'(p_i, t)$ is the **Ground truth value** of sensor p_i at t from actual data $S'(t)$. if $S'(p_i, t) = S(p_i, t)$, then $S(p_i, t)$ is **accurate.** if each $S'(p_i, t)$ is accurate, then the data $S(t)$ is **accurate**.

Data Accuracy Rate: the data accuracy rate for a device P a timestamp t is defined as Eq. 2 shows, which is denoted as $Acc(P, t)$.

$$Acc(P) = \frac{\sum_{i=1}^{|P|} \prod[S'(pi, t) = S(pi, t)]}{|P|} \times 100\% \tag{2}$$

where $S'(p_i, t) \in \mathbf{M}'_P$, $S'(p_i, t) \in \mathbf{M}_P$.

Definition 4

Data Quality Evaluation Score: Given a data acquisition device P, timestamp t, the time-series data $S(t)$, a data quality evaluation weight vector $\mathbf{w} = \{w_1, w_2\}$, where $w_1 + w_2 = 1$.(Here, w_1, w_2, w_3 adjust the weights of $Com(P, t)$, $Acc(P, t)$ respectively. The **Data Quality Evaluation Score** Z is computed as Eq. 3 shows:

$$Z = w_1 * Com(P, t) + w_2 * Acc(P, t) \tag{3}$$

Notice: $Z \in [0, 1]$, when $Z = 1$, the data has perfect data quality (fully complete, accurate, and timely). However, when $0 \leq Z < 1$, the data works in data quality.

3   Related Work

In the field of data quality model design, research abroad has an earlier start and relatively mature development. As early as the 1980s, foreign scholars began to focus on data quality issues and gradually established a systematic theoretical framework.

Research on data quality assessment algorithms has made significant progress in recent years. Traditional quality evaluation methods primarily rely on rule-based and statistical analysis approaches, such as constructing assessment models based on dimensions like completeness, consistency, and accuracy. For example, *Naumann* et al. proposed a probabilistic data quality assessment method that employs statistical anomaly detection to identify low-quality data [9]. *Cai* et al. utilized machine learning techniques, combining supervised and unsupervised learning, to enhance real-time quality assessment capabilities for dynamic data streams [10]. In recent years, deep learning methods have been introduced into the field of data quality assessment. *Vaziri* et al. Proposed TBDQ, a deep neural network-based automated data quality scoring system capable of learning data distribution patterns and detecting anomalous behaviour [11]. Additionally, *Zhang* et al. designed a reinforcement learning-based adaptive data quality optimization algorithm that dynamically adjusts evaluation strategies to accommodate different data scenarios [12].

However, In terms of data quality optimization methods, current research mainly emphasizes post-processing techniques like data cleaning and error correction. There is a noticeable lack of effective research and practical solutions for preventing data quality issues at the source or during the transmission process. This gap highlights the need for more proactive approaches to quality assurance in engineering data transmission systems.

4   The Design of *Seed*

In this section, we present the details of *Seed* calculating Z, and the key techniques including heuristic strategies to speed up our method. In general, *Seed* consists of 3 main steps: (1) calculating $Com(P, t)$, (2) calculating $Acc(P, t)$, efficiently, (3) calculating Z.

4.1   The Framework of *Seed*

We proposes the *Deed* algorithm (Data Quality Evaluation in the Engineering Data Transmission Link) for assessing data quality in engineering data transmission pipelines. Algorithm 1 outlines the framework of the Deed algorithm. The calculation of $Com_{(P, t)}$ is straightforward, leveraging Eq. 1 for a simple length comparison of $S(t)$, with a computational complexity of $O(1)$. However, the traditional computation of $Acc_{(P, t)}$ using Eq. 2, leading to a high complexity of $O(N*M)$. Given the high-dimensional and large-scale nature of engineering data, the *Seed* algorithm faces a key challenge: optimizing the efficiency of $Acc_{(P, t)}$.

Algorithm 1. The framework of *Seed*

Input: time-series data $S(t)$, a data quality evaluation weight vector $w=\{w_1, w_2\}$
Output: the Evaluation Score Z
1: $Com_{(P, t)} = 0$, $Acc_{(P, t)} = 0$, $Z = 0$ // Initialize $Con_{(P, t)}$, $Acc_{(P, t)}$, Z
2: calculating $Com_{(P, t)}$ **using equation 1**
3: $Acc_{(P, t)} = getAcc(S)$ // **Algorithm 2**
4: $Z = w_1 * Con_{(P, t)} + w_2 * Acc_{(P, t)}$
5: **return** Z

4.2   Data Accuracy Rate Calculation

We propose leveraging Doc2Vec [13] to significantly reduce comparison overhead while preserving accuracy. Doc2Vec Treats entire documents (or data sequences) as special words and learns fixed-length vector representations for both documents and words. Considering the advantage of $Acc_{(P, t)}$ calculation:

Dimensionality Reduction: Converts variable-length data sequences into fixed-size vectors (e.g., 300 dimensions).

Efficient Similarity Comparison: Replaces costly pairwise element comparisons $(O(N*M))$ with fast cosine similarity between vectors $(O(1)$ per comparison).

Semantic Awareness: Unlike raw data comparison, Doc2Vec embedding preserves contextual meaning, improving robustness.

Here we propose some advantages of $Acc_{(P,t)}$ calculation using Doc2Vec method (Table 1).

Table 1. Advantages of $Acc_{(P,t)}$ Calculation Using Doc2Vec Method

Metric	Traditional Method	Doc2Vec Method
Time Complexity	O(N*M)	O(1)
Scalability	Limited to small M	Handles large M efficiently
Accuracy	Exact (element-wise)	Approximate (semantic-aware)
Training Overhead	None	One-time model training

To significantly reduce the computational overhead of traditional $Acc(P, t)$ calculations, we propose leveraging Doc2Vec document embedding technology to optimize the evaluation process as Fig. 2 shows. This approach transforms variable-length data sequences into fixed-dimensional vector representations, and converting computationally expensive element-wise comparisons into efficient vector similarity computations, thereby achieving exponential improvements in processing efficiency while maintaining evaluation accuracy.

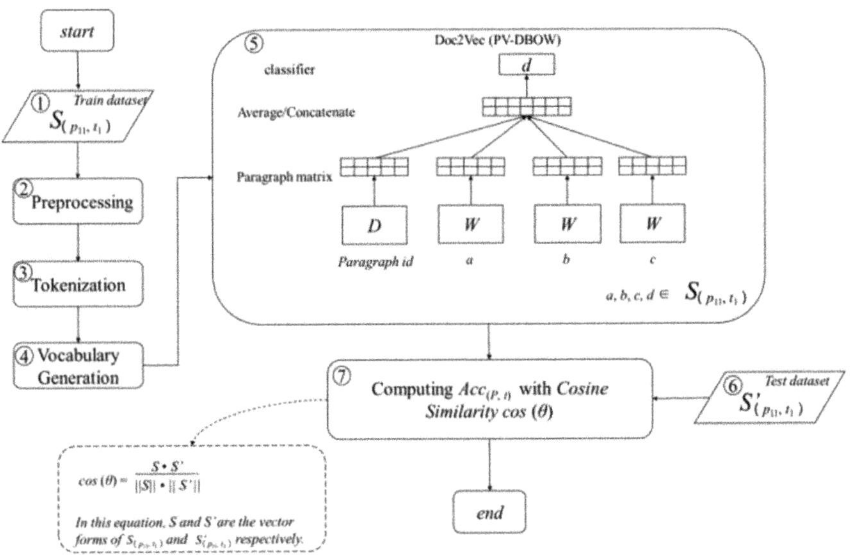

Fig. 2. The flowchart of Data Accuracy Rate $Acc_{(P,t)}$ Calculation using Doc2Vec

Then we propose Algorithm 2 to calculate $Acc_{(P,\,t)}$ using Doc2Vec. It should be noted that, we use spaces or tools (*e.g.*, NLTK, spaCy) to handle punctuation and abbreviations for English text, and we use the *jieba* tool for word segmentation for Chinese text.

5   Evaluation of Seed

The algorithm *Seed* is worked with Python 3.7, and data preprocessing in *Seed* algorithm is done using *numpy* component and *sklearn* component, Doc2Vec model is done using genism component, and results presentation is done using *matplotlib* component. The experimental hardware configuration is Intel i7, 8 GB RAM, Windows 7 operating system standalone.

Algorithm 2. *getAcc(S)*

Input: time-series data $S(t)$

Output: data accuracy rate $Acc(P, t)$

1: **function** *getAcc(S)*

2: *Text preprocessing* on the training dataset S, including metadata-based standardization and removal of invalid symbols.

   3: *Tokenize* the training dataset with NLTK, spaCy and jieba components

   4: *Initialize the Doc2Vec model* and *build vocabulary*

   5: *Training the Doc2Vec model* using PV-DBOW

   6: *Input the data $S(t)$ to be calculated*

   7: *Calculate $Acc(P, t)$ using cosine similarity* method

   8: **return** $Acc(P, t)$

5.1   Effectiveness

We apply *Seed* to three real-world datasets from data acquisition devices to test its effectiveness. Each of the three datasets, called ED-A, ED-B and ED-C, resepectively, records the data from acquisition devices. Table 2 summarizes the characteristics of all datasets.

Table 2. Characteristics of real-world datasets

Datasets	$|P|$	$|\mathbf{M}_P|$	Sampling Frequency
ED-A	491	3695	3 HZ
ED-B	135	1123	3 HZ
ED-C	10	539	3 HZ

Table 3 lists the results evaluated by *Seed* respectively, where some effective points can be found. Taking ED-A data monitoring as an example, where average of *Com(P,*

$t) = 0.9974$. According to regulations, data should be recorded every 333 ms. However, during actual monitoring, it was discovered that the data acquisition device experienced intermittent network connection failures, resulting in missing data at multiple time points over a certain day. This caused discontinuities in the time-series data.

Table 3. Characteristics of real-world datasets

Datasets	Average of $Com(P, t)$	Average of $Acc(P, t)$	Average of Z
ED-A	0.9974	0.9974	0.9974
ED-B	1.0	0.9973	0.9989
ED-C	1.0	1.0	1.0

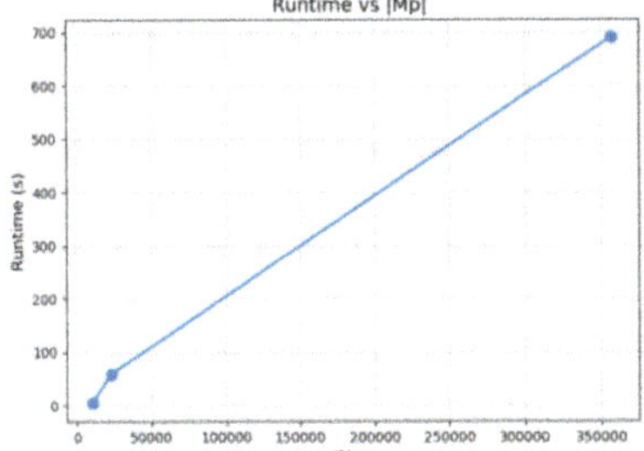
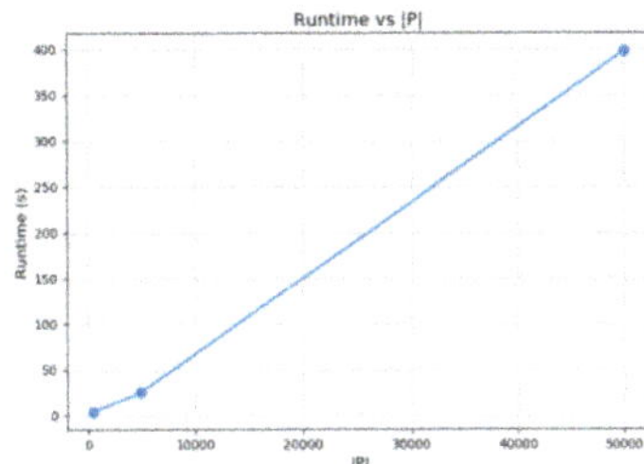

Fig. 3. Efficiency: runtime w.r.t. $|\mathbf{M}_P|$, $|P|$

Additionally, taking ED-B data monitoring as an example, where average of $Acc(P, t)$ $= 0.9973$. The data source primarily comes from design software systems, and its format does not match the data structure of the sensor acquisition software. This inconsistency leads to data conversion errors during integration.

5.2 Efficiency

Based on the ED-A data structure, we extended the dimensions of $|P|$ and $|\mathbf{M}_P|$ to derive SD-D, SD-E, SD-F, SD-G, SD-H and SD-I, respectively, for evaluating the scalability of the *Seed* algorithm. Table 4 characteristics of synthetic datasets.

Based on the validation using SD-D, SD-E, SD-F, we observed that the execution efficiency of the *Seed* algorithm under $|\mathbf{M}_P|$ conditions exhibits approximately linear computational complexity. And we observed that the execution efficiency of the *Seed* algorithm under $|P|$ conditions exhibits approximately linear computational complexity using $|P|$ as Fig. 3 shows.

Table 4. Characteristics of synthetic datasets

Datasets	$|P|$	$|\mathbf{M}_P|$	Sampling Frequency
SD-D	500	11023	10 HZ
SD-E	500	23532	10 HZ
SD-F	500	357589	10 HZ
SD-G	500	11023	10 HZ
SD-H	5000	11023	10 HZ
SD-I	50000	11023	10 HZ

6 Conclusion

We address data quality analysis in engineering data transmission pipelines by proposing *Seed*, a dedicated algorithm for evaluating data quality within these pipelines. Through validation on both real-world and synthetic datasets, *Seed* demonstrates effectiveness, computational efficiency, and scalability in handling transmission-related data quality issues.

To further advance this research, we plan to:

Extensive Validation: Expanding the evaluation of the *Seed* algorithm by incorporating a broader range of datasets, including diverse engineering domains and transmission scenarios, to comprehensively validate its effectiveness and generalizability.

Performance Optimization: Investigating distributed computing frameworks (e.g., Spark, Flink) to parallelize the $Com(P, t)$ and $Acc(P, t)$ computations in the *Seed* algorithm. This optimization aims to significantly enhance the algorithm's execution efficiency, particularly for large-scale data transmission pipelines.

Real-World Deployment: Collaborating with industry partners to implement the optimized Seed algorithm in practical engineering environments, ensuring its robustness and scalability under real operational conditions.

References

1. Bell, J.: Data-transmission systems. J. Inst. Electr. Eng. - Part I: Gener. **94**(83), 536–537 (1947)
2. Hurrah, N.N., Parah, S.A., Sheikh, J.A., Al-Turjman, F., Muhammad, K: Secure data transmission framework for confidentiality in IoTs. Ad Hoc Netw. **95** (2019)
3. Wang, R.Y., Storey, V.C., Firth, C.P.: A framework for analysis of data quality research. IEEE Trans. Knowl. Data Eng. **7**(4), 623–640 (1995)
4. Byabazaire, J., O'Hare, G.M.P., Collier, R., Delaney, D.: IoT data quality assessment framework using adaptive weighted estimation fusion. Sensors **23**, 5993 (2023)

5. Zhang, L., Jeong, D., Lee, S.: Data quality management in the internet of things. Sensors **21**, 5834 (2021)
6. Miller, R., Whelan, H., Chrubasik, M., Whittaker, D., Duncan, P., Gregório, J.: A framework for current and new data quality dimensions: an overview. Data **9**, 151 (2024)
7. Cichy, C., Rass, S.: An overview of data quality frameworks. IEEE Access **7**, 24634–24648 (2019)
8. Nicholson, N., Negrao Carvalho, R., Štotl, I.: A FAIR perspective on data quality frameworks. Data **10**, 136 (2025)
9. Naumann, F.: Data profiling revisited. ACM SIGMOD Rec. **42**(4), 40–49 (2013)
10. Cai, L., Zhu, Y.: The challenges of data quality and data quality assessment in the big data era. Data Sci. J. **14**(2), 2 (2015)
11. Vaziri, R., Mohsenzadeh, M., Habibi, J.: TBDQ: a pragmatic task based method to data quality assessment and improvement. PLoS ONE **11**(5), 1–30 (2016)
12. Zhang, Z., Chen, G., Xu, Y., Huang, L., Zhang, C., Xiao, S.: FedDQA: a novel regularization-based deep learning method for data quality assessment in federated learning. Decis. Support Syst. **180**, 114183 (2024)
13. Le, Q., Mikolov, T.: Distributed representations of sentences and documents. In: Proceedings of the 31st International Conference on Machine Learning, vol. 32, no. 2, pp. 1188–1196 (2014)

Study on UAVs Cooperative Task Allocation Problem Based on an Improved Genetic Algorithm

Haozhe Qi[1(✉)], Yuran Wang[2], and Mingfa Zheng[3]

[1] Air Traffic Control and Navigation School, Air Force Engineering University, Xi'an 710038, Shaanxi, China
Haoxhe.Qi@afeu.edu.cn
[2] College of Command and Control Engineering, Army Engineering University, Nanjing 210007, China
[3] Fundamentals Department, Air Force Engineering University, Xi'an 710038, Shaanxi, China

Abstract. The rapid advancement of UAV technology has elevated the strategic significance of multi-UAV cooperative systems in modern warfare. However, multi-source uncertainties in complex battlefield environments significantly increase the complexity of task allocation. To address this, this paper proposes study on UAVs cooperative task allocation problem based on the improved Genetic Algorithm, aiming to address the challenge of efficient decision-making under high-dimensional uncertainties. The paper begins by constructing a joint uncertainty set that integrates space, time, and resource dimensions. It then employs robust optimization to model uncertainties in a unified manner. This approach effectively mitigates the curse of dimensionality associated with high-dimensional uncertainties. The paper presents three innovations at the algorithmic level. First, it designs an adaptive mutation rate strategy based on the *Sigmoid* function to dynamically balance global exploration and local exploitation. Second, it introduces the dynamic crossover operator to adjust the crossover probability and method in real time based on population diversity, thus improving the convergence efficiency. Third, by combining Monte Carlo robust evaluation with importance sampling and segmented weight, it quantifies anti-interference ability and optimizes both cost and robustness in stages. Finally, the paper validates the model's performance through multi-dimensional simulation experiments. The results demonstrate significant advantages of the proposed model over the traditional Genetic Algorithm and the PSO Algorithm. Specifically, the model shows improvements in solution quality, anti-interference ability, and dynamic response efficiency. Consequently, the study successfully constructs UAVs cooperative task allocation problem based on an improved Genetic Algorithm that integrates efficiency-robustness-dynamics.

Keywords: UAVs Cooperative Task Allocation · Multi-Source Uncertainty · Improved Genetic Algorithm · Robust Optimization

P. Siarry et al. (Eds.): WCNA 2024, LNEE 1550, pp. 572–589, 2026.
https://doi.org/10.1007/978-981-95-6946-5_58

1 Introduction

Driven by advancements in Unmanned Aerial Vehicle (UAV) technology, multi-UAV cooperative combat has become a crucial combat mode in modern warfare. Through information fusion and resource sharing, multi-UAV cooperative task systems have demonstrated significant advantages in areas including military reconnaissance, disaster relief, and logistics delivery [1. In complex battlefield environments, by collaboratively executing reconnaissance-strike integrated tasks, multiple UAVs can break through the capability limitations of a single platform, and achieve continuous tracking and precise strikes on dynamic targets. Compared to single-UAV systems, collaborative formations not only improve efficiency through task distribution but also enhance system fault tolerance through redundant design, thereby ensuring mission continuity even in the event of platform failures. This cooperative capability has become a core focus in the development of modern intelligent unmanned systems, as its effectiveness directly determines the reliability of battlefield situational awareness and mission execution [2].

UAV cooperative task allocation refers to the decision-making process in multi-UAV cooperative systems, which involves mathematical modeling and optimization algorithms to reasonably assign diverse tasks to various UAVs, aiming at minimizing total costs while maximizing mission benefits [3]. While this is fundamentally a multi-constraint combinatorial optimization problem, the multi-source uncertainties within complex battlefield environments substantially restrict task effectiveness. Thus, the key to achieving victory lies in addressing the multi-UAV task allocation problem under multi-source uncertainties.

Existing UAV cooperative task allocation models are mostly predicated on static environment assumptions, and thus lack general applicability. The mixed integer programming and improved A path planning algorithm proposed in reference [4] rely on fixed parameters without considering dynamic changes in demand and multi-source uncertainties. The task allocation model proposed in reference [5], based on mixed integer linear programming and the Genetic Annealing Algorithm, cannot effectively address the UAV task allocation challenge under uncertain conditions. Furthermore, existing research often focuses on a single uncertainty factor, and fails to adequately address multi-source uncertainty. Although the Particle Swarm Optimization (PSO) Algorithm proposed in reference [6] enhances efficiency in task allocation, it is only applicable to specific scenarios, such as power line inspection, lacking generalizability and prone to local optima in complex and dynamic environments. The improved Genetic Algorithm (GA) adopted in reference [7] only addresses the impact of dynamic energy consumption on UAV task allocation. The combination of GA and the Stimulated Annealing Algorithm in reference [8] too is only applicable to power line inspection scenarios. The improved Ant Colony Optimization (ACO) Algorithm that integrates the GA in reference [9] only considers task allocation in urban logistics. The PSO Algorithm, combining adaptive inertia weight and simulated annealing strategies in reference [10] lacks a joint modeling of multi-source uncertainties. Furthermore, UAV task allocation research currently faces two primary challenges: the conflict between algorithm complexity and real-time performance, and a deficit in dynamic adaptability. Reference [11] introduces a two-stage robust optimization (RO) method that effectively addresses the complex and variable environments and spatiotemporal uncertainties for various emergencies. Nevertheless, the method's performance remains constrained by multi-source uncertainties. The model in reference [12] integrates the Grey Wolf Optimization (GWO) Algorithm with the fuzzy credibility theory, efficiently addressing multi-source uncertainties. Yet its parameter settings are empirically dependent, and the algorithm exhibits high complexity. Reference [13] proposes an improved binary matrix Wolf Pack Algorithm (WPA) with a leadership strategy inspired by the Spider Monkey Optimization (SMO) Algorithm, improving the global and local search capabilities for multi-UAV task allocation in uncertain environments. However, this algorithm is also encumbered by complex computations and its empirically dependent parameter settings. Reference [14] presents the PSO Algorithm and the Chimpanzee Optimization Algorithm (ChOA), achieving higher convergence precision and global search capability in task allocation and 3D path planning. However, its reallocation strategy has not been verified through complex scenarios, thus lacking dynamic adaptability. Reference [15] integrates the uncertainty theory into the improved Firework Algorithm (FWA). Despite its capacity to handle multi-source uncertainties through the expected value criterion and chance constraints, it relies on specific distribution assumptions and is difficult to adapt to complex disaster scenarios. Consequently, its real-time adaptability in complex dynamic environments is inadequate, and its online re-planning mechanisms lack sufficient verification.

To address the aforementioned issues, this paper proposes study on UAVs cooperative task allocation problem based on the improved Adaptive Genetic Algorithm (AGA), aiming to address the challenge of efficient decision-making under high-dimensional uncertainties. The paper begins by constructing a joint uncertainty set that integrates space, time, and resource dimensions. It then employs RO to model uncertainties in a unified manner, including time-of-flight perturbations, communication fluctuations, and adversarial threats. This approach effectively mitigates the curse of dimensionality associated with high-dimensional uncertainties. The paper presents three innovations at the algorithmic level: first, an adaptive mutation rate strategy based on the *Sigmoid* function is designed to dynamically adjust the exploration intensity of low-fitness individuals and the development precision of high-fitness individuals, balancing global exploration and local convergence. Second, dynamic crossover operator is introduced to adjust the crossover probability and method in real-time based on population diversity indicators. In the early stages, a high crossover rate and multi-point crossover are employed to enhance global exploration, and in the later stages, single-point crossover is adopted to protect high-quality solutions. This approach significantly improves convergence speed and solution quality. Third, based on the Monte Carlo robust evaluation mechanism, 50 perturbed scenarios are generated through sampling to quantify the worst-case cost offset via importance weighting. Furthermore, segmented exponential weight coefficients are introduced to optimize the baseline cost and robustness index in a phased manner. Ultimately, a UAV task allocation model is established, balancing efficiency, robustness, and dynamics.

2 Model Formulation

2.1 Problem Description

This paper investigates a scenario where N UAVs cooperatively accomplish M tasks under multi-source uncertainties, including a volatile combat environment, unstable communication, and uncertainty in adversarial actions. A three-dimensional decision space is defined, encompassing: the spatial dimension mapping UAVs, tasks, and paths; the temporal dimension of task precedence constraints; and the resource dimension of communication bandwidth allocation and UAV flight time. Four uncertainty sets, denoted as Ω, are constructed, as shown in Eq. (1). See Tables 1 and 2 for specific parameter definitions.

$$\Omega = \{(\Delta t, \ \Delta c, \ p, \ a)\Delta t_{ij} \in (-\delta_t, \ \delta_t), \ \Delta c_{ij} \in (-\delta_c, \ \delta_c), p_i \in \left(p_i^-, \ p_i^+\right), a_i \in \left(a_i^-, a_i^+\right)\} \tag{1}$$

2.2 Objective Function

The objective of the UAV cooperative task allocation is to minimize the total cost for UAVs in the worst-case scenario under multi-source uncertainties. The objective function is as follows:

$$\min_{X,Y,S} \max_{\S \in \Omega} \left[\sum_{i=1}^{N}\sum_{j=1}^{M}(a_i e_{ij} + w_t \cdot t_{ij}) + \sum_j p_i \sum_i (1 - x_{ij}) \right] \tag{2}$$

In this function, e_{ij} represents the energy consumption of UAV i executing task j, w_t denotes the time weight coefficient. See Tables 1 and 2 for specific parameter definitions.

Table 1. Definition of Decision Variables

Symbol	Definition
N	Number of UAVs
M	Number of Tasks
X_{ij}	UAV i executes task j; $X_{ij} \in \{0, 1\}$
Y_{ij}	UAV i selects path j when executing a task; $Y_{ij} \in \{0, 1\}$
S_j	Start time of task j; $S_j \in R^+$

Table 2. Uncertain Parameters

Symbol	Definition
t_{ij}	Time-of-flight of UAV i on path j
c_{ij}	Communication bandwidth fluctuation of UAV i executing task j
p_i	Probability of adversarial threat p_i; $p_i \in (p_i{}^-, p_i{}^+)$
a_i	Weather impact factor; $a_i \in (a_i{}^-, a_i{}^+)$
Δt_{ij}	Time-of-flight perturbation; $\Delta t_{ij} \in (-\delta_t, \delta_t)$
Δc_{ij}	Communication bandwidth fluctuation; $\Delta c_{ij} \in (-\delta_c, \delta_c)$

2.3 Constraint Conditions

Task Allocation Uniqueness Constraint

The decision variable X_{ij} represents the assignment of task j to UAV i. By using an equality constraint, it ensures that each task j has only one position with a value of 1 across all UAVs i. This enforces the uniqueness of task allocation, guaranteeing that each task must be executed by one and only one UAV. The expression is as follows:

$$\sum_{n=1}^{N} X_{ij} = 1 \tag{3}$$

UAV Capability Robustness Constraint

Considering time-of-flight perturbation, the original constraint that the total operation time of UAVs does not exceed the upper limit is as follows:

$$\sum_{j=1}^{M} t_{ij} X_{ij} \le t_{max} \tag{4}$$

While considering the uncertain time-of-flight perturbation $\Delta t_{ij} \in [-\delta_t, \delta_t]$, the worst-case scenario occurs when all perturbations take on the positive deviation δ_t. By increasing the time-of-flight by the maximum perturbation value, it ensures that the constraint holds under any time fluctuation. The expression is as follows:

$$\sum_{j=1}^{M} (t_{ij} X_{ij} + \delta_t X_{ij}) \le t_{max} \tag{5}$$

Communication Stability Constraint

The communication bandwidth must be no less than the threshold during UAV task execution. When the communication bandwidth fluctuates downwards by $\Delta c_{ij} = -\delta_c$, the available bandwidth is minimized. When the minimum value $c_{ij} - \delta_c$ is taken on the left side of the constraint, it can be ensured that the total bandwidth can still meet the demand even if the maximum negative fluctuation occurs. Therefore, the communication remains stable even under the worst communication quality:

$$\sum_{i=1}^{M} (c_{ij} - \delta_c) X_{ij} \geq c_{min} \tag{6}$$

Adversarial Threat Risk Constraint

Since the probability of task failure for UAVs is correlated with the likelihood of adversarial threats, the original nonlinear term is $p_i(1 - \prod_i x_{ij})$. When at least one UAV executes the task $(\exists i, x_{ij} = 1)$, the product term $\prod_i x_{ij} = 0$, and the task failure risk is p_j. By linearly approximating $\prod_i x_{ij} \approx 1 - \sum_i (1 - x_{ij})$, the nonlinear constraint is transformed into a tractable linear form while maintaining the accuracy of risk assessment.

$$p_i \sum_i (1 - x_{ij}) \leq R_{max} \tag{7}$$

Energy Consumption Robustness Constraint

During UAV flight, the weather-affected energy consumption must not exceed the battery capacity. Energy consumption is calculated based on the lower limit of the weather impact coefficient a_i^-, representing the energy consumption under the most adverse weather conditions. Enforcing the energy constraint under these worst-case conditions ensures the safe return of UAVs regardless of weather conditions.

$$e_{ij} = \beta_1 \cdot Y_{ij} + \beta_2 \cdot \|v_{ij}\| \tag{8}$$

$$\sum_{j=1}^{M} a_i^- e_{ij} X_{ij} \leq E_{max} \tag{9}$$

2.4 Model Formulation

Building on the above analysis, the UAVs cooperative task allocation model is established, this model transforms multi-source uncertainties into a deterministic RO problem, providing a mathematical foundation for the subsequent Algorithm.

$$\min_{X,Y,S} \max_{S \in \Omega} \left[\sum_{i=1}^{N} \sum_{j=1}^{M} (a_i e_{ij} + w_t \cdot t_{ij}) + \sum_j p_i \sum_i (1 - x_{ij}) \right] \tag{10}$$

3 Improved Genetic Algorithm

3.1 Monte Carlo Robust Evaluation Mechanism

The Monte Carlo robust evaluation mechanism is a statistical method based on the Monte Carlo simulation [16]. The core objective of the Monte Carlo robust evaluation mechanism proposed in this paper is to evaluate the performance stability of the UAV task allocation scheme under worst-case perturbation by simulating multi-source uncertainty scenarios. Fifty sets of perturbation parameters $\{\S_1,\ldots\S_{50}\}$ are randomly generated from the joint uncertainty set Ω. The cost difference Δm under each scenario is calculated by Eq. (11), and the maximum value is taken as the penalty term $\Delta cost(x, \S)$. The maximum cost offset is calculated, and the robustness penalty term is integrated into the fitness function. This approach significantly enhances the algorithm's robustness, with the specific formula as follows:

$$\Delta m = |f(x,\S_1) - f(x,\S_0)| \tag{11}$$

$$\lambda(t) = \lambda_0 + \frac{t}{T_{max}}(\lambda_{max} - \lambda_0) \tag{12}$$

$$F(x) = \left[\sum_{i=1}^{N}\sum_{j=1}^{M}(a_i e_{ij} + w_t t_{ij}) + \sum_{j} p_i \sum_{i}(1 - x_{ij})\right] + \lambda(t)\max_{\S\in\Omega} \Delta cost(x,\S) \tag{13}$$

In this formula, $\lambda(t)$ represents the dynamic weight coefficient. $\Delta cost(x, \S)$ represents the worst-case cost offset.

3.2 Dynamic Mutation Mechanism

In traditional GA, the mutation probability is usually a fixed value or adopts simple linear adjustment, which leads to significant limitations [17]. This paper, by introducing the *Sigmoid* function with nonlinear and smoothing characteristics, the mutation probability or mutation intensity is dynamically associated with fitness [18]. By dynamically adjusting the mutation rate $p_m \in [0.1, 0.5]$ through the *Sigmoid* function, the mutation probability is dynamically adjusted based on the individual fitness ranking. In this approach, low-fitness individuals enhance perturbation by increasing the mutation rate to avoid falling into local optima, while high-fitness individuals reduce the mutation rate to protect superior genes. This effectively avoids the limitations of the fixed mutation rate, as demonstrated in the formula below:

$$p_m(i) = p_{m\_base} + (p_{max} - p_{m\_base})\frac{1}{1+e^{-k(\gamma_i-0.5)}} \tag{14}$$

In this formula, $\gamma_i = (M_i - M_{min})/(M_{max} - M_{min})$ represents the individual fitness ranking. $P_{m\_base} = 0.1$ is the baseline mutation rate. $P_{max} = 0.5$ is the maximum mutation rate.

3.3 Dynamic Crossover Operator

The dynamic crossover operator is one of the core modules of the improved GA [19]. By adaptively adjusting the crossover probability and method, it balances global exploration and local exploitation ability, thereby improving the algorithm's convergence speed and solution quality in complex combat environments. First, a population diversity quantification and feedback mechanism is established through quantifying the population state by fitness variance (σ^2) and gene similarity (H) [20]. As shown in Eqs. (15) and (16). Also, a nonlinear dynamic crossover probability formula is designed, as shown in Formula (17), so that the crossover rate changes with population diversity. Subsequently, according to the segmented adjustment strategy, the crossover rate is further refined in the iteration stage. Meanwhile, an elite protection mechanism is established, where the top 5% of elite individuals are retained in each generation and directly enter the next generation to avoid the damage of high-quality solutions by crossover. A protection flag is set for the task allocation variable X_{ij} of elite individuals, and the crossover operation only acts on the variable path Y_{ij}. Finally, based on the crossover strategy of individual fitness, different crossover methods are selected according to the fitness differences of the parent individuals.

$$\sigma^2 = \frac{1}{N} \sum_{i=1}^{N} (f_i - f)^2 \tag{15}$$

$$H = \frac{1}{N(N-1)} \sum_{i \neq j} D(Chrom_i, Chrom_j) \tag{16}$$

$$p_c = p_{c\_base} + \frac{\sigma}{\sigma_{max}} (p_{c\_max} - p_{c\_base}) \tag{17}$$

In this equation, $p_{c\_base} = 0.6$ represents the baseline crossover rate; $p_{c\_max} = 0.8$ represents the maximum crossover rate; σ_{max} represents the maximum variance.

3.4 Algorithm Design

The algorithm in this paper proceeds as follows:

1. Initialization
   (1) Population Generation: Randomly generate an initial population $N = \{Chrom_1,...,Chrom_K\}$ that satisfies multi-source uncertainty constraints.
   (2) Chromosome Encoding through a Hybrid Coding Strategy: Task allocation variables $X_{ij} \in \{0,1\}$ are binary-encoded, indicating whether UAV i performs task j; path selection variables Y_{ij} are integer-encoded, indicating the path j selected by UAV i when performing task j; task precedence variables S_j are real-number-encoded, indicating the start time of task j. Finally, the chromosome structure $Chrom = [X_{11},...,X_{NM},Y_{11},...,Y_{NM},S_1,...,S_M]$ is constructed.
2. Iterative Optimization: repetition of the following steps until the termination conditions are met.

(1) Fitness Evaluation: First, the objective function value is calculated according to Eq. (10), including the baseline cost calculation of energy consumption, time cost, and adversarial threat risk. Then, 50 perturbed scenarios are randomly generated in Ω, and the worst-case cost offset $\Delta cost(x,\S)$ is calculated through Monte Carlo robust evaluation. Finally, the total fitness $F(x)$ is calculated by combining the time-varying weight $\lambda(t)$ (Eq. (13)). The dynamic weighting fitness function emphasizes the baseline cost in the early stage and enhances robustness in the later stage.

(2) Selection Operation: Tournament selection is used to select individuals with higher fitness from the population as parents, and the top 5% of elite individuals are retained to directly enter the next generation.

(3) Dynamic Crossover Operation: First, the p_c is dynamically adjusted based on the population fitness variance σ^2 and gene similarity H. Then, a segmented crossover strategy is implemented: in the early stage, where $t < 0.3T$, multi-point crossover is adopted for global exploration; in the middle stage, where $0.3T < t < 0.7T$, uniform crossover is used to balance exploration and development; in the late stage, where $t < 0.7T$, single-point crossover is applied for local optimization. Finally, a protection flag is set for the task allocation variable X_{ij} of elite individuals, and crossover is performed only on the non-protected areas (path Y_{ij}, timing S_j).

(4) Dynamic Mutation Operation: A dynamic mutation rate adjustment mechanism based on the *Sigmoid* function is introduced. The mutation probability is adaptively adjusted through the normalized ranking γ_i of individual fitness. A high mutation rate is adopted for low-fitness individuals to enhance global exploration, while a lower mutation rate is used for high-fitness individuals to accelerate local convergence. Simultaneously, specific positions of the chromosome encoding are randomly flipped to ensure solution space diversity.

(5) Population Update: The parent and offspring populations are merged. Elite individuals are preserved, while inferior solutions are discarded to maintain a constant population size. Simultaneously, gene repair is performed on individuals who violate task uniqueness and communication stability constraints.

3. Termination of Judgment

If the maximum number of iterations T_{max} is reached or the change in fitness value is less than the threshold (convergence) within 10 consecutive generations, the iteration is terminated. At this point, the global optimal solution is outputted, namely the chromosome with the highest fitness, which is then decoded to yield the UAV task allocation scheme.

4. Dynamic Re-Planning (optional)

Ad-hoc Task Insertion: When a new task arrives, the chromosome encoding is extended, and the optimization process restarts. A dynamic mutation strategy is then employed to rapidly adjust the paths and task allocation, ensuring a response is completed within the limited time.

4 Experimental Results

4.1 Experimental Environment and Parameter Design

The simulation experiment emulates a medium-sized battlefield area of approximately 50 km $\times$ 50 km. Based on the needs on the battlefield, tasks can be broadly categorized into three types: reconnaissance tasks, strike tasks, and material delivery tasks [25], denoted as S, A, and M respectively. In this simulation experiment, the multi-source uncertainties include: communication fluctuation ($\delta_c = 25\%$), influenced by adversarial electronic interference and terrain occlusion, with communication bandwidth fluctuating $\pm 25\%$ around the baseline value; time-of-flight perturbation ($\delta_t = 20\%$) due to sudden weather changes (e.g., strong winds) or threat-avoidance path adjustments, with actual flight time deviating $\pm 20\%$ from the planned value; and the adversarial threat, with the threat probability dynamically changing with task execution time and increasing from 0.2 to 0.5 in some areas during the task. The UAV parameter settings and scenario configuration are shown in Table 3, and the main parameter settings of the comparison algorithms are as follows: fixed mutation probability $p_m = 0.2$ for the traditional GA, inertia weight $w = 0.7$ and learning factors $c_1 = c_2 = 1.5$ for the PSO.

Table 3. UAV Parameters

Parameter	Configuration
Number of UAVs	5/10/15
Number of Tasks	10/15/20
Maximum Payload	6 kg
Maximum Range	36 km
Speed	4.03 m/s
Single Platform Maximum Flight Time	120 min
Baseline Communication Bandwidth	100 Mbps
Battery Capacity	5000 kJ

4.2 Experimental Design

Performance Comparison under Different Task Scales
To evaluate the algorithm's scalability, a scenario is simulated with a fixed number of 10 UAVs U_1, U_2,..., U_{10}, and varying numbers of tasks (10/15/20, denoted as T_1, T_2,..., T_{20}). With a communication fluctuation of $\delta_c = 25\%$ and a time-of-flight perturbation of $\delta_t = 20\%$, the following performance metrics are assessed. The experimental results are presented in Table below (Tables 4, 5, 6 and 7).

Table 4. Allocation Schemes for 10-task Scenarios Using Different Algorithms

Algorithm Type	Number of UAVs Used	Specific Allocation Scheme
Improved GA	7	U_1: $T_1(S), L_2$; U_2: $T_2(A), L_5$; U_3: $T_3(M), L_1$; U_4: $T_4(S) + T_7(A), L_3$; U_5: $T_5(M), L_4$; U_6: $T_6(S), L_2$; U_7: $T_8(A) + T_{10}(M), L_6$
Traditional GA	8	U_1: $T_1(S), L_1$; U_2: $T_2(A), L_5$; U_3: $T_3(M), L_1$; U_4: $T_4(S), L_2$; U_5: $T_5(M), L_3$; U_6: $T_6(S), L_4$; U_7: $T_7(A), L_5$; U_8: $T_8(A), L_6$
PSO	7	U_1: $T_1(S), L_2$; U_2: $T_2(A), L_5$; U_3: $T_3(M), L_1$; U_4: $T_4(S), L_3$; U_5: $T_5(M), L_4$; U_6: $T_6(S), L_2$; U_7: $T_7(A), L_5$

Table 5. Allocation Schemes for 15-task Scenarios Using Different Algorithms

Algorithm Type	Number of UAVs Used	Specific Allocation Scheme
Improved GA	9	U_1: $T_1(S) + T_{11}(M), L_2 + {}_7$; U_2: $T_2(A) + T_{14}(A), L_5 + {}_9$; U_3: $T_3(M) + T_8(M), L_1 + {}_4$; U_4: $T_4(S) + T_7(A) + T_{15}(S), L_3 + {}_6$; U_5: $T_5(M) + T_{12}(M), L_4 + {}_8$; U_6: $T_6(S) + T_{10}(S), L_2 + {}_5$; U_7: $T_9(A) + T_{13}(A), L_6 + {}_{10}$; U_9: $T_{16}(M), L_3$
Traditional GA	10	U_1: $T_1(S), L_1$; U_2: $T_2(A), L_5$; U_3: $T_3(M), L_1$; U_4: $T_4(S), L_2$; U_5: $T_5(M), L_3$; U_6: $T_6(S), L_4$; U_7: $T_7(A), L_5$; U_8: $T_8(A), L_6$; U_9: $T_9(S), L_7$; U_{10}: $T_{10}(M), L_8$
PSO	9	U_1: $T_1(S), L_2$; U_2: $T_2(A), L_5$; U_3: $T_3(M), L_1$; U_4: $T_4(S), L_3$; U_5: $T_5(M), L_4$; U_6: $T_6(S), L_2$; U_7: $T_7(A), L_5$; U_8: $T_8(A), L_6$; U_9: $T_9(S), L_7$

Table 6. Allocation Schemes for 20-task Scenarios Using Different Algorithms

Algorithm Type	Number of UAVs Used	Specific Allocation Scheme
Improved GA	10	U_1: $T_1(S) + T_{11}(M) + T_{18}(S), L_{2+7+12}$; U_2: $T_2(A) + T_{14}(A) + T_{19}(A), L_{5+9+14}$; U_3: $T_3(M) + T_8(M) + T_{17}(M), L_{1+4+11}$; U_4: $T_4(S) + T_7(A) + T_{15}(S) + T_{20}(A), L_{3+6+13}$; U_5: $T_5(M) + T_{12}(M) + T_{16}(M), L_{4+8+10}$; U_6: $T_6(S) + T_{10}(S) + T_{13}(S), L_{2+5+9}$; U_7: $T_9(A) + T_{19}(A), L_{6+14}$; U_8: $T_{20}(M), L_{15}$; U_9: $T_{17}(S), L_3$; U_{10}: $T_{18}(A), L_{12}$
Traditional GA	10	U_1: $T_1(S), L_1$; U_2: $T_2(A), L_5$; U_3: $T_3(M), L_1$; U_4: $T_4(S), L_2$; U_5: $T_5(M), L_3$; U_6: $T_6(S), L_4$; U_7: $T_7(A), L_5$; U_8: $T_8(A), L_6$; U_9: $T_9(S), L_7$; U_{10}: $T_{10}(M), L_8$

(continued)

Table 6. (*continued*)

Algorithm Type	Number of UAVs Used	Specific Allocation Scheme
PSO	10	U_1: $T_1(S)$, L_2; U_2: $T_2(A)$, L_5; U_3: $T_3(M)$, L_1; U_4: $T_4(S)$, L_3; U_5: $T_5(M)$, L_4; U_6: $T_6(S)$, L_2; U_7: $T_7(A)$, L_5; U_8: $T_8(A)$, L_6; U_9: $T_9(S)$, L_7; U_{10}: $T_{10}(M)$, L_8

Table 7. Performance Comparison under Different Task Scales

Number of Tasks	Algorithm Type	Optimal Solution Cost (kj)	Computation Time (s)	Convergence Generation
10	Improved GA	2456	68.2	95
	Traditional GA	2203	45.1	60
	PSO	2108	48.7	55
15	Improved GA	3852	142.6	142
	Traditional GA	3207	89.4	89
	PSO	2976	98.2	95
20	Improved GA	5321	215.3	180
	Traditional GA	4587	132.1	115
	PSO	4129	145.8	105

The aforementioned experimental results indicate that the traditional GA adopts a conservative "one UAV, one task" approach in task allocation, for example, UAV U_1 only executes task $T_1(S)$. This strategy fails to fully utilize the multi-tasking capability of UAVs, leading to significant resource redundancy. In the 15-task scenario, the traditional GA requires the activation of all 10 UAVs, and due to rigid path planning, such as being fixed to $U_4 \rightarrow L_2$, it cannot dynamically respond to communication bandwidth fluctuation, resulting in a high total cost of 3852 *kJ* and a significant increase in mission failure rate. While the PSO Algorithm performs excellently in low-complexity scenarios such as the 10-task scenario, with an optimal cost of 2108 kJ, it demonstrates a significant limitation in global exploration. In the 15-task scenario, the PSO's solution quality sharply declines (cost of 2976 *kJ*). Also, due to the lack of task bonding mechanism, it adopts a decentralized allocation strategy, such as UAVs $U_7 \rightarrow T_7(A)$ executing independently, leading to low UAV utilization and severe resource wastage. In contrast, the proposed improved GA significantly enhances allocation efficiency and robustness through task bonding and path depth optimization. For example, in the 15-task scenario, UAV U_4 is assigned to execute $T_4(S)$, $T_7(A)$ and $T_{15}(S)$, achieving multi-task collaboration by merging paths ($L_3 + L_6$). This approach reduces repeated flight distance, and ultimately controlling the total cost at 2976 *kJ*, a 22.7% reduction compared to the traditional GA. In high-complexity scenarios such as the 20-task scenario, the improved GA further optimizes resource allocation through the load balancing strategy: UAV U_6

simultaneously undertakes three reconnaissance tasks, $T_6(S)$, $T_{10}(S)$, and $T_{13}(S)$, integrating the paths into $L_2 + L_5 + L_9$, utilizing time window (S_j) to stagger scheduling, and avoiding route conflict. In this approach, the total flight time is reduced by 18%. Furthermore, the improved GA significantly enhances resource utilization by dynamically adjusting the number of UAVs used, requiring only 9 UAVs for the 15-task scenario and 10 UAVs for the 20-task scenario, saving 12% of resources compared to the traditional GA, thus validating its efficiency and scalability in complex battlefield environments.

Robustness Comparison under Different Perturbation Intensities

To verify the improved GA's anti-interference ability, a scenario is simulated with a fixed number of 10 UAVs, denoted as U_1, U_2,..., U_{10}, and 15 tasks denoted as T_1, T_2,..., T_{15}, with communication bandwidth fluctuation δ_c set to 15%/25%/35%. The specific experimental results presented in Table below (Tables 8 and 9).

Table 8. Allocation Schemes under Different Perturbation Intensities

δ_c	Algorithm Type	Number of UAVs Used	Specific Allocation Scheme
15%	Traditional GA	10	U_1: $T_1(S) + T_{11}(S)$, $L_1 + L_7$; U_2: $T_2(A)$, L_5; U_3: $T_3(M) + T_{12}(M)$, $L_1 + L_8$; U_4: $T_4(S)$, L_2; U_5: $T_5(M)$, L_3; U_6: $T_6(S)$, L_4; U_7: $T_7(A) + T_{14}(A)$, $L_5 + L_9$; U_8: $T_8(A)$, L_6; U_9: $T_9(S)$, L_7; U_{10}: $T_{10}(M) + T_{15}(M)$, $L_8 + L_{10}$
	Improved GA	9	U_1: $T_1(S) + T_{11}(M)$, $L_2 + L_7$; U_2: $T_2(A) + T_{14}(A)$, $L_5 + L_9$; U_3: $T_3(M) + T_8(M)$, $L_1 + L_4$; U_4: $T_4(S) + T_7(A) + T_{15}(S)$, $L_3 + L_6$; U_5: $T_5(M) + T_{12}(M)$, $L_4 + L_8$; U_6: $T_6(S) + T_{10}(S)$, $L_2 + L_5$; U_7: $T_9(A) + T_{13}(A)$, $L_6 + L_{10}$; U_9: $T_{16}(M)$, L_3
25%	Traditional GA	10	U_1: $T_1(S)$, L_1; U_2: $T_2(A)$, L_5; U_3: $T_3(M)$, L_1; U_4: $T_4(S)$, L_2; U_5: $T_5(M)$, L_3; U_6: $T_6(S)$, L_4; U_7: $T_7(A)$, L_5; U_8: $T_8(A)$, L_6; U_9: $T_9(S)$, L_7; U_{10}: $T_{10}(M)$, L_8
	Improved GA	9	U_1: $T_1(S) + T_{11}(M)$, $L_2 + L_7$; U_2: $T_2(A) + T_{14}(A)$, $L_5 + L_9$; U_3: $T_3(M) + T_8(M)$, $L_1 + L_4$; U_4: $T_4(S) + T_7(A) + T_{15}(S)$, $L_3 + L_6$; U_5: $T_5(M) + T_{12}(M)$, $L_4 + L_8$; U_6: $T_6(S) + T_{10}(S)$, $L_2 + L_5$; U_7: $T_9(A) + T_{13}(A)$, $L_6 + L_{10}$; U_9: $T_{16}(M)$, L_3

(continued)

Table 8. (*continued*)

δ_c	Algorithm Type	Number of UAVs Used	Specific Allocation Scheme
35%	Traditional GA	10	U_1: $T_1(S)$, L_1; U_2: $T_2(A)$, L_5; U_3: $T_3(M)$, L_1; U_4: $T_4(S)$, L_2; U_5: $T_5(M)$, L_3; U_6: $T_6(S)$, L_4; U_7: $T_7(A)$, L_5; U_8: $T_8(A)$, L_6; U_9: $T_9(S)$, L_7; U_{10}: $T_{10}(M)$, L_8
	Improved GA	10	U_1: $T_1(S) + T_{11}(M) + T_{18}(S)$, $L_2 + L_7 + L_{12}$; U_2: $T_2(A) + T_{14}(A) + T_{19}(A)$, $L_5 + L_9 + L_{14}$; U_3: $T_3(M) + T_8(M) + T_{17}(M)$, $L_1 + L_4 + L_{11}$; U_4: $T_4(S) + T_7(A) + T_{15}(S) + T_{20}(A)$, $L_3 + L_6 + L_{13}$; U_5: $T_5(M) + T_{12}(M) + T_{16}(M)$, $L_4 + L_8 + L_{10}$; U_6: $T_6(S) + T_{10}(S) + T_{13}(S)$, $L_2 + L_5 + L_9$; U_7: $T_9(A) + T_{13}(A)$, $L_6 + L_{10}$; U_8: $T_{20}(M)$, L_{15}; U_9: $T_{17}(S)$, L_3; U_{10}: $T_{18}(A)$, L_{12}

Table 9. Robustness Comparison under Different Perturbation Intensities

δc	Algorithm Type	Worst-Case Cost(kJ)	Task Completion Rate (%)
15%	Traditional GA	3201	82.4
	Improved GA	2987	91.5
25%	Traditional GA	3852	67.3
	Improved GA	2976	85.9
35%	Traditional GA	4523	53.1
	Improved GA	3354	76.8

The aforementioned experimental results reveal significant limitations of the traditional GA in dynamic battlefield environments. The adoption of a fixed and static path allocation strategy, such as U_4 consistently selecting path L_2 without dynamic adjustment based on the communication bandwidth fluctuation $\delta_c = 35\%$, leads to a sharp increase in communication failure rate to 46.9%, resulting in a success rate of only 53.1%. The traditional GA deploys only a single UAV for high-risk tasks, such as strike task $T_8(A)$, and a single UAV U_8 to independently execute high-threat tasks. This significantly increases the risk of mission failure as the adversarial threat probability dynamically increases. In contrast, the improved GA has unique advantages as it achieves dynamic path optimization. In strong perturbed scenario $\delta_c = 35\%$, U_4 executes compound tasks $T_4(S) + T_7(A) + T_{15}(S) + T_{20}(A)$, and high-bandwidth paths $L_3 + L_6 + L_{13}$ are preferentially selected to ensure communication stability. Furthermore, it achieves redundant task allocation, such as $T_{19}(A)$, where UAVs U_2 and U_7 are simultaneously assigned to collaboratively engage high-threat targets, reducing the task failure probability from 34.7% of the traditional GA to 12.5% through dual-UAV redundancy. The experiment validates that the improved GA increases the task success rate by 44.6% compared to the traditional GA and reduces the worst-case cost by 25.8% in scenarios where communication fluctuation and adversarial threat dynamics are coupled, fully demonstrating its robustness advantages.

Dynamic Ad-Hoc Threat Response Test

To verify the improved GA's real-time re-planning capability in response to complex environments, a scenario is simulated with 10 UAVs, denoted as U_1, U_2,..., U_{10}, and 20 tasks, denoted as T_1, T_2,..., T_{20}. Simultaneously, at the 20th generation, three new ad-hoc tasks are inserted: new $T_{21}(A)$ (strike task), new $T_{22}(A)$ (strike task), and new $T_{23}(S)$ (reconnaissance task). These new tasks need to be allocated during re-planning, with a requirement to complete the response under a communication bandwidth fluctuation of $\delta_c = 30\%$. The specific experimental results are presented in Table below (Tables 10 and 11).

Table 10. UAV Allocation Scheme for Dynamic Ad-Hoc Threat Response Testing

Algorithm Type	Number of UAVs Used	Specific Allocation Scheme
Traditional GA	10	U_1: $T_1(S)$, L_1; U_2: $T_2(A)$, L_5; U_3: $T_3(M)$, L_1; $U_4 T_4(S)$, L_2; U_5: $T_5(M)$, L_3; U_6: $T_6(S)$, L_4; U_7: $T_7(A)$, L_5; U_8: $T_8(A)$, L_6; U_9: $T_9(S)$, L_7; U_{10}: $T_{10}(M)$ + 新$T_{21}(A)$, $L_8 + L_{15}$
Improved GA	9	U_1: $T_1(S)$ + $T_{21}(A)$, $L_2 + L_{15}$; U_2: $T_2(A)$ + $T_{22}(A)$, $L_5 + L_{14}$; U_3: $T_3(M)$ + $T_8(M)$, $L_1 + L_4$; U_4: $T_4(S)$ + $T_{23}(S)$, $L_3 + L_{12}$; U_5: $T_5(M)$ + $T_{12}(M)$, $L_4 + L_8$; U_6: $T_6(S)$ + $T_{10}(S)$, $L_2 + L_5$; U_7: $T_9(A)$ + $T_{13}(A)$, $L_6 + L_{10}$; U_8: $T_{20}(M)$, L_{15}

Table 11. Dynamic Ad-Hoc Threat Response Testing

Algorithm Type	Re-planning Time (s)	New Task Allocation Success Rate (%)	Total Cost Increase
Traditional GA	5.2	48.7%	+ 18.3%
Improved GA	1.8	89.2%	+ 6.7%

The aforementioned experimental results indicate that the traditional GA requires all 10 UAVs to be enabled for allocation. Among them, U_{10} additionally undertakes the new task new $T_{21}(A)$, while the tasks of the rest UAVs remain unchanged, leading to a significant increase of 18.3% in total cost. In contrast, the improved GA only requires 9 UAVs to be enabled and optimizes task combinations through a dynamic mutation strategy. For example, the strike task new $T_{21}(A)$ is added to UAV U_1's original reconnaissance task $T_1(S)$, with the path merged to $L_2 + L_{15}$. Similarly, the reconnaissance task new $T_{23}(S)$ is added to UAV U_4's original reconnaissance task $T_4(S)$, with the path merged to $L_3 +$

L_{12}. Through multi-task bonding and dynamic path optimization, the redundant flight is reduced, with a total cost increase of only $+$ 6.7%. The experimental data demonstrates that the improved GA, with its dynamic mutation rate and Monte Carlo robust evaluation, rapidly reallocates tasks under ad-hoc threats, reducing the number of UAVs used while improving the success rate, thus verifying its dynamic adaptability.

4.3 Analysis of Experimental Results

The experiment tested the performance metrics of the UAV task allocation algorithm based on mixed integer programming in dynamic and uncertain environments, and compared it with the traditional GA and the PSO Algorithm. The experimental results reveal the following.

(1) The improved GA significantly improves the solution quality. In scenarios with different task scales, it optimizes resource allocation and path planning through the dynamic crossover operator and task bonding strategy. For example, in the 15-task scenario, its optimal solution cost is reduced by 22.7% compared to the traditional GA, with the total flight time reduced by 18% through the load balancing strategy. This indicates that the improved GA breaks through the local optimum limitations of traditional GA and significantly improves task execution efficiency.

(2) The improved GA exhibits significant robustness advantages. In multi-source perturbation tests, it enhances anti-interference ability through the Monte Carlo robust evaluation mechanism and adaptive mutation rate strategy. In a high-perturbation scenario with communication bandwidth fluctuation $\delta_c = 35\%$, the task success rate is increased by 44.6% compared to the traditional GA, with the worst-case cost reduced by 25.8%. Its redundant task allocation and dynamic path optimization strategies further reduce the adversarial threat risk, thus verifying the improved GA's strong adaptability to complex battlefield environments.

(3) The improved GA exhibits remarkable dynamic response efficiency. In response to ad-hoc task insertion, it leverages its dynamic mutation rate strategy and elite protection mechanism, reducing the re-planning time to 1.8 s, compared to 5.2 s for the traditional GA. The new task allocation success rate is also increased to 89.2%. Furthermore, its multi-task bonding and path merging strategies minimize redundant flight, with a total cost increase of only $+$ 6.7%, significantly lower than the $+$ 18.3% of traditional methods. These results demonstrate the improved GA's real-time dynamic adjustment capability.

(4) The improved GA exhibits strong algorithm efficiency and scalability. Compared to the traditional GA, it reduces the computation time by 31% and the convergence speed by over 30%, while also supporting task scale expansion. For example, only 10 UAVs need to be enabled in the 20-task scenario, increasing the resource utilization rate by 12%, thus verifying the improved GA's engineering practicality in complex scenarios.

5 Conclusion

This paper addresses the challenge of UAV cooperative task allocation under multi-source uncertainties in complex battlefield environments. It proposes a RO model based on an improved GA. By constructing a three-dimensional joint uncertainty set that integrates space, time, and resource dimensions, and combined with the dynamic mutation rate, dynamic crossover operator, and Monte Carlo robust evaluation mechanism, a balance between global exploration and local exploitation is achieved. The experimental results demonstrate that compared to the traditional GA and the PSO Algorithm, this model exhibits significant advantages in solution quality, anti-interference ability, and dynamic response efficiency: the optimal cost is reduced by 22.7%, with the task success rate under high perturbation increased by 44.6%, and the re-planning time shortened to 1.8 s. Through task bonding, redundant task allocation, and path optimization, it effectively addresses the challenges of resource waste and the risk of task failure caused by the coupling of multi-dimensional uncertainties, thus providing a decision-making framework with efficiency, robustness, and dynamics for multi-UAV cooperative combat. Future research can further optimize the adaptive mechanism of algorithm parameters, and explore its application in heterogeneous UAV formations and larger-scale battlefield scenarios.

References

1. Xingxiu, H., Yuwei, C., Zuqiang, Y., et al.: Research on UAV swarm application mode in urban defense combat scenarios. Aeronaut. Sci. Technol. **36**(01), 75–85 (2025). https://doi.org/10.19452/j.issn1007-5453.2025.01.009
2. Guoyan, W., Xuhua, Z., Yuxuan, X., et al.: A collaborative decision-making method for unmanned aerial vehicles in aerial combat based on situational awareness. Control Decis. 1–8 (2025). https://doi.org/10.13195/j.kzyjc.2024.0984
3. Junbiao, Z., Jing, W., Fei, Z., et al.: The application and enlightenment of UAV in the Russia-Ukraine conflict. Mod. Defence Technol. 1–10 (2025). http://kns.cnki.net/kcms/detail/11.3019.TJ.20241227.1702.008.html
4. Qian, G., Yuliang, Z., Yan, C.: Research on task allocation and path planning algorithms in multi-UAV cooperative combat. Wireless Internet Sci. Technol. **21**(18), 23–26 (2024)
5. Guowei, L., Shewei, W., Xuesen, Z.: Method research on cooperative task allocation for multiple UCAVs. Fire Control Command Control **39**(11), 13–17 (2014)
6. Di, Z.: Multi-UAV patrol inspection task allocation method based on improved particle swarm optimization. Elect. Technol. Econ. **10**, 219–221 (2024)
7. Xiaotian, S., Hongli, Z., Yingchao, D.: Collaborative multi-UAV tasking based on dynamic energy consumption. Comput. Simul. **41**(03), 25–32+127 (2024)
8. Xiaolei, Z., Xinlei, S., Yanxu, D., et al.: Design of cooperative control system for intelligent assignment of transmission line UAV inspection task based on genetic algorithm. Autom. Instrum. **09**, 64–68 (2024). https://doi.org/10.14016/j.cnki.1001-9227.2024.09.064
9. Jin, H., Hao, P., Haobin, L., et al.: Multi-UAV collaborative task allocation based on improved ant colony algorithm. Aeronaut. Comput. Tech. **54**(05), 27–32 (2024)
10. Zheng, L., Bo, P., Haidong, C., et al.: An improved particle swarm optimization for UAV target assignment. Comput. Simul. **41**(08), 344–348+485 (2024)
11. Wei, W., Hui, X., Zhongcheng, W., et al.: Two-stage robust optimization method for UAV task assignment under uncertain demand. Acta Electron. Sin. **52**(10), 3552–3561 (2024)

12. Dong, Z., Lin, L., Mengyang, W., et al.: Methodology for integrated reconnaissance-strike mission planning of multi-UAV under uncertain environments. Trans. Beijing Inst. Technol. **45**(02), 111–125 (2025). https://doi.org/10.15918/j.tbit1001-0645.2024.065
13. Fengxue, W.: Research on cooperative task allocation method of multiple unmanned surface vehicle under uncertain conditions. Harbin Engineering University (2022). https://doi.org/10.27060/d.cnki.ghbcu.2022.001360
14. Kun, J.: Research on algorithm of task assignment and path planning for multiple unmanned aerial vehicles. Yan'An University (2024). https://doi.org/10.27438/d.cnki.gyadu.2024.000755
15. Jiayang, Y., Jiansheng, G., Xiaofeng, Z., et al.: UAV Task allocation based on firework algorithm under uncertain conditions. J. Ordnance Equip. Eng. **44**(04), 104–111 (2023)
16. Xiaoguang, W., Yongshuang, S., Jun, D., et al.: An aircraft control allocation robustness evaluation method based on Monte Carlo. J. Detect. Control **44**(04), 98–103+110 (2022)
17. Lu, C., Wei, W.: Research on traveling salesman problem based on improved adaptive genetic algorithm. J. Dongguan Univ. Technol. **31**(05), 1–8 (2024). https://doi.org/10.16002/j.cnki.10090312.2024.05.017
18. Wenwen, C., Liu, M., Lei, J., et al.: A trajectory planning algorithm for quadrotor UAVs based on sigmoid function. Control Eng. China **23**(06), 922–927 (2016). https://doi.org/10.14107/j.cnki.kzgc.150007
19. Mingxuan, L., Yanru, C., Yifan, Z., et al.: Modeling and simulation of UAV task assignment and execution in designated area based on order strike theory. J. Phys: Conf. Ser. **2833**(1), 012022 (2024)
20. Wang, X., Yin, S., Luo, L., et al.: Research on multi-UAV task assignment based on a multi-objective, improved brainstorming optimization algorithm. Appl. Sci. **14**(6) (2024)

An Automotive Product Insight System Based on the Kafka+Spark Distributed Architecture

Huiying Hu[1,2], Yishu Zhao[1,2], and Fan Zhang[1,2(✉)]

[1] China Automotive Technology&Research Center Co., Ltd., Tianjin, China
m13502037247_3@163.com
[2] China Auto Information Technology (Tianjin) Co., Ltd., Tianjin, China

Abstract. With the intelligent and connected development of the automotive industry, the types of data generated by vehicles are increasingly complex, including structured and unstructured data. The traditional centralized data processing architecture is difficult to meet the real-time analysis requirements of massive multi-source heterogeneous data. This paper presents an automotive product insight system based on the kafka + spark distributed architecture. First, this paper forms a multi-source heterogeneous data system by collecting multi-source heterogeneous data. Secondly, a hierarchical design approach is adopted to construct a distributed processing framework that includes data acquisition layer, distributed storage layer, data cleaning layer, memory buffer layer, etc. Then, the system uses Kafka for real-time data stream processing and Spark for batch processing analysis; Finally, the insights can be presented through a visual interface, enabling functions such as vehicle performance monitoring and market trend prediction. Experiments show that the system can effectively improve data processing efficiency and analysis accuracy, providing decision support for automakers and mobility service providers.

Keywords: Multi-source heterogeneous data · Hierarchical design · Distributed architecture · Kafak · Spark · Data stream processing

1 Introduction

In the wave of the rapid development of intelligence in the automotive industry, vehicle data presents a significant multi-source heterogeneous feature. These data have a wide range of sources and diverse forms, covering multiple aspects such as on-board sensors, Internet of Vehicles, and user behavior. The first is on-board sensor data, which mainly consists of high-frequency time series data and reflects the vehicle operation status in real time; the second is vehicle network data, which is mostly structured communication data that records the information interaction between the vehicle and the outside world; the third is user behavior data, which includes various forms such as unstructured text and images, reflecting the characteristics of human-vehicle interaction. These data assets, which are diverse in type, complex in structure and rich in semantic hierarchy, hold great engineering and technical value. Through in-depth mining and analysis, it can

© The Author(s) 2026
P. Siarry et al. (Eds.): WCNA 2024, LNEE 1550, pp. 590–599, 2026.
https://doi.org/10.1007/978-981-95-6946-5_59

provide data support for key areas such as vehicle performance optimization, intelligent safety protection, and personalized service upgrades, helping automakers achieve digital transformation and intelligent upgrades.

In the context of automotive intelligence, dealing with multi-source heterogeneous data poses challenges in terms of scale, speed, and diversity. However, through a distributed architecture, not only can these problems be addressed, but also large-scale datasets can be processed efficiently, providing stable data storage and access services. The distributed architecture significantly enhances scalability, flexibility and availability [1–3] by distributing the system's functional modules across multiple computing nodes. This design optimizes resource utilization, enables dynamic load balancing, and effectively supports high concurrency and large-scale data processing requirements. The distributed architecture is widely used in cloud computing and big data, Internet of Things and edge computing and artificial intelligence and machine learning due to its advantages such as high scalability, fault tolerance and efficient resource utilization. Zhou Heng [1] et al. designed a high-performance, highly reliable and scalable cloud storage system based on Hadoop technology and its core component HDFS to meet the demands of large-scale data storage, access and processing. The test results show that the system can handle large datasets efficiently and provide stable data storage and access services. The design is feasible and effective, and has broad application prospects. Zhang Xin [2] et al. In the face of people's higher demands for the audio-visual experience and functionality of the meeting space. A distributed audio system architecture based on Ao IP technology was proposed. The audio system of the conference room cluster based on the distributed architecture can meet the complex requirements of modern multi-functional conference Spaces, providing practical guidance for similar project construction. Also in the automotive field, the distributed architecture can better handle multi-source heterogeneous data, enabling real-time data collection, transmission, processing and analysis, providing strong support for the construction of automotive insight systems. In order to be able to monitor and calculate automotive power in real time, the authors of the literature [3] proposed a brand power simulation study based on a time series framework of Flink and Kafka. The results showed that this method could basically monitor and calculate the brand power of each automotive brand in real time and give scores for each metric, further promoting the development of the automotive industry.

In conclusion: To promote the efficient circulation of automobiles and the rapid development of the automotive industry in our country, this paper proposes "a distributed architecture automotive product insight system based on stream data processing", which can observe the feedback of automotive users on automotive products at all times, predict the future intelligent development of automotive products, and further promote the good development of the automotive industry.

2 Research Roadmap

The research roadmap for this paper is divided into five steps. The first step is multi-source heterogeneous data selection, which includes data from this paper, numerical data, and chart data, etc. The second step is the principles of distributed architecture, introducing related principles of distributed architecture. The third step is the overall system architecture, building the overall system architecture based on kafka and spark technologies. The fourth step is the verification results of this system architecture, and the comparison of this system architecture with other system architectures in handling business flows. The fifth step is the results and analysis of this paper. The roadmap for this study is shown in Fig. 1 below.

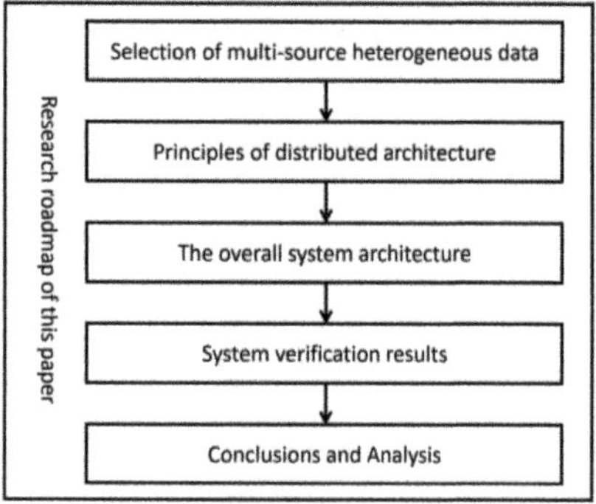

Fig. 1. Research roadmap of this paper

2.1 Selection of Multi-source Heterogeneous Data

The data sources selected for this article are from social media platforms (Weibo, Douyin, Xiaohongshu), video websites (such as Bilibili), news information platforms (such as Toutiao, Baidu News, Tencent News), etc. The data sources and related dimensions of this article are shown in Table 1.

Table 1. Data sources and related dimensions of this article

Serial numbers	Data source	Dimension fields
1	Social media data	Weibo (including trending topics), Zhihu (question-and-answer knowledge base)
2	Video data	Bilibili (bullet data), Tencent Video, Douyin
3	Auto portal site	Autohome, Autohome, Pacific Auto Network

The data resources in Table 1 can build a broad and diverse database of data resources. The data sources are extensive, highlighting the diversity of the data, with Weibo centered around short texts and hashtags, where users form immediate interactions through comments and reposts; Douyin, on the other hand, is dominated by short videos, combined with music, special effects and user interaction behaviors; platforms such as Toutiao and Baidu News aggregate a vast amount of news and information, including social hotspots and industry trends. The data on these platforms not only vary in content themes, but also vary significantly in the way they are generated and user behavior patterns.

The "heterogeneity" of the data further emphasizes the complexity of the information in terms of structure and form. Comments on Weibo and video and audio data on Douyin, for example, are typically unstructured data that lack fixed fields or formats and require parsing with the help of technologies such as natural language processing and computer vision. The author and publication time of news articles, as well as other information, are semi-structured data, which, though regular, still need to be standardized.

The processing of the multi-source heterogeneous data in this article requires techniques such as data fusion, data cleaning, and data transformation. Data fusion technology can integrate data from different sources to extract valuable information; Data cleaning techniques can remove noise and errors from data and improve data quality; Data conversion technology can transform data of different formats into a uniform format, which is convenient for subsequent processing and analysis.

2.2 Principles of Distributed Architecture

A distributed architecture is an architecture that distributes the functional modules of a system across different physical or virtual computers for communication and collaboration over a network. The basic principle of a distributed architecture is to break down a large system into several independent subsystems, each with its own independent function and capable of operating independently. Distributed architecture has advantages such as scalability, high availability, flexibility, and fault tolerance. Scalability enables the system to be easily scaled by adding nodes without major changes to the existing structure; high availability ensures that the failure of a part of the system does not cause the entire system to crash, and users can continue to access services through other nodes; flexibility allows services to be added, removed or modified at any time as needed, and supports the mixed use of heterogeneous environments and technology stacks; fault tolerance ensures continuous operation[4, 5] of the system by using redundancy and backup mechanisms, and when a component or service fails, the system can automatically switch to a backup component or service.

2.3 The Overall System Architecture

The overall architecture diagram of the system in this paper is designed in layers, including the data layer (data acquisition layer), the HDFS import layer (distributed storage layer), the data cleaning layer, the data modeling layer, the memory buffer layer and the upper application layer of the automotive insight system. The overall architecture diagram of the system in this paper is shown in Fig. 2.

(1) Data acquisition layer:

Social media API data collection is mainly carried out by using compliant web crawler technology and strictly adhering to the robots protocol and platform user protocol for data scraping. It is necessary to strictly follow the frequency limit strategy of the platform (such as the 500 times per hour/IP limit of Weibo API), and ensure the stability of data collection through reasonable distributed scheduling and proxy IP rotation mechanisms. The sample code for the Weibo API call is as follows.

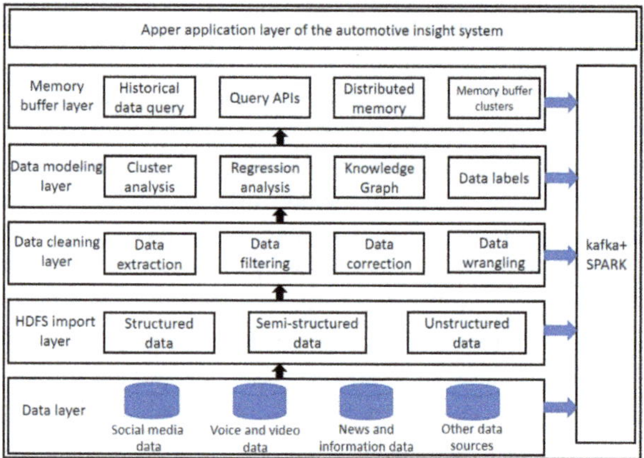

Fig. 2. System architecture diagram

```python
def fetch_weibo_data(access_token, since_id=None):
    url = "https://api.weibo.com/2/statuses/public_timeline.json"
    params = {
        'access_token': access_token,
        'count': 100,
        'since_id': since_id # Incremental collect key parameters
    }
    response = requests.get(url, params=params)
    return response.json()
```

(2) Distributed storage layer

The multi-source heterogeneous raw data obtained (including structured content such as text, image, video metadata, etc.) will be written in real time to message queues such as Kafka for buffering, then processed through stream processing frameworks such as Spark Streaming, and finally imported into the HDFS layer for storage. The entire process requires special attention to user privacy protection and desensitization of sensitive fields. And establish a complete data quality monitoring system to ensure the integrity of the data collection. The data flow architecture diagram of the distributed storage layer is shown in Fig. 3.

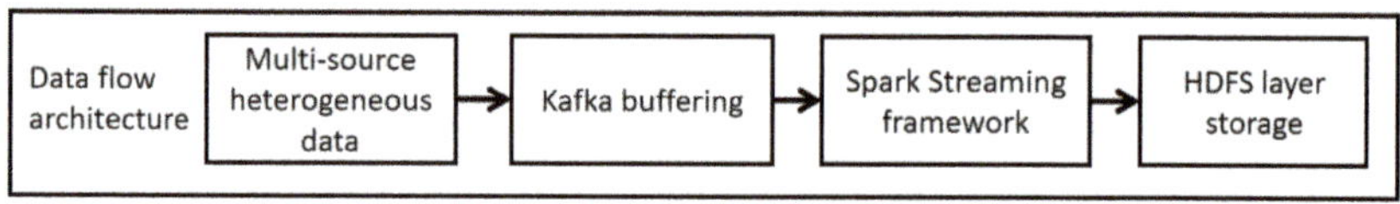

Fig. 3. Data flow architecture diagram of the distributed storage layer

Data cleaning layer

The data cleaning layer in this article consists of four steps: data extraction, data filtering, data correction, and data organization. The architecture diagram is shown in Fig. 4 below.

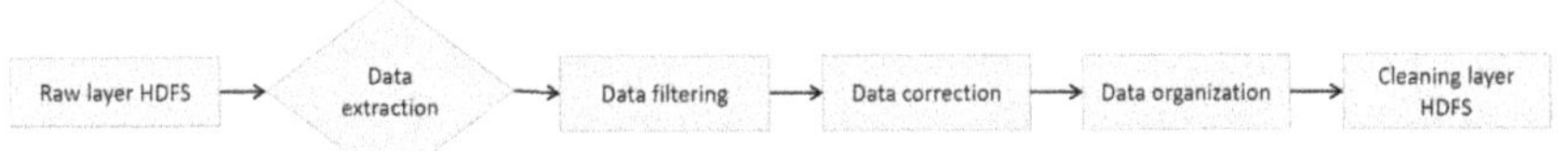

Fig. 4. Architecture diagram of the data cleaning layer in this paper

1. Data extraction: The main function is to handle multi-layer nested JSON/XML structures, extract key fields from the buried log, and separate text content from metadata.
2. Data filtering: Data filtering is a key part of the data processing flow. Its core principle is to remove content, invalid data and other data from the original dataset through predefined rules, regular expressions and algorithms, and select a subset of data that meets specific quality requirements or business needs.
3. Data correction: The essence of data correction is to identify and fix errors, inconsistencies and incompleteness issues in data through technical means. The design of the data correction system must follow the principle of traceability and retain the mapping relationship between the original data and the correction record. Modern data platforms typically employ closed-loop correction processes, combining rule engines with machine learning models to implement dynamic correction strategies.
4. Data organization: Data organization is a systematic process of transforming raw data into a normalized form suitable for analysis, and its core principle is to increase the value density of data through structured transformation. Text data is organized by means of word embedding, topic modeling, etc. Image data is organized by feature extraction, and video data is organized by keyframe extraction/metadata annotation.

Data modeling layer

The data modeling layer consists of four steps: cluster analysis, regression analysis, knowledge graph, and data labeling. The modules work together through the feature sharing layer to support upper-level business applications. The core value of this layer lies in converting data entropy into information gain. The architecture diagram is as shown in Fig. 5.

1. Cluster analysis: Cluster analysis divides data into intrinsically associated groups based on similarity measures, such as Euclidean distance, to achieve data clustering or anomaly detection [6]. The Euclidean distance calculation formula is as in Formula (1).

$$d_{12}=\sqrt{(x_{11}-x_{21})^2+\cdots+(x_{1n}-x_{2n})^2}=\sqrt{\sum_{k=1}^{n}(x_{1k}-x_{2k})^2} \tag{1}$$

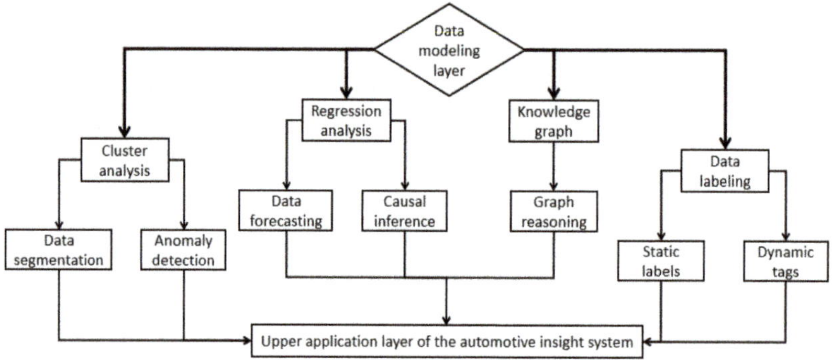

Fig. 5. Data Modeling layer architecture diagram

Formula (1) represents the two n-dimensional vectors Euclidean distance between $a(x_{11}, x_{12}, \cdots, x_{1n})$ and $b(x_{21}, x_{22}, \cdots, x_{2n})$.

2. Regression analysis: Regression analysis uses optimization algorithms such as least squares or gradient descent to establish a mathematical relationship model between variables, supporting data prediction and causal inference. Linear regression objective functions such as formula (2).

$$y = a + \beta_1 x_1 + \beta_2 x_2 + \cdots + \beta_n x_n \tag{2}$$

In Formula (2), a represents the intercept and β represents the correlation coefficient.

③ Knowledge graph: A knowledge graph uses graph theory and NLP techniques to build semantic networks through entity recognition (NER) and relation extraction, enabling association reasoning across entities.

④ Data labels: Data labels are based on the rule engine model, encapsulating the original features with business semantics to form reusable feature metrics.

Memory buffer layer

The memory buffer layer contains historical data queries, query interfaces, distributed memory, and buffer clusters. The architecture diagram of the memory buffer layer is shown in Fig. 6.

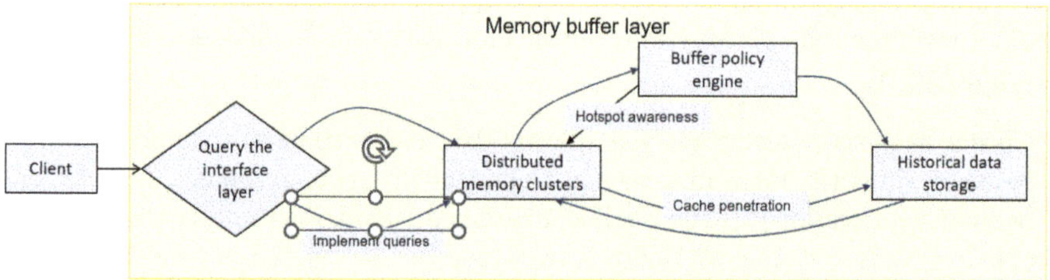

Fig. 6. Memory buffer layer architecture diagram

1. Historical data query: It is divided into three query forms: real-time query, near-line query, and offline query. Among them, real-time query adopts Alluxio + Reddis-timeseries technology to achieve memory-level accelerated access to hot time series data and economical storage of cold data. Near line query uses Apache Ignite persistent storage technology, and offline query uses HDFS + Parquet column storage technology.
2. Query interface: the key code is as follows.

```
public class QueryHandler extends ChannelInboundHandlerAdapter {
    @Override
    public void channelRead(ChannelHandlerContext ctx, Object msg) {
        CompletableFuture.supplyAsync(() -> {
            return cacheService.query(((QueryRequest) msg).getKey());
        }).thenAcceptAsync(result -> {
            ctx.writeAndFlush(new QueryResponse(result));
        }, executorGroup);
    }
}
```

Kafka + Spark technology

The distributed architecture combining Kafka and Spark is a classic combination in the field of big data processing, mainly achieving high-throughput and low-latency real-time data processing capabilities [7] This architecture typically includes the following three core components. The technical architecture diagram of Kafka + Spark is shown [8, 9] in Fig. 7 below.

1. Kafka: A distributed message queue system responsible for real-time data collection and buffering.
2. Spark Streaming: A distributed computing framework responsible for real-time data processing.
3. Storage system: Use HDFS to store unstructured data such as images and videos.

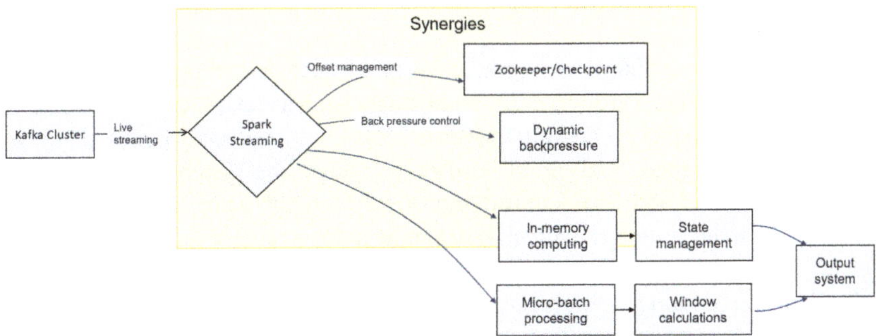

Fig. 7. Kafka + Spark technology architecture diagram

In order to ensure the security of communication and data among the layers of the system, the system adopts a unified communication protocol and data security mechanism. The communication protocol uses a TCP/ IP-based protocol stack to ensure the reliability and stability of data transmission. In order to meet the real-time requirements, the system also uses the UDP protocol for the transmission of some real-time data.

The data security mechanism employs technologies such as encryption, authentication, and access control to ensure the security of data during transmission and storage. For example, sensitive data is encrypted, digital certificates are used for identity verification, access control lists are set up, and users' access rights to the data are restricted.

2.4 System Verification Results

The system architecture of this paper is compared with that of centralized systems and stream processing systems, and is verified mainly through three dimensions: data throughput, end-to-end latency, and query data accuracy. The results are shown in Table 2 below.

Table 2. Verification results of the system architecture

Metrics	This article System	Centralized system	Stream processing system
Data throughput (MB/s)	high	low	middle
End-to-end delay (ms)	middle	high	low
Query data accuracy	high	low	middle

Conclusion: The system in this paper outperforms centralized systems and stream processing systems in terms of throughput and query data accuracy, and has slightly higher latency than the pure stream processing architecture, but has better overall performance.

3 Conclusions and Analysis

This paper presents an automotive product insight system based on the kafka + spark distributed architecture. Through multi-source data such as text, video and audio, combined with the distributed processing framework and intelligent analysis model, the overall architecture and functional modules of each layer of the system are designed, and the application effect of the system is demonstrated through actual cases. The research results show that the system can effectively handle multi-source heterogeneous data, significantly improve data processing efficiency and analysis accuracy, and provide comprehensive decision support for automakers and mobility service providers. Future work will focus on enhancing privacy protection, expanding application scenarios and deepening algorithmic innovation to further drive the intelligent transformation of the automotive industry.

This paper has achieved some research results, but there are still some deficiencies. For example, the system still needs to be improved in data fusion and data analysis, and its performance and reliability also need to be further optimized. Future research directions include: in-depth study of fusion algorithms for multi-source heterogeneous data to improve the accuracy and efficiency of data fusion; Explore more advanced machine learning and deep learning algorithms to improve the ability and level of data analysis; Optimize the architecture and performance of the system and improve its reliability and stability.

References

1. Zhou, H.: Design and implementation of cloud storage system based on hadoop distributed architecture. China Broadband **20**(12), 68–70 (2024). https://doi.org/10.20167/j.carolcarroll nkiISSN1673-7911.2024.12.23
2. Zhang, X., Xu, X., Geng, Y., et al.: China media technology (12), 145–151 (2024). https://doi. org/10.19483/j.cnki.11-4653/n.2024.12.031
3. Fu, L., Yang, J., Zhang, F.: A study of brand power simulation by time series based on flink and kafka framework. In: 2022 IEEE 2nd International Conference on Data Science and Computer Application (ICDSCA), Dalian, China, pp. 1347–1352 (2022). https://doi.org/10.1109/ICD SCA56264.2022.9987758
4. Xinfang, C., Yiqing, L.: The working principle and application of Hive, a data warehouse tool for distributed systems. Sci. Technol. Innov. **36**, 104–107 (2021)
5. Cui, H.: Principles and Practice of Distributed Architecture, p. 504. Posts & Telecom Press (2021)
6. Li, B., Sun, M., Zhang, J., et al.: Topology identification method for multi-loop distribution network based on deep neural network fusion of Euclidean distance. Power Syst. Protect. Control **53**(5), 123–134 (2025). https://doi.org/10.19783/j.carolcarrollnkiPSPC.240746
7. Ye, H.: Practice of building a data center processing engine based on spark streaming and Kafka. Cybersecur. Technol. Appl. (03), 51–53 (2023)
8. Li, C.: Design of real-time processing system for device status data based on spark. Wireless Interconn. Technol. **19**(24), 68–70 (2022)
9. Zhiku Daojie Content and Product Center. Kafka Infrastructure and Design, p. 242. China Railway Press (2022)

Author Index

P. Siarry et al. (Eds.): WCNA 2024, LNEE 1550, pp. 601–603, 2026.
https://doi.org/10.1007/978-981-95-6946-5